Springer Water

The book series Springer Water comprises a broad portfolio of multi- and interdisciplinary scientific books, aiming at researchers, students, and everyone interested in water-related science. The series includes peer-reviewed monographs, edited volumes, textbooks, and conference proceedings. Its volumes combine all kinds of water-related research areas, such as: the movement, distribution and quality of freshwater; water resources; the quality and pollution of water and its influence on health; the water industry including drinking water, wastewater, and desalination services and technologies; water history; as well as water management and the governmental, political, developmental, and ethical aspects of water.

More information about this series at http://www.springer.com/series/13419

Martina Zeleňáková ·
Katarzyna Kubiak-Wójcicka ·
Abdelazim M. Negm
Editors

Management of Water Resources in Poland

Springer

Editors
Martina Zeleňáková
Institute of Environmental Engineering
Technical University of Košice
Košice, Slovakia

Katarzyna Kubiak-Wójcicka
Nicolaus Copernicus University
Toruń, Poland

Abdelazim M. Negm
Faculty of Engineering
Zagazig University
Zagazig, Egypt

ISSN 2364-6934 ISSN 2364-8198 (electronic)
Springer Water
ISBN 978-3-030-61964-0 ISBN 978-3-030-61965-7 (eBook)
https://doi.org/10.1007/978-3-030-61965-7

This Springer imprint is published by the registered company Springer Nature Switzerland AG
The registered company address is: Gewerbestrasse 11, 6330 Cham, Switzerland

Preface

The issue of water resources management is extremely broad and covers several related areas. There are several publications that discuss water resource management in Poland, but focus only on selected water management issues. Hence, the idea of a book on management of water resources in Poland, which in an exceptionally broad way, would present the most important issues in the field of water management in quantitative and qualitative terms. In this volume, only the most important aspects of water resources management in Poland are discussed, indicating specific problems and possibilities of solving or eliminating them.

This book consists of 21 chapters that form a coherent and logical entity. This is the result of a teamwork of 32 scientists from various institutions and research centers, who in their scientific research address the issues of determining water resources and their importance and use in various sectors of the national economy (agriculture, industry, municipal economy, transport).

The scientific material presented in this book is a coherent collection of information that is extremely useful to both practitioners and experts in water resource management. We hope that professional, designers, employees and graduate students dealing with water management, but also the society will find useful information on water resource management in this book. Educating and informing the public are extremely important because it largely contributes to increasing public awareness of water use and taking further measures to reduce water consumption. Involving society in positive actions for water resources protection contributes to the increase of water resources protection.

Water is an important element of the landscape, which requires protection against excessive use. Ensuring sustainable development in water management requires first of all knowledge of its resources, which will allow for planning protection as well as for its rational use. We keep in mind that we must protect these resources and leave them for future generations. The basis of proper management is the issue of proper administration through a reliable diagnosis of the existing state and forecasting future changes.

Systematic investigation and evaluation of surface water and groundwater in Poland and other countries are the basic responsibility of the state as a necessary requirement to ensure the conditions necessary for sustainable development.

This volume consists of five parts. The book certainly does not exhaust the entirety of the issues contained in the title, which is why each chapter contains references extending the problems discussed.

The first part consists of two chapters, and the first one is titled "Introduction to the "Management of Water Resources in Poland"" which was written by the volume's editors. Its main goal is to familiarize the reader with the research issues that have been discussed in this volume, while the next chapter is titled "Water Resources and Management of Poland in SCOPUS Database." It provides an overview of the trends and statistics of water resources (e.g., surface water and groundwater) in Poland, based on the SCOPUS database. Moreover, the chapter includes sufficient information on the institutional affiliations, countries and funding agencies and sponsors, concerning the management of water bodies in Poland. Some recommendations are considered to maintain water availability in the country.

The second part discusses "Water Resources in Poland." The first chapter in this part presents the institutions responsible for the administration and management of water resources, which is the basic responsibility of the state. The chapter "Administration of Water Resources Management: Key Facts About Water Resources in Poland" was presented by Katarzyna Kubiak-Wójcicka from Nicolaus Copernicus University in Toruń, Faculty of Earth Sciences and Spatial Management. The second chapter "Water Resources in Poland and Their Use" was presented by Zdzisław Michalczyk and Joanna Sposób from the Marie Curie-Skłodowska University in Lublin, Faculty of Earth Sciences and Spatial Management. It is dedicated to the general presentation of the state of surface (mainly rivers) and underground water resources and their use in the national perspective.

In the next chapter, Adam Choiński from Adam Mickiewicz University in Poznań, Faculty of Geographical and Geological Sciences, and Rajmund Skowron from Nicolaus Copernicus University in Toruń, Faculty of Earth Sciences and Spatial Management, presented "Water Resources of Stagnant Waters." It discusses the surface water resources that are accumulated in the largest lakes and artificial water reservoirs in Poland and their transformation. Supplement to the information on lake water resources, the chapter titled to "Natural and Anthropogenic Lakes of River Valleys" is introduced and is written in cooperation with authors from three scientific centers. These are Jarosław Dawidek from the Marie Curie-Skłodowska University in Lublin, Faculty of Earth Sciences and Spatial Management, Beata Ferencz from the University of Life Sciences in Lublin, Faculty of Animal Sciences and Bioeconomy, and Katarzyna Kubiak-Wójcicka from Nicolaus Copernicus University in Toruń, Faculty of Earth Sciences and Spatial Management. This chapter discusses the problem of genetic types of lakes, with particular emphasis on lakes located in river valleys.

The chapter entitled "Studies on, the Use and Protection of Springs in Poland" was prepared by Paweł Jokiel from the University of Łódź, Faculty of Geography Sciences and Zdzisław Michalczyk, from the Marie Curie-Skłodowska University in Lublin, Faculty of Earth Sciences and Spatial Management. The authors discussed sources located throughout Poland. These are spots of groundwater outflow to the surface of the area with unique features that should be legally protected. Source protection measures are to ensure the preservation of high-quality groundwater flowing out of the sources as well as the natural and landscape values of their surroundings.

The last chapter in this part of the book is entitled "Groundwater Resources of Poland." The authors of the chapter are Arkadiusz Krawiec and Andrzej Sadurski from Nicolaus Copernicus University in Toruń, Faculty of Earth Sciences and Spatial Management. The authors discussed the status of groundwater resources in Poland. Fresh groundwater is an abundant resource and of high quality, which is why it is the main source of water supply for the population, which is constantly monitored. Mineral and thermal waters that occur at greater depths are used for therapeutic, recreational and heating purposes.

In the third part of the book, "Change of Flow in Polish Rivers," there are nine chapters that deal with the long-term and seasonal variability of rivers in Poland. The first chapter in this part was written by Dariusz Wrzesiński from Adam Mickiewicz University in Poznań, Faculty of Geographical and Geological Sciences, and is entitled "Flow Regime Patterns and Their Changes." The author characterized the types of hydrological regime of major Polish rivers and the detection of changes in the hydrological cycle. He drew attention to the problem of availability and quantity of water resources, which is affected by the seasonal flow of the river and the stability of the hydrological regime.

The second chapter in part three is entitled "Flow Seasonality in Two Big Polish Rivers – The Vistula and the Oder" by Paweł Jokiel and Przemysław Tomalski from the University of Łódź, Faculty of Geography Sciences. The authors compared the long-term flow seasonality along the two largest rivers in Poland, the Vistula and the Oder.

The chapter "Low-Flows in Polish Rivers" is authored by Edmund Tomaszewski from the University of Łódź, Faculty of Geography Sciences, and Katarzyna Kubiak-Wójcicka Nicolaus Copernicus University in Toruń, Faculty of Earth Sciences and Spatial Management. This chapter draws attention to the spatial, seasonal and long-term variability of low flows in selected Polish rivers. Low flows in rivers affect the river regime as well as the availability, dynamics and management of water resources.

Additionally, the chapter titled "Dynamics, Range, and Severity of Hydrological Drought in Poland" was presented by Edmund Tomaszewski and Malwina Kozek from the University of Łódź, Faculty of Geography Sciences. The study includes parameters describing the duration, severity, extent and identification of periods with different patterns of hydrological drought development. This problem is extremely important in reducing losses caused by periods with long-term water shortages.

Further discussed are the issues of "Flood Marks in Poland and Their Significance in Water Management and Flood Safety Education" by Marcin Gorączko from the University of Science and Technology in Bydgoszcz, Faculty of Civil and Environmental Engineering and Architecture. This chapter presents an overview flood registered in Poland. They are regarded as an important source of hydrological information to study the extent and effects of the most catastrophic floods in history.

Flood issues are presented in the chapter titled "Flood Potential of Polish Rivers" written by Artur Magnuszewski from the University of Warsaw, Faculty of Geography and Regional Studies. The different reaction of rivers to precipitation in Poland was discussed, which was measured using the flood potential index k proposed by J. Françou. In case of Poland, the flood potential indicator map was used to prepare flood risk management plans for the Vistula.

Moreover, the chapter titled "Flood Risk Management System in Poland" was written by Renata Graf from Adam Mickiewicz University in Poznań, Faculty of Geographical and Geological Sciences. The key elements of flood risk management in Poland were discussed by determining the distribution and level of aggregate risk in several categories: health and life, natural environment, cultural heritage and economic activity.

The last chapter in part three is titled "Water Management in the Pomeranian Rivers Estuary Zone on the Background of Hydro-meteorological Conditions" and was written by Joanna Fac-Beneda, Izabela Chlost and Alicja Olszewska from the University of Gdańsk, Faculty of Oceanography and Geography. The authors discuss the issues of river runoff in the contact zone between the Baltic Sea and the mainland on the example of the Pomeranian Lake District.

The fourth part of the book is entitled "Economic Importance of Water Resources" and consists of four chapters.

The chapter titled "Exploitation of Rivers in Poland for Electricity Production – Current Condition and Perspectives for Development" was written by Katarzyna Kubiak-Wójcicka from Nicolaus Copernicus University in Toruń, Faculty of Earth Sciences and Spatial Management, and Lech Szczęch from Military University of Technology in Warsaw, Faculty of Mechanical Engineering. The chapter provides an overview of the past and current use of the potential of hydropower and the importance of hydropower in the Polish energy system.

Next, the issues of the current state, use and significance of inland waterways in Poland were discussed in the chapter titled "Waterways in Poland – The History, Present State and Future" which was written by Marcin Gorączko from the University of Science and Technology in Bydgoszcz, Faculty of Civil and Environmental Engineering and Architecture, and Katarzyna Kubiak-Wójcicka from Nicolaus Copernicus University in Toruń, Faculty of Earth Sciences and Spatial Management.

Next to the above chapter, the chapter titled "The Effects of Plant Irrigation in Poland" was written by Renata Kuśmierek-Tomaszewska and Jacek Żarski from the University of Science and Technology in Bydgoszcz, Faculty of Agriculture and Biotechnology. The chapter discusses the importance of plant irrigation in Poland

and their impact on production effects, which depend on weather, climate and soil factors.

The last chapter in this part of the book entitled "Power Plant Open Cooling System in the Context of the Objectives of the Water Framework Directive" was written by Leszek Pająk and Agnieszka Operacz, from the University of Agriculture in Kraków, Faculty of Environmental Engineering and Land Surveying, and Barbara Tomaszewska from AGH University of Science and Technology in Kraków, Faculty of Geology, Geophysic and Environmental Protection. This chapter discusses the impact of cooling water discharges on achieving environmental objectives resulting from the WFD used in a conventional power plant.

The last part of the book entitled "Conclusions" was written by the editors. The last chapter contains "Updates, Conclusions, and Recommendations for "Management of Water Resources in Poland"" and closes the volume of the book with the main conclusions and recommendations of the volume, as well as an update of some arrangements.

Special thanks to all who contributed in making this high-quality volume a real source of knowledge and the latest findings in the field of water resources in Poland. We would love to thank all the authors for their invaluable contributions. Without their patience and effort in writing and revising the different versions to satisfy the high-quality standards of Springer, it would not have been possible to produce this book and make it a reality. Much appreciation and great thanks are also owed to the reviewers and the editors of the Earth and Environmental Sciences series at Springer for the constructive comments, advice and critical reviews. Acknowledgments are extended to include all members of the Springer team who have worked long and hard to produce this volume.

The volume editor would be happy to receive any comments to improve future editions. Comments, feedback, suggestions for improvement or new chapters for next editions are welcome and should be sent directly to the volume editors.

Zagazig, Egypt
Toruń, Poland
Košice, Slovakia
September 2020

Abdelazim M. Negm
Katarzyna Kubiak-Wójcicka
Martina Zeleňáková

Contents

Part I
Introducing the Book

Chapter 1
Introduction to the "Management of Water Resources in Poland"

Katarzyna Kubiak-Wójcicka, Martina Zeleňáková, and Abdelazim M. Negm

Abstract This chapter presents the main features of the book "**Management of Water Resources in Poland**" and related current problems and research topics implemented by scientists dealing with water resources in Poland. The discussed research issues were divided into 5 thematic blocks. These are: Introduction (I part), Water Resources in Poland (II part), Change of Flow in Polish Rivers (III part), Economic Importance of Water Resources (IV part) and Conclusions (V part). The main technical elements of each chapter are presented under the appropriate topic.

Keywords Management · Surface and underground water resources · Flow variability · Extreme phenomena · Modeling · Hydropower · Inland waterways · Irrigation · Pollution · Poland

1.1 Poland: A Brief Background

The total area of the country according to the administrative division amounts to 312,722 km^2 and includes the land area (including inland waters) of 311,895 km^2 as well as part of internal waters – 827 km^2. The internal water include part of the Vistula Bay including waters of ports, a part of Lake Nowowarpieńskie and a part of Szczecin Bay including Świna and Dziwna as well as Kamieński Bay including waters of ports, Oder between the Szczecin Bay and waters of Szczecin port as well

K. Kubiak-Wójcicka
Department of Hydrology and Water Management, Faculty of Earth Sciences and Spatial Management, Nicolaus Copernicus University, Lwowska 1, 87-100 Toruń, Poland
e-mail: kubiak@umk.pl

M. Zeleňáková
Department of Environmental Engineering, Faculty of Civil Engineering, Technical University, Košice, Slovakia
e-mail: martina.zelenakova@tuke.sk

A. M. Negm (✉)
Water and Water Structures Engineering Department, Faculty of Engineering, Zagazig University, Zagazig 44519, Egypt
e-mail: Amnegm85@yahoo.com; Amnegm@zu.edu.eg

M. Zeleňáková et al. (eds.), *Management of Water Resources in Poland*, Springer Water,
https://doi.org/10.1007/978-3-030-61965-7_1

as waters of the Gulf of Gdańsk and ports bordering on territorial sea waters. The length of the sea border is 440 km, which is 12.5% of the length of the state border [1]. Almost 50% of the country's surface is 100–200 m above sea level, while 3% of the country's surface is above 500 m amsl. The highest peak in Poland is Rysy (2499 m amsl).

The most important elements of the hydrographic network in Poland are: rivers, lakes, artificial water reservoirs and canals. Poland is located almost entirely in the basin area of the Baltic Sea (99.7% of the country's area) which includes the basins of the 2 largest rivers, the Vistula and Oder. The remaining 0.3% of the country's territory is occupied by rivers falling within the Black Sea (0.2%) and North Sea (0.1%) basin systems. Most rivers in Poland flow north-west, according to the slope of the country's surface.

Almost 88% of Poland's total area lies in the basin of the two largest Polish rivers: the Vistula and the Oder. The Vistula river basin (without delta) covers the area of 194.0 thousands km^2, of which 168.9 thousands km^2 in Poland. The length of the Vistula is 1022 km, while the average flow is 1044 $m^3 . s^{-1}$. The Oder river basin covers an area of 119.1 thousands km^2, of which 106.0 thousands km^2 in Poland. The length of this river is 840 km (including 726 km in Poland and the border section 187 km). The average Oder flow is 522 $m^3 . s^{-1}$ [2].

Rivers in Poland are fed directly by precipitation and indirectly by snow thaws. High water levels in Polish rivers occur mainly in spring (February–April). The second high water level is noted in summer, most often on mountain rivers, as a result of intensive July rainfall. The lowest water levels take place in early autumn [3, 4]. On the Baltic coast, high water levels are caused by the storms surges [5].

Most of the lakes in Poland are glacial lakes. Small and shallow lakes predominate among Polish lakes [6]. The largest lakes in terms of area include Śniardwy (113.4 km^2), Mamry (102.8 km^2). The deepest lakes include Lake Hańcza with a depth of 108.5 m, which is located in the Suwałki Lake District. The largest artificial reservoirs include the Solina Reservoir located in a mountain area with a total capacity of 472.4 million m^3 and the Włocławski Reservoir in a lowland area with a capacity of 453.6 million m^3.

In the years 1951–2013 the average annual precipitation for Poland amounted to 618 mm and ranged from 480 mm in Leszno to over 1000 mm in Zakopane [7]. In general, precipitation in the central lowland of Poland is the lowest [8]. For example, the average annual precipitation in the period 1951–2015 at the station in Toruń was 527 mm, while in Warsaw 521 mm [9]. The highest rainfall amounts occur in the summer months (most often in July), while the lowest in the winter months (in January or February).

The average annual air temperature in Poland in the years 1951–2015 ranges between 6.4 °C (Suwałki, north-eastern Poland) and 8.9 °C (Słubice, western Poland). In general, the western part of Poland is warmer than the eastern part. In high parts of the Tatra Mountains in southern Poland, the average annual air temperature is much lower and amounts to −0.5 °C at the Kasprowy Wierch station [10].

1.2 Main Themes of the Book

The book includes the following parts: Introduction (I part), Water Resources in Poland (II part), Change of Flow in Polish Rivers (III part), Economic Importance of Water Resources (IV part) and Conclusions (V part). The second part is discussed in seven chapters. Part III is presented in 8 chapters, while Part IV is discussed in 4 chapters. In the sections below, the main features of each of them will be presented under the appropriate topic.

1.3 Water Resources in Poland

The thematic block dedicated to "**Water Resources in Poland**" presents the size of Poland's water resources and their economic use nationwide. The current state of water resources is the effect of water management over many years in accordance with the applicable water resource management system and the organizational structure of water management in Poland.

The chapter entitled "**Water Resources and Management of Poland in SCOPUS database**" provides a short overview of the state of water management in Poland. The focus was on research funded by grants, leading international journals, institutional affiliations as well as recognized sponsors and projects available in the SCOPUS database, covering water aspects in Poland. Detailed studies are presented in the following chapters.

The chapter "**Administration of Water Resources Management: Key Facts about Water Resources in Poland**" discusses the legal status and institutions responsible for administration and management of water resources. The organizational structure of water management in Poland in 2001–2017 did not support the sustainable development of water management. This was mainly due to two institutions dealing with water management, which operated within administrative boundaries and drainage basins. This led to confusion and conflicts in terms of competence. The water management reform was carried out in 2017, as part of which the State Water Management of Polish Waters was established in 2018. The new institution focuses exclusively on the catchment area, however, an assessment of its activities can be summarized in a few years.

Chapter entitled "**Water Resources in Poland and their use**" introduces the reader to a general presentation of the amount of surface and underground water resources and their economic use in Poland. The authors of the chapter emphasize that water resources in Poland are extremely low compared to other European countries. On average, in the years 1901–2017, the annual outflow through Polish rivers was 61.1 km^3 per year, which gives on average about 1600 m^3 per capita per year. Annual outflow of Polish rivers in particular years and decades of the period 1901–2017 fluctuated in a very broad range from 37.5 km^3 in 1954 to 89.9 km^3 in 1981, averaging 61.1 km^3. Water resources are shrinking as a result of abstraction for economic use (surface water) and supplying the population with drinking water (mainly groundwater). That is why it is so important to protect their resources and preserve their good quality.

Subsequently, chapters are presented that discuss water resources in Poland, divided into surface and underground.

The chapter "**Water Resources of Stagnant Waters**" presents standing water resources and their regional distribution. In total, there are 7081 lakes in Poland with an area of over 1 ha. According to the authors' calculations, the water resources accumulated in the lakes in Poland amount to 19.7349 km^3, of which the lakes located in the Masurian Lake District (51.27%), the Pomeranian Lake District (36.12%) and the Greater Poland-Kujawskie Lake District have the largest capacity (11.93%). Apart from natural lakes, water resources are collected in artificial reservoirs, of which there are 101 in Poland (with a capacity of over 1 million m^3). In total, there are only 11 artificial reservoirs and 26 natural lakes in Poland with a capacity of over 100 million m^3). The authors point out the problem of water resources reduction/decrease resulting from the reduction of the surface and their volume. This process is determined by two basic factors: fluctuations in water levels in the lake and reservoir filling with sediments. The problem of lake disappearance is particularly evident in the case of small water bodies. The chapter "**Natural and Anthropogenic Lakes of Rivers Valleys**" discusses the genetic types of lakes occurring in river valleys. These objects are an important and underestimated element of small retention. The lakes of river valleys are very diverse reservoirs. High variability of morphometric, hydrochemical and hydrobiological as well as ecological and landscape parameters makes them attractive. In addition, they are young reservoirs subject to dynamic disappearance processes. Although they constitute a very large group of lakes, their recognition is relatively low.

In the chapter "**Studies on, the Use and Protection of Springs in Poland**" special attention was paid to groundwater outflows. Sources are connections between underground and surface links of the water circulation system, and its existence is proof of unity between the two links, not their separateness. Their presence, type of drainage and drainage regime indicate the conditions of the water cycle in the underground phase of the water cycle. In Poland, the most numerous springs occur in mountainous and upland areas, and the largest source discharges occur in parts of the Tatra Mountains built of carbonate rocks. The water in the springs is usually very high quality and requires special protection.

From the point of view of supplying people with water, the chapter "**Groundwater Resources of Poland**" is an important element. Groundwater in Poland is the main source of drinking water supply for urban and rural populations. This is due to large documented groundwater resources and their high quality. Over 70% of water is used to collectively supply the population with groundwater intakes, and this demand is growing every year. Groundwater is of interest due to its special properties, e.g. curative, thermal or highly mineralized (brine) waters. Waters with special properties are used in health resorts for therapeutic purposes, in geothermal power plants or recreational facilities. These waters require special protection and monitoring. As part of the sustainable management of groundwater resources, particular attention is paid to the rational use of resources and their protection in order to maintain water quality.

Summing up this part of the volume, it should be pointed out that Poland's water resources are small and therefore require proper management.

1.4 Change of Flow in Polish Rivers

Part two of the volume is entitled "**Change of Flow in Polish Rivers**". The chapter "Flow Regime Patterns and their Changes" characterizes the hydrological regime of the Polish rivers. The article distinguishes and characterizes five types of hydrological regime of rivers in Poland. In the lowland areas, three simple nival regimes are dominant: poorly, moderately and well-formed, while in the highlands and in the mountains two complex regimes: nival-pluvial and pluvial-nival can be distinguished. Their spatial differentiation results from the diversity of environmental conditions of the basin, which leads to the appearance of various forms of river supply, seasonality of the river flow and its variability. The paper also describes how contemporary changes in climatic conditions and human impact on water relations affect the characteristics of the river flow regime. Multiannual and seasonal changes in river flows were found, which are of particular importance in the spring and summer seasons due to the adopted criteria of the typology of the river regime. Studies confirmed that the destabilization of the flow regime characterisitcs of many rivers in Poland may be affected by changing climatic conditions. It is caused by the varied intensity of macro-scale air circulation types, such as the North Atlantic Oscillation, whose impact is not strong but noriceable and statistically significant, especially in the winter season.

The purpose of the chapter entitled "**Flow Seasonality in Two Big Polish Rivers – the Vistula and the Oder**" is to determine the seasonality of flows of the two largest Polish rivers: the Vistula and the Oder. For this purpose, the authors used the seasonality index (IS), half-life date (PK), half-flow date (TPO) and seasonality factor (GMO) for 14 hydrological stations located along the Vistula and 11 hydrological stations located along the Oder in 1951–2016. The highest features of IS are for the upper part of rivers. Then the seasonality index decreases sharply due to the sum of flows from river inflows, which are characterized by different outflow conditions. A clear correlation was found between the concentration date (PK) and the half-flow date (TPO) for both rivers. Similarity of PK indicators in individual years for the Vistula and Oder rivers in the analyzed multi-year period occurred in the lower and middle rivers. In the upper sections, the differences between the rivers are greater, and over the past 30 years the tendency to an earlier concentration date of the Oder has been increasing. Changes in the IS flow seasonality index showed high variability and depended on the catchment size. In contrast to PK, relatively the greatest similarities between the Vistula and the Oder were recorded for their upper sections. With the increase in the catchment, the seasonality of the Vistula flows became much higher than in the Oder, and the differences were most noticeable in the 1960s and 1970s. In recent years their importance has been decreasing.

The chapter entitled "**Low-Flows in Polish Rivers**" discusses the spatial, seasonal and long-term variability of low flows in Poland in the years 1951–2015. The progression dynamics of low flow episodes are characterized by high seasonal and long-term variability. Its basis is determined by the strong diversity of hydrometeorological conditions generating individual episodes and the modification of the role of anthropogenic and physiographic factors (mainly hydrogeological conditions and lakes). Long-term trends of selected features of low flows, such as: minimum flows, duration of low flows and intervals between low flows, covered only a part of the studied catchments and seem to be caused by natural and anthropogenic factors. The assessment of the likelihood of occurrence of maximum low flow episodes allowed to indicate areas with a high risk of water shortage dangerous for water ecosystems and water management operations. The analysis of selected features of extreme low flow events has identified a group of determinants that should be taken into account when planning the local exploitation of water resources, as well as in long-term water management strategies nationwide.

The next chapter is entitled "**Dynamics, Range, and Severity of Hydrological Drought in Poland**". The study shows that the use of low flows as indicators of the development of hydrological drought seems promising. The adopted identification and separation criteria allowed for clear isolation of mild and severe hydrological droughts in Poland in 1985–2014. The transformation of parameters describing the duration and relative deficit of the stream flow and the share of the catchment in the entire study area enabled the assessment of hydrological droughts in terms of their duration, severity and range. It also made it possible to indicate periods of varying intensity of the assessed parameters. Assessment of seasonal and long-term variability, as well as analysis of genetic relationships between selected hydrological drought estimators, provided new and valuable cognitive information and allowed to identify a group of factors determining drought development. The analysis proves that both the date of seasonal concentration and the intensity of hydrological drought in Poland show significant and multidirectional variability.

The rest of the work presents a chapter entitled "**Flood Marks in Poland and Their Significance in Water Management and Flood Safety Education**". This chapter provides an overview of floods registered in Poland. They are treated here as an important source of hydrological information to study the extent and effects of the most catastrophic floods in history. In many cases, these objects provide the only objective information about these events. Both typical and unique flood markings in Poland were discussed, including the place and method of their installation in river valleys. Particular attention was paid to research projects aimed at creating a nationwide database available to all Internet users and updated with their participation. This is a new approach that allows more effective search for such objects in the field and their full inventory and archiving. The paper pointed out that flood markings should play an important role in flood education. Especially for communities living in flood plains that are threatened by floods.

The next chapter of the "**Flood Potential of Polish Rivers**" presents the spatial diversity of the flood potential indicator on Polish territory. This measure can also be applied to other geographical areas and can be a useful measure to compare, for example, EU countries. In the case of Poland, the flood potential indicator map was used to prepare flood risk management plans for the Vistula. It was shown that floods generated in the Carpathian Mountains spread in the middle course of the Vistula. This means that the middle course of the Vistula can spread floods of a magnitude characteristic for mountain tributaries. To obtain consistent results, it is recommended to use a common maximum flow record length for all hydrological meters analyzed. The presented method of calculating the flood indicator can be used to draw up flood risk management plans. The indicator quantifies the catchment's potential to achieve maximum flow at flood peak.

"**Flood Risk Management System in Poland**" is discussed in the next chapter. The key element of flood risk management in Poland is determining the distribution and level of aggregate risk in the following categories: health and life, natural environment, cultural heritage and economic activity. An effective management system based on flood risk management strategies and plans will contribute to reducing the negative effects of floods and will contribute to improving public safety.

The purpose of risk management is to reduce the likelihood of flood risk and loss by introducing appropriate mechanisms and instruments. The new risk management system in Poland, also based on non-technical methods of flood prevention, includes initial risk assessment, preparation of flood risk maps and flood risk maps as well as flood risk management plans. Flood risk management plans are a planning document setting out the main objectives, including, for example, reducing the risk, reducing vulnerability to flood in threatened areas and improving the ability to reduce flood risk. Flood risk management strategies implemented in Poland include: "Leading floods away from people", "Leading people away from floods" and "Learning how to live with a flood". Non-technical methods of reducing flood risk include: administrative and social activities Flood risk prevention focuses on designing flood management plans taking into account economic development and climate.

The chapter entitled "**Water Management in the Pomeranian Rivers Estuary Zone on the Background of Hydro-Meteorological Conditions**" the issues of the contact zone between sea and land. Some coastal estuaries lead directly to the sea, while others lead to large coastal lakes, creating different hydrographic systems. Reconstruction and construction of new levees, engineering channels and construction of flow regulation devices, strengthening of river banks, construction of retention reservoirs, as well as increasing retention capacity of catchments will help prevent floods. The hierarchical organisation of outflow is characteristic for the South Baltic Lake Districts, including the Pomeranian catchment. The water cycle of Pomeranian catchments depends more on the zonal factors, and less on local factors. The morphological contrast between the raised and diversified relief of the lake district and the coastal plains and the range of the Baltic Sea influence are of vital importance in shaping the climate conditions of Pomerania. Atmospheric precipitation determines the water cycle in Pomerania. The main alimentation zone of surface and underground waters in the Pomeranian Lake District and the coastal region are highly elevated plateaus. Water resources of the studied rivers were determined on the basis of the unit outflow and the runoff layer are measures. All rivers have a unit outflow higher than the average for Poland estimated at 5.6 $dm^2 \cdot s^{-1} \cdot km^{-2}$. Air pressure changes, wind waves and hydrological factors associated with the seasonal outflow of land waters cause changes in the level of the southern Baltic waters include water accumulation and depressions. The coastal lakes that often are connected with the sea by engineered canals, play a unique role in the potamic outflow of the Pomeranian rivers. These lakes are somehow an extension of the main drainage base on the continent, creating a regional base. The high content of substances in the coastal waters of the Southern Baltic is mainly due to the rivers carrying pollution load. Delivery of pollutants to the Baltic Sea, takes place via rivers, but the nutrient load in rivers contains only a particular part of nutrients from various sources in the basin get to the rivers and then to the receiver.

1.5 Economic Importance of Water Resources

The next part of the work consists of four chapters. The first of them concerns "**Exploitation of Rivers in Poland for the Purposes of Electricity Production – Current Condition and Perspectives for Development**". According to the findings of the European Union, the energy industry in Poland is forced to increase electricity production from renewable energy sources. The use of renewable energy sources in Poland does not exceed 8–10% of electricity generated. Environmental pollution and decreasing fossil fuel resources convince to increase the share of wind, solar and water energy. The construction of new hydropower plants is expensive, but on the other hand, the durability of these energy sources is incomparably greater than that of photovoltaics or wind turbines. Poland has the potential and a good 100-year tradition in the use of hydropower. It is only necessary to draw the attention of decision-makers to this underdeveloped area of the Polish energy sector. The potential for increasing electricity production from hydropower plants is seen in the hydrotechnical development of the lower Vistula. This would increase the production of renewable electricity by 4.5 TWh/year. This is a great potential that has not been used, which requires support from the Polish state.

One of the basic goals of European transport policy is to ensure sustainable development in the field of transport. The next chapter was devoted to "**Waterways in Poland – the History, Present State and Future**". Currently, the role of inland navigation in Poland is negligible. Poland has a favorable arrangement of inland roads. Polish waterways are elements of three international waterways: E30 – connecting the Baltic Sea with the Danube, E40 – connecting the Baltic Sea with the Black Sea and E70 – connecting the Baltic Sea with the Atlantic Coast. However, the main factor determining the efficiency of inland navigation is the correspondence between ship sizes and waterways. Only about 6% of Polish waterways are suitable for modern navigation on international waterway standards (categories IV and V). The remaining 94% (3441 km) of waterways meet regional standards (categories I,

II and III). This sector of the economy requires not only significant expenditure from the state budget, but also public acceptance of the scope of large-scale activities and their impact on the environment. The lack of a forward-looking plan for the modernization of waterways and the construction of infrastructure developed in consultation with various social groups only leads to the intensification of existing conflicts. It seems to be of utmost importance in terms of the program for the development of inland waterways in Poland should be the results of scientific research, which should form the basis for shaping the transport policy of the country and the region when developing strategic plans and making decisions about their implementation.

A separate problem raised in this part of the work concerns "**The Effects of Plant Irrigation in Poland**". The chapter discusses the importance of plant irrigation in Poland. The production effects of the treatment have been characterized depending on the weather, climate and soil factors. The results of irrigation of the main crops are presented. Actions for the development of plant irrigation in Poland were indicated, taking into account, among others, the economic aspect. Plant irrigation in Poland is complementary. The effects of production depend to a large extent on soil conditions and precipitation during the period of high water demand of plants. The production and economic efficiency of agricultural crops condition the development of irrigation systems, and thus the water consumption in this area.

In Poland, droughts are weather phenomena occurring quite often, but irregularly. Their occurrence in the period of high demand for water, e.g. cereal plants and rapeseed (May–June) in the years 1981–2010 in Pomerania and Kujawy amounted to 23.3–30.0%. Therefore, irrigation of plants in this country is an emergency procedure applicable to supplement periodic rainfall shortages in relation to the water demand of plants. The authors developed prognostic formulas that determine the variability of production effects during irrigation in a given area in subsequent growing seasons, resulting from uneven rainfall. Currently, the development of irrigation in Poland concerns primarily intensive production of vegetables and fruits on a large scale, because irrigation profitability rates are much lower in the production of agricultural plants compared to horticultural plants. Growth prospects for plant irrigation are determined by two main factors: economic efficiency and availability of water sources.

The last chapter in this part of the work concerns "Power Plant Open Cooling System in the Context of the Objectives of the Water Framework Directive". It discusses the problem of using water for cooling in conventional power plants in the light of the application of the provisions of the Water Framework Directive and Water Management Plans in River basin management plans (RBMPs) were first developed in Poland in 2009 and were approved in 2011. The plans are a strategic document that forms the basis for making decisions determining the state of water resources, improving the process of achieving or maintaining good status of waters and related ecosystems, and to determine the need for rational water management principles in the future. For conventional coal or lignite power plants, the impact of conventional power plants on surface waters is usually limited to an increase in their temperature a few degrees Celsius. The authors point out that each case is different and many aspects related to the aquatic environment should be considered. This analysis should be carried out to demonstrate all the pros and cons of the solutions being considered in order to meet applicable formal and legal requirements. To assess the possibility of avoiding the negative impact of discharged waste water on the environmental objectives of the WFD, the feasibility of implementing alternative solutions, including economic, energy and environmental considerations, should be assessed.

The book ends with conclusions and recommendations, which are contained in Chap. 19.

Acknowledgements The authors would like to acknowledge the authors of the chapters for their efforts during the different phases of the book including their inputs in this chapter.

References

1. Environment (2018) Statistical analyses. Statistics Poland. https://stat.gov.pl/obszary-tematyczne/srodowisko-energia/srodowisko/ochrona-srodowiska-2018,1,19.html
2. Kubiak-Wójcicka K (2020) Assessment of water resources in Poland. In Water resources in Poland: management, Springer Water Series, Springer (Under production)
3. Dynowska I (1971) Types of River Regimes in Poland. Zeszyty Naukowe UJ, CCLXVIII, Prace Geogr 28, p 150 (In Polish)
4. Wrzesiński D, Sobkowiak L (2018) Detection of changes in flow regime of rivers in Poland. J Hydrol Hydromech 66(1):55–64. https://doi.org/10.1515/johh-2017-0045
5. Kundzewicz ZW, Dobrowolski A, Lorenc H, Niedźwiedź T, Pińskwar I, Kowalczak P (2012) Floods in Poland. In: Kundzewicz ZW, Special IAHS (eds) changes in flood risk in Europe. Publication Press, CRC Press, Taylor and Francis, pp 319–334
6. Choiński A (2007) Physical limnology of Poland. UAM Science Publishing, Poznań (In Polish)
7. Szwed M (2019) Variability of precipitation in Poland under climate change. Theoret Appl Climatol 135:1003–1015. https://doi.org/10.1007/s00704-018-2408-6
8. Ilnicki P, Farat R, Górecki K, Lewandowski P (2015) Long-term air temperature and precipitation variability in the Warta River catchment area. J Water Land Dev 27 (X–XII):3–13. https://doi.org/10.1515/jwld-2015-0019
9. Kubiak-Wójcicka K (2019) Long-term variability of runoff of Vistula River in 1951–2015. "Air and water—components of the environment" conference proceedings, Cluj-Napoca, Romania, pp 109–120. https://doi.org/10.24193/AWC2019_11
10. Owczarek M, Filipiak J (2016) Contemporary changes of thermal conditions in Poland, 1951–2015. Bull Geogr Phys Geogr Ser 10:31–50. https://doi.org/10.1515/bgeo-2016-0003

Chapter 2
Water Resources and Management of Poland in SCOPUS Database

Mahmoud Nasr, Michael Attia, and Abdelazim M. Negm

Abstract Recently, Poland has faced significant issues in the sector of water resource, availability, and management because of the sharp increase in various domestic, agricultural, and industrial activities. This chapter gives an overview of the status, challenges, and management of water in Poland. The major water resources, e.g., rivers, lakes, and dams/reservoirs in Poland are also listed. Moreover, the chapter mentions the grant-funded researches, leading international journals, institutional affiliation, and acknowledged sponsors and projects available in the SCOPUS database, covering the water aspects in Poland. The findings of this work would support local residents, investors, government officials, industrialists, and academics, dealing with water concerns in Poland.

Keywords Poland · SCOPUS database · Water resources

2.1 Introduction

Poland is a Baltic Sea country located in Central Europe, having a population of 37,921,592 capita in 2018 [1]. Recently, the water consumption profile in Poland has enlarged due to national economic development, population growth, and

M. Nasr (✉)
Sanitary Engineering Department, Faculty of Engineering, Alexandria University, Alexandria 21544, Egypt
e-mail: mahmmoudsaid@gmail.com; mahmoud-nasr@alexu.edu.eg; mahmoud.nasr@ejust.edu.eg

Environmental Engineering Department, Egypt-Japan University of Science and Technology (E-JUST), Alexandria 21934, Egypt

M. Attia
Irrigation Engineering and Hydraulics Department, Faculty of Engineering, Alexandria University, Alexandria 21544, Egypt
e-mail: eng.michael_george@yahoo.com

A. M. Negm
Water and Water Structures Engineering Department, Faculty of Engineering, Zagazig University, Zagazig 44519, Egypt
e-mail: amnegm85@yahoo.com; amnegm@zu.edu.eg

M. Zeleňáková et al. (eds.), *Management of Water Resources in Poland*, Springer Water,
https://doi.org/10.1007/978-3-030-61965-7_2

climatic change [2]. The increases in social practices, agricultural activities, and industrialization have also contributed to the rise in water consumption [3, 4].

The status of water in Poland can be described as follows [5]:

- In 2017, water precipitation in depth = 600 mm per year, corresponding to an annual volume of 187.6 billion m^3.
- In 2017, renewable water resources = 61 billion m^3/year, providing a yearly water resource of 1585 m^3 per capita. However, the average freshwater availability in Poland is considered one of the lowest values in Europe.
- In 2016, the percentages of water utilization were 19% for municipal use, 71% for industrial use, and 10% for agricultural use. These percentages were covered by a total water withdrawal of 10.58 billion m^3.
- In 2015, about 1.7% of the population in Poland had no access to safe drinking water.

This chapter provides an overview of the trends and statistics of water resources (e.g., surface water and groundwater) in Poland. Moreover, the chapter includes sufficient information on the institutional affiliations, countries, and funding agencies and sponsors, concerning the management of water bodies in Poland.

2.2 Rivers in Poland

The river system in Poland is recognized as an essential source of drinking, domestic, and agricultural purposes [6]. The river system has also played an important role in the growth and development of Poland's economy and civilization. Vistula River is the most important river in Poland [7], having a total length of about 1022 km. The river is associated with a basin area in Poland, equivalent to 168,868 km^2. Warta River is the second largest river in Poland, with a total length of 795 km and a basin area = 54,520 km^2. Another important river in Poland is known as Oder, representing a length of 726 km and a basin area = 106,043 km^2 in the country. These rivers are followed by Bug, Narew, and Noteć, having lengths of 590, 443, and 391 km in Poland. However, the rivers have recently received untreated and/or partially treated wastewater; hence, several parts of the river water became unfit for direct potable use.

2.3 Lakes in Poland

Lakes are recognized as large water bodies (either fresh- or salt-water) surrounded by land. Śniardwy Lake is the most important lake in Poland, with an area of 113.8 km^2 and a maximum depth of 23 m. This lake is followed by Lake Mamry with an area of 104 km^2 and Łebsko Lake (71.4 km^2). Other lakes in Poland include Lake Jamno, Lake Wigry, Lake Gopło, Lake Orzysz, Lake Karwowo, Lake Karaś, Lake

Licheńskie, Lake Ełk, Lake Gosławskie, Lake Leśnias, Nyskie Lake, and Lake Słupca [8–10].

2.4 Dams in Poland

Generally, dams are constructed for water storage in associated reservoirs, the control of disasters (i.e., floods and droughts), and navigation purposes. The major dams in Poland can be described as follows [11], viz., Name (capacity in 10^6 m^3 and maximum depth in m): Leśna (16.8 and 35.8), Pilchowice (50.0 and 46.7), Złotniki (12.1 and 27.5), Otmuchów (130.5 and 18.4), Kozłowa Góra (17.6 and 6.5), Porąbka (27.2 and 21.2), Rożnów (159.3 and 31.5), Czchów (12.0 and 9.5), Goczałkowice (161.3 and 13.0), Tresna (96.1 and 23.8), Solina (472.4 and 60.0), Włocławek (453.6 and 12.7), Nysa (124.7 and 13.3), Sulejów (84.3 and 11.3), Słup (38.7 and 19.1), Chańcza (24.2 and 12.8), Mietków (71.9 and 15.3), Bukówka (16.8 and 22.4), Dobromierz (11.4 and 26.7), Dobczyce (141.7 and 27.9), Jeziorsko (202.0 and 11.5), Dzierżno Małe (12.6 and 13.1), Czorsztyn (231.9 and 54.5), Nielisz (28.5 and 8.6), Wióry (35.0 and 23.4), and Sulejów (84.3 and 11.3).

2.5 Reservoirs in Poland

Generally, the reservoirs are used for the aims of water storage and supply, irrigation, hydroelectric production, and land protection against flooding. Poland includes a number of reservoirs such as Czorsztyn Reservoir in Dunajec; Dobczyce Reservoir in Raba; Goczałkowice Reservoir in Vistula; Jeziorsko Reservoir in Warta; Koronowo Reservoir in Brda; Nysa Reservoir in Nysa Kłodzka; Otmuchów Reservoir in Nysa Kłodzka; Rożnów Reservoir in Dunajec; Solina Reservoir in San; Turawa Reservoir in Mała Panew; Włocławek Reservoir in Vistula; Zegrze Reservoir in Narew; Żywiec Reservoir in Soła.

2.6 Water Management in Poland

The "Water Law" act of 1974 and a series of laws, regulations, and guidelines have been established to handle the issues of water management in Poland [12]. For instance, a framework for the river basin management was proposed in the late 1980s, followed by the initiation of Regional Water Management Boards (RWMBs) in 1991. With a further modification of the sub-basin boundary agreement in 1999, the RWMBs became responsible for the planning and co-ordination of the polish river basins and lakes. The management of waterworks and wastewater treatment systems was undertaken by counties and municipalities, whereas the water use tasks were considered and regulated by provincial governments. In 2004 (after Poland's entry into the EU), the EU Water Framework Directive (WFD) comprised an official agenda for Polish water management, in which ten river basin districts were defined. Further, the National Water Management Authority (i.e., a sector in the Ministry of Environment) was established in 2006 to share the competent authority for water resources management with the RWMBs. For example, the national authority arranges and regulates the river basin management plans (RBMPs), whereas the Boards make reporting, provide the required data, and establish consultations. However, these management regimes have recently experienced some water-related issues, which were beyond the jurisdiction of Poland's Ministry of Environment. Currently, the Ministry of Environmental Protection, Natural Resources and Forestry is responsible for the water framework directive in Poland (i.e., measures and implications). Moreover, some individual and small pilot water management projects are carried out to reduce the water and wastewater treatment costs.

2.7 Poland's Water Statistics from SCOPUS Resource Library

Figure 2.1 shows the number of published documents for the study period from 2001 to 2019, handling and addressing the major water issues in Poland [13]. Figure 2.1a represents the records retrieved from the SCOPUS database using "Water", "Quality", and "Poland" as the keywords for search. A total number of documents of about 503 was reported during 2001–2010, which increased by two-fold during 2011–2019. Furthermore, the documents were searched in the SCOPUS database by using "Water", "Resource", and "Poland" as keywords (Fig. 2.1b), depicting 220 documents during 2001–2010 and 545 documents during 2011–2019. The entire number of publications using the search keywords "Water", "Management", and "Poland" was 335 documents during 2001–2010, and it increased to 777 documents during 2011–2019 (Fig. 2.1c). The recent improvement in the number of published articles indicates that the title of "Water Quality, Resource, and Management in Poland" has become a crucial field of investigation.

The search signified different types of documents, including article (~84%), conference paper (~10%), review (~3%), and book chapter (~1%). The articles covered various subject areas such as Environmental Science; Earth and Planetary Sciences; Agricultural and Biological Sciences; Engineering; Social Sciences; Chemistry. The articles have been published in several international journals, including Ecological Indicators; Smart Innovation, Systems and Technologies; Science of the Total Environment; Environmental Management; Environmental Pollution. The aims and scopes covered by these journals include the minimization of environmental pollution related to human health, and the use and conservation of natural resources. The journals were managed by several publishers, viz., Springer,

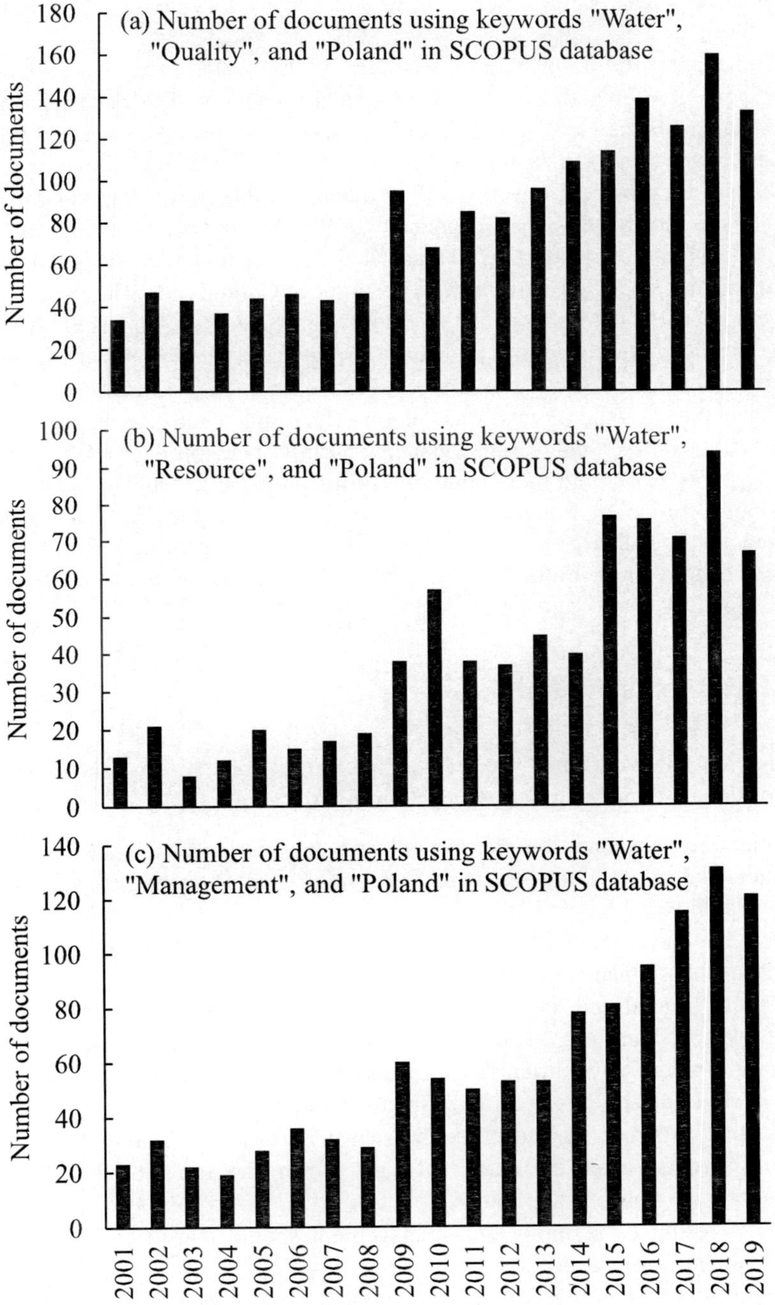

Fig. 2.1 Increasing number of published documents retrieved from SCOPUS database from 2001 to 2019 using research keywords **a** "Water", "Quality", and "Poland", **b** "Water", "Resources", and "Poland", and **c** "Water", "Management", and "Poland"

Taylor & Francis, Nature Publishing Group, Sage Publications, Wiley-Blackwell, and Elsevier.

The authors of publications associated with "Water Quality, Resource, and Management in Poland" are affiliated to Polish Academy of Sciences; Uniwersytet im. Adama Mickiewicza w Poznaniu; Uniwersytet Warminsko-Mazurski w Olsztynie; University of Lodz; AGH University of Science and Technology; University of Silesia in Katowice; Uniwersytet Przyrodniczy w Poznaniu; Gdańsk University of Technology; Szkoła Główna Gospodarstwa Wiejskiego; Polish Geological Institute – National Research Institute. The main funding sponsors acknowledge in the documents were Narodowe Centrum Nauki; European Commission; European Regional Development Fund; Ministry of Higher Education; Scientific Committee on Antarctic Research; Narodowe Centrum Badań i Rozwoju; Ministerstwo Nauki i Szkolnictwa Wyższego; Komitet Badań Naukowych; National Centre for Atmospheric Science; European Social Fund. Poland was the top country that contributed to the subject of "Water Resources, Quality, and Management in Poland" in the SCOPUS database during 2001–2019, with more than 90% of the published documents. Based on the previous information, Poland country was followed by Germany, United Kingdom, United States, Netherlands, and Czech Republic, revealing the importance of Poland water resources on the Baltic region and other countries with similar climatic regimes.

2.8 Recommendations

Poland has relatively small water resources; hence, some recommendations should be considered to maintain water availability in the country:

- Scientific, academic, environmental, and engineering contributions should be interacted to solve the recent and upcoming challenges essential to the water availability in Poland.
- Feasible and cost-effective frameworks for the long-term management of water resources in Poland are required.
- Regular spatial and temporal surveys on various physicochemical parameters of water bodies (e.g., temperature, pH, turbidity, solids, dissolved oxygen, organic matter, and nutrients), as well as pathogenic *Escherichia coli*, *Salmonella typhimurium*, and *Vibrio cholera*, should be assessed.
- Adaptation strategies and proper wastewater treatment technologies should be considered to avoid the release of wastes from the agricultural and industrial sectors into Polish water bodies.
- Water resources planning and management should consider additional water supply infrastructures.
- More stakeholders should contribute to the Regional Water Management Boards, dealing with water resources planning and management, flood and erosion control, and water purification.

- The amendments of the "Water Law" act signed by Poland should harmonize with the EU legislations and international agreements.

2.9 Conclusions

This chapter aims at providing an overview of the essential water resources in Poland. It can be concluded that:

- Poland comprises several freshwater resources retained in surface water bodies and within soil layers.
- In Poland, the main four rivers are Vistula, Warta, Oder, and Bug; the essential three lakes are Śniardwy, Mamry, and Łebsko; the main dams are Solina, Pilchowice, Czorsztyn-Niedzica, Swinna Poreba, and Roznow; the major reservoirs are Czorsztyn, Dobczyce, Goczałkowice, Jeziorsko, and Koronowo.
- In 2017, renewable water resources in Poland reached 61 billion m^3 per year, providing an annual water resource of 1585 m^3 per capita.
- Based on the SCOPUS database, the total number of published documents that handled the water issues in Poland during 2011–2019 was approximately two-fold that during 2001–2010.
- Proper awareness, frameworks, and scientific meetings that address the environmental and health issues should be delivered to residents and farmers.
- The government should give significant actions for rainwater harvesting and wastewater reuse to overcome future water demands.
- The findings obtained from this chapter would support the government, policymakers, and scientists dealing with all problems of water resources in Poland.

Acknowledgements The first author would like to acknowledge Nasr Academy for Sustainable Environment (NASE).

References

1. [Online]. Available https://www.worldometers.info/world-population/poland-population/. Accessed 25 Dec 2019
2. [Online]. Available https://stat.gov.pl/en/basic-data/. Accessed 25 Dec 2019
3. Gilewski P, Nawalany M (2018) Inter-comparison of rain-gauge, radar, and satellite (IMERG GPM) precipitation estimates performance for rainfall-runoff modeling in a mountainous catchment in Poland. Water 10(11):1665
4. Jaiswal M, Hussain J, Gupta S, Nasr M, Nema A (2019) Comprehensive evaluation of water quality status for entire stretch of Yamuna river, India. Environ Monit Assess 191(4):208
5. [Online]. Available https://www.worldometers.info/water/poland-water/. Accessed 25 Dec 2019

6. Bajkiewicz-Grabowska E, Markowski M, Golus W (2020) Polish rivers as hydrographic objects. In: Korzeniewska E, Harnisz M (eds) Polish river basins and lakes—part I. The handbook of environmental chemistry, vol 86. Springer, Cham
7. Cerkowniak G, Ostrowski R (2020) Simple approach to long-term morphodynamics of the river delta applied to the Vistula river outlet. J Waterw Port Coast Ocean Eng 146(1):05019002
8. Bajkiewicz-Grabowska E (2020) Geoecosystems of Polish lakes. In: Korzeniewska E, Harnisz M (eds) Polish river basins and lakes—part I. The handbook of environmental chemistry, vol 86. Springer, Cham
9. Choiński A, Ptak M (2020) Occurrence, genetic types, and evolution of lake basins in Poland. In: Korzeniewska E, Harnisz M (eds) Polish river basins and lakes—part I. The handbook of environmental chemistry, vol 86. Springer, Cham
10. Tandyrak R, Grochowska J, Parszuto K, Augustyniak R, Łopata M (2020) Environmental conditions in Polish lakes with different types of catchments. In: Korzeniewska E, Harnisz M (eds) Polish river basins and lakes—part I. The handbook of environmental chemistry, vol 86. Springer, Cham
11. Chmiel S, Sposób J, Mięsiak-Wójcik K, Michalczyk Z, Głowacki S (2020) The effect of a dam reservoir on water trophic status and forms of river transport of nutrients. In: Korzeniewska E, Harnisz M (eds) Polish river basins and lakes—part I. The handbook of environmental chemistry, vol 86. Springer, Cham
12. [Online]. Available https://www.un.org/esa/earthsummit/pold-cp.htm#chap18. Accessed 25 Dec 2019
13. Scopus, 12 2019. [Online]. Available https://www.scopus.com/search/form.uri?display=basic. Accessed 25 Dec 2019

Part II
Water Resources in Poland

Chapter 3
Administration of Water Resources Management: Key Facts About Water Resources in Poland

Katarzyna Kubiak-Wójcicka

Abstract This chapter reviews and evaluates institutional solutions related to water management. It indicates which institutions have been responsible for water management in Poland in the last dozen or so years and what institutional changes have been introduced since 2018. The study identifies the main problems that over recent years have resulted from water management by various institutions responsible for key issues regarding water quantity and quality. The main problems related to proper water management in Poland include lack of uniform interpretation of water law, lack of a coherent information system, and access to information by both water management employees and water users. In order to deal with these problems, first of all, there should be conducted a series of training for both employees and stakeholders regarding a uniform interpretation of water law. A full assessment of the water management reform implementation in Poland and of all the changes introduced after 2018 will be possible in about 2 or 3 years.

Keywords Water management · Water law · Institution · Organizational chart · Water resource · Governance · Poland

3.1 Introduction

Water is a resource directly relevant to the development of society. It has always been the basis of civilisation. Unlimited use of water resources by humans and their activity has led to their exhaustion or degradation as well as transformation of water conditions.

The extent of these changes and problems caused by them are dependent on various factors, which include the amount of water resources, their availability, and management. Hence, in a number of scientific publications there appear considerations on water resources management in different countries, e.g. the Netherlands [1], Italy [2], Spain [3, 4], Switzerland [5], Ukraine [6, 7] and many others [8–10].

K. Kubiak-Wójcicka (✉)
Department of Hydrology and Water Management, Faculty of Earth Sciences and Spatial Management, Nicolaus Copernicus University, Lwowska 1, 87-100 Toruń, Poland
e-mail: kubiak@umk.pl

M. Zeleňáková et al. (eds.), *Management of Water Resources in Poland*, Springer Water,
https://doi.org/10.1007/978-3-030-61965-7_3

These studies provide a picture of systemic solutions to water management at the national or regional level, and above all, how they work in practice. Guidance on the strengths and weaknesses of these systems, based on different experiences, show that they are not only important for the areas affected by water scarcity, e.g., Qatar [11], Iran [12], but also regions that have large water resources, such as Canada [13, 14]. The issues of the amount of water, and more and more often emerging issues of its quality, have led to the examination of water resources in many aspects. The key to solving the problems is the appropriate management of water resources [15–17]. In recent years, researchers have been discussing the importance of water management in the face of climate change [18–20], as well as the scale and form of social participation in water resources management [21–25].

The only possibility for rational water resources use is integrated water resources management (IWRM). According to Biswas [26], this was already known at the beginning of the twentieth century, but the return to it occurred only in the 1990s. One definition of integrated water management has been presented by Global Water Partnership [27] as "the process promoting the harmonious development and management of water, space and other resources, in order to maximise social and economic benefits while maintaining healthy ecosystems."

The IWRM concept has been debated by many scholars [28–30]. A new management concept is an adaptive comanagement (ACM). The term "comanagement" indicates the collaboration of a wide range of actors from government and civil society in sharing managing power and responsibilities across local, regional, and national levels [31–33]. In Poland, the problem of integrated systems in water management was considered by many authors [34–37].

This chapter presents water resources management in Poland over the last twenty years. Based on the analysis of official documents (legal acts), literature review, interviews with employees at various levels of offices related to water management and the author's own experiences, the case of water management in Poland has been described. The interviews focus on the determination of roles and efficiency of the organizations responsible for water management at the governmental and municipal levels. The surveys were conducted with 10 persons with at least 5 years of professional experience.

Poland is a country where water resources are one of the poorer in Europe. In the years 1951–2015 the average annual surface runoff from the territory of Poland, including its tributaries from abroad, was 61 km^3. The average specific runoff in the largest basin of the Vistula is between 5.3 and 10 $dm^3 \cdot s^{-1} \cdot km^{-2}$ [38]. This gives the annual water resource size of 1600 m^3 per capita, while in most European countries, freshwater resources are at a level of approximately 5000 m^3 per capita. In addition, surface water resources of Poland are highly volatile in terms of time and space, which causes periodic water excesses and deficits in rivers [39]. Therefore, the management of water resources is extremely difficult. On May 1st 2004, Poland became a member of the European Union and hence must apply the rules of the Frame Water Directive, which obliges the member states to rational use and protection of water resources, following the paradigm of the balanced development. The current scheme of water management is the result of the transposition of rules of the WFD to the Polish code of laws, mainly through the Water Law act. The integrated water resources management (IWRM) in Poland faces significant problems, which will be presented in this article. Moreover, Poland is in the course of political changes related to government change which occurred at the end of 2015. Significant changes in the scheme of organizational water management were initiated in January 2018.

3.2 History of Water Management in Poland Until 2001

Formal water management in Poland dates back to the early twentieth century when the first bill of Water Law of 19 September 1922 was passed. The bases of the draft were two documents: the Austrian Water Act of 30 May 1869 from which the ownership matters were taken, and the Prussian Water Law of 7 April 1913 from which the issues of water use were taken.

The term "water management" was used for the first time in Poland at the end of the 1920s during the discussion of the subject and organizational assumptions of the First Polish Hydrotechnical Convention. After the Second World War, the issues of water management in Poland were initially entrusted to the Ministry of Transport (Department of Waterways), and in 1948 water management was handed over to the Ministry of Shipping. In 1960, the Central Office for Water Management was established, which was to deal with the entirety of water management in Poland.

Subsequent Water Law acts were passed on May 30th, 1962 and October 24th, 1974 and were related to the socialist system in Poland. Water management in that period was based on the administrative division of the state, according to the division into voivodeships. Changes in water management are related to system transformation in Poland which started in 1989. The administration, maintenance and operation of the State Treasury assets (i.e., rivers of special importance, waterways, water reservoirs, and water infrastructure) were dealt with by the Regional Water Management Authorities. On February 1st 1991, the Ministry of Environment Protection, Natural Resources, and Forestry appointed 7 Regional Water Management Authorities, that borders followed the hydrographic (and not the administrative division) of the country.

Regional Water Management Directorates which were established to maintain waters owned by the State Treasury and Regional Water Management Authorities were functioning in parallel. Pursuant to the Ordinance of the Minister of Environment of November 29, 1999, regarding the reorganization and scope of operation of regional water management boards, changes were introduced in the administrative structure of water management. On January 1, 2000, regional directorates were merged with regional management boards, summing up their existing tasks and assuming further reform of water management based on the new Water Law of 2001. Changes in the field of water management resulted from the administrative reform of the Polish state, which came into force on January 1, 1999. As a result of the reform, a three-level structure of territorial division was introduced, divided into voivodships, districts and communes. According to this division, 16 voivodships (instead of 49), 308 districts and 2489 communes were created [40]. This division, since its introduction, has undergone minor modifications.

3.3 Water Management in Years 2001–2017

3.3.1 Water Management Organization Structure

The new Water Law act of 18 July 2001, valid until 2017, had a significant impact on the shape of the water management system in Poland.

Water management in Poland was carried out both via the state and local governments. The authority responsible for the proper management of inland waters in Poland, according to the provisions of the Water Law act of 2001, was the President of National Water Management Authority, directors of national parks and Voivodeship Marshals. According to the act, water is the property of the State Treasury, other legal entities, and private persons. Waters that belong to the State Treasury or local government units are the public waters [41].

There was no single ministry which would deal with water management as a whole, which means that competences in water management were divided among different sectors of the national economy. Management of water resources and the related infrastructure was carried out by the Ministry of the Environment, Ministry of Naval Management and Inland Navigation, Ministry of the Interior, Ministry of Agriculture and Rural Development (Fig. 3.1).

The Prime Minister exercised direct supervision over the Government Plenipotentiary for the "Program for the Oder – 2006". The plenipotentiary was appointed by the Act of 6 July, 2001 on establishing a long-term program "Program for the Oder – 2006" (Journal of Laws No. 98, item 1067), which came into force in 2002. The program included the reconstruction and modernization of the Oder Water System, which was destroyed during the flood on the Oder River in 1997. The function of the plenipotentiary was held by the voivode of Lower Silesia. The purpose of this program was to modernize the Oder River to a modernly developed ecological corridor in accordance with the principles of sustainable development. As part of this program, specific statutory tasks were implemented in the area of construction of a flood protection system, protection of the natural environment and water purity, elimination of flood damage and preventive spatial development and restoration of eco-systems. The Government Plenipotentiary for the "Program for the Oder – 2006" served until the end of 2014. He was dismissed by the Council of Ministers, and all his activities were then coordinated by the President of the National Water Management Authority (KZGW) by the end of 2015.

A key role in the management of inland waterways is played by the Ministry of the Environment, which has an extensive organisational structure. The Minister of

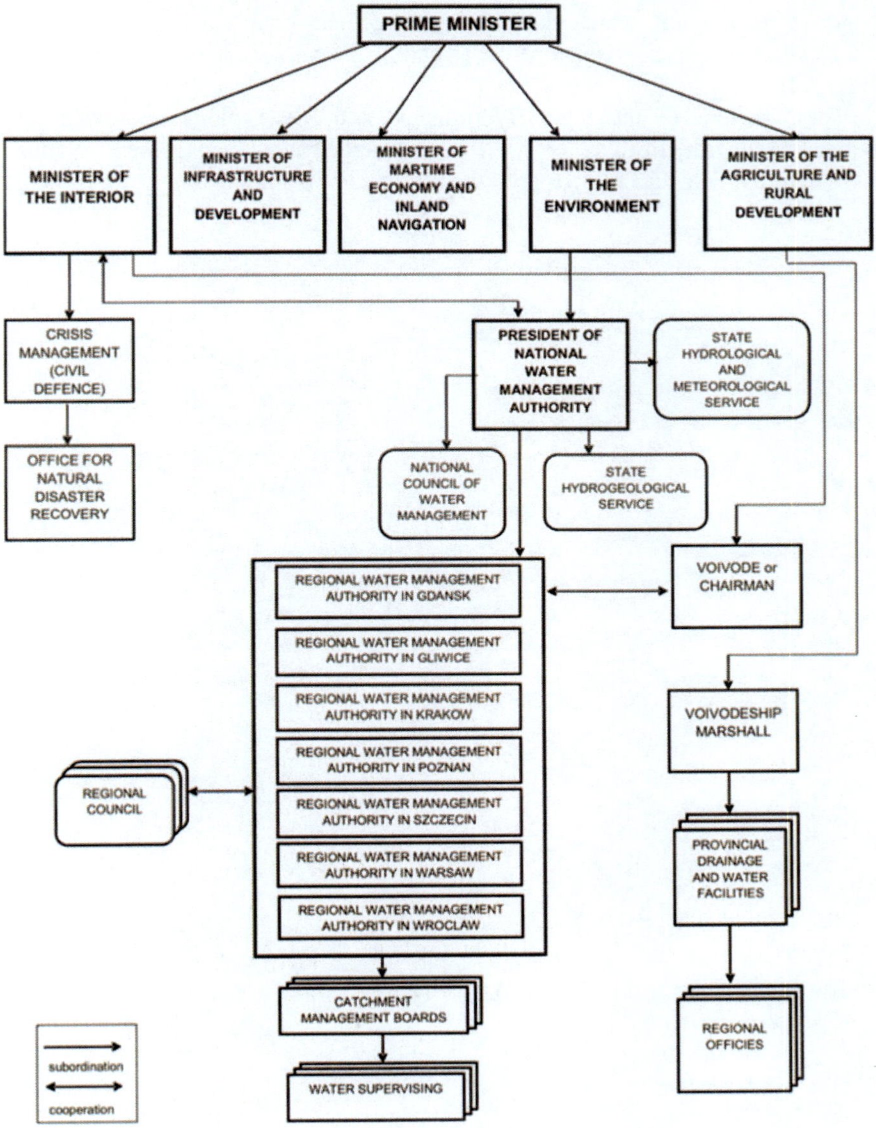

Fig. 3.1 Organisation of water management in Poland (until December 2017) [42]

the Environment manages the government administration departments and oversees the President of the National Water Management Authority (KZGW). In turn, the President of the KZGW plays a paramount function for the Regional Water Management Authorities, the State Hydrological and Meteorological Service and the State Hydrogeological Service. Regional water management authorities are engaged in the maintenance of water and water facilities in individual drainage basins of the water regions, within which catchment management boards are delimited. Local units called water supervisors operate as part of the catchment management boards (Fig. 3.2).

The functions of the State Hydrological and Meteorological Service were performed by the Institute of Meteorology and Water Management, National Research Institute, which supervised the activities of 6 administrative departments.

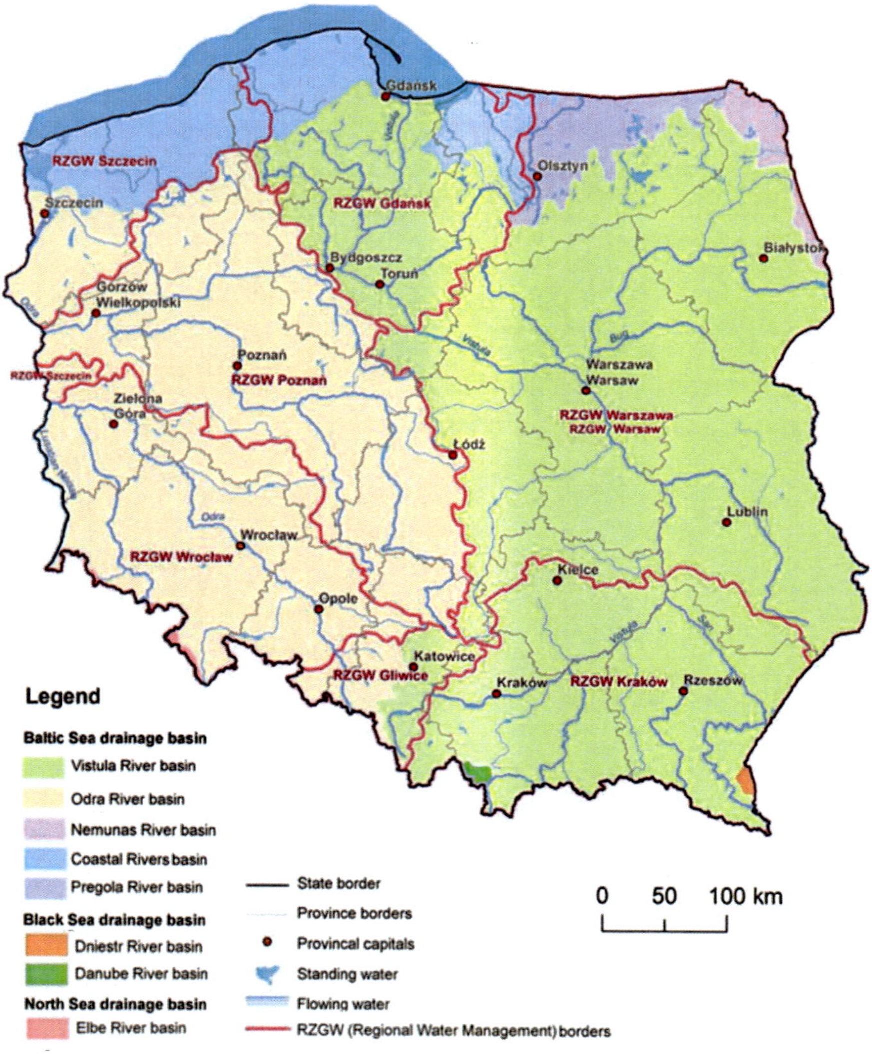

Fig. 3.2 Administrative and hydrographical division of Poland

The Minister of the Environment also held direct authority over the General Inspector of Environmental Protection (GIOŚ), the General Directorate for Environmental Protection (GDOŚ) and the Polish Geological Institute (PGI State Research Institute). The GIOŚ performs its tasks using 16 branches of Voivodship Inspectorates for Environmental Protection, which conduct State Environmental Monitoring (including, among others, surface water quality) and supervision of entities using the environment. In turn, the GDOŚ supervises the tasks of 16 Regional Directors of Environmental Protection, which implement tasks related to environmental policy in the field of nature conservation management, supervision of the investment process and provision of information about the environment in the area of the voivodship.

The President of the National Water Management Authority (KZGW – Krajowy Zarząd Gospodarki Wodnej) performed ownership rights in relation to public waters owned by the state, in relation to the waters essential for the development of water resources and flood protection, particularly groundwater and surface inland waters. The inland surface waters are in mountain streams and their sources, in natural watercourses, from the sources to the mouth, with an average multi-year discharge equal to or greater than 2.0 $m^3 \cdot s^{-1}$ at the cross-estuary, lakes and artificial water reservoirs through which referred streams flow, as well as in inland waterways and border watercourses. Administering waters took into account the division of the country into drainage basins and water regions.

In addition to the President of the KZGW, the unit responsible for water management was the provincial (*voivodeship*) marshal. As part of the government, this task was performed by the voivodeship government in relation to the water essential for regulating water relations for agriculture in order to improve the soil capacity and facilitate its cultivation. Water management was carried out in line with the state administrative division, within administrative units.

The main bodies responsible for water management at the regional level were Provincial Drainage and Water Facilities, which were subject to the marshal of the province. They performed tasks in the field of water management in agriculture and flood protection. Therefore, they were subject to the Minister of Agriculture and Rural Development.

The Ministry of Naval Management and Inland Navigation carried out the activities related to the administration and operation of inland waterways.

In turn, the Ministry of the Interior was taking over some of the tasks in crisis situations in the field of water management during, e.g., floods or droughts.

The authorities responsible for water management in the context of local government were a municipality leader called *wójt* (mayor or president) and the municipality council as well as the governor and district council. Their tasks were mainly related to water and sewage infrastructure (municipal council) and the issuance of water supply consents.

The commune's own tasks include issues related to water supply, sewerage, water supply, and sewage disposal. Detailed legal regulation in this respect is the Act of June 7, 2001, on collective water supply and collective sewage disposal. Collective supply of inhabitants with drinking water is the commune's own task – this results from the content of art. 7 par. 3.1 point 3 of the Act of 8 March 1990 on municipal self-government (Journal of Laws 2013 item 594 as amended) and art. 3 section 1 of the Act of June 7, 2001, on collective water supply and collective sewage disposal (Journal of Laws from 2006 No. 123, item 858, as amended). Communes are responsible for providing the community with water supply and sewage disposal which forces the creation of water and sewage companies or entrusts certain obligations to a water and sewage company. In the last 20 years, we have had in Poland the unique development of infrastructure investments, including those related to the concentration of expenditures from local government funds. Although various aid funds from the European Union were widely used, e.g., the Operational Program Infrastructure and Environment, as well as the National Fund for Environmental Protection and Water Management (NFOŚiGW) however, also the own share of communes was high enough. A significant part of the infrastructure was created in rural areas. In the case of low-income communes, where the settlement network is dispersed, and thus the cost of building the infrastructure is high, these investments have no chance for implementation without the support of the state.

As a result of such division of competence, water management takes place in a river basin and administrative system (Table 3.1).

Poland's accession to the European Union on 1 May 2004 meant that Polish regulations and legislation should be aligned to the EU. The most important EU directives on water policy in the European Union is the Water Framework Directive (WFD), which was approved by the European Parliament and the Council of the European Union as the Directive 2000/60/EC, which came into force on 22 Dec. 2000 and introduction to the new EU Water Framework Directive (accessed 24 July 2006). Transposition of the WFD regulations into Polish legislation was primarily through the Water Law and its implementing legislation. Also, the WFD has reflected also in the Environmental Protection Law and the Act on collective water supply and discharge of wastewater with the implementing legislation to these laws.

The EU water policy is based on the principles of Integrated Water Resources Management (IWRM). These principles include treatment of the drainage basin as a primary area of any planning and decision-making action, socialisation of the decision-making process, integrated approach to surface water and groundwater, treatment of water as a fundamental factor influencing the functioning of ecosystems, as well as implementation of economic mechanisms in water management.

As a result of Poland's accession to the European Union in 2004, in order to align the law with the WFD, the Polish Parliament introduced amendments to the existing Water Law act of 18 July 2001. Due to complicated nature of the changes introduced to the existing Water Law, it has been decided to create a new Water Law act, which initially was supposed to come into force in 2015. That act assumed changes in organizational scheme of water management in Poland and in the matter of water usage fees following the rule "the one that uses the water pays." The act has not been introduced because of political changes that took place. In October 2015, as a result of election to the parliament, after 8 years the ruling party stepped down. The existing ministry division has been preserved with one exception. A new Ministry of Naval Management and Inland Navigation has been created. Its competences were previously (until 2015) in the domain of the Ministry of Infrastructure.

Table 3.1 List of water management authorities in Poland (until December 31, 2017)

Authority	Competences	Area
Regional Water Management Authority	Managing waters with discharge greater than 2 m^3 per second (water management plans)	Basin
Provincial Drainage and Water Facilities	Managing waters with discharge less than 2 m^3 per second and waters agriculturally significant, lakes, ditches, and canals	Voivodeship
Provincial Environment Protection Inspector	Water quality in rivers and lakes	Voivodeship
Regional Environment Protection Management	Nature protection as a part of assessments of environment impact, protection, and management of various sorts of nature protection	Voivodeship
Local municipal government, city mayor or president	Issuing water law consents for special water use	Administrative area of a commune or a city
Provincial crisis management centre	Facilitation of cooperation of all governmental and municipal administration units in the scope of prevention of environmental diseases or environmental threats, e.g., floods Monitoring of threat degree, e.g. from surface waters	Voivodeship
Institute of Meteorlogy and Water Management National Research Institute	Conducting systematic measurements and observations with the use of the network of meteorological and hydrological stations Creation and distribution of hydrological and meteorological forecasts	Administrative departments

In the meantime, the new government has created a new Water Law Act that was announced on July 20, 2017. The Act introduced significant changes in the field of water management, which entered into force on January 1, 2018.

3.3.2 Assessment of the State of Water Management in Poland

The decentralized (or polycentric) system of water management that existed in years 2001–2017 in Poland has faced criticism from society, scientists and users of water resources in the recent years. Among the causes of poor water, management state were scattered management at central and regional levels and lack of strict definition of competences, rules and cooperation forms between state and municipal authorities that were engaged in water management. Based on document analysis and authors own observations, endogenous, and exogenous factors of integrated water management following the idea of the Framework Water Management have been identified. Strong and weak sides as well as opportunities and threats to decentralised water management in Poland have been analysed. The main administrators of waters in Poland were: Voivodeship Marshalls, which did their duties through the regional drainage managements (16 voivodeships) and Country Water Management Authority, which did its duties though 7 Regional Water Management Authorities.

In light of the above mentioned division of water management, SWOT analysis of the scheme of water management in Poland has been conducted (Table 3.2).

Both endogenous and exogenous factors listed in the SWOT analysis are mutually correlated. For example, lack of mutual information exchange between the institutions responsible for water management and condition of geographic environment influence overall knowledge about conditions and users in a whole basin. On the other hand, the participation of all stakeholders in the decision making process related to water management in a region may lead to a slowdown of projects implementation due to a long decision making and agreement procedures. The consequences may be incomplete exploitation of the EU-provided funding.

In order to verify endogenous and exogenous factors of integrated water management in Poland, there have been 10 interviews conducted with persons employed by the institutions responsible for water management in Poland. Those were: Regional Water Management Authorities, Provincial Drainage, and Water Facilities Authorities and civilian officers responsible for issuing water law consents in district offices. During the interviews, attention was paid at practical functioning of the organizational scheme of water management in Poland and to pointing out the strengths and weaknesses of the water management system in Poland. The results were grouped into a few topics. Five issues, which were important from the point of view of state and municipal administrations and may comprise a source of conflict have been noted.

- Lack of Information Flow Between Local and National Government Units

The division of competence at government level between different ministries, as well as vertically in different areas of activity (basin, region, country) mean that the proper implementation of some tasks in water management becomes almost impossible. The main source of conflict, therefore, is the lack of information flow (bad system of communication and coordination between various levels of administration).

- Lack of Cooperation on Planning River Basin Management

The implementation of the Water Framework Directive into national law meant that public participation in water management finally became possible. During the determination of drainage basin management plans, public consultations are organised with the invited local government institutions responsible for water management in the area and environmental groups. Residents of the region (local community) can

Table 3.2 Endogenous and exogenous factors water resources management in Poland on the basis of the SWOT method

Endogenous factors	
Strengths	Weaknesses
– Separation of authority at state and municipal levels (decentralisation) – Management based on catchment of large rivers (hydrographic division) – Fulfilment of tasks at state, regional and local levels – Plans and planning documentation at the state level and for the EU	– Fragmented small watercourse management (administrative borders) – Lack of mutual information exchange between the institutions responsible for water management and geographic environment state – Overlapping competences of the employees of the institutions related to water resources management – Lack of a coherent database containing information on the geographic environment that would be available to all institutions responsible for water management, to the society and to water users – Lack of coherent interpretation of provisions of Water Law and other acts – Setting public consultations for local groups at locations far from the place of residence of the people interested in a given problem and lack of information about such consultation at the local level – Poor hydrotechnic infrastructure state on small watercourses resulting from lack of funding in the budgets of small municipal units (e.g., communes)
Exogenous factors	
Opportunities	Threats
– Comprehensive knowledge about conditions and users in the whole catchment – Increased involvement of local societies – Enhancement of hydrotechnic facilities – Easier implementation of large investments – Enhancement of the natural environment state with cooperation of local, municipal and state authorities – Inland navigation development – Development of tourism and recreation – Complete implementation of the assumptions of the WFD	– Competition between organisations in the area of decision making, which makes cooperation more difficult – Too long decision making and agreement procedures, which results in a slowdown in projects implementation – Too many individuals taking part in decision making – Reduction of biodiversity and fragmentation of biologically active areas – Incomplete usage of funding from the EU – Penalties imposed by the EU due to incompliance with the WFD

– and in principle should – participate in these consultations, but information about meetings does not always reach them on time, or they do not know where to look for such information, or the consultations are located in the significant distance from the place of residence. However, there is a chance that with time, social awareness in this regard will be higher, and interest will increase.

- Inconsistency of Databases and Unified Information System as Well as Access to Information

In Poland, there was no integrated information system of water management. Every Regional Water Management Authority had its own system, which was not consistent with that in other water regions. This prevented potential users and decision-makers from using full information about the state of the environment in drainage basins, analysing and assessing the environment as well as examining their impact on the state of various alternative investment and organisational solutions. On the other hand, the lack of direct access to basic hydrological data by users, but also local governments, shows that the information contained in different types of studies is inconsistent.

- No Precise Description of Powers and Tasks for Individual Institutions

This is evident in the context of issuing water supply consent for special use of water. Each of the institutions involved in the procedure had different guidelines, and thus in many issued decisions, there is no relevant hydrological information, or the decisions were issued illegally. It was, therefore, necessary to introduce a provision relating to the substantive scope to be met by water supply consent.

- Inconsistent Interpretation of Laws by Individual Institutions

According to the interviewees, this factor was the major source of conflicts. Often different interpretation of the regulation by different institutions led to issuing different decisions in the same subject. It seems essential to introduce appropriate training for employees of institutions on each level, starting from communes up to voivodeship (regional) and as well within Regional Water Management Authorities and Provincial Drainage and Water Facilities authorities. Such training may comprise a point of experience and knowledge exchange in the area of proper management and may be a stage for determination of proper task coordination at various levels.

3.4 Water Management After 2018

Water management in Poland is still carried out both through the state and local governments. The management of water resources after the introduction of changes in the organizational structure is carried out mainly by the Ministry of the Environment and the Ministry of Maritime Economy and Inland Navigation.

The main entity responsible for national water management is Państwowe Gospodarstwo Wodne Wody Polskie. It operates based on the provisions of the Act of 20 July 2017 – Water Law (Journal of Laws, items 1566 and 2180), and the statute granted by the Regulation of the Minister of the Environment of December 28, 2017 (Journal of Laws 2017, item 2506).

Changes in the organizational structure in water management were introduced on January 1, 2018. As a result of these changes, the Boards of Melioration and Water Facilities were liquidated. The entire water management was taken over by the "Wody Polskie", which was established based on the existing Regional Water Management Boards. It is a state legal entity (Article 9, point 14 of the Act of August 27, 2009 on Public Finance, Journal of Laws of 2016, item 1870, as amended), which includes the following organizational units (Figs. 3.3 and 3.4):

1. The National Water Management Authority with headquarters in Warsaw;
2. Regional Water Management Authorities with headquarters in Białystok, Bydgoszcz, Gdańsk, Gliwice, Kraków, Lublin, Poznań, Rzeszów, Szczecin, Warsaw, and Wrocław;
3. 50 catchment management boards;
4. 330 water supervising committees.

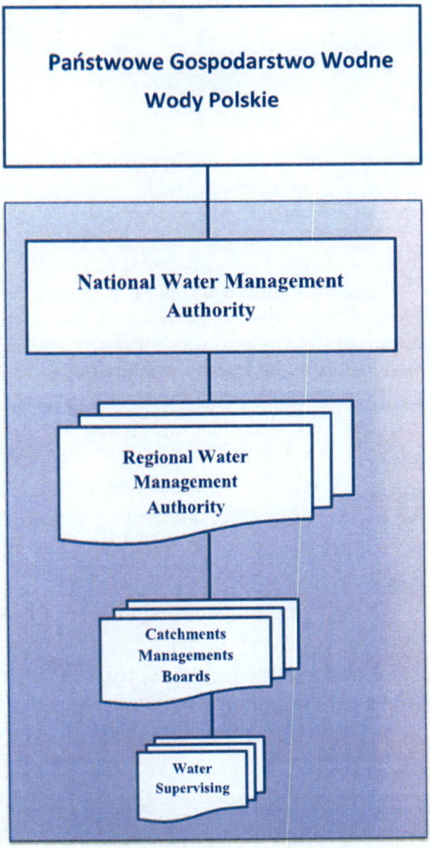

Fig. 3.3 Organizational structure of Państwowe Gospodarstwo Wodne Wody Polskie

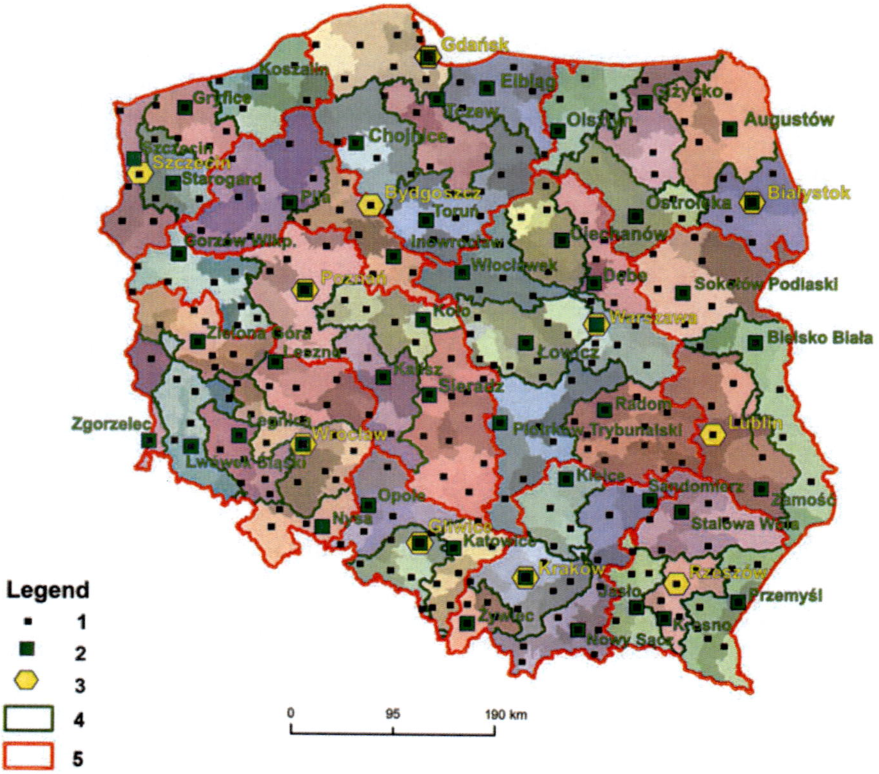

Fig. 3.4 Map of areas of operation of Regional Water Management Authorities, Catchments Management Boards and Water Supervising (as of July 6, 2018) [43]. Explanations: 1 – Water Supervisings headquarters, 2 – Catchments Management Board headquarters, 3 – Regional Management Authorities headquarters, 4 – boundaries of Catchments Management Boards, 5 – boundaries of Regional Water Management Authorities

Regional Water Management Authorities were established based on the previously existing 7 RZGW (Gdańsk, Szczecin, Poznań, Warsaw, Kraków, Gliwice, Wrocław). As a result of the reduction of the area of operation of the existing RZGW units, additional 4 new units have been created, i.e., Bydgoszcz, Lublin, Rzeszów, and Białystok.

The new institution responsible for water management in Poland has taken over from local governments the tasks of issuing water law consents for special use of water and establishing fees for water usage. These tasks are currently carried out by the management of the catchment. Collective water supply and collective sewage disposal is still the communes' own task. The changes concerned issues related to the setting of water tariffs charged by municipal or commune's water and sewage companies. The rates of water fees are subject to approval by the Director of Regional Water Management Authority and not, as before, by the municipal council.

Supervision over the Institute of Meteorology and Water Management from January 9, 2018 is held by the Minister of Maritime Economy and Inland Navigation, and not as previously the Minister of the Environment.

Despite the establishment of a new institution dealing with water management in Poland, which is Państwowe Gospodarstwo Wodne Wody Polskie, already in the first months of its functioning, there were problems with the implementation of its tasks. Starting from technical issues related to the creation of new regional water management boards and river basin management boards, appointing new directors, employing staff from former melioration boards and from local governments, creating new posts for issues related to the timely processing of applications for the issue of water law consents. Some of these problems have already been resolved and resulted from the lack of a transitional period when transferring assets, documentation, and arrangements that were previously in the remit of the melioration boards or local governments. The problems related to the interpretation of the provisions of the Water Law Act, as well as the lack of a common database and a unified information system remain valid. The solution to these problems should in the first place be the creation of a training system for the employees, but also for the stakeholders. This will avoid conflicts between officials and the public. Doubts about the special use of water, management principles in catchment terms should be determined and consulted as part of regular meetings of officials with the local community. The creation of a shared database is necessary, but it requires a systematic approach that takes into account the interests of all institutions dealing with water management in Poland. This requires a solution at the central level, taking into account the scope of data and the form of making it available. Proper assessment of the functioning of the current structure of institutions responsible for water management in Poland should be prepared in a few years.

3.5 Discussion

The problem of proper water management is not Poland-specific. It pertains to other countries as well [44–47]. Lack of coordination between state government and municipal governments, data sharing and informing the society are the problems of, among others, Canada [48] and Germany [49]. German experience shows [50, 51] that determination of introduction of the WFD requires an analysis of the efficiency of new law regulations introduced in the administration. Similar conclusions were made on the implementation of the first stage of the WFD introduction in Sweden [52].

Integrated water management is characterized by the consistency between various aspects of water management [53]. As indicated in this paper, integrated water management in Poland did not fully comply with the assumptions.

Water resource management in Poland at the catchment level was introduced yet in the 1990s. However, the other matters of integrated water resource management like participation of society and all stakeholders have been introduced by the Framework Water Directive. An important part of water management on the state or municipal level is the issue of interactions between all users of a basin. According to van den Brink and Meijerink [54], it is important to shape dependencies between various entities. Relatively independent actors have to work together in one way or another while possessing different bits of information, representing different interests and pursuing different interests through separate, often conflicting courses of action. In addition, as indicated by Bakker and Cook [48], an active approach to sharing the experience among various level officials will allow introducing efficient innovations throughout the whole country.

As written by McDonnell [55], independent on the water management system type, decentralised or governmental, conscious decision making in water management should be based on thorough and up to date information, which must be available to all the stakeholders within a basin [56]. Only providing access to historical and current hydrological data allows for a full understanding of water management in environmental, social, cultural, and economic aspects. According to Franzen et al. [57], access to information should be adapted to various integration levels, from passive information access towards higher integration levels, such as consultations and common planning, up to local involvement. Information and consultation are a passive form of participation, while the third level encourages active involvement.

There is a need for cooperation between various levels of authority, political sectors, and public and private entities on local, regional and country levels. It is possible through the application of a coordinator role, whose responsibility is cooperation between various organisations involved in the creation and implementation of climate adaptation strategies. Most often, the governmental agencies are the ones who play a key role in adaptive projects implementation. Establishing connections between various levels of state administration, politics, science, and private investors may consist in, e.g., initiating meetings, establishing workgroups, including additional people into existing networks [58].

Integrated water management underlines the joint functioning of various institutions responsible for water management within a river basin. The analysis showed that in practice such activities in Poland were far from being integrated, and the levels of mutual cooperation differed.

The introduction of the new Water Law Act of July 20, 2017 and organizational changes in water management in Poland after January 1, 2018 led to the creation of one institution responsible for the management of waters in scope catchments. The current structure of water management in Poland has been functioning for less than two years, and it is difficult to assess the extent to which the situation in water management has improved. However, there is a chance that thanks to such an organization of water management in Poland, with fewer institutions deciding on the fate of water management, we get a quicker response and exchange of information, better communication between the watercourse administration and stakeholders and public participation.

3.6 Conclusions

The organizational structure of water management, which operated in 2001–2017 in Poland, was not conducive to the sustainable development of water management. It resulted mainly from 2 institutions dealing with water management, which functioned in the administrative and sub-basin divisions. Despite the introduction of the WFD and its functioning in Poland for over 15 years, there were many problems that were not completely resolved. The new institution dealing with water management, Państwowe Gospodarstwo Wodne Wody Polskie, has been operating since 2018 and its focus is only on the catchment area. It is subordinate directly to the Ministry of the Environment. Despite the reform, some of the problems related to water management are still valid. The most important issues to be dealt with in the near future are the training of employees at various levels and stakeholders in the interpretation of water law. Systematic meetings of officials responsible for water management with the local community, directly in the field, constitute an important element of integrated water resources management. The lack of a common database and access to information for all authorities at local, regional, and public level causes misunderstandings. This database should be implemented as part of interdisciplinary national projects, which should be coordinated at the government level. The problems mentioned above are not new. They have already appeared before. As the time of functioning of the new institution responsible for water management is short, there is a hope that in the coming years these problems will be solved in a systematic manner, which will allow for the integrated management of water resources. Therefore, the time for a full assessment of the activity is still ahead, in the future.

3.7 Recommendations

This chapter presents institutions responsible for water management in Poland and their responsibilities. In the future, the focus should be on research showing current water management in Poland in the opinion of officials, local government officials, stakeholders and society. It will be particularly important to pay attention to the development of effective communication schemes between various institutions. The method of communication plays a key role in making decisions, especially in crisis situations.

References

1. Dieperink C, Raadgever GT, Driessen PPJ, Smit AAH, van Rijswick HFM (2012) Ecological ambitions and complications in the regional implementation of the Water Framework Directive in the Netherlands. Water Policy 14:160–173. https://doi.org/10.2166/wp.2011.223
2. Carrozza Ch (2011) Italian water services reform from 1994 to 2008: decisional rounds and local modes of governance. Water Policy 13(6):751–768. https://doi.org/10.2166/wp.2011.030
3. Śánchez-Martínez MT, Salas-Velasco M, Rodríguez-Ferrero N (2012) Who manages Spain's water resources? The political and administrative division of water management. Int J Water Resour Dev 28(1):27–42. https://doi.org/10.1080/07900627.2012.640610
4. Bielsa J, Cazcarro I (2015) Implementing integrated water resources management in the Ebro River Basin: from theory to facts. Sustainability 7:441–464. https://doi.org/10.3390/su7010441
5. Metz F, Glaus A (2019) Article integrated water resources management and policy integration: lessons from 169 years of flood policies in Switzerland. Water 11:1173. https://doi.org/10.3390/w11061173
6. Dombrowsky I, Hagemann N, Houdret A (2014) The river basin as a new scale for water governance in transition countries? A comparative study of Mongolia and Ukraine. Environ Earth Sci 72(12):4705–4726. https://doi.org/10.1007/s12665-014-3308-4
7. Yara O, Uliutina O, Golovko L, Andrushchenko L (2018) The EU Water Framework Directive: challenges and prospects for implementation in Ukraine. Eur J Sustain Dev 7(2):175–182. https://doi.org/10.14207/ejsd.2018.v7n2p175
8. Biswas S, Vacik H, Swanson ME, Sirajul Haque SM (2012) Evaluating integrated watershed management using multiple criteria analysis—a case study at Chittagong Hill Tracts in Bangladesh. Environ Monit Assess 184:2741. https://doi.org/10.1007/s10661-011-2148-x
9. Frey K, Ramirez DRC (2018) Multi-level network governance of disaster risks: the case of the Metropolitan Region of the Aburra Valley (Medellin, Colombia). J Environ Plann Manag. https://doi.org/10.14256/JCE.2063.2017
10. Zhou W, Mu R (2019) Exploring coordinative mechanisms for environmental governance in Guangdong-Hong Kong-Macao Greater Bay Area: an ecology of games framework. Sustainability 11:3119. https://doi.org/10.3390/su11113119
11. Shomar B, Darwish M, Rowell C (2014) What does integrated water resources management from local to global perspective mean? Qatar as a case study, the very rich country with no water. Water Resour Manage 28:2781–2791. https://doi.org/10.1007/s11269-014-0636-9
12. Shahraki AS, Shahraki J, Monfared SA (2019) An integrated water resources management considering agricultural demands and the assessment of different scenarios in Hirmand Catchment, Iran. Water Resour 46(308). https://doi.org/10.1134/S0097807819020143
13. De Loë R (2008) Invited commentary: Canada needs a national water strategy. Can Water Resour J 33(4):309–314
14. Hurlbert M, Andrews E (2018) Deliberative democracy in Canadian watershed governance. Water Altern 11(1):163–186
15. Araral E, Wang Y (2013) Water Governance 2.0. A review and second generation research agenda. Water Resour Manag 27(11):3945–3957. https://doi.org/10.1007/s11269-013-0389-x
16. Danilov-Danilyan VI, Demin AP, Pryazhinskaya VG, Pokidysheva IV (2015a) Markets of water and water management services in the world and the Russian Federation: Part I. Water Resour 42(2):260–268
17. Danilov-Danilyan VI, Demin AP, Pryazhinskaya VG, Pokidysheva IV (2015b) Markets of water and water management services in the world and the Russian Federation: Part II. Water Resour 42(3):378–388
18. Huntjens P, Pahl-Wostl C, Grin J (2010) Climate change adaptation in European river basins. Reg Environ Changes 10:263–284. https://doi.org/10.1007/s10113-009-0108-6
19. Greckseh K (2013) Adaptive capacity and regional water governance in north-western Germany. Water Policy 15(5):794–815. https://doi.org/10.2166/wp.2013.124

20. Bozzola M, Swanson T (2014) Policy implications of climate variability on agriculture: water management in the Po river basin, Italy. Environ Sci Policy 43:26–38. https://doi.org/10.1016/j.envsci.2013.12.002
21. Priscoli JD (2004) What is public participation in water resources management and why it is important? International Water Resources Association. Water Int 29(2):2–7
22. Bruch C, Jansky L, Nakayama M, Salewicz KA (eds) (2005) Public participation in the governance of international freshwater resources. Unites Nations University Press, Tokyo-New York-Paris
23. Drenovak-Ivanovic M (2015) The development of the right to public participation in environmental matters as a new concept of administrative decision making in Serbia. Transylvanian Rev Adm Sci 44E(2015):74–90
24. Warner JF (2006) More sustainable participation? Multi-stakeholder platforms for integrated catchment management. Water Resour Dev 22(1):15–35. https://doi.org/10.1080/07900620500404992
25. Watson N (2015) Factors influencing the frames and approaches of host organizations for collaborative catchment management in England. Soc Nat Resour Int J 28(4):360–376. https://doi.org/10.1080/08941920.2014.945059
26. Biswas AK (2008) Integrated water resources management: Is it working? Int J Water Resour Dev 24:5–22. https://doi.org/10.1080/07900620701871718
27. Global Water Partnership. Effective water governance. Stockholm, Sweden (2002). https://dlc.dlib.indiana.edu/dlc/bitstream/handle/10535/4995/TEC+7.pdf?sequence=1
28. Blomquist W, Schlager E (2005) Political pitfalls of integrated watershed management. Soc Nat Resour 18:101–117. https://doi.org/10.1080/08941920590894435
29. Gallego-Ayala J (2013) Trends in integrated water resources management research: a literature review. Water Policy 15:628–647. https://doi.org/10.2166/wp.2013.149
30. Suhardiman D, Clement F, Bharati L (2015) Integrated water resources management in Nepal: key stakeholders' perceptions and lessons learned. Int J Water Resour Dev 31:284–300. https://doi.org/10.1080/07900627.2015.1020999
31. Folke C, Hahn T, Olsson P, Norberg J (2005) Adaptive governance of social-ecological systems. Annu Rev Environ Resour 30:441–473. https://doi.org/10.1146/annurev.energy.30.050504.144511
32. Armitage D, de Loë R, Plummer R (2012) Environmental governance and its implications for conservation practice. Conserv Lett 5:245–255. https://doi.org/10.1111/j.1755-263X.2012.00238.x
33. Becker G, Huitema D, Aerts JCJH (2015) Prescriptions for adaptive comanagement: the case of flood management in the German Rhine basin. Ecol Soc 20(3). https://doi.org/10.5751/ES-07562-200301
34. Miłaszewski R (ed) Zintegrowane systemy zarządzania w gospodarce wodnej. Rozprawy Naukowe, 69, Politechnika Białostocka, Białystok (2000)
35. Gromiec M (2011) Projekt polityki państwa a diagnoza stanu gospodarki wodnej. Rocznik Ochrony Środowiska 13:283–301
36. Iwanicki J, Kindler J, Kundzewicz ZW (2014) Zagrożenia związane z wodą. Nauka 1/2014:63–76. https://www.pan.poznan.pl/nauki/N_114_06_Iwanicki.pdf
37. Kindler J, Iwanicki J, Kundzewicz ZW, Matczak P, Miłaszewski R, Żelazo J (2014) Institutional hazards. Nauka, 1/2014:173–195. https://www.portalwiedzy.pan.pl/images/2014/nauka_20141/N114-11-Kindler.pdf
38. Kubiak-Wójcicka K (2019) Long-term variability of runoff of Vistula river in 1951–2015. In: Air and water—components of the environment" conference proceedings, Cluj-Napoca, Romania, pp 109–120. https://doi.org/10.24193/AWC2019_11
39. Environment. Central Statistical Office, Warsaw (2015)
40. Główny Urząd Statystyczny (GUS). https://stat.gov.pl/
41. Kubiak-Wójcicka K, Marszelewski M (2012) Definitions and evolutions of the terms "flowing and stagnant waters" in the context of the proprietorship of the lakes in Poland. Limnol Rev 12:189–195. https://doi.org/10.2478/v10194-012

42. Projekt polityki wodnej państwa do roku 2010 (z uwzględnieniem etapu 2016) (2010) Krajowy Zarząd Gospodarki Wodnej. https://gwppl.org/data/uploads/warsztaty-konsultacje/dokumenty_krajowe/Projekt_Polityki_wodnej_panstwa_do_roku_2030_z_uwzglednieniem_etapu_2016.pdf
43. https://www.kzgw.gov.pl/index.php/pl/jednostki-organizacyjne/mapa-obszarow-dzialania
44. Ayre M, Nettle R (2015) Doing integration in catchment management research: insights into a dynamic learning process. Environ Sci Policy 47:18–31. https://doi.org/10.1016/j.envsci.2014.10.011
45. Ćosić Flajsig G, Belaj M, Carleuša B (2017) Combined approach to surface water management. Gradevinar 69(8) 617–631. https://doi.org/10.14256/JCE.2063.2017
46. Fan L, Wang H, Lai W, Wang Ch (2015) Administration of water resources in Beijing: problems and countermeasures. Water Policy 17:563–580. https://doi.org/10.2166/wp.2014.407
47. Thiel A (2015) Constitutional state structure and scalar re-organization of natural resource governance: the transformation of polycentric water governance in Spain, Portugal and Germany. Land Use Policy 45:176–188. https://doi.org/10.1016/j.landusepol.2015.01.012
48. Bakker K, Cook C (2011) Water governance in Canada: innovation in the context of fragmentation. Int J Water Resour Dev 27:275–289
49. Kampa E, Kranz N, Hansen W (2003) Public participation in River Basin management in Germany: from borders to natural boundaries. Report of Workpackage 4 of the Harmoni COP Project. Ecologic, Institute for International and European Environmental Policy, Berlin
50. Albrecht J (2013) The Europeanization of water law by the Water Framework Directive: a second chance for water planning in Germany. Land Use Policy 30:381–391. https://doi.org/10.1016/j.landusepol.2012.04.009
51. Voulvoulis N, Arpon KD, Giakoumis T (2017) The EU Water Framework Directive: from great expectations to problems with implementation. Sci Total Environ 575:358–366. https://doi.org/10.1016/j.scitotenv.2016.09.228
52. Keskitalo ECH (2015) Actors' perceptions of issues in the implementation of the first round of the Water Framework Directive: examples from the water management and forestry sectors in Southern Sweden. Water 7:2202–2213. https://doi.org/10.3390/w7052202
53. Benson D, Lorenzoni I (2016) Climate change adaptation, flood risks and policy coherence in integrated water resources management in England. Reg Environ Changes. https://doi.org/10.1007/s10113-016-0959-6
54. van den Brink M, Meijerink S (2006) Implementing policy innovations. Resource dependence, struggle for discursive hegemony and institutional inertia in the Dutch river policy domain. Working Paper Series. https://hdl.handle.net/2066/45540
55. McDonnell RA (2008) Challenges for integrated water resources management: how do we provide the knowledge to support truly integrated thinking? Int J Water Resour Dev 24:131–143. https://doi.org/10.1080/07900620701723240
56. Mees H, Suykens C, Crabé A (2017) Evaluating conditions for Integrated Water Resource Management at sub-basin scale. A comparison of the Flemish sub-basin boards and Walloon river contracts. Environ Policy Gov 27:59–73. https://doi.org/10.1002/eet.1736
57. Franzen F, Hammer M, Balfors B (2015) Institutional development for stakeholder participation in local water management—an analysis of two Swedish catchments. Land Use Policy 43:217–227. https://doi.org/10.1016/j.landusepol.2014.11.013
58. Stiller S, Meijerink S (2016) Leadership within regional climate change adaptation networks: the case of climate adaptation officers in Northern Hesse, Germany. Reg Environ Changes 16:1543–1555. https://doi.org/10.1007/s10113-015-0886-y

Chapter 4
Water Resources in Poland and Their Use

Zdzisław Michalczyk and Joanna Sposób

Abstract Water resources in Poland are directly related to the features of the natural environment, particularly the hydrogeological and climatic conditions. The territory of Poland is characterised by relatively low water resources, and the outflow of rivers is among the lowest in Europe. In the context of the occurring climatic changes, measures aimed at water retention in the catchment are recommended. Changes in the conditions of water circulation should be towards its retention in the soil and bedrock, leading to a decrease in surface runoff and an increase in exploitable resources in catchments. In addition to the construction of retention reservoirs, this purpose can be met by changes in the land use structure, application of suitable agrotechnical measures, as well as the construction of corrective steps slowing down water outflow. All water resources are subject to strong human pressure leading to their quantitative and qualitative transformations. They result from an increase in the water needs accompanying an increase in the standard of life and development of industrial infrastructure and raw material extraction. It is important to reduce the water consumption at every stage of its use: in households, industry and agriculture. The level of exploitation, especially on the regional scale, should not endanger the stability of low flows and natural functioning of hydrogenic ecosystems. The water safety of the country requires the maintenance of the quantity and quality of water resources on a good level and effective protection against the effects of drought and floods.

Keywords Water resources · Water use · Water protection · Poland

Z. Michalczyk (✉) · J. Sposób
Department of Hydrology and Climatology, Faculty of Earth Sciences and Spatial Management, Maria Curie-Skłodowska University, 2 d Kraśnicka Ave, 20-718 Lublin, Poland
e-mail: zdzislaw.michalczyk@poczta.umcs.lublin.pl

J. Sposób
e-mail: joanna.sposob@poczta.umcs.lublin.pl

M. Zeleňáková et al. (eds.), *Management of Water Resources in Poland*, Springer Water,
https://doi.org/10.1007/978-3-030-61965-7_4

4.1 Introduction

Water – as a chemical compound of an atom of oxygen and two atoms of hydrogen – occurs in nature in three states of concentration. It participates in physical, chemical, and biological processes occurring in nature. It is used in the economy and in recreation. It is the basic component of the natural environment with unique and irreplaceable properties. It is also necessary for the functioning of life. Water determines the physical and chemical processes occurring in the natural environment, as well as the development of the life and health of all ecosystems. It affects the development of agriculture and food production. It is used for transport. It participates in industrial processes and is necessary for maintaining personal hygiene. Its resources in the hydrosphere, atmosphere, lithosphere, and biosphere are constant on the global scale. They are theoretically unlimited, although acceptable by living organisms, they are always subject to reduction by human pressure. Water existed on our planet long before any life form appeared. It alone created the environment of development of biological and biophysical processes leading to the creation of life, its maintenance, and development. The origin of water in the atmosphere and the history of its precipitation on the Earth's surface resulting in the development of the hydrosphere are still uncertain. The erosion of rocks and deposition of sediments suggest deep changes in the manner of interaction between the atmosphere and hydrosphere in the prehistoric times, determining the conditions of life on Earth. Water is one of the treasures of nature with the slow renewal of resources, forcing its rational and economical use. The occurring periods of both water excess and deficits are dangerous for the economic growth of the region and country. The right to access to safe, clean, and affordable drinking water and sanitary infrastructure is commonly considered the fundamental human right.

The problem of access to water of suitable quality is currently becoming one of the most critical challenges for humanity, particularly in the context of the observed climate changes [1, 2]. According to the United Nations Organisation, approximately 2 billion people on Earth currently have no access to safe drinking water. Lack of access to good water caused the proclamation of the years 2005–2015 as the International Water for Life Decade by the General Assembly of the United Nations. On 22 March 2018, members of the UN General Assembly inaugurated the Water Decade, officially described as the International Decade of Action "Water for Sustainable Development", with emphasis on sustainable development and development of the integrated process of management of water resources [3].

The undertaken measures aimed at better management of freshwater resources, and providing humanity with good quality and healthy water. This requires the protection of water sources, and construction of systems of storage and transport of suitably treated water, as well as the provision of use of sanitary infrastructure.

4.2 Water Resources in Selected European Countries

The abundance of water in a given area depends on the factors of the geographic environment such as climate, geological structure, land relief, and land use. Water resources of an area are particularly dependent on climatic conditions, shaped by the amount of solar radiation, distance from seas and oceans, land relief, including height above sea level, and course of mountain ranges. In reference to Europe, it should be emphasised that it is located in only one climatic zone, and is surrounded by seas having a warming effect on the continent. The mountains have latitudinal distribution facilitating airflow of both marine and oceanic air.

Atmospheric precipitation and the related outflows are distributed over continents, including Europe and therefore Poland, very unevenly. On the Earth, dry areas occupy almost ¼ of the land surface, and a constant deficit of freshwater is recorded on more than half of the surface area of continents.

The assessment of water resources considers the absolute amount of precipitation water is reaching the surface of a given country, the amount of outflow and potentially inflow of water, as well as the amount of water per capita in a year. The assessment of particular values is imprecise. The starting material comprised data collected in materials of the Central Statistical Office [4], including absolute values of atmospheric precipitation, evapotranspiration, own resources of the country, and water inflow in selected countries. Based on hydrometeorological data and comparison of the surface area and size of the population, precipitation and outflow indices were calculated, expressed in millimetres of the water layer, as well as annual water resources per one resident.

Among the selected European countries, atmospheric alimentation somewhat exceeding 1400 mm·year^{-1} was determined in Switzerland and Slovenia. In Croatia, Austria, UK, Norway, and Ireland, it is maintained at a level of 1000–1300 mm. In Poland, Romania, Hungary, and several other countries (Fig. 4.1), it approximates 600–700 mm·year^{-1}. In reference to the climatic conditions of the countries, and particularly to high potential evaporation, the amount of outflow (own resources of a country) is low, because it does not reach 200 mm. In countries with higher precipitation, the amount of outflowing water is maintained at a level of 600–1000 mm per year.

The absolute amount of water resources is equivalent to the amount of mean outflow from a given country. The outflowing water can originate from internal resources, i.e. from atmospheric alimentation of the territory of the country, and outflow from outside its borders. Discharge of transit rivers (having its springs in another country) increases the water resources of the country, sometimes very radically. Figure 4.2 presents a comparison of water resources developing in the territory of the country and total renewable resources, i.e. with water outflow from parts of catchments located outside of the national borders. A particularly favourable situation concerns Slovakia, Bulgaria, Hungary, Netherlands, Serbia, and Croatia. Their resources are small, and the rivers flowing through the territories carry high amounts

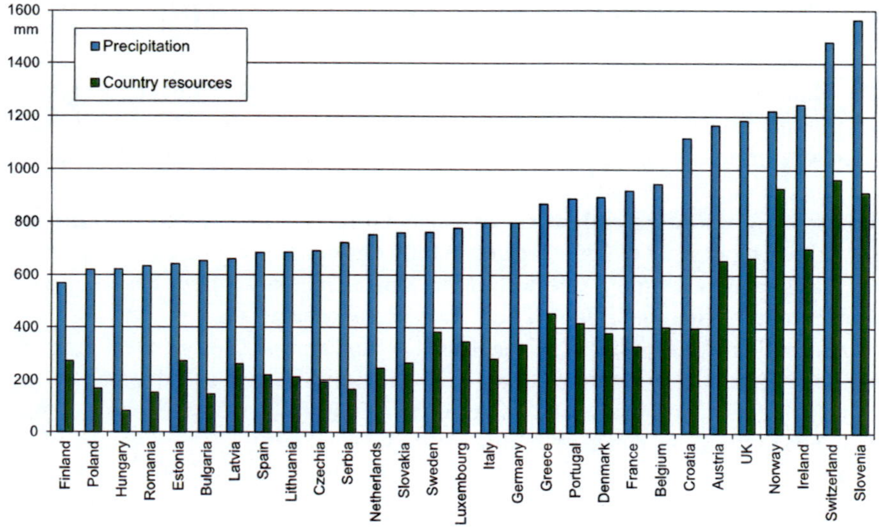

Fig. 4.1 Mean annual values of precipitation and water resources (total runoff) from the territories of selected European countries (OECD Environmental Data. Compendium 2017, after [4])

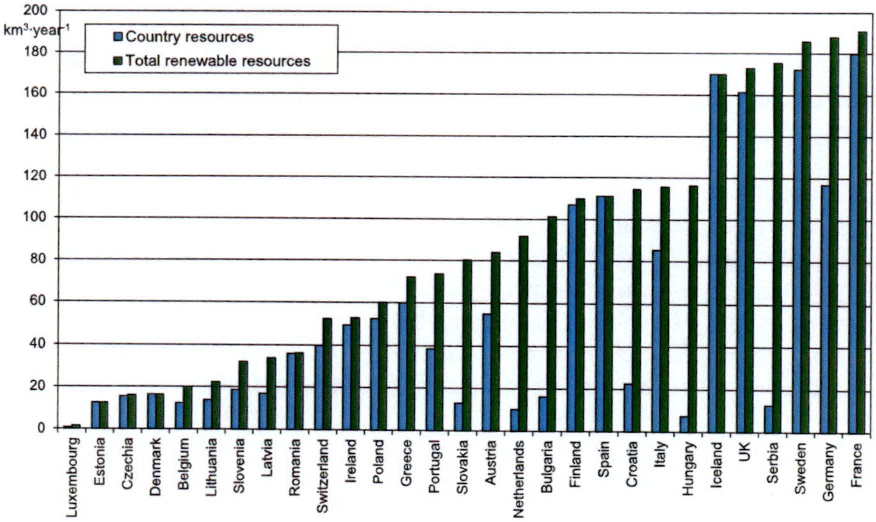

Fig. 4.2 Mean annual resources of surface waters in selected European countries (OECD Environmental Data. Compendium 2017, after [4])

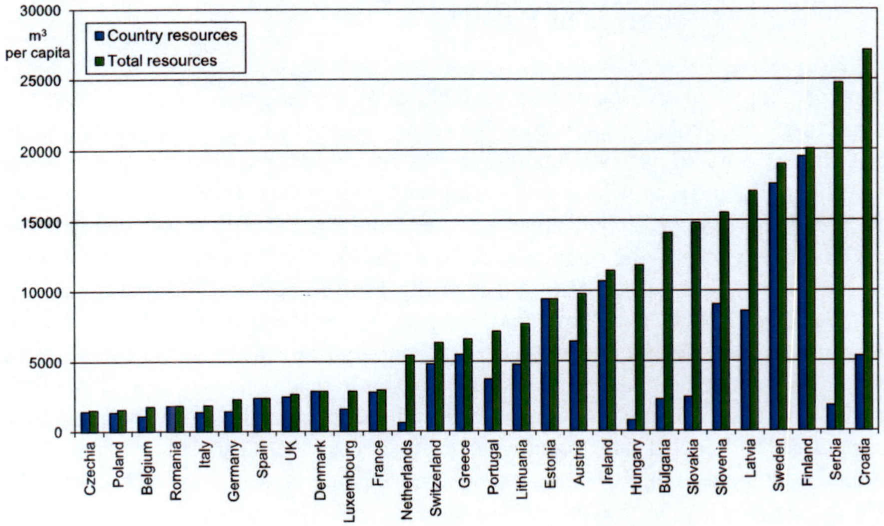

Fig. 4.3 Mean annual water resources per capita in selected European countries (OECD Environmental Data. Compendium 2017, after [4])

of water. Water resources of Poland, both own and total, are small, also in reference to smaller neighbouring countries.

The relative amount of water resources of a country is provided by the water abundance index, expressing the amount of water per capita in a year. Its amount can be referred only to own resources, or to total renewable resources (Fig. 4.3). Among European countries, Poland together with the Czechia and Belgium are included in a group of countries with the smallest water resources – approximately 1500 m^3 per capita. It should be emphasised that the level of resources of 1000–2000 m^3 per capita is considered very small. The highest index of total water resources concerns Finland and Sweden. For Europe, the index is 4500 m^3 per year.

Water resources on particular continents per capita in a year are mostly different, averaging 18.1 $m^3{\cdot}d^{-1}$ per resident [5, 6]. This results both from the conditions of the natural environment and from the number of residents. They are the highest in Australia and Oceania (199.6 $m^3{\cdot}d^{-1}$ per resident), twice lower in South America (88.4), and three times lower in North America (65.6). On the remaining continents, they are considerably lower: in Asia (9.5), Europe (10.9), and Africa (12.2 $m^3{\cdot}d^{-1}$ per resident).

4.3 Water Resources of Poland

Water resources of an area depend on features of the geographic environment, and particularly the geological structure, land relief, soils, land use, atmospheric precipitation, and evapotranspiration, and from the retention capacity of the ground. The territory of Poland is a lowland area – areas with a height from 0 to 300 m amsl occupy 91.3% of the territory. Upland areas, elevated from 300 to 500 m amsl, occupy 5.6%, and mountain areas – the remaining 3.1% of the area of Poland. The highest point in the country is Rysy (2499 m amsl) in the Tatra Mountains, and the lowest the depression in Żuławy (−1.8 m amsl). Despite the small elevation of the territory of Poland above sea level (173 m amsl), the land relief of Poland shows considerable variability, with a belt arrangement of geomorphological regions. The system of the main features of land relief, arranged in latitudinal zones, results from the geological structure. The Sudetes and Carpathians occupy the southern part of the country. The Sudeten mountain ranges with a character of tectonic horsts, developed during the Hercynian orogeny, are built of metamorphic stones. The Carpathians, developed in the Alpine folding, are built of flysch formations (sandstones, slates, conglomerate rocks) with low permeability. The highest part of the Carpathian range are the Tatra Mountains, built in the eastern part of massive granitoids, and in the western part of Mesozoic carbonate formations with high water drainage. A belt of uplands is located to the north, mainly built of Mesozoic carbonate deposits: limestones, dolomites, and margles. The Świętokrzyskie Mountains are also built of Palaeozoic carbonate rocks as well as sandstones and quartzites. The area of uplands is connected with a belt of extensive lowlands with diverse morphology, shaped by the Pleistocene ice sheet. Post-glacial formations occur on the surface, particularly moraine clays and sandy formations constituting reservoirs of groundwaters with moderate and locally high water capacity.

The land use structure of the territory of Poland is dominated by arable land and forests (Table 4.1). They are areas with good conditions for precipitation water retention, i.e. areas where groundwater resources develop, feeding rivers, springs, and water intakes. Areas with low permeability and built-up areas occupy approximately a dozen percent of the area. The surface area of rivers, ponds, ditches – land under waters slightly exceeds 2.0% of the area of Poland.

The factor directly affecting water resources are climatic conditions, particularly precipitation and air temperature, determining the amount of evapotranspiration. Despite a relatively small area, the basic parameters describing climate features show high spatial and seasonal variability. In the territory of Poland, mean precipitation in the years 1901–2000 equalled 628 mm, and outflow 175.2 mm, providing the outflow coefficient of 0.278 [7]. In a year with an average humidity, the amount of precipitation in the territory of the country varied from approximately 500 mm in the lowland belt to more than 1000 mm in the mountains. The highest monthly precipitation and evapotranspiration totals in Poland occur in July. The lowest precipitation and evapotranspiration occur in January and February.

Table 4.1 Land use structure in 2017 [4]

	Total area	Agriculture land							
		Total	Arable lands	Orchards	Meadows	Pastures	Agricultural built-up areas	Agricultural lands under ponds	Woody and bushy lands
thousads hectares	31,268.0	18,810.1	13,684.3	294.7	2243.7	1589.5	547.7	211.7	238.6
%	100.00	60.15	43.76	0.94	7.18	5.08	1.75	0.68	0.76

	Forest, woody and bushy lands			Lands under waters				Ecological arable lands	Miscellaneous and waste lands
	Total	Forests	Woody and bushy lands	Total	Internal waters	Flowing waters	Standing waters		
thousads hectares	9513.2	9382.1	131.3	650.6	79.2	514.0	57.4	43.1	550.2
%	30.43	30.01	0.42	2.08	0.25	1.65	0.18	0.14	1.76

	Built-up and urbanised areas									
	Total	Industrial					Transport areas			Minerals
		Residential areas	Industrial	Other build-up	Urbanized non-residential	Recreation and leisure	Roads	Railways	Others	
thousads hectares	1700.6	340.4	123.6	152.4	56.8	66.3	808.3	102.6	13.8	28.4
%	5.44	1.10	0.40	0.49	0.18	0.21	2.59	0.33	0.05	0.09

The outflow module is the most accurate estimate of water resources. Its value constitutes the difference between atmospheric precipitation and evapotranspiration. The amount of outflow is calculated based on multiannual observations and hydrometric measurements performed in river channels where both very high and extremely low flows can occur. Precipitation falling on the land surface in the country determines the development of own water resources. In the environmental cycle, precipitation water can directly flow to rivers, return to the atmosphere in the process of evaporation and transpiration, or can be retained in the soil cover, and then retained in the bedrock. The contribution of precipitation water in particular phases determines the rate of water inflow to river channels.

Almost the entire area of Poland is located in the catchment of the Baltic Sea (99.7% of the area), and only small fragments are incorporated in the catchment of the North Sea (0.1%) and Black Sea (0.2% of the territory of the country). The surface of catchments of the Oder and Vistula Rivers constitute 89.6% of Poland's territory, and the remaining part is occupied by those of rivers of the Pomeranian and Masurian Lake Districts [8], draining water directly to the Baltic Sea or to the catchment of the Neman and Pregolya River [9]. Water resources of the territory of Poland are associated with the amount of mean outflow from the area of the country, i.e. 312,677 km^2. They develop as a result of alimentation with atmospheric precipitation as own renewable resources. Own resources are increased by internal resources, i.e. by the volume of water introduced through rivers to the territory of Poland, where part of catchments is located outside the national borders. After adding the parts of catchments located in the neighbouring countries, the area from which Polish rivers collect water occupies 351,208 km^2, i.e. it is 12.3% larger than the territory of Poland.

Amounts of outflowing water are calculated based on daily water gauge observations and flow measurements performed in the scope of the network of the Institute of Meteorology and Water Management. The data are collected in water gauge profiles established in river channels. The number of water gauges and way of registration of water stages changes with technical progress. At the beginning of the current century, daily water stages and discharges were determined in 611 water gauge stations [7] distributed on Polish rivers. Their catchment areas occupy from several tens to 194 thousand km^2.

Discharge of Polish rivers fluctuates in annual and multi-annual periods. Cyclical occurrence of years with high and low discharges is registered, and seasonal discharge fluctuations are observed. According to calculations performed by IMGW, mean total water outflow in Polish rivers in the twentieth century was 61.5 km^3, whereas outflow from the area of the country equalled 53.9 km^3 [7]. The difference between mean river outflow and outflow from the area of Poland equals 7.6 km^3. It results from the outflow of water introduced by rivers inflowing to the territory of Poland, i.e. they are resources developed outside of the national borders. Total annual outflows showed a series of several years with high or low discharges, and following annual values showed a significant level of autocorrelation. This points to significant environmental conditions of development of water resources, and particularly to the way of alimentation of rivers with a high contribution of the groundwater component. The outflow module of Polish rivers in the twentieth century was 1950 $m^3 \cdot s^{-1}$, and from the territory of Poland 1710 $m^3 \cdot s^{-1}$ [7]. Out of the entire volume of waters outflowing in rivers, approximately 55% concerns the catchment of the Vistula River, and 25% the catchment of the Oder River. The rivers of Przymorze and the Vistula Lagoon carry respectively 9.5 and 5.9% of waters of Poland [10].

Annual outflow of Polish rivers in particular years and decades of the period 1901–2017 fluctuated in a very broad range, averaging 61.1 km^3. The lowest and highest annual outflows occurred in the second half of the previous century: lowest – 37.6 km^3 in 1954, and highest – 89.9 km^3 in 1981. No significant differences were determined in the frequency and duration of dry and wet periods in the first and second half of the twentieth century. The driest periods were 1932–1937 and 1989–1993 [7, 11]. High values of outflowing water were determined in the 1970s and at the turn of the centuries. Outflow in the second half of the twentieth century was higher by 1.5% than the average calculated from the entire period. In recent years, river outflow fluctuated in a vast range (Fig. 4.4), from 41 $km^3 \cdot year^{-1}$ (years 2015–2016) to almost 87 km^3 (2010).

A measure of variability of water resources in the area is unitary outflow expressed in unitary outflow in one second from one square kilometre. Mean unitary outflow from the multi-annual period 1951–2000 in the Vistula catchment was 5.56 $dm^3 \cdot s^{-1} \cdot km^{-2}$. It was somewhat higher than in the catchment of the Oder River (4.83 $dm^3 \cdot s^{-1} \cdot km^{-2}$), whereas in the territory of the country it is higher, equalling 5.64 $dm^3 \cdot s^{-1} \cdot km^{-2}$ [7]. The lowest unitary outflow concerns catchments of rivers of lowlands in the central parts of Kujawy and Wielkopolska and the highest in the Tatra Mountains. In the lowland area, mean unitary outflows vary from 3 to 4 $dm^3 \cdot s^{-1} \cdot km^{-2}$, whereas, in the Upper Noteć River catchment, they do not reach 3

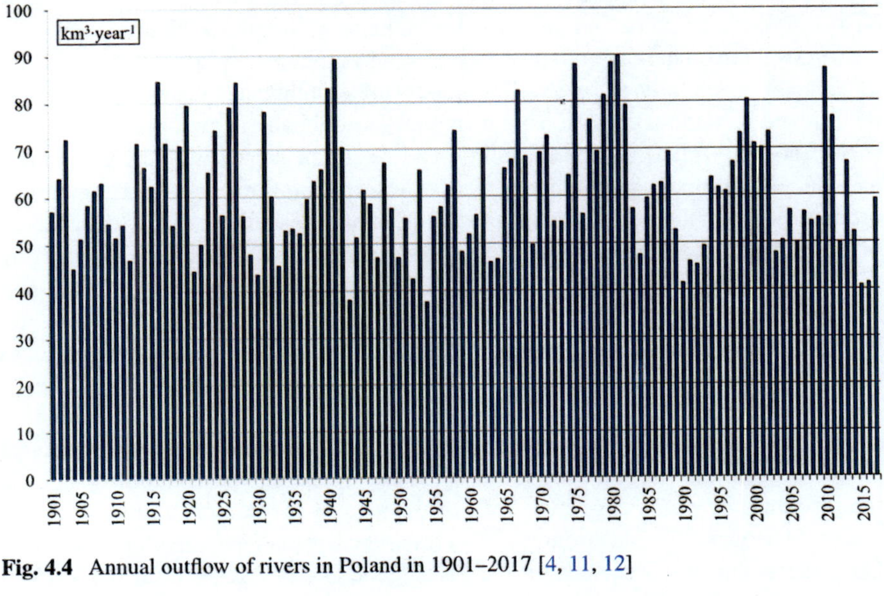

Fig. 4.4 Annual outflow of rivers in Poland in 1901–2017 [4, 11, 12]

$dm^3 \cdot s^{-1} \cdot km^{-2}$. In the Masurian Lake District, they increase to 6 $dm^3 \cdot s^{-1} \cdot km^{-2}$, and in the Pomeranian Lake District to 10 $dm^3 \cdot s^{-1} \cdot km^{-2}$. In the zone of uplands, values of unitary outflows consequently increase with height from 4 to 8 $dm^3 \cdot s^{-1} \cdot km^{-2}$. In the Carpathian Foothills, Carpathians, and Sudeten Foothills and Sudeten, outflow increases with height, from 6 $dm^3 \cdot s^{-1} \cdot km^{-2}$ in the boundary zone of the mountain area to 15–30 $dm^3 \cdot s^{-1} \cdot km^{-2}$ in the Carpathians and Sudeten. In the Tatra Mountains, mean unitary outflow exceeds 40 $dm^3 \cdot s^{-1} \cdot km^{-2}$.

In the southern part of Poland, the height of the outflow layer increases with land elevation as a consequence of an increase in atmospheric precipitation in the mountains. In the Tatra Mountains, the annual outflow layer exceeds 1000 mm, in the Carpathians and Sudeten it equals 300–500 mm. In the zone of Uplands of Central Poland, the outflow layer is 150–220 mm. It is somewhat higher in the Lake District, where it exceeds even 300 mm. The lowest outflow, below 100 mm, is observed in the zone of Lowlands of Central Poland, and in some of its parts, it does not even reach 70 mm (part of Kujawy and central Wielkopolska).

The value of river discharge, mainly dependent on the amount of atmospheric precipitation and the surface area of the catchment, changes seasonally, which is directly related to the type of alimentation of rivers. Outside the mountain area, the highest discharges occur in spring as a result of the release of water retained in the snow cover. In the mountains, the highest river discharges occur during intensive rainfalls at the beginning of summer. High variability of discharges in rivers makes it challenging to manage the entire mass of outflowing water, and causes problems with water excess or deficits. A particularly difficult situation occurs both during flood flows and low flows. In the remaining period, rivers are relatively evenly fed from the groundwater resources. Discharges of the largest Polish rivers in their mouth sections, together with the length of the rivers and size of their catchment areas, are presented in Table 4.2. Polish rivers with the most abundant water resources are the Vistula and Oder Rivers, and their tributaries, namely Narew and Warta.

Atmospheric waters are retained in rocks, lakes, ponds, and reservoirs, and part of them flows directly to rivers. The amount of groundwater resources is mainly determined by natural conditions: the amount of atmospheric precipitation, type of rocks and land relief, soil permeability, and retention capacity of the catchment, as well as anthropogenic factors: water bodies, meliorations and river regulation, urbanisation, and the related change of land use structure and others. The distribution of groundwater resources in the territory of Poland refers to conditions resulting from the geological structure, and particularly the range and thickness of aquifers and their water capacity. Water resources of a given area can occur in rocks of different ages, and be renewed through infiltration of precipitation waters. The exploitable resources of the main aquifers occurring in the territory of Poland and their area of occurrence are presented in Table 4.3.

In the territory of Poland, groundwaters of the first aquifer occur in rocks of different ages. The largest area is occupied by waters of the Quaternary aquifer the resources of which exist over 75% of the area of the country (Table 4.3), and collect 51% of exploitable resources. Over a somewhat smaller area, often underlying Quaternary deposits, wet rocks of the Palaeogene and Neogene occur (61%). Wet

Table 4.2 Principal rivers in Poland [4]

River	Recipient	Basin area (km^2)	River length (km)	River discharge ($m^3 \cdot s^{-1}$)
Vistula	Baltic sea	193 960	1 022	1 080.0
Dunajec	Vistula	6 796	249	85.5
Wisłoka	Vistula	4 110	173	35.5
San	Vistula	16 877	458	129.0
Wieprz	Vistula	10 497	349	36.4
Pilica	Vistula	9 258	333	47.4
Narew	Vistula	74 527	499	313.0
Bug	Narew	38 712	774	155.0
Bzura	Vistula	7 664	173	28.6
Drwęca	Vistula	5 697	231	30.0
Brda	Vistula	4 665	245	28.0
Oder	Baltic sea	119 074	840	567.0
Nysa Kłodzka	Oder	4 570	189	37.7
Barycz	Oder	5 547	136	18.8
Bóbr	Oder	5 874	279	44.8
Nysa Łużycka	Oder	4 403	246	31.0
Warta	Oder	54 520	795	216.0
Noteć	Warta	17 302	391	76.6
Parsęta	Baltic sea	3 084	143	29.1
Łyna	Pregolya	7 126	264	34.7

Table 4.3 Main muliti-aquifer formations in Poland and related disposable groundwater resources [13]

Resources	Area	Disposable resources	Disposable resources in main groundwaters reservoirs	Approved resources (as of 31st December 2001)
Resources ($km^3 \cdot rok^{-1}$)		12.5	7.35	16.17
Multiaquifer formation	Thousands km^2	%		
Quaternary	234	65.0	51.3	65.9
Neogene	191	11.0	5.5	10.1
Cretaceous	70	13.0	23.1	13.5
Jurassic	60	5.0	11.7	10.5
Triassic	15	3.0	7.1	
Older formations		3.0	2.3	

Cretaceous, Jurassic, and Triassic formations occur over a total of 46% of the area of the country (Cretaceous 22%, Jurassic 19%, Triassic 5%), and collect 41.9% of exploitable resources occurring in the main groundwater reservoirs (Table 4.3).

According to materials presented in the Hydrogeological atlas of Poland [14], total infiltration of precipitation was evaluated as 100 $km^3 \cdot year^{-1}$, and the amount of renewable groundwater resources within useful aquifers was estimated for approximately 18 ± 3 $km^3 \cdot year^{-1}$. In balance calculations concerning the possibilities of use of groundwater, exploitable resources of useful aquifers should not exceed 15 $km^3 \cdot year^{-1}$. In periods of moderate and low discharges, rivers are fed from groundwater resources, and floods occur during the runoff of precipitation and meltwaters. Random occurrence of such events forces undertaking measures for regulating river discharge. In addition to the construction of retention reservoirs, the regulation should cover measures aimed at water retention in catchments, i.e. mainly in agricultural and forest areas.

An essential role in water retention and development of water resources is played by lakes, ponds, and retention reservoirs. There are 7081 lakes (larger than 1 ha) in Poland, with a surface area of 2813.8 km^2 [15, 16]. Their number in the last century considerably decreased, and Majdanowski [17] provides their number of 9296 and total surface area of 3169.3 km^2. K. Dębski [18] mentions that lakes in Poland occupy the area of approx. 3400 km^2 and their total volume is estimated for 33 km^3. Whereas, the capacity of soil and groundwaters in the water balance for Poland is evaluated for 36.6 km^3 [18]. Total static resources of lakes in Poland are 18.747 km^3, 99% of which is located inside the range of Weichselian glaciation [19]. A large amount of surface water is retained in wetlands, which is estimated for 15.5 km^3 [20]. The basic parameters of the largest lakes in Poland are presented in Table 4.4.

To increase surface retention, dam reservoirs are constructed, potentially substantially contributing to a reduction of the irregularity of outflow. Retention reservoirs increase exploitable water resources, particularly in the period of the lowest discharge in the river, or when the periodical demand increases. Several tens of thousands of different hydrotechnical constructions have been built on Polish rivers and lakes.

Table 4.4 The largest lakes in Poland [4]

Lake	Basin	Area (km^2)	Depth (m)	Capacity (mln m^3)
Śniardwy	Pisa	113.4	23.4	660.2
Mamry	Węgorapa	102.8	43.8	1009.8
Łebsko	Łeba	71.4	6.3	117.5
Dąbie	Odra	56.0	4.2	168.0
Miedwie	Płonia	35.3	43.8	681.7
Jeziorak	Iławka	32.2	12.0	141.6
Niegocin	Pisa	26.0	39.7	258.5
Gardno	Łupawa	24.7	2.6	31.0

Table 4.5 The largest artificial water reservoirs in Poland [4]

Reservoir	River	Area (km^2)	Total capacity (hm^3)
Solina	San	22.0	472.4
Włocławek	Wisła	75.0	453.6
Czorsztyn	Dunajec	12.3	231.9
Jeziorsko	Warta	42.3	202.8
Goczałkowice	Wisła	32.0	161.3
Rożnów	Dunajec	16.0	159.3
Dobczyce	Skawa	10.7	141.7
Otmuchów	Nysa Kłodzka	20.6	131.5
Nysa	Nysa Kłodzka	20.7	124.7
Turawa	Mała Panew	20.8	106.2

hm^3 – cubic hectometres, 1,000,000 m^3

According to the Central Statistical Office [4], a total of 32,272 of objects of so-called small retention exist in Poland, concentrating water with a volume of 0.826 km^3. An increase in retention resources was obtained through damming 360 lakes, construction of 4176 small water reservoirs, 8317 ponds, and 18,760 damming structures on rivers. From the point of view of water management, the largest 60 reservoirs are of vital importance for hydrology and water management. They have a substantial effect on the level of exploitable resources, including 10 with a volume of more than 100 hm^3 (Table 4.5). The total volume of the reservoirs exceeds 4 km^3, which constitutes approximately 6.5% of annual outflow.

Water resources occurring in rivers, lakes, ponds, retention reservoirs, rocks, and in wetland areas are subject to seasonal fluctuations. They are usually interrelated, and a disturbance in some places of occurrence of water in the environment results in consequences in others. The economic use of water requires its supply during the entire period of work, i.e. also in the case of low and high resources. It is possible to increase water resources to a certain extent through regulating river discharge, water retention in reservoirs of different ranks, and through other measures favouring water retention in the catchment. In the case of firm regulation of river discharge, the amount of water resources available for use can be determined by the amount of outflowing water. In the case of cyclical occurrence of dry and humid periods, the amount of water resources of the country, and therefore also the possible level of use of the resources for economic purposes, should be determined based on discharges with a high probability of occurrence. Economic balancing must consider so-called inviolable (biological) river discharges, determined for Polish rivers at a level of 15 km^3. Kaczmarek [21] adopted the amount of outflowing water that can be supplied to users over 95% of time in each year as the amount of exploitable resources. The outflow of Polish rivers at such a level of probability was determined as 22 km^3. A similar amount of exploitable resources is obtained by adopting their value as groundwater outflow in the year with the lowest outflow. After subtracting the value of biological discharge from exploitable resources, we obtain the amount of exploitable resources of Poland for irreclaimable use equalling 7 km^3·year^{-1}. The value can be increased to 9–10 km^3·year^{-1} after regulating discharges of Polish rivers, i.e. after an increase in reservoir and soil retention.

4.4 Water Resources Balance and Use

The amount of water in the natural environment changes in a seasonal and annual scale. Performing its balance requires the determination of the time and areas for which the value of water inflow and outflow is assessed. The water balance equation components may be expressed as a mean depth of water over the basin or water body (mm), as a volume of water (m^3), or in the form of discharge (m^3·s^{-1}) [22]. Therefore, it is possible to present components of water circulation for different areas and time intervals. It is the most efficient, due to the unambiguously determined water runoff area, to prepare water balances for catchments. Balances for other areas are also based on calculations performed in catchments within the analysed area. Basic information on water resources of the area is included in normal water balance, prepared based on data from the multi-annual period in which precipitation is balanced by outflow and evapotranspiration. In the case of presentation of parameters for shorter periods, they should consider the change in retention in the balancing period, described as the difference between the amount of resources at the end and beginning of balancing.

In an average year, atmospheric precipitation supplies approximately 190 km^3 of water to the territory of Poland, and approximately 7 km^3 inflows from the neighbouring countries. In the years 1901–2000, mean precipitation in the territory of Poland was 628 mm, and outflow 175.2 mm [7]. Approximately 450 mm corresponds to evapotranspiration, i.e. field evaporation. In a year with average humidity, the amount of precipitation in the territory of the country varies from approximately 500 mm in the lowland belt to more than 1000 mm in the mountains, which directly affects the amount of regional water resources. In an average year [23], mean river outflow is dominated by water supply from groundwater resources (95 mm, 55%). The remaining part is accounted for by surface runoff (76 mm, 45%). The percent contribution of both components varies in a regional scale. The mountain areas surface runoff is prevalent, and on uplands – groundwater outflow.

Satisfying municipal, industrial, and agricultural water needs of the population and economy of Poland occurs from groundwater and surface water resources. They develop from atmospheric precipitation, within particular regions and places of occurrence of water which in water management should be treated jointly. A separate calculation of surface water and groundwater resources is only justified in valleys of large transit rivers, i.e. in situations when flowing water resources development in different precipitation regions. Water intake for economic purposes is only possible in the case of relevant values of its discharge occurring in river channels or in rock pores and fissures, with the maintenance of biological flow.

In the second half of the previous century, water use rapidly increased as a result of the "planned" urbanisation and industrialisation of the country. In 1965, controlled water intake was 7.6 km^3, and from 1977 it regularly exceeded 14 $km^3 \cdot year^{-1}$ [24]. The occurrence of low flows at the turn of the 1980s and 1990s, and the resulting periodical water shortages and its heavy pollution, determined the directional transition to economic water management, as well as more effective sewage treatment. The decrease in the amount of collected water, however, was to the greatest extent determined by the economic crisis and the related restructuring of the industry, and liberalisation of prices of water at the beginning of the 1990s. These factors caused considerable limiting of water use for different purposes and led to substantial changes in the structure and amount of water use in Poland. The approach of society and state administration to the resources of the natural environment and increase in the standard of life of residents also changed.

The country's economy is particularly supplied with surface waters that currently satisfy the needs of approximately 83.6%. Groundwaters, together with waters from drainage of mines, accounting for 16.5% [4]. In 2016, 70.81% of total water intake was used for production purposes, 19.34% for municipal and irrigation purposes, and 9.85% for supplementation of water in fish ponds. Amounts of water intake and use in different sectors of the national economy in three-time intervals were compared in Table 4.6. Values from 1980 were adopted as the reference level.

Data concerning 1980 include information on the highest level of water use that in the 1960s annually exceeded 14 km^3, including groundwaters of approximately 2.0 km^3. At the turn of the 1980s and 1990s, the approach to water resources changed. This was undoubtedly partially determined by changes in the economic system, as well as the occurrence of a series of very dry years which caused serious problems in the municipal economy, industry, and agriculture. At the end of the twentieth century, the water needs of the country stabilised at the annual level of 11 km^3. Over the last decade, due to the introduction of new production technologies, the amount of water intake for the purposes of the industry has slightly decreased. Water intake for the purposes of agriculture and municipal economy has been maintained on a similar level. Total annual water intake in the years 1980–2016 decreased from 14.2 km^3 to 10.6 km^3. Moreover, the amount of water collected for the agricultural irrigation purposes is dependent on the humidity of a given period. Over the last twenty years, atmospheric alimentation was relatively high, which positively translated into the resources of shallow groundwaters. Data of the Central Statistical Office [4] show that the total water intake in the years 1980–2016 for agriculture and forestry for the purpose of irrigations and for supplementation of water in fish ponds decreased from 1.323 to 1.043 km^3. In the same period, the amount of groundwater intake for the purposes of the economy decreased from 1.958 to 1.688 km^3.

The use of groundwaters accounts for approximately 10% of their exploitable resources. Nonetheless, areas exist, particularly in the vicinity of large cities, where overdrying of the ground occurred, and therefore the possibilities of agricultural land use decreased. Reserves of groundwaters usually exist in non-urbanised areas not affected by human pressure.

Table 4.6 Water withdrawal for national economy and population purposes [4]

Water withdrawal	1980		2000		2016	
	hm^3	%	hm^3	%	hm^3	%
Total	14,183.6	100.0	11,048.5	100.0	10,581.4	100.0
Surface waters	11,899.0	83.89	9150.6	82.82	8840.8	83.55
Groundwaters	1958.3	13.81	1747.3	15.82	1687.9	15.95
Waters from mines drainage – used for production	326.2	2.30	150.6	1.36	52.8	0.50
Production purposes	10,137.6	71.47	7637.9	69.13	7492.8	70.81
Sufarce waters	9168.5	64.64	7221.5	65.36	7228.7	68.32
Groundwaters	642.9	4.53	265.8	2.42	211.3	2.00
Water from mine and building constructions drainage – used for production	326.2	2.30	150.6	1.36	52.8	0.50
Irrigation in agriculture and forestry and filling and completing fishponds	1323.4	9.33	1060.6	9.60	1042.7	9.85
Surface waters	1323.2	9.33	1060.6	9.60	1039.9	9.82
Groundwaters					2.7	0.02
Exploitation of water supply network – withdrawal in water intakes	2722.6	19.20	2350.1	21.27	2045.9	19.34
Surface waters	1407.2	9.92	868.5	7.86	572.2	5.41
Groundwaters	1315.4	9.28	1481.5	13.41	1473.8	13.93

hm^3 – cubic hectometres, 1,000,000 m^3

According to the materials of the Ministry of the Environment, the total registered intake of groundwaters in the area of the country as at the end of 2013 was more than 2.6 km^3·year^{-1} [25]. This value includes:

- registered intake in more than 18,500 groundwater intakes functioning for the purposes of provision of the population and industry with water to the amount of more than 1.6 km^3·year^{-1},
- intake of more than 0.95 km^3·year^{-1} in the scope of meliorations of mines exploiting more than 90 largest deposits in Poland,
- intake from meliorations of inactive mines in the Upper Silesian Coal Basin with a value of more than 0.08 km^3·year^{-1}.

In the Vistula River catchment area, according to the state from 2013, groundwater intake exceeded 1.25 km^3·year^{-1}, and in the Oder River catchment 1.31 km^3·year^{-1}.

A systematic increase in water intake should be expected in the upcoming years, particularly for agriculture, up to a level of 13.4 km^3 in 2025 [26]. An increase in water use, also in agriculture (irrigation) and fish farming, must consider the rules of the basic economic calculation. It is particularly important in lowland areas of

Table 4.7 Primary melioration [4]

Year	Rivers and canals		Embankments		Usable capacity of water reservoirs in dam^3	Drainage pump stations	
	Lenght (km)	Of which regulated (km)	Lenght (km)	Protected area in thousands ha		Number	Area of interaction in thousands ha
1990	72,577	37,923	8148	1004.3	163,408	609	571.4
2003	73,812	39,972	8450	1074.9	261,334	574	601.4
2016	75,297	43,442	8451	1091.2	279,955	579	616.1

the country experiencing structural water deficits. The intensification of agriculture requires securing water supply in the period of growth of plants, which occurs through land melioration. The measures involve the application of the regulation of water outflow and inflow, in accordance with the production requirements of crops. Incomplete functioning of melioration facilities leads to negative changes in the environment, and particularly to the overdrying of the ground, which is extremely dangerous for organic soils. Their dehydration reduces the retention resources of the catchment and launches an intensive process of mineralisation of organic substance, degrading natural water resources.

The area of meliorated arable land is slowly increasing. This results from the intensification of agricultural production, currently also related to the profitability of agricultural and breeding production. Inconsiderable changes also occur in the state of the environment covered by so-called basic meliorations (Table 4.7). The value is too low for overtaking the rapidly developing surface runoff. It is estimated that the basic melioration requires the reconstruction or modernisation of 15,551 km of rivers, 3,657 km of embankments, and 1,443 thousand ha of arable land in need of modernisation [4]. It should be emphasised that the existing hydrotechnical infrastructure can mitigate the effects of drought only to a low degree. The most efficient direction of action is works aimed at a decrease and slowing down of water outflow from small catchments, an increase in forest cover, maintenance and reconstruction of wetlands, and water retention in the existing small reservoirs in upper parts of the catchment. It should be emphasised that as a result of inappropriately performed meliorations, particularly in periods of hydrological drought, the area of wetlands and peat bogs is subject to continuous and permanent reduction.

Approximately 8.5 thousand km of anti-flood embankments have been constructed in Poland, protecting around 4% of the territory of the country. It is estimated that 40% of the embankments are in a good technical state, and the remaining ones require renovation or modernisation. The existing embankments protect approximately 75% of areas threatened with floods [27].

4.5 Summary and Conclusions

The territory of Poland is characterised by relatively low water resources at the European scale. This results from the amount of precipitation and evapotranspiration, and their seasonal variability. Mean outflow in Polish rivers in the period 1901–2000 was 61.5 km^3, and that from the area of Poland equalled 53.9 km^3. The value corresponds to the layer of outflow of 175.2 mm and unitary outflow of 5.56 $dm^3 \cdot s^{-1} \cdot km^{-2}$. The outflow of rivers, expressed both in the outflow layer and per resident, is among the lowest in Europe.

Water resources of Poland are directly related to the features of the natural environment, particularly the hydrogeological and climatic conditions. They determine the amount of resources and their temporal and spatial variability. This is of particular importance in the observed climate changes that will affect the renewability of groundwaters and surface waters, especially in the conditions of increasing uncertainty of precipitation. Particular importance is gained by groundwater resources, showing higher inertia in the conditions of changes in precipitation and evapotranspiration. They can satisfy the water needs of the population and economy with respect for the rules of sustainable use of waters.

In the context of the occurring climatic changes, measures aimed at water retention in the catchment are recommended. Changes in the conditions of water circulation and outflow should be towards its retention in the soil and bedrock, leading to a decrease in surface runoff and an increase in exploitable resources in catchments. In addition to the construction of retention reservoirs, this purpose can be met by changes in the land use structure – particularly through forestation of areas with high slope inclination, application of suitable agrotechnical measures, and observing the term of their performance, as well as the construction of corrective steps slowing down water outflow.

All resources, of both surface and groundwaters, are subject to strong human pressure leading to their quantitative and qualitative transformations. They result from an increase in the water needs accompanying an increase in the standard of life and development of industrial infrastructure and raw material extraction. It is important to reduce the consumption levels at every stage of its use, i.e. in households, and particularly in the industry and agriculture. The level of exploitation, especially on the regional scale, should not endanger the stability of low flows and natural functioning of hydrogenic ecosystems. The water safety of the country requires the maintenance of the quantity and quality of water resources on a good level and effective protection against the effects of drought and floods.

4.6 Recommendation

In the aspect of climate change and low water resources, actions aiming at improvement of water balance structure of in Poland, mainly reduction of water losses and increase of retention, are necessary. Changes of water outflow conditions show cause increase of retention possibilities of soils, as well as aeration zone and groundwater retention, which will led to decrease of surface runoff and increase of water disposable resources in the catchment. Indicated direction of changes, construction of water reservoirs and correction sills slowing down the outflow, can be archived by land use structure changes (afforestation of areas of high gradient of slopes and reduction of surface runoff from urbanized areas). In order to water resources protection, rational management of available resources in catchments, elimination or limitation of influence of point, linear and diffuse sources of pollution, decision processes simplification in spatial management and exploitation of water-economic systems are necessary.

References

1. Vörösmarty CJ, Green P, Salisbury J, Lammers RB (2000) Global Water resources: vulnerability from climate change and population growth. Science 289(5477):284–288. https://doi.org/10.1126/science.589.5477.284
2. Oki T, Kanae S (2006) Global hydrological cycles and world water resources. Science 313(5790):1068–1072. https://doi.org/10.1126/science.1128845
3. www.unic.un.org.pl/dekada-wody/miedzynarodowa-dekada-wody-2018%E2%80%932028/3220. Accessed 24 Nov 2018
4. Ochrona Środowiska (Environmental protection). Wyd. GUS, Warszawa (in Polish) (2017)
5. Shiklomanov IA (2000) Appraisal and assessment of world water resources. Water Inter 25(1):11–32. https://doi.org/10.1080/02508060008686794
6. Kowalczak P (2007) Konflikty o wodę (Water conflicts). Wyd. Kurpisz Przeźmierowo (in Polish)
7. Fal B, Bogdanowicz E (2002) Zasoby wód powierzchniowych Polski (Surface water resources in Poland). Wiad IMGW 3:3–37 (in Polish)
8. Bajkiewicz-Grabowska E (2020) Geoecosystems of Polish lakes. In: Korzeniewska E, Harnisz M (eds), Polish river Basins and lakes—Part I. Handbook Environ Chem 86, 57–67. Springer, Cham. https://doi.org/10.1007/978-3-030-12123-5_3
9. Bajkiewicz-Grabowska E, Markowski M, Golus W (2020) Polish rivers as hydrographic objects. In: Korzeniewska E, Harnisz M (eds), Polish river Basins and lakes—Part I. Handbook Environ Chem 86:27–55. Springer, Cham. https://doi.org/10.1007/978-3-030-12123-5_2
10. Gutry-Korycka M, Sadurski A, Kundzewicz ZW, Pociask-Karteczka J, Skrzypczyk L (2014) Zasoby wodne a ich wykorzystanie (Water resources and their use). Nauka 1:77–98 (in Polish)
11. Fal B (1993) Zmienność odpływu z obszaru Polski w bieżącym stuleciu (Changeability of outflow from the Poland territory in present century). Wiad IMGW 16:3–20 (in Polish)
12. Biuletyn Państwowej Służby Hydrologiczno-Meteorologicznej (Bulletin of the State Hydrological and Meteorological Service). IMiGW-PIB, Warszawa (2017) (in Polish)
13. Szczepański A (2008) Współczesne problemy gospodarki wodnej i zasobów wód (Present problems of water management and water resources). In: Kotarba MJ (ed) Przemiany środowiska naturalnego a rozwój zrównoważony (Changes of natural environment and sustainable development), pp 53–64. Wydawnictwo TBPŚ GEOSFERA Kraków (in Polish)
14. Paczyński B (ed) (1995) Atlas hydrogeologiczny Polski (Hydrogeological atlas of Poland). Wyd. PIG Warszawa (in Polish)
15. Choiński A (2007) Katalog jezior Polski (Catalogue of lakes in Poland). Wyd. UAM Poznań (in Polish)
16. Choiński A (2017) Geneza i rozmieszczenie jezior (Origin and location of lakes). In: Jokiel P, Marszelewski W, Pociask-Karteczka J (eds), Hydrologia Polski (Hydrology of Poland), pp 223–229. Wyd. Naukowe PWN Warszawa (in Polish)

17. Majdanowski S (1954) Jeziora Polski (Lakes of Poland). Przegl Geogr 26(2):17–50 (in Polish)
18. Dębski K (1970) Hydrologia (Hydrology). Wyd. Arkady Warszawa (in Polish)
19. Borowiak D (2017) Zasoby i bilans wodny jezior (Resources and water balance of lakes). In: Jokiel P, Marszelewski W, Pociask-Karteczka J (eds), Hydrologia Polski (Hydrology of Poland), pp 229–234. Wyd. Naukowe PWN Warszawa (in Polish)
20. Kowalewski G (2017) Rola mokradeł w obiegu wody (The role of wetlands in water circulation). In: Jokiel P, Marszelewski W, Pociask-Karteczka J (eds), Hydrologia Polski (Hydrology of Poland), pp 264–269. Wyd. Naukowe PWN Warszawa (in Polish)
21. Kaczmarek Z (1978) Zasoby wodne Polski i zasady ich racjonalnego użytkowania (Water resources in Poland and rules of their rational use). Nauka Polska 8:43–54 (in Polish)
22. Sposób J (2011) Water balance in terrestrial ecosystems. In: Gliński J, Lipiec J (eds), Encyclopedia of agrophysics, pp 955–959. Springer Science and Business Media. https://doi.org/10.1007/978-90-481-3585-1
23. Mikulski Z (1998) Gospodarka wodna (Water management). Wyd. PWN Warszawa (in Polish)
24. Wilgat T (1984) Ochrona zasobów wodnych Polski (Protection of water resources in Poland). Wyd. PWN Warszawa-Łódź (in Polish)
25. Gospodarowanie wodami w Polsce w latach 2012–2013 (Water management in Poland in 2012–2103) Ministerstwo Środowiska Warszawa (2014) (in Polish)
26. Kaczmarek Z (2005) Gospodarka wodna w Polsce u progu XXI wieku (Water management in Poland on the threshold of 21st century). Monografie Komitetu Inżynierii Środowiska PAN 32:27–40 (in Polish)
27. Strategia gospodarki wodnej (Strategy of water management). (In:) Światowy dzień wody (2005) (World water day 2005), 7–32. Wyd. KGW PAN Warszawa (in Polish)

Chapter 5
Water Resources of Stagnant Waters

Adam Choiński and Rajmund Skowron

Abstract Total water resources of lakes in Poland for 7081 lakes (above 1 ha) equal 19.7349 km^3 (Masurian Lake District – 51.27%, Pomeranian Lake District – 36.12%), Wielkopolska-Kujawy Lake District – 11.93%, and area south of the delineated range of the last glaciation – 0.68%). Retention for the area north of the line delimiting the maximum range of the last glaciation is 162 mm. It constitutes only 10% of mean annual precipitation, and the value is approximately 3.5 times smaller than the amount of mean outflow from the territory of Poland.

Keywords Water resources of lakes and dam reservoirs · Changes in the surface area of lakes · Poland

5.1 Introduction

From the geological point of view, the age of lakes is exceptionally short. This particularly refers to postglacial lakes, predominating in Poland. Two usually co-occurring factors are considered to be responsible for the process of the decline of lakes, and therefore also their water resources. The first one is water level fluctuations, contributing to changes in the surface area. A decrease in the water level results in the intensification of the sedimentation process. Accumulation of sediments, i.e. sedimentation within the lake basin, determines the degree of shallowing, and therefore leads to a decrease in water resources.

Causes for water level fluctuations include: short- or long-term climate changes, determining changes in water supply to the lake from aquifers; deforestation of the lake's catchment, effect of local factors, for example determining changes in the

A. Choiński (✉)
Institute of Physical Geography and Environmental Planning, Adam Mickiewicz University, Krygowskiego 10, 61-680 Poznań, Poland
e-mail: choinski@amu.edu.pl

R. Skowron
Faculty of Earth Sciences and Spatial Management, Nicolaus Copernicus University, Lwowska 1, 87-100 Toruń, Poland
e-mail: rskowron@umk.pl

M. Zeleňáková et al. (eds.), *Management of Water Resources in Poland*, Springer Water,
https://doi.org/10.1007/978-3-030-61965-7_5

erosion base, and variable time of the lake's incorporation into the hydrographic network; performance of different hydrotechnical works, both within the lake – outflow regulation, and within its catchment, e.g. meliorations, as well as natural factors that can contribute to water level increase, such as landslides or construction of beaver dams.

The second factor contributing to the decline of lakes, i.e. aggregation of sediments, can occur through: accumulation of biogenic mass, i.e. sedimentation; precipitation of chemical compounds; sedimentation of clastic sediments supplied by streams, colluvia and deluvia deposited in the lake; effect of aeolian processes.

The course of evolution of a given lake basin depends on many factors, among others: the genetic type of the basin; location in a given climate zone; individual features of the basin determining the morphometric parameters; size of the catchment and the ratio of its size to the surface area of the lake; stability of climate which determines water balance; effect of human pressure of water relations in the lake's catchment.

Three stages of development of lakes are usually designated. The first one is the youth stage, characterised by the basin's shape approximate to the original one, with no evident changes resulting from sedimentation processes. The second stage, i.e. maturity stage, is characterised by obscuring of depth contrasts as a consequence of deposition of a high amount of sediments. The final stage, i.e. the old age stage, is associated with complete disappearance and leads to filling the entire basin with sediments. The evolution of lake basins in some cases does not always follow the above pattern. Sometimes, with time, lakes not only fail to disappear, but even increase their surface area, and therefore their water volume. Moreover, in lakes with a very variable basin, their certain fragments can be in different stages of development due to their different depths.

In reference to stagnant waters, water resources and their variability can be considered in three aspects, i.e. in reference to lakes (natural water objects >1 ha), tarns (natural water objects <1 ha), and artificial reservoirs.

5.2 Water Resources of Lakes

Total water resources of lakes in Poland were estimated based on the Catalogue of Polish Lakes [1], including information concerning surface area for 7081 lakes (above 1 ha). The calculation procedure involved adding up known lake water volumes for more than 30% of lakes and adding the total water volume of unmeasured lakes. It was calculated as a product of their surface area and mean depth. Lake water resources calculated this way equal 19.7349 km^3. Water resources within particular lake districts are as follows: Masurian Lake District – 10.1183 km^3 (51.27%), Pomeranian Lake District – 7.1292 km^3 (36.12%), Wielkopolska-Kujawy Lake District – 2.3535 km^3 (11.93%), and area south of the delineated range of the last glaciation – 0.1339 km^3 (0.68%). Approximate values were also added in earlier Polish publications [2].

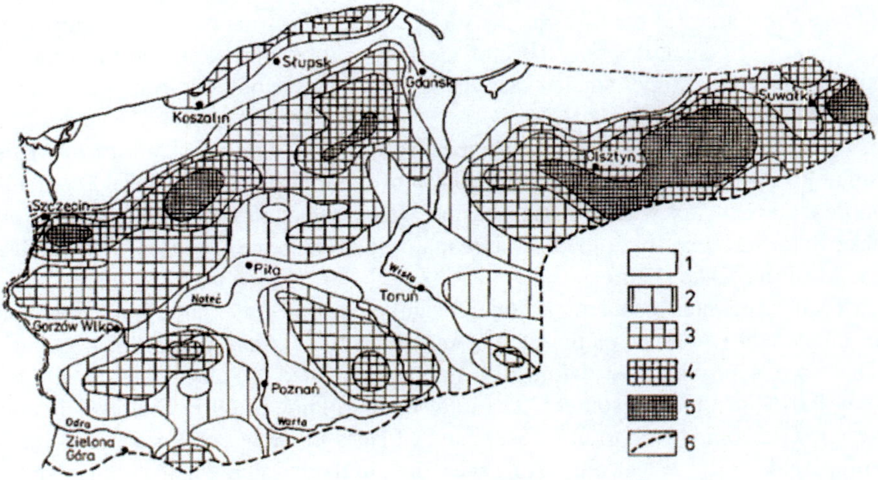

Fig. 5.1 Variability of the retention index of lake waters (mm): 1 – below 50, 2 – 50–100, 3 – 100–250, 4 – 250–500, 5 – above 500, 6 – delineated the range of the last glaciation

Figure 5.1 presents regional variability of water resources of lakes expressed as the retention index in mm. The map of retention capacities was prepared based on the outlines of maps at a scale of 1:50,000 for more than 200 sheets of topographic maps. The middle of each field was ascribed a retention value providing the basis for interpolation. Retention capacity was expressed in mm, permitting its comparison with other indices. The determined retention capacity for the area north of the line delimiting the maximum range of the last glaciation is 162 mm. An evident two-fold character of the analysed phenomenon is observed on the map, i.e. the Masurian and Pomeranian Lake District have considerably larger resources than the Wielkopolska-Kujawy Lake District. The zone of maximum resources has an evident latitudinal orientation and reaches the highest values, usually south of the I degree watershed, both in the case of the Pomeranian and Masurian Lakelands. Based on delineation of [3], the greatest resources within the Masurian Lake District occur in the following catchments: Węgorapa (404 A) – 1976 mm, Biała Hańcza (407) – 958 mm, and Pisa (227 A) – 893 mm, and in the Pomeranian Lake District in catchments of: Drawa (119 D) – 389 mm, Wda (237 B) – 196 mm, and in the Przymorze catchment from Wieprza to Słupia (310) – 184 mm. Within the Wielkopolska-Kujawy Lake District, lake catchments retain considerably less waters, and the most abundant ones include Vistula from Bzura to Skrwa (233 A) – 173 mm, Warta from Wełna to Obra (118 G) – 141 mm, and Noteć to Gwda (119 A) – 116 mm. Zones with the lowest retention capacity, i.e. up to 50 mm, are located in valleys of the largest rivers with the adjacent areas, and coasts: Gdańsk, East Baltic, Szczecin, and the southern part of the Koszalin Coast. Mean retention capacity determined within the three lake districts is exceptionally variable. In the case of the Masurian Lake District, it is 283 mm, for the Pomeranian Lake District the index equals 150 mm, and for the Wielkopolska-Kujawy Lake District only 58 mm. The division of the total volume of water resources of the lakes by the surface area of Poland provided a water layer of 63 mm. It constitutes only 10% of mean annual precipitation, and the value is approximately 3.5 times smaller than the amount of mean outflow from the territory of Poland.

Table 5.1 presents lakes with the highest volumes, i.e. above 100 million m^3. This small group including only 26 lakes (0.4% of the total number of lakes) contains more than 6 km^3 of waters in their basins, which constitutes more than 30% of all lake water resources in Poland. A large majority of them, i.e. 16 lakes, is located in the Masurian Lake District.

Total lake water resources in Poland can be described as scarce in comparison to lake water resources of the globe. According to [5], they equal 176 400 km^3. Therefore, approximately 20 km^3 of waters accumulated in Polish lakes constitutes only 0.01% of global resources. The total water volume of all Polish lakes is also very low in comparison to the most abundant lakes in the world. For example, it is comparable to the water volume of Lake Como in Italy, occupying the 66th position in the ranking of the lakes the most abundant in water resources in the world [6].

5.3 Water Resources of Tarns

Postglacial tarns are defined as small, usually closed-drainage water bodies. They usually have a round shape, whereas their maximum depths exceptionally exceed 3 m. In the summer period in the case of natural decrease in groundwater level, they frequently remain without water. Therefore, depending on their location, they can be of permanent, periodic, or episodic character. Their permanent disappearance is usually associated with extensive melioration works. Tarns occur both north and south of the maximum range of the last glaciation. They are the most characteristic of zones of moraine plateaus and ground moraine areas. The spatial distribution of tarns was analysed by Choiński [7] within the area north of the maximum range of the Baltic glaciation. The number of tarns was determined based on the analysis of topographic maps at a scale of 1:50 000, edited in the mid-1970s. Tarns were classified as natural water bodies not exceeding 1 ha. The background of more than 200 maps at a scale of 1:50 000 was covered with a grid of primary fields with an area of 100 km^2. The determination of the number of tarns in primary fields permitted spatial presentation of the analysed phenomenon and interpolation (Fig. 5.2). The analysed area is characterised by a variable concentration of occurrence of tarns. In the Pomeranian Lake District, their largest clusters are located in the Kashubian and Drawa Lake Districts, in the eastern part of the Krajeńskie Lake District, and the western part of the Myśliborskie Lake District. A higher concentration of tarns also occurs over an extensive section of the watershed zone of coastal rivers and rivers flowing south of them. A small number of tarns occurs in the area of the West Pomeranian Coast, in zones directly adjacent to river valleys, and outwash plains.

In the Masurian Lake District, evident variability is observed between its northern and southern part, where the concentration of tarns is considerably lower. Their

Table 5.1 Lakes with the highest water volume (in million m^3) based on data of the Institute of Inland Fisheries in Olsztyn

No	Lake	Volume (million m^3)	Area (ha)	Location (according to [4])
1	Mamry	1 003.4	10 282.4	Poj. Mazurskie (Kraina Wielkich Jezior Mazurskich)
2	Miedwie	681.7	3 527.0	Poj. Pomorskie (Równina Pyrzycko-Starogardzka)
3	Śniardwy	660.2	11 340.4	Poj. Mazurskie (Kraina Wielkich Jezior Mazurskich)
4	Wigry	336.7	2 118.2	Poj. Mazurskie (Równina Augustowska)
5	Drawsko	333.4	1 871.5	Poj. Pomorskie (Poj. Drawskie)
6	Niegocin	258.5	2 600.0	Poj. Mazurskie (Kraina Wielkich Jezior Mazurskich)
7	Tałty-Ryńskie	248.4	1 831.2	Poj. Mazurskie (Kraina Wielkich Jezior Mazurskich)
8	Wdzydze	220.8	1 455.6	Poj. Pomorskie (Równina Charzykowska)
9	Lubie	169.9	1 439.0	Poj. Pomorskie (Poj. Drawskie)
10	Łańskie	168.0	1 042.0	Poj. Mazurskie (Poj. Olsztyńskie)
11	Roś	152.9	1 887.7	Poj. Mazurskie (Kraina Wielkich Jezior Mazurskich)
12	Rajgrodzkie	142.6	1 503.2	Poj. Mazurskie (Poj. Ełckie)
13	Jeziorak	141.6	3219.4	Poj. Mazurskie (Poj. Iławskie)
14	Pluszne	134.9	903.3	Poj. Mazurskie (Poj. Olsztyńskie)
15	Powidzkie	134.8	1 174.7	Poj. Wielkopolskie (Poj. Gnieźnieńskie)
16	Charzykowskie	134.5	1 363.8	Poj. Pomorskie (Poj. Charzykowskie)
17	Narie	124.6	1 240.1	Poj. Mazurskie (Poj. Ełckie)
18	Dadaj	120.8	976.8	Poj. Mazurskie (Poj. Olsztyńskie)
19	Żarnowieckie	120.8	1 431.6	Poj. Pomorskie (Wysoczyzna Żarnowiecka)
20	Hańcza	120.4	311.4	Poj. Mazurskie (Poj. Wschodniosuwalskie)
21	Łebsko	117.5	7 140.0	Poj. Pomorskie (Pobrzeże Słowińskie)

(continued)

Table 5.1 (continued)

No	Lake	Volume (million m^3)	Area (ha)	Location (according to [4])
22	Pile	115.2	980.1	Poj. Pomorskie (Poj. Szczecineckie)
23	Nidzkie	113.9	1 818.0	Poj. Mazurskie (Równina Mazurska)
24	Mokre	107.3	841.0	Poj. Mazurskie (Poj. Mrągowskie)
25	Siecino	104.4	729.7	Poj. Pomorskie (Poj. Drawskie)
26	Gołdapiwo	101.5	862.5	Poj. Mazurskie (Kraina Wielkich Jezior Mazurskich)

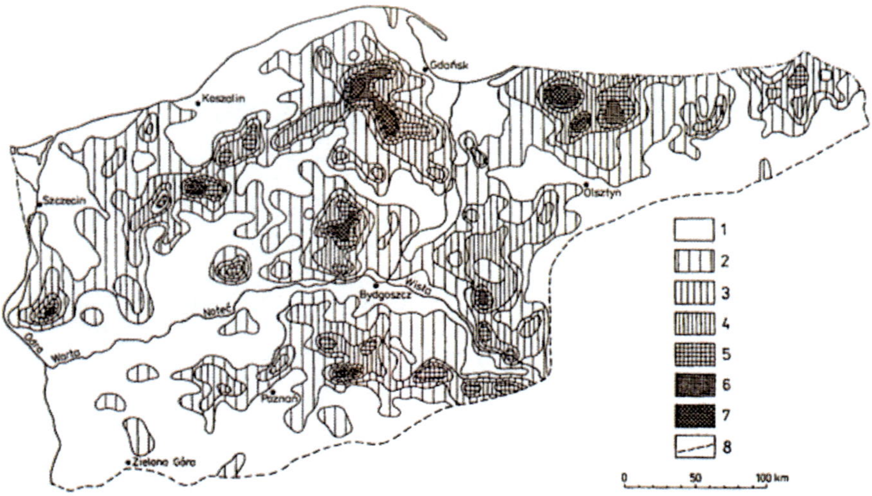

Fig. 5.2 Density of occurrence of tarns per 100 km^2: 1 – 0–50, 2 – 51–100, 3 – 101–150, 4 – 151–200, 5 – 201–250, 6 – 251–300, 7 – above 300, 8 – range of the last glaciation (according to [8])

largest clusters are located north of the Olsztyn Lake District, and on the Dobrzyń and Lithuanian Lake Districts.

Due to high dispersion of tarns, the Wielkopolska-Kujawy Lake District differs from the lake districts mentioned earlier. Tarns occur much more seldom there, and their concentrations are considerably lower, whereas in the eastern part of the Lake District they are substantially more abundant than in the western part. In the case of the Wielkopolska-Kujawy Lake District, a considerably higher effect of human pressure is observed in comparison to the Pomeranian and Masurian Lake Districts. Table 5.2 presents the number of tarns within the designated lake districts [7].

Table 5.2 The occurrence of tarns within the designated lake districts (according to [7])

Lake district	Number of tarns	Share (%)	Surface area (km^2)	Mean density per 100 km^2
Pomorskie	36 657	44.6	48 330	75.8
Mazurskie	28 603	34.8	35 928	79.6
Wielkopolsko-Kujawskie	16 916	20.6	31 747	53.3
Summary	82 176	100.0	116 005	70.8

A total of 82 176 tarns were identified in the analysed area. Their highest number (as well as that of lakes with an area >1 ha) occurs in the Pomeranian Lake District, i.e. 36 657, constituting 44.6% of their total abundance. Their highest mean density, i.e. 79.6 tarns/100 km^2, occurs in the Masurian Lake District, although in terms of mean density of tarns, the Masurian and Pomeranian Lake Districts are similar. The Wielkopolska-Kujawy Lake District considerably differs from them in this aspect. The mean density of tarns in the Lake District is lower by approximately 30%. The mean density of tarns throughout the analysed area is 70.8 per 100 km^2. The highest density of tarns reached more than 300 per 100 km^2. Zones of the type occurring in the Kashubian, Drawa, Krajeńskie, Gniezno Lake Districts, and north of the Olsztyn Lake District. An evident tendency of dependency on the number of tarns on the type of substrate is observed. In zones covered by sands and gravels of fluvioglacial accumulation, the number of tarns is small and does not exceed 50 per 100 km^2. In zones of occurrence of glacial tills, it is several times higher.

The determination of the number of tarns permits the estimation of their total surface area. It is possible with the assumption of an average surface area of a tarn. In the case above, the adopted average surface area was 0.5 ha. Therefore, with such an assumption, the total surface area of tarns was estimated for 41 088 ha. The area is approximately 4 times larger than the largest Polish Lake Śniardwy.

A procedure analogical to the one described above can be applied in the estimation of total water resources of tarns. Assuming its previously determined total surface area of 41 088 ha and mean depth of 0.5 m, the total water volume in tarns can be estimated for 0.20544 km^3. The value constitutes 1.0% of water resources of lakes larger than 1 ha. For particular lake districts, the contribution is as follows: Pomeranian – 1.28%, Masurian – 0.71%, and Wielkopolska-Kujawy – 1.80%. Total resources accumulated in tarns determine the retention index within the analysed area equal to 1.8 mm.

Sebzda [9] analysing topographic maps at a scale of 1:50 000 from the 1970s determined the number of tarns in the area south of the range of the Baltic glaciation, i.e. over an area of 195 836 km^2. Only 10 676 tarns were identified there, corresponding to a mean density of only 5.5 per 100 km^2. With assumptions analogical to the previous ones, their total surface area can be estimated for 5 338 ha. It is therefore approximate to the surface area of Lakes Jeziorak and Jamno and constitutes 1.9% of Polish lakes larger than 1 ha. Their water resources can be estimated for 0.02669 km^3, which corresponds to only 0.13% of lake water resources in Poland. In the case of spatial distribution of tarns, their considerably lower densities of occurrence are observed south of the range of the last glaciation. This evidence suggests their high susceptibility to decline, depending on their age.

5.4 Water Resources of Dam Reservoirs

No favourable conditions for the construction of water reservoirs occur in Poland, particularly those with considerable surface areas. This results from small water resources in rivers, considerable variability of outflow, and unfavourable natural conditions for their construction. The largest reservoirs in terms of surface area as well as volume are approximate to natural lakes. The reservoir in Solina on the San River is the largest in terms of water resources. It accumulates 0.47 km^3 of waters, which constitutes approximately half of the water resources of Mamry, the lake with the richest water resources in Poland. In terms of surface area, the reservoir on the Vistula River above the barrage in Włocławek is the largest. Its surface area of 70.4 km^2 is comparable with that of Lake Łebsko, occupying the third position in this aspect in Poland.

Wiśniewski [10] listed dam reservoirs (more than 1 million m^3) and determined their number for 101 (Table 5.3). Only 11 among them have resources exceeding 100 million m^3. It is worth emphasising that 26 natural lakes are identified to have water resources exceeding 100 million m^3.

The location of dam reservoirs in Poland is presented in Fig. 5.3. Notice the very evident concentration of reservoirs in mountain and sub-mountain regions, in the Pomeranian Lake District, and on the Małopolska Upland. In the Masurian Lake District and the belt of lowlands, dam reservoirs occur sporadically. Rivers the best developed in terms of the number of reservoirs include Bóbr, Radunia, Nysa Łużycka, Nysa Kłodzka, Rega, Brda, and Soła.

It is worth emphasising that human activity in Poland in reference to the construction of large dam reservoirs dates back 170 years. The oldest reservoir with a volume of more than 1 million m^3 is Mylof on the Brda River, launched in 1848. Several reservoirs in the Sudetic region and the Pomerania come from the first quarter of the twentieth century. The construction of more than half of the reservoirs (with a volume of more than 1 million m^3) dates back to the period after World War 2.

The total volume of dam reservoirs (more than 1 million m^3) is approximately 3.5 km^3. This constitutes approximately 18% of water resources of Polish lakes and approximately 6% of the volume of waters outflowing annually from the territory of Poland. Their total surface area is approximately 500 km^2, i.e. around 18% of the area of lakes in Poland.

Due to the inconsiderable economic importance of artificial reservoirs with small surface areas, they are not included in any comparisons. Obtaining an answer to the question, however, is equivalent to the determination to what degree man managed to increase the system of stagnant surface waters in comparison with what nature created. The analysis of the above issue was performed for the zone south of the

Table 5.3 Artificial water reservoirs in Poland with water volume exceeding 1 million m^3 (elaboration based on [10] and Yearbooks of the Central Statistical Office)

No	Reservoir	River	Year of launching	Total volume (million m^3)	Surface area (km^2)
1	Solina	San	1968	472.0	21.1
2	Włocławek	Wisła	1970	408.0	70.4
3	Czorsztyn-Niedzica	Dunajec	1997	231.9	12.3
4	Jeziorsko	Warta	1986	202.8	42.3
5	Goczałkowice	Mała Wisła	1956	166.8	32.0
6	Rożnów	Dunajec	1941	166.6	16.0
7	Dobczyce	Raba	1986	125.0	10.7
8	Otmuchów	Nysa Kłodzka	1933	124.5	19.8
9	Nysa	Nysa Kłodzka	1972	113.6	20.4
10	Turawa	Mała Panew	1948	106.2	20.8
11	Tresna	Soła	1967	100.0	10.0
12	Dębe	Narew	1963	94.3	30.3
13	Dzierżono Duże	Kłodnica	1964	94.0	6.2
14	Sulejów	Pilica	1973	88.1	19.8
15	Koronowo	Brda	1960	80.6	15.6
16	Siemianówka	Narew	1995	79.5	32.5
17	Mietków	Bystrzyca	1986	70.5	9.2
18	Pilchowice	Bóbr	1912	54.0	2.4
19	Dziećkowice	Przemsza	1976	52.5	7.1
20	Klimkówka	Ropa	1994	43.5	3.1
21	Słup	Nysa Szalona	1978	38.6	4.9
22	Pławniowce	Potok Toszewski	1976	29.1	2.4
23	Porąbka	Soła	1936	28.4	3.7
24	Poraj	Warta	1978	25.1	5.5
25	Chańcza	Czarna Staszowska	1984	24.5	4.7
26	Rybnik	Ruda	1972	22.0	4.7
27	Przeczyce	Czarna Przemsza	1963	20.7	5.1
28	Leśna	Kwisa	1906	18.0	1.4
29	Bukówka	Bóbr	1987	16.8	2.0
30	Żur	Wda	1929	16.0	3.0
31	Besko	Wisłok	1978	16.0	1.3
32	Kozłowa Góra	Krynica	1937	15.8	5.8

(continued)

Table 5.3 (continued)

No	Reservoir	River	Year of launching	Total volume (million m^3)	Surface area (km^2)
33	Dzierżono Małe	Drama	1938	14.1	1.3
34	Złotniki	Kwisa	1934	12.4	1.2
35	Czchów	Dunajec	1949	12.0	2.5
36	Pogoria III	Pogoria	1974	12.0	2.0
37	Łąka	Pszczynka	1986	12.0	4.2
38	Pierzchały	Pasłęka	1916	11.5	2.4
39	Dobromierz	Strzegomka	1986	11.3	1.0
40	Myczkowce	San	1961	10.9	2.0
41	Rosnowo	Radew	1922	8.8	1.9
42	Brzeg Dolny	Odra	1958	8.0	2.1
43	Lubachów	Bystrzyca	1917	8.0	0.5
44	Sromowce Wyżne	Dunajec	1994	7.4	0.9
45	Brody Iłżeckie	Kamienna	1964	7.3	2.6
46	Mosty	Kanał Wieprz-Krzna	1969	6.9	3.9
47	Żelizna	Kanał Wieprz-Krzna	1971	6.9	3.5
48	Słupca	Meszna	1965	6.4	2.6
49	Zemborzyce	Bystrzyca	1974	6.3	2.8
50	Jastrowie	Gwda	1931	6.2	1.5
51	Gródek	Wda	1923	5.5	1.0
52	Niedalino	Radew	1913	5.5	0.9
53	Strzegomino-Konradów	Słupia	1924	5.0	1.0
54	Mylof	Brda	1848	5.0	1.2
55	Borowo	Drawa	b.d	1.9	b.d
56	Wisła-Czarne	Mała Wisła	1973	4.9	0.4
57	Niedów	Witka	1962	4.9	1.9
58	Raduszew Stary	Bóbr	1935	4.7	1.9
59	Rejowice	Rega	1924	4.6	2.2
60	Łączany	Wisła	1958	4.5	b.d
61	Opole-Podedworze	Kanał Wieprz-Krzna	1970	4.5	2.8
62	Zahajki	Kanał Wieprz-Krzna	1968	4.4	2.4

(continued)

Table 5.3 (continued)

No	Reservoir	River	Year of launching	Total volume (million m^3)	Surface area (km^2)
63	Wąglanka-Miedzna	Wąglanka	1979	4.2	1.8
64	Dychów	Bóbr	1936	4.1	1.0
65	Ptusza	Gwda	1933	4.0	2.0
66	Podgaje	Gwda	1930	3.9	1.2
67	Nielisz	Wieprz	1976	3.8	3.8
68	Kamienna	Drawa	1939	3.5	b.d
69	Pogoria I	Pogoria	1943	3.4	0.7
70	Straszyn	Radunia	1910	3.4	0.7
71	Bledzew	Obra	1909	3.0	3.2
72	Przewóz	Wisła	1955	2.5	b.d
73	Dąbie	Wisła	1961	2.5	b.d
74	Paprocany	Gostynia	b.d	2.5	b.d
75	Bielkowo	Radunia	1924	2.4	0.6
76	Krzywaniec	Bóbr	b.d	2.4	b.d
77	Smukała	Brda	1951	2.2	b.d
78	Tryszczyn	Brda	1960	2.2	b.d
79	Dobrzyca	Gwda	1912	2.2	0.9
80	Zesławice	Dłubnia	b.d	2.2	b.d
81	Zatonie	Plebanka	b.d	2.0	b.d
82	Krzynia	Słupia	b.d	2.0	b.d
83	Brąsław	Łyna	1936	1.9	0.6
84	Rzeszów	Wisłok	1974	1.8	b.d
85	Husynne	Udal	b.d	1.8	b.d
86	Wrzeszczyn	Bóbr	b.d	1.7	b.d
87	Likowo	Rega	1926	1.6	1.5
88	Łapino	Radunia	1927	1.6	0.4
89	Cedzyna	Lubrzanka	b.d	1.6	b.d
90	Bielawa	Brzęczek	b.d	1.5	b.d
91	Gołuchów	Ciemna	b.d	1.4	b.d
92	Stronie Śląskie	Morawka	b.d	1.4	b.d
93	Czaniec	Soła	b.d	1.3	b.d
94	Wrzosy	Juszka	b.d	1.3	b.d
95	Bytów	Bytowa	b.d	1.2	b.d

(continued)

Table 5.3 (continued)

No	Reservoir	River	Year of launching	Total volume (million m^3)	Surface area (km^2)
96	Chechło	Chechło	b.d	1.2	b.d
97	Wapienica	Wapienia	b.d	1.1	b.d
98	Środa	Moskawa	b.d	1.1	b.d
99	Kaczorów	Kaczawa	b.d	1.0	b.d
100	Drzewica	Drzewiczka	b.d	1.0	b.d
101	Zielona	Mała Panew	1925	1.0	b.d

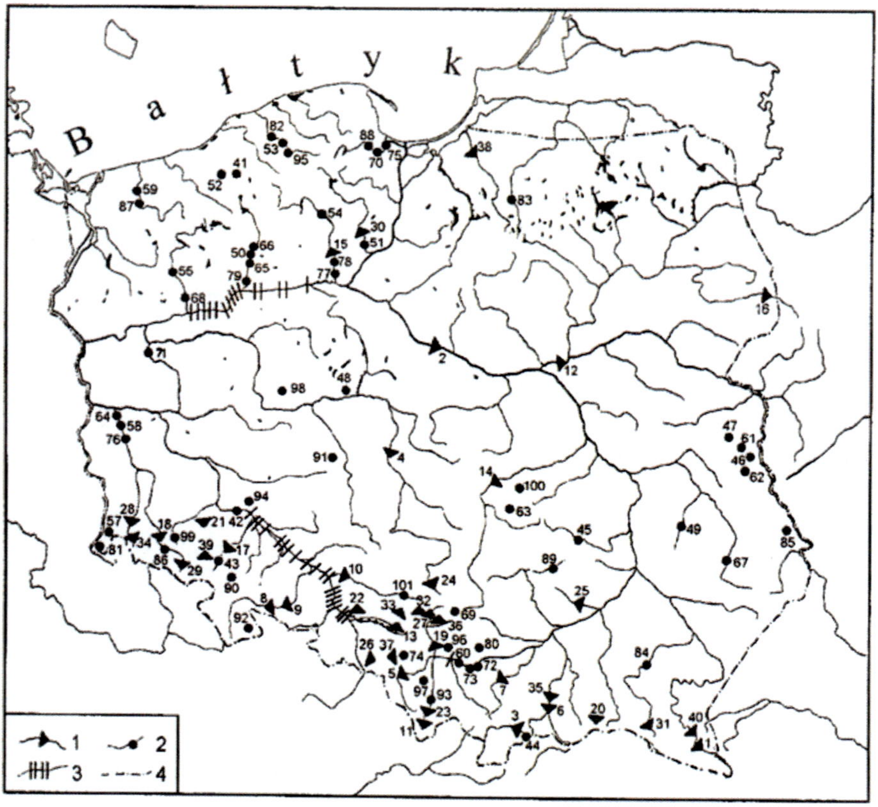

Fig. 5.3 Distribution of dam reservoirs in Poland according to [10] – numbering of reservoirs as in Table 5.3

range of the last glaciation, approximately overlapping with the Oder and Vistula River catchments. Materials used for calculations were hydrographic maps of Poland at a scale of 1:50 000 and topographic maps at the same scale (both cartographic documents were largely edited after 2000). A total of 712 maps were used for the analysis. Within the Oder River catchment, measurements were performed by Choiński [11], and the surface area of the Vistula River catchment was determined based on calculations of Lipiecka [12].

The result of the above analysis is the map of the percent share of dam reservoirs (Fig. 5.4). The applied ranges of the percent share correspond with those presented by Majdanowski [13] on the map of lake density of Poland. The procedure permitted direct comparison of the north and south Poland. The surface area of the Oder River catchment south of the range of the last glaciation is 58 618.6 km^2. Within the catchment area, the total surface area of artificial water reservoirs was estimated for 402.06 km^2. The value is equivalent to 0.69% of the cover of the area with dam reservoirs. The analogical range of the Vistula River has a surface area of 139 012.8 km^2. The total surface area of dam reservoirs within its boundaries was determined for 524.85 km^2, equivalent to 0.38% of share in the total area. The surface area of the western part of the country is therefore considerably more abundant in

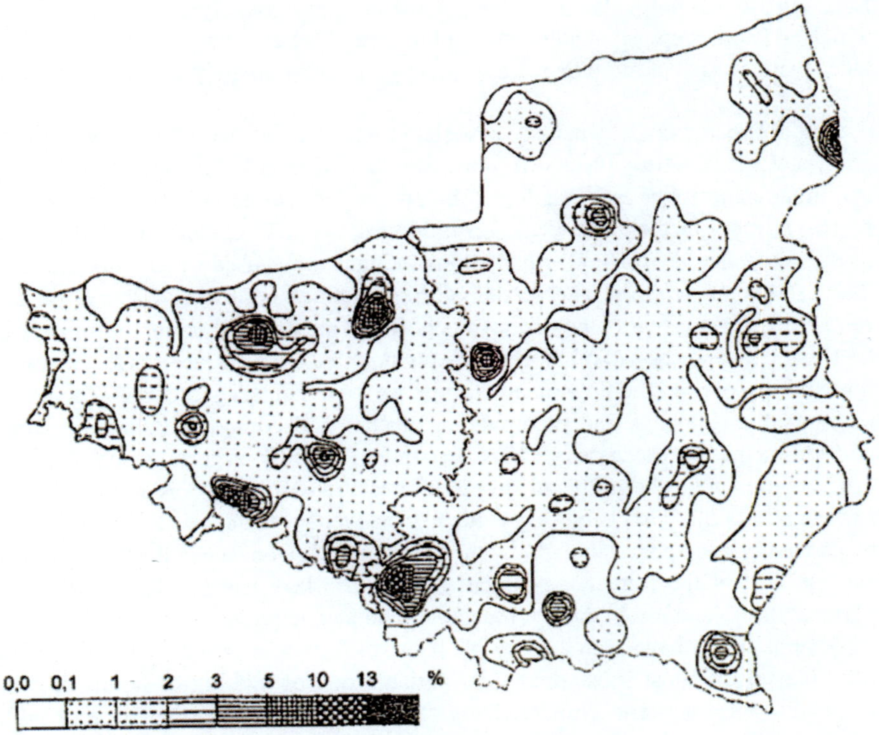

Fig. 5.4 Percent share of dam reservoirs south of the range of the last glaciation ranges in accordance with those adopted by Majdanowski [13]

dam reservoirs in comparison to the eastern part. The area south of the range of the last glaciation occupies 197 631.4 km^2, and the occurring dam reservoirs have a surface area of 926.91 km^2. The value constitutes 33% of the surface area of Polish lakes and is equivalent to the index of cover with dam reservoirs of 0.47%. It is, therefore, a very substantial value in comparison to the mean lake density of Poland equal to 0.9%. The number of dam reservoirs in the analysed area can be estimated for a dozen thousand. It should be mentioned that in the lake district zone, covered by the range of the last glaciation, many dam reservoirs are also located. The areas should be therefore considered in the total balance for the country.

Despite the known total surface area of dam reservoirs (approximately 1000 km^2), it is difficult to adopt their mean depth due to high depth variability. Their total water resources, however, are at a level of several km^3 and can constitute approximately ¼ of surface waters.

5.5 Changes in the Surface Area of Lakes and Their Water Resources

Lake Districts in north Poland cover an area of approximately 110 thousand km^2, which constitutes approximately 35% of the area of the country. They include 6 793 lakes (with an area of more than 1 ha), occupying approximately 2.77 thousand km^2 [1].

The decline of lakes is usually associated with a decrease in their surface area and water volume. This results from the fact that lakes in Poland are usually small and in the majority of cases shallow. The process of decline of lakes is determined by two primary factors: lake water level fluctuations and aggradation of sediments in the lake basin. Water level fluctuations in lakes are caused by climate changes, deforestation of the direct catchment, incorporation of lakes to the system of surface runoff, and various hydrotechnical works [14]. An increase in bottom sediments can be a result of deposition of mineral sediments supplied by rivers, aeolian processes, or those related to overland flow (deluvia) or mass movements, aggradation of the mass of biogenic sediments, and precipitation of chemical compounds. Sedimentation and sedimentation processes in lakes occur very rapidly, especially that over the last centuries in northern Poland strong deforestation took place. A decrease in the surface area of lakes has been very evident even over several decades. The process is considerably affected particularly by the location of a given lake in the hydrographic system, size of the (direct) catchment feeding the lake, surface formations (e.g. permeability), land use in the catchment, and human activity.

Several factors have been responsible for transformations of lakes. The following ones played the most substantial role: variable rate of melting of lumps of dead ice, short- and long-term climate changes, variable time of incorporation of lakes into the surface outflow system, catchment deforestation (concerning both the entire and direct catchment), hydrotechnical measures (from the mid-eighteenth century) and extensive melioration works. The rate of transformations of the lake basin from the moment of its origin to modern times, as a consequence leading to a decrease in their surface area, was characterised by high fluctuations. It was determined by two primary factors, namely: lake water level fluctuations (decrease), and filling the basin with accumulated biogenic and terrestrial sediments. Although in the process of evolution lakes underwent several stages of similar transformations, they are in different stages of development of their basins.

An attempt of determination of changes occurring in lake basins in the twentieth century was presented based on the example of selected lakes in Poland described in the relatively rich literature.

According to research conducted by various authors, a decrease in the water level in many lakes by 05–0.7 m caused an evident decrease in their surface area, sometimes even by 25–30% [15]. In some lakes, the water level decreased by even 2.5 m (Lake Miedwie) and more (Lake Gopło), causing a decrease in their surface area by more than 50% [16–18]). Choiński and Madalińska [19], analysing bathymetric plans of lakes in the Masurian and Pomeranian Lake District, concluded that over 60–70 years in the twentieth century, the surface area decreased by several percent, and the volume by several tens percent.

In a group of 18 lakes located in the Wielkopolskie Lake District throughout approximately 50 years, their considerable transformation was observed. Research evidenced a considerable decrease in the surface area by 14.8% (i.e. 172.6 ha). As a result of changes in the surface area and shallowing of lakes, water resources accumulated in the lakes decreased by a total of 7.7%, which constitutes 3.6 million m^3 [20]. It should be remembered, however, that the transformations are a natural element of their evolution. In the case of the analysed lakes, evolution is accelerated by human activity and climate changes.

Mięsiak-Wójcik et al. [21] presented changes in the surface area of stagnant waters in the western part of West Polesie (East Poland). The research was based on the analysis of the content of archival topographic maps of Poland at a scale of 1:10 000, presenting the situation from the early 1980s, and orthophoto maps and satellite images from the years 2010–2014. The objective of the analysis was the determination of the direction of transformations, as well as the presentation of results at variance with the opinion commonly adopted in Europe on a progressing decrease in the surface area of water bodies. The number of objects and their surface area increased over the last three decades. This resulted from the co-occurrence of natural factors and human activity, both destructive and aimed at the restoration of postglacial areas unique at the European scale.

Based on analyses of bathymetric plans of Lake Jamno from 1889 and 1960, [15] obtained an image of transformations of not only the surface area but also water resources. Throughout 71 years, the surface area of the lake decreased by approximately 150 ha, and the volume of the lake decreased by 9.247 thousand m^3, i.e. by 22.7%.

Water level fluctuations in 24 lakes of the Pomeranian Lake District were presented by Dąbrowski [22]. The author concluded that in addition to the primary elements of the climate, they are also largely determined by a group of individual features of lake catchments.

Very considerable changes occurred within the range of Lake Gopło as a result of regulatory measures and melioration works performed in the Noteć River valley in the years 1775–1878. The measures caused a decrease in the water level in the lake by 3.2–3.3 m, from 80.20 m amsl to below 77.00 m amsl [17]. A consequence of the changes is the division of former Lake Gopło into the modern lake and several smaller ones (Szarlej, Tryszczyn, Łunin, Gocanowskie, and Mielno) (Fig. 5.5). For Lake Gopło, changes in the volume of the lake in the analysed period of 246 years decreased from 235.8 to 76.7 million m^3. Therefore, the present lake only fills the deepest parts of the former Lake Gopło, constituting only 23.4% of its original surface area, and only 32.5% of its volume.

In the Pomeranian Lake District, the effect of all processes occurring in lakes is an evident succession of vegetation in lakes [23]. A good indicator showing an increase in vegetation succession is the index of overgrowing of the shoreline ($ha \cdot km^{-1}$). The value of the index from the turn of the 1950s and 1960s (preparation of bathymetric plans) is 0.98 $ha \cdot km^{-1}$, and that from the turn of the first and second decade of the twenty-first century (60 years of difference) already reaches 1.14 $ha \cdot km^{-1}$. This is strong evidence of an increase in overgrowing of lakes and vegetation succession irrespective of the type of vegetation [24].

The performed calculations with the application of a digital terrain model showed that the surface area of Lake Gopło in the period from the second half of the eighteenth century to the end of the first decade of the nineteenth century (1772–2010) decreased from 9 245.2 to 2 163.9 ha, i.e. by 76.6%.

Lake Ostrowskie (Gnieźnieńskie Lake District) also decreased its surface area but to a considerably lower degree. In the years 1887–2010, the total decrease in the surface area was 74.9 ha, i.e. 23.5%. Simultaneously with a decrease in the surface area of the analysed lakes, their water resources were also reduced. Lake Gopło decreased its volume by more than 67%, and Lake Ostrowskie by 21% [18].

Based on cartographic and teledetection materials, and geodesic measurements, changes in the morphometry of Lake Ostrowskie were traced throughout the last 123 years. As a result of a decrease in the water level, caused by various factors, the modern lake was divided into two separate basins (eastern and western). It caused a decrease in its surface area by 242.0 ha, and volume by 28.9%, whereas over the last 28 years by as much as 18.3% [25].

The effect of extensive regulation works conducted in the area of Wielkopolska and Kujawy generally involving drying of wetlands, straightening of river channels, channelling many sections of rivers, and incorporation of closed-drainage lakes to the outflow system was a decrease in the water level on many lakes by an average of 0.6–0.8 m, and in extreme cases even by 1.1–1.4 m. Such a considerable interference caused a decrease in their surface area by a dozen, and in extreme cases even by 80% [16].

The problem of the decline of lakes is exceptionally complex, particularly in the Wielkopolska-Kujawy Lake District, located in an area exceptional at the scale of the country due to low supply of atmospheric precipitation and therefore small water

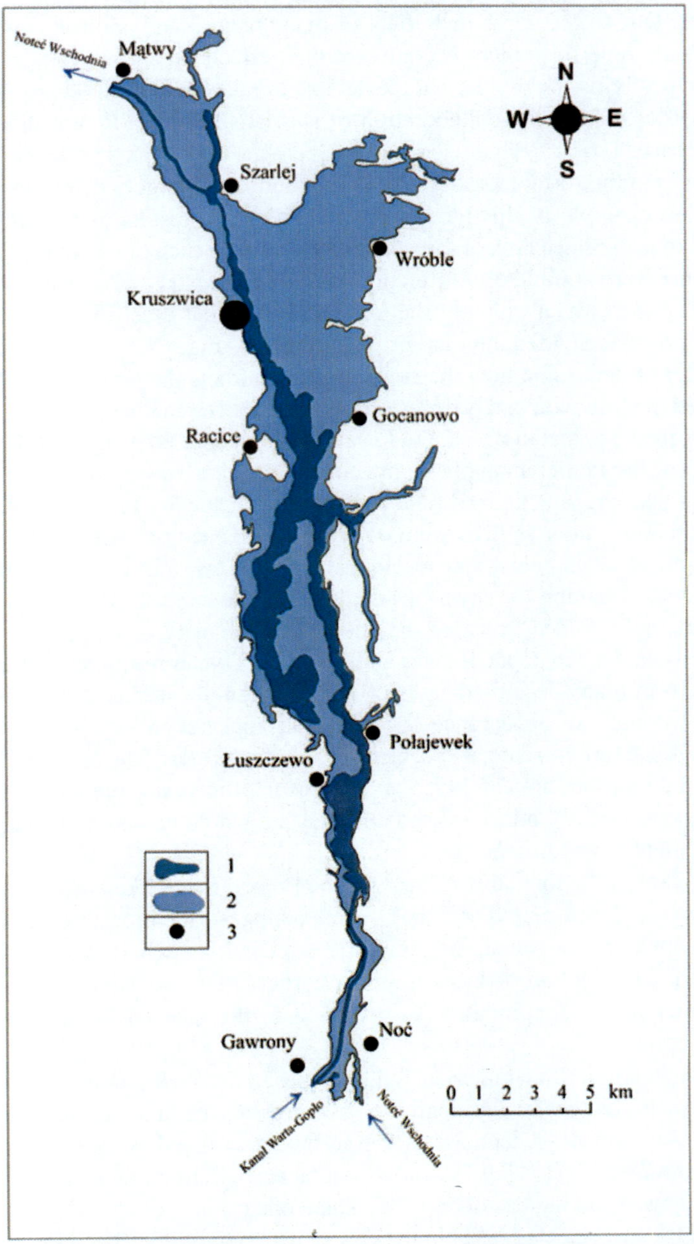

Fig. 5.5 Situational draft and changes in the range of Lake Gopło: 1 – today (77.00 m above sea level, 2 – rage of the lake at datum 80.00 above sea level (1772), 3 – localities

resources. Due to the combined effect of many natural and anthropogenic factors, it is difficult to perform hierarchization of their effects on a given lake. Due to this, attempts to solving the problem of the decline of lakes in the Wielkopolska-Kujawy Lake District require the collaboration of specialists from different disciplines of Earth sciences [26].

Changes in the surface area of lakes are undoubtedly largely determined by local factors. An example of this type of determination are results of research by [27]. Changes in the surface area of three lakes located near each other (in the Oder River catchment) were completely different. Over approximately 170 years, the surface area of one lake was absolutely stable, that of the other drastically decreased, and the third one was divided into three smaller basins – Fig. 5.6.

In order to determine how the course of time affects the longevity of lakes [28] performed the following analysis. In zones located between the ranges of subsequent phases of the last glaciation, i.e. the Leszno and Poznań, Poznań and Pomneranian, and north of the Pomeranian phase, the number of lakes was determined along with their total water resources, and total surface area. Knowing the surface area of the designated areas, three indices were determined for each of them, i.e. lake density, mean depth of lakes, and water resources in mm. The values were referred to the surface area, obtaining the layer of lake waters. The analysis shows that the lakes are increasingly "older" towards the south. Lake density consequently decreases southwards, and mean depth of lakes and the layer of water resources decreases [28].

Apart from many described cases of a decrease in the surface area of lakes, also cases of complete disappearance occur. The disappearance of Lake Jelenino near Szczecinek can be described as spectacular [29]. At the end of the eighteenth century, the lake had a surface area of 495.2 ha. As a result of intensive melioration, its basin lost water entirely. If the lake existed nowadays, it would be one of the largest lakes in Poland in terms of surface area.

A decrease in the total area of lakes in the Pomeranian Lake District throughout 40–50 years in the twentieth century by 9.69% suggests the process of their decline. This also concerns the remaining regions (Lake Districts) where the decline is even greater [1]. The decline of lakes is justly particularly associated with a decrease in the water level, accumulation of sediment in the lake basin, and progressing eutrophication.

Based on cartographic materials for 256 lakes in the Wielkopolska Lake District, [30] analysed changes in the degree of overgrowing of lakes throughout the last 60 years. As a result, a decrease in the surface area of lakes by 0.6% was determined (from 28 152.6 to 27 983.7 ha). Simultaneously, the surface area occupied by emergent vegetation increased by 1.7%. The tendency is confirmed by overgrowing presented for two selected lakes in the region with polymictic character [31].

Based on cartographic materials and aerial photographs, [14] analysed changes in the degree of overgrowing of lakes for 893 lakes located in lake districts in Poland (Pomeranian, Masurian, and Wielkopolska Lake District). The lakes were selected based on the existing bathymetric plans and information on their overgrowing and depth relations. Over the last 60 years, a decrease in the surface area of lakes by 1.9% was observed (from 140 975.0 to 138 273.7 ha). The surface area of lakes occupied

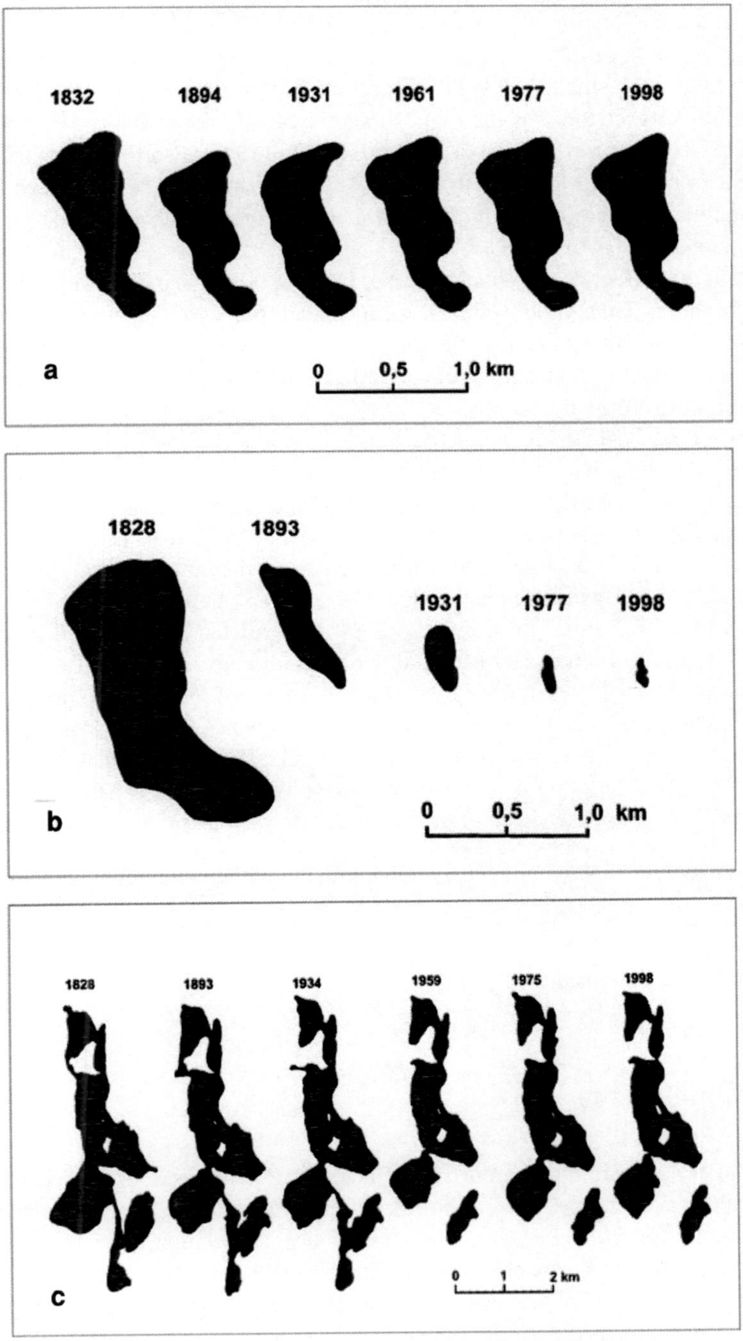

Fig. 5.6 Variability of the surface area of lakes located near each other; A – Lake Cisie – example of stability of surface area; B – Lake Grążyk – an example of drastic disappearance; C – a system of Lakes Chobienicko-Wielkowiejskie-Kopanickie – development of new lakes as a result of a decrease in the surface area of larger lakes

by emergent vegetation also decreased by 0.27% (from 11 219.0 to 10 637.2 ha). The surface area of lakes occupied by emergent vegetation averages 7.69%. In the case of lakes with small (below 80 ha) and medium surface areas (80 ÷ 200 ha), the degree of overgrowing was the highest, and equalled, respectively: 14.3 and 9.6%. Fast rate of overgrowing of the lakes is related to the amount and quality of biogenic substances migrating to lakes from the catchment area, the resistance of lakes to degradation, the contribution of the littoral zone in the surface area of lakes, and age of the lake.

A tangible effect of the changes in the lake basin was usually a shift of isobaths from the shores to the middle of the lake, and development of new peninsulas, islands, and in some cases the division of the original lake into several smaller ones [32]. An example is Lake Jamno where in the period 1889–1960 a decrease in the surface area by 6.5% and volume by 22.7% was determined [15]. Similar situations of changes were observed in other regions of the Polish Lowland ([25, 33–36]).

Similar changes also occur in lakes in Lithuania, Latvia, Estonia, and Finland. In Lake Luupuvesi (central Finland), the range of macrophytes increased from 96 ha in 1953 to 355 ha in 1996 [37]. Research conducted in shallow lakes: Engure (Latvia) and Võrtsjärv (South Estonia) confirm an increase in the range of macrophytes in the second half of the twentieth century [38].

The range of changes in water resources accumulated in lake basins, tarns, and artificial reservoirs can be insubstantial, which is undoubtedly particularly related to the process of decline of lakes. The parameter largely reflecting the predisposition of particular lakes to decline is mean depth. The value of the parameter for lakes in Poland is very variable, and in the Pomeranian Lake Districts it equals 6.84 m, in the Masurian Lake District – 7.45 m, in the Wielkopolska-Kujawy Lake District – 5.70 m, and in lakes south of the range of the last glaciation – 4.34 m [8]. Assuming that the mean depth of Polish lakes is 7.02 m, and mean annual accumulation of sediments is approximately 1 mm, the range of loss of water resources can be concluded. Considering only the sedimentation process in lake basins, the loss of resources is 0.014% annually, which permits the estimation of the prospective age of lakes for approximately 7 thousand years [8].

5.6 Conclusions

The total water resources comprised of lakes in Poland were estimated based on the Catalogue of Polish Lakes (Choiński 2006a) including information concerning surface area for 7081 lakes (above 1 ha). The calculation procedure involved summating known lake water volumes for more than 30% of lakes, and adding the total water volume of unmeasured lakes. This was calculated as a product of their surface area and mean depth equalling to 19.7349 km^3. Water resources within particular lake districts are as follows: Masurian Lake District – 10.1183 km^3 (51.27%), Pomeranian Lake District – 7.1292 km^3 (36.12%), Greater Poland-Kujavian Lake

District – 2.3535 km^3 (11.93%), and area south of the delineated range of the last glaciation – 0.1339 km^3 (0.68%).

Retention capacities were prepared based on the outlines of maps at a scale of 1:50,000 for more than 200 sheets of topographic maps. The determined retention capacity for the area north of the line delimiting the maximum range of the last glaciation is 162 mm. The greatest resources within the Masurian Lake District occur in the following catchments: Węgorapa (404 A) – 1976 mm, Biała Hańcza (407) – 958 mm, and Pisa (227 A) – 893 mm; and in the Pomeranian Lake District in catchments of: Drawa (119 D) – 389 mm, Wda (237 B) – 196 mm, and in the Przymorze catchment from Wieprza to Słupia (310) – 184 mm. Within the Greater Poland-Kujavian Lake District, lake catchments retain considerably less waters, and the most abundant ones include: Vistula from Bzura to Skrwa (233 A) – 173 mm, Warta from Wełna to Obra (118 G) – 141 mm, and Noteć to Gwda (119 A) – 116 mm. Zones with the lowest retention capacity, i.e. up to 50 mm, are located in valleys of the largest rivers with the adjacent areas, and coasts: Gdańsk, East Baltic, Szczecin, and the southern part of the Koszalin Coast. Mean retention capacity determined within the three lake districts is exceptionally variable. In the case of the Masurian Lake District, it is 283 mm, for the Pomeranian Lake District the index equals 150 mm, and for the Greater Poland-Kujavian Lake District only 58 mm. The division of the total volume of water resources of the lakes by the surface area of Poland provided a water layer of 63 mm. It constitutes only 10% of mean annual precipitation, and the value is approximately 3.5 times smaller than the amount of mean outflow from the territory of Poland.

5.7 Recommendations

Determining the volume of available water resources is of key importance in water resources management and also in defining the thermal budget of water, flood protection capabilities, fish farming and agriculture. The issues discussed should be of interest to a wide spectrum of researchers, decision makers and policy planners.

References

1. Choiński A (2006) Katalog jezior Polski (Catalogue of Polish Lakes). Adam Mickiewicz University Press, Poznań, p 600 (in Polish)
2. Jańczak J (1984) Wstępna ocena zasobów wodnych jezior Polski. Czasopismo Geograficzne 54(4):441–451
3. Hydrographic Division of Poland (1980) Part. II, Map 1: 200 000, IMiGW, Warszawa
4. Kondracki J (2009) Geografia regionalna Polski (Regional geography of Poland). Wydawnictwo Naukowe PWN, Warszawa, p 441 (in Polish)
5. World Water Balance and Water Resources of the Earth (1978) UNESCO, Paris
6. Choiński A (2000) Jeziora kuli ziemskiej. Wydawnictwo Naukowe PWN, Warszawa

7. Choiński A (1999) Oczka wodne w Polsce w strefie zasięgu zlodowacenia bałtyckiego, Acta Universitatis Nicolai Copernici, Geografia, XXIX, Toruń, 320–326
8. Choiński A (2007) Limnologia fizyczna Polski (Physical limnology of Poland), UAM Science Publishing, Poznań, p 548 (in Polish)
9. Sebzda P (2002) Oczka wodne w Polsce poza zasięgiem zlodowacenia bałtyckiego. IGFiKŚP UAM, Poznań (maszynopis)
10. Wiśniewski RJ (1998) Zbiorniki zaporowe. In: Dobrowolski KA, Lewandowski K (eds) Ochrona środowisk wodnych i błotnych w Polsce. Stan i perspektywy, Oficyna Wydawnicza, Instytut Ekologii PAN, Dziekanów Leśny
11. Choiński A (2006) Percentage of the area of artificial water bodies in the Oder catchment south of the maximum limit of the Baltic Glaciation. Limnol Rev 6:47–50
12. Lipiecka M (2007) Sztuczne zbiorniki wodne w dorzeczu Wisły na południe od linii zasięgu zlodowacenia bałtyckiego. IGFiKŚP, UAM, Poznań (maszynopis)
13. Majdanowski S (1954) Jeziora Polski. Przegląd Geograficzny 26:17–50
14. Skowron R, Jaworski T (2017) Changes in lake area as a consequence of plant overgrowth in the South Baltic Lakelands (Northern Poland). Bull Geogr Phys Geogr Ser 12:19–30. https://doi.org/10.2478/11383
15. Choiński A (2001) Analysis of changes in the area and water volume of Lake Jamno. Limnol Rev 1:41–44
16. Kaniecki A (1997) Influence of XIXth centuries—the meliorations on change of level of waters. In: Choiński A (ed) Influence of human impact on lake. UAM, Poznań-Bydgoszcz, pp 67–71 (in Polish)
17. Dorożyński R, Skowron R (2002) Changes of the basin of Lake Gopło caused by melioration work in the 18th and 19th centuries. Limnol Rev 2:93–102
18. Skowron R, Piasecki A (2012) Zmiany zasobów wodnych oraz geometrii niecek jeziora Gopło i Ostrowskiego w wyniku wpływu antropopresji. In: Grześkowiak A, Nowak B (eds) Anthropogenic and natural transformations of lakes, Poznań, pp 95–97
19. Choiński A, Madalińska K (2002) Changes in lake percentage in Pomeranian Lakeland catchments adjacent to the Baltic since the close of the 19th century. Limnol Rev 2:63–68
20. Choiński A, Ptak M, Ławniczak AE (2016) Changes in water resources of Polish lakes as influenced by natural and anthropogenic factors. Pol J Environ Stud 25(5):1883–1890. https://doi.org/10.15244/pjoes/62906
21. Mięsiak-Wójcik K, Turczyński M, Sposób J (2014) Natural and anthropogenic changes of standing water bodies in West Polesie (East Poland). In: 2nd international conference—water resources and wetlands. 11–13 Sept 2014, Tulcea (Romania). https://www.limnology.ro/water2014/proceedings.html
22. Dąbrowski M (2004) Trends in changes of lake water levels in the Pomerania Lakeland. Limnol Rev 4:75–80
23. Ptak M (2013) Changes in the area and bathymetry of selected lakes of the Pomeranian Lake District. Prace Geograficzne 133:61–76 ((in Polish))
24. Piasecki A, Skowron R (2014) Changing the geometry of basins and water resources of Lakes Gopło and Ostrowskie under the influence of anthropopressure. Limnol Rev 14(1):33–43. https://doi.org/10.2478/limre-2014-0004
25. Kunz M, Skowron R, Sz S (2010) Morphometry changes of Lake Ostrowskie (the Gniezno Lakeland) on the basis of cartographic, remote sensing and geodetic surveying. Limnol Rev 10(2):77–85
26. Marszelewski W, Ptak M, Skowron R (2011) Anthropogenic and natural conditionings of disappearing lakes in the Wielkopolska-Kujawy Lake District, (in Polish). Roczniki Glebozawcze 62(2):283–294
27. Choiński A., 2009. Changes in the area of lakes from the Obra River drainage basin taking place from the begiming of the 19th century. Limnol Rev 9(4):159–164
28. Choiński A (2017) Jeziora i zbiorniki wodne w Polsce. In: Jokiel P, Marszelewski W, Pociak-Karteczka J (eds) Hydrologia Polski, PWN, pp 223–229

29. Ptak M, Choiński A, Strzelczak A, Targosz A (2013) Disappearance of Lake Jelenino Since the end of the 18th century as on effect of antropogenic transformation of the natural environment. Pol J Environ Stud 22(1):191–196
30. Skowron R., Piasecki A., 2015. The spatial analysis of overgrowing the lakes—on example of the Wielkopolska Lake District. In: Doganovsky AM, Naumenko MA, Isaev DI, Grześ M, Glazik R, Skowron R (eds), Modern problems of Hydrology, Sankt Petersburg, RSHU, pp 58–68
31. Ławniczak AE (2010) Overgrowing of two polymictic lakes in Central-Western Poland. Limnological Rev 3–4:147–156
32. Marszelewski W (2005) Zmiany warunków abiotycznych w jeziorach Polski Północno-Wschodniej (Changes of the abiotic conditions in the lakes of North-East Poland). Nicolaus Copernicus University Press, Toruń, p 288
33. Dąbrowski M (2001) Changes in the water level of lakes in Northeastern Poland. Limnol Rev 2:85–92
34. Nowacka A, Ptak M (2007) Zmiany powierzchni jezior na pojezierzu Wielkopolsko-Kujawskim w XX wieku (Changes in the surface of lakes in the Wielkopolsko-Kujawskie Lakeland in the twentieth century). Badania Fizjograficzne Nad Polską Zachodnią, Geografia Fizyczna 58:149–157
35. Ptak M (2010) Percentage of the area covered by forest and change surface lakes in the middle and lower Warta River Basin from the end 19th century. In: Ciupa T, Suligowski TR (eds) Woda w badaniach geograficznych. Jan Kochanowski University Press, Kielce, pp 151–158
36. Ptak M, Ławniczak A (2012) Changes in water resources in selected lakes in the middle and lower catchment of the River Warta. Limnol Rev 12:35–44
37. Valta-Hulkkonen K, Kanninen A, Pellikka P (2004) Remote sensing and GIS for detecting changes in the aquatic vegetation of rehabilitated lake. Int J Remote Sens 25:5745–5758
38. Brižs J (2011) Dynamics of emergent macrophytes for 50 years in the coastal Lake Engure, Latvia. Proc Latvian Acad Sci 65:170–177

Chapter 6
Natural and Anthropogenic Lakes of River Valleys

Jarosław Dawidek, Beata Ferencz, and Katarzyna Kubiak-Wójcicka

Abstract This chapter discusses the forms and ways of functioning of the lakes in river valleys in Poland. Due to the specific nature of these objects, which are characterized by a distinct separation from natural and artificial lakes, the knowledge gathered so far concerning the appropriate nomenclature and classification has been systematized. The paper presents bathymetric plans of selected objects which origin is related to both natural and anthropogenic lakes. These objects are an important and underestimated element of small retention.

Keywords Floodplain lakes · Lakes typology · Potamophase · Limnophase · Poland

6.1 Introduction

The issue of the forms of occurrence and functioning of lakes located in river valleys is relatively rarely addressed in the literature, despite the high hydrological and ecological significance, as well as their unique function in river valleys landscape. The chapter will present the classification and types of floodplain lakes observed in Polish river valleys, as well as the nomenclature used. Moreover, selected examples of lakes occurring in river valleys – natural, quasi-natural and transformed by

J. Dawidek (✉)
Department of Hydrology and Climatology, Faculty of Earth Sciences and Spatial Management, Maria Curie-Skłodowska University, 2 cd Aleja Kraśnicka, 20-718 Lublin, Poland
e-mail: jaroslaw.dawidek@poczta.umcs.lublin.pl

B. Ferencz
Department of Hydrobiology and Protection of Ecosystems, University of Life Sciences, 13 Akademicka St., 20-950 Lublin, Poland
e-mail: beata.ferencz@up.lublin.pl

K. Kubiak-Wójcicka
Department of Hydrology and Water Management, Faculty of Earth Sciences and Spatial Management, Nicolaus Copernicus University, Lwowska 1, 87-100 Toruń, Poland
e-mail: kubiak@umk.pl

M. Zeleňáková et al. (eds.), *Management of Water Resources in Poland*, Springer Water,
https://doi.org/10.1007/978-3-030-61965-7_6

human activity – will be discussed. The environment of river valleys is characterized by a wide variety of processes and forms resulting from them. Limnic forms, although are frequently observed in some valleys, are generally poorly recognized, and their retention and ecological potential are underestimated in the context of river bed regulation and hydrotechnical interventions in valleys. The floodplain lakes, however, are an essential element of small retention of valley waters. Due to the diversity of types: genetic, balance, hydrochemical, hydrobiological and other, they are also a very attractive subject of research. The specificity of lakes located in river valleys is often characterized by a distinct separation from the environment of river waters, which encourages research and practical solutions aimed at the protection or renaturalization of the limnic environment.

6.2 Definition of Floodplain Lakes

In world literature, the floodplain lakes are called differently. Examples of terms are: "neck cutoff" [1–3], oxbow lake [4–6], fluvial lake [7, 8], or more rarely "river lake" [9], hydro-biological "limnocren" [10, 11], hydrological "floodplain lake" [12–16]. In the Polish literature, especially in reference to lakes, the creation of which is the result of regulatory and hydrotechnical works in river valleys, apart from the general and imprecise name "old riverbed," there are also names: quasi-ox-bows [17], anthropogenic reservoirs, post-training reservoirs [18]. Sometimes local names are used, which are derived from river names (e.g., in the Vistula valley – "wiśliska", in the Odra valley – "odrzyska", Warta – "warciska", Bug – "bużyska"). A separate problem is the definition of a floodplain lake, which is not precise. A general definition of a lake assumes the natural origin of a basin in which water accumulates (from surface and groundwater supply) and the advantage of recharge over losses. However, an important criterion assuming the existence of turbulent flow of river waters through a basin during potamophase, as well as autonomous areas with a laminar character of water movement and stagnation zones, should be added. The number of floodplain lakes within particular genetic groups, presented in Fig. 6.1, depends on macroscale features of the natural environment (e.g., geological structure, relief, climate and others) and human pressure, while the way of functioning is favored by the local conditions of their catchment areas. The ecological separateness of reservoirs is commonly observed, which often determines the uniqueness of floodplain lakes.

6.3 Materials and Methods

The analyses of natural floodplain lakes was based mostly on published papers concerning various river systems. Fluvial, morphological, hydrochemical and hydrobiological aspects were analyzed in order to quantify the main factors determining floodplain lakes heterogeneity. A representative group of temperate zone floodplain

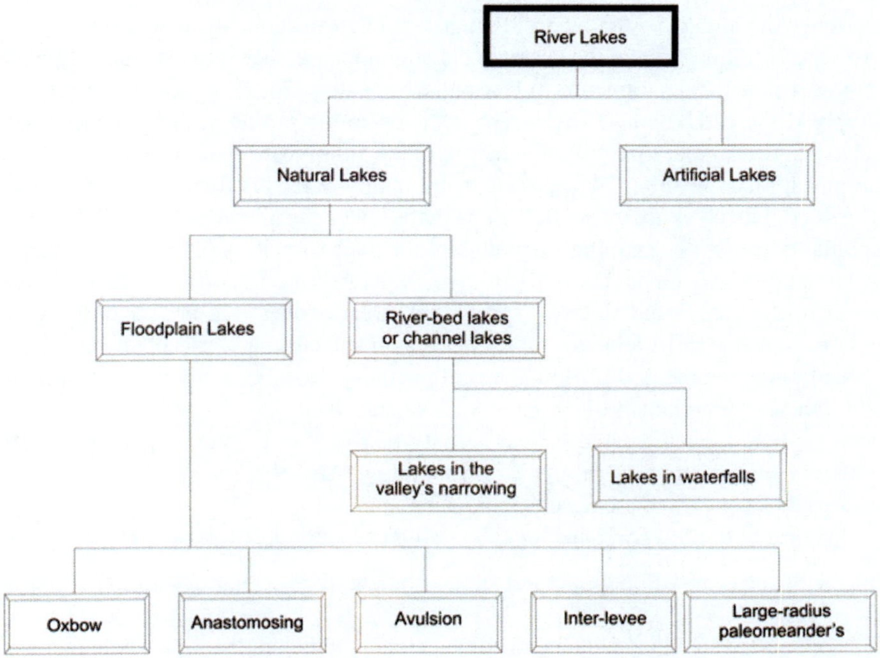

Fig. 6.1 Lakes in the River valeys

lakes located in the Bug river valley, one of very few large, quasi-natural European rivers in Eastern Poland was selected. A functional analysis of lake basins allowed the construction of uniform floodplain lakes classifications.

Bathymetric research was conducted in all seasons, at different water levels of the floodplain lakes. Measurements were made using echo sounder or probe weights (in lakes where intense bottom vegetation was observed) and a GPS receiver. In addition crevasses that connect floodplain lakes with the parent river were identified, and water distribution was determined during the spring potamophase periods.

6.4 Typology of Floodplain Lakes

6.4.1 Natural Lakes

The floodplain lakes are characterized by cyclical changes in the basin's supply forms. In general, there are four phases of the cycle: filling, flooding, drainage and isolation [19, 20]. There is also a division into two basic functional periods called limnophase and potamophase or inundation and isolation phase [21–23]. In the limnophase, the lake has its catchment that can be determined. It is also possible to determine the range of supply and the type and intensity of hydrochemical and hydrobiological processes for this phase of the cycle. In the potamophase, there is a clear convergence of physicochemical parameters of lake and river waters [24, 25], and the area of basin supply is the catchment of the parent river (up to the profile at the place of water supply to the lake). During high recharge of the valley with river waters, lakes may disappear for some time. The lakes are then a hollow in the flooded floodplain terrace. The floodplain lakes are most often found in the lower and middle sections of the river, and the criterion determining the potential for their formation is the predominance of lateral erosion processes over depth one in the river bed. Floodplain lakes do not occur in the crenal and rhithral (spring and stream zones) but appear only in the potamal (river zone). Detailed studies of erosion and cutting of meanders have been carried out since the end of the nineteenth century [26], but initially, they were an addition to comprehensive studies of valleys and rivers [26, 27]. Already then, the need to divide lakes created by the activity of rivers was noticed. Due to a large number of lakes and a large area of meandering lakes, the first typologies of river valley lakes concerned these lakes.

According to [28], for a lake to be classified as a floodplain lake, it must:

(a) be located on a floodplain and shows some apparent relationship to the major stream,
(b) be located within a distance of fifteen stream widths of the major stream,
(c) contain water,
(d) include at least one segment with a crescent-shaped channel (called the "fundamental component"),
(e) have a size of the same general magnitude as the major stream, and
(f) be independent of other topographic influences, which means that each oxbow lake must be free-standing.

Division of lakes by morphometry

The simplest floodplain lakes typologies were based on shoreline shape and development. Three types of lakes were distinguished in terms of oxbow lakes morphometry [29].

- simple oxbow lakes are reservoirs with a compact, curved and not dismembered basin with at least one arm through which river water is supplied (Fig. 6.2a),
- compound oxbow lakes with a dismembered basin, many channels and a developed shoreline (Fig. 6.2b),
- complex oxbow lakes with a complex basin shape, which was created by connecting several floodplain lakes into one body of water (Fig. 6.2c).

Another concept of typological division of meandering lakes by [28] also includes three lake categories:

- open oxbow lakes (Fig. 6.3a)
- normal oxbow lakes (Fig. 6.3b)
- closed oxbow lakes (Fig. 6.3c).

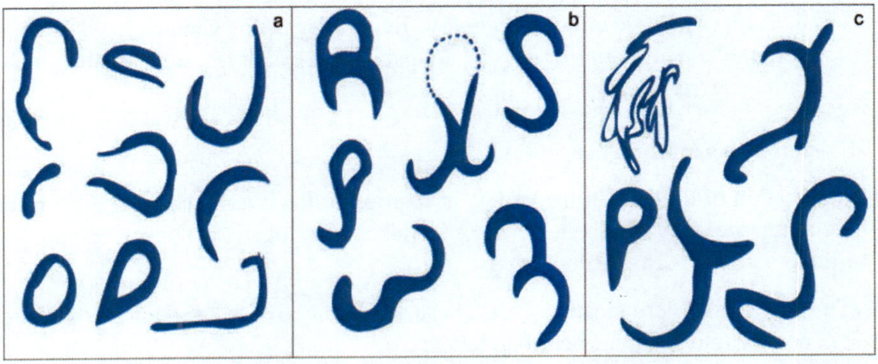

Fig. 6.2 Oxbow lakes typology according to [28]: **a** simple, **b** compound, and **c** complex

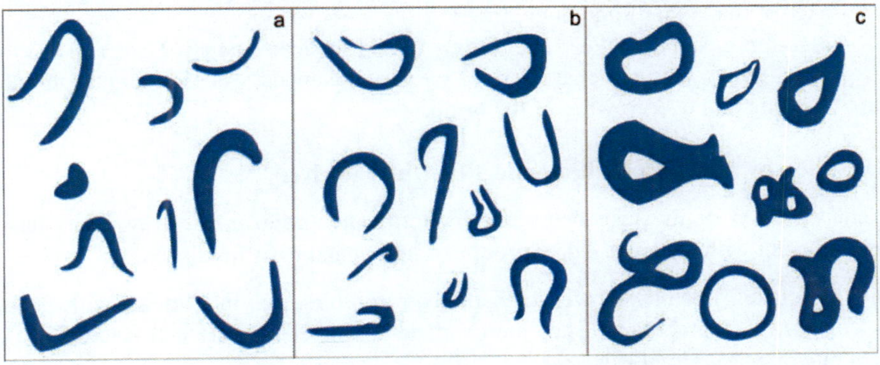

Fig. 6.3 Oxbow lakes typology according to [28]: **a** open, **b** normal, and **c** closed

The criterion determining the lake's belonging to the group was the index of shoreline development, allowing to determine the degree of closure (or opening) of the lake basin. The open lakes had arms exposed in the direction of river waters supplying them and resembled the elongated letter 'c' in shape. Normal lakes were typical moon-shaped meandering lakes. Closed lakes, on the other hand, were water reservoirs with connected arms (Fig. 6.3c).

Division by ratio of bathymetric parameters

Floodplain lakes are generally characterized by their considerable length and narrow width. The basis for the next classification of floodplain lakes in the period of limnophase according to [19] was the index being the quotient of the length and width of the reservoir. They separated two groups of lakes:

- channel lakes, in case of which the size of the index was higher than five. Lakes of this type were characterized by rapid water exchange, mainly based on fluvial supply and short retention time.

– dish lakes with the index size below five. They were characterized by slow water movement in the basin and longer water exchange time (mainly based on atmospheric supply).

Division by location in the valley

The location of the lake in the valley, in relation to the river bed or the edge of the lagoon terrace, may also be a criterion for lakes classification. Four different types of floodplain lakes can be separated:

– the lakes of modern meander belt. These are mainly oxbows and inter-levee lakes, generally referred to as near-bed lakes.
– under-edge lakes. They are located on the opposite side of the floodplain terrace in relation to the above mentioned lakes. They are located in sub-edge depressions. They generally have a small surface area and the depth of the basin.
– lakes of the higher floodplain terrace
– large-radius oxbow lakes. They rarely occur, far from the river bed and are the oldest lakes. They are characterized by very advanced processes of atrophy due to silting and overgrowing of the basin.

Division by manner of lake connection with the river

The criterion of the lake belonging to a specific group is the way of its basin connecting with the river. Three groups of lakes can be distinguished:

– lotic lakes, open lakes, well oxygenated reservoirs, rich in nutrients with a relatively fast metabolism. They are characterized by an apparent slowdown in the rate of succession [30].
– lentic lakes, closed [31] with a very short duration of potamophase, which favors the formation and maintenance of oxygen deficits. Denitrification and sulfate reduction processes often dominate in bottom zones. The circulation of matter in such lakes takes place mainly on the basis of indigenous substances.
– semiopen, semi-lotic lakes, characterized by a wide range of hydrological changes, which results in differentiation of aerobic conditions and water trophy [32], as well as a large mosaic of habitats.

In each of the above mentioned groups, there are also three forms that take into account the shape of the lake. Straight, complex and irregular lakes differ in the degree of complexity of the shoreline of the basins. In the case of straight reservoirs, there is usually one formed basin, typically curved (croissant), being a part of a cutoff river meander. The basins of complex lakes are usually connected through channels. Sometimes it is one basin with a developed system of arms. Irregular lakes have the most complex shape. These are usually several basins connected or a system of irrigated inter-splitter depressions.

The method of connecting the lake with the river determines the dynamics of the ecosystem and the physicochemical composition of the water. This issue is widely discussed in the literature [16, 23, 33–36].

Division by the origin

The oxbow lakes are the most numerous group of floodplain lakes and generally have a relatively large surface area and basin capacity. They are formed in the valleys of meandering rivers. They are always located in the meandering belt of the channel zone of the river valley bottom. They are a fragment of an old river bed, cut off from the present one with a near-bed embankment. The flow of flooding waters through lower and higher flood terraces is the basic factor shaping the surface of inter-levee lakes. Erosion, transport and deposition of sediments directly determine the possibility of such reservoirs creation and functioning. Lithologically diversified forms of dumping, of different durability and size, are most often created under the conditions of free-flowing or standing water. Meander drops are accumulation forms of the largest size and high stability. Lakes can be formed in places where embankments separate depressions, with favorable arrangement of supplying them valley waters. Flows of water in the out-of-bed zone determine the recharge of the basins of inter-splitter lakes during potamophase. The recharge of the lake basin results from the occurrence of erosive forms created during flooding flows. The shape of the lakes refers to the form of an inter-levee depression, which was filled with water. Inter-levee lakes are characterized by a clear loss of water in the limnophase period, a small maximum depth and a relatively small area.

Avulsion occurs in fragments of the valley where a typical free migration of meanders is replaced by the process of abandoning of longer sections of the river bed. Sections of valleys, which are characterized by an increase in the longitudinal slope of the river bed, are predisposed to the occurrence of avulsion. Especially in the gorges zones, before the topographical obstacle of water outflow, the longitudinal slope of the river decreases, but after its crossing, it increases significantly. It is the increase in the gradient of the longitudinal slope during flooding that favors the abandonment of the old bed. Increased slope of the water table also favors an increase in the role of deep erosion at the expense of side erosion, which results in a weakening of the river's tendency to meandering. The modeling processes then took place within abandoned, straightened, narrower and deeper beds. Under the conditions of existence of stabilized channel drains delimiting the shoreline zone, alternating outflows and depressions occurred in the longitudinal profile. A system of linear cascades of basins was created, which were formed in the deep zones and isolated from each other by shallows. The underground recharge may also be the factor favoring the preparation and stabilization of the deepest zones of the basins of the avulsion lakes. Especially in gorges areas, water-bearing layers, the drainage of which allows to maintain full watering of the lake basins throughout the year, can be often intersected.

Anastomotic lake basins were shaped by the processes resulting from the adaptation of the flow of flooding waters to the decreasing slope of the valley bottom (most often sections located before its narrowing or topographic obstacle). The lower transport efficiency of the bed (resulting from the lower velocity of flowing water) led to swelling of water on the floodplain terrace during flooding. At that time, alternative (supplementary) beds were created. Such organization of the outflow, in conditions of the high variability of flows, helped to increase the transport efficiency of the beds and faster outflow of river waters throughout the year. Significant lengthening of the basins (beds) in relation to their width does not favor the limnic functioning of these forms. Only a few basins can be considered anastomotic lakes. The condition of the functioning of the flow of potamic waters through the basin must be met while maintaining the zones of standing waters or waters with very slow motion.

Division by supply direction

The origin of the floodplain lake determines the manner and rate of supplying the basin with the parent river water, and during the potamophase, most of the floodplain lakes have a flowing character. Depending on the direction of supplying waters in relation to the slope of the valley bottom, there are four hydrological types of floodplain lakes [37]. Confluent supply is mentioned when the direction of flowing river water (supplying the lake basin) is consistent with the longitudinal gradient of the river (Fig. 6.4a). This type of lakes is characterized by the highest dynamics of water exchange in the period of potamophase.

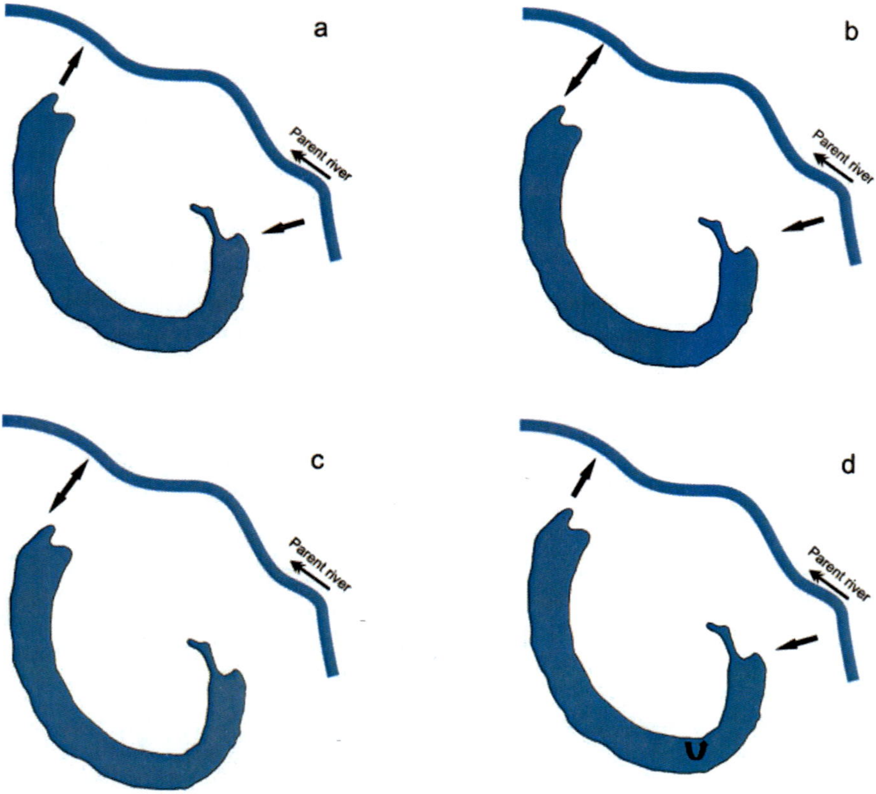

Fig. 6.4 Hydrological types of floodplain lakes: **a** confluent, **b** contrafluent–confluent, **c** – contrafluent, and **d** profundal

The inflow of river water in floodplain lakes with a contrafluent type of supply takes place through one arm of the lake, located on the lower side of the river water (Fig. 6.4c). In the reservoirs with this type of supply, the process of slower (than in confluent lakes) water exchange with the river is observed. In deep lake basins (e.g., avulsion lakes), apart from confluent supply, constantly high underground water supply is observed. These lakes belong to the profundal hydrological type (Fig. 6.4d). There were also cases of river water inflow through two arms of the floodplain lake, first from the lower and then from the upper water side. In the first phase of potamic recharge, the lake water masses are pushed into the central part of the basin, and after connecting the basin with the river by the second arm, they change direction in accordance with the inclination of the river bed. It is a complex contrafluent–confluent way of lakes supplying (Fig. 6.4b).

Confluent, (Latin *confluentia* "floating," *confluere* "floating together") recharge of floodplain lakes takes place consequently in relation to the slope of the valley. The supply of confluent lake basins, from the upper side of the river water, results from the local relations between the shape of the surface of the modern meander belt and the geometry of the river bed. Potamophase in such lakes begins when the flow of the river, which corresponds to the shore point, is exceeded. The direction of flooding waters in the valley and river waters is consequent, and the amount of supply to the basins of floodplain lakes is a result of the elevation of the ceiling of the crevasse splay (limnological effective rise – LER). Limnological Effective Rise (Fig. 6.5) is, therefore, a layer of water measured on the crevasse glyph that feeds the basin (most commonly given in centimeters). In the case of contrafluent lakes, a subsequent or obsequent direction of feeding waters in relation to the river bed is observed, which results from the system of local landform features. Contrafluent supply is activated when the relative height of the dike isolating the lake basin from the river, from the lower water side is lower than from the upper water side.

Mixed hydrological type occurs in the case of supplying the floodplain lakes with two arms (contrafluent–confluent). At the beginning inflow occurs from the lower water side, and after some time also through the second arm (from the upper water

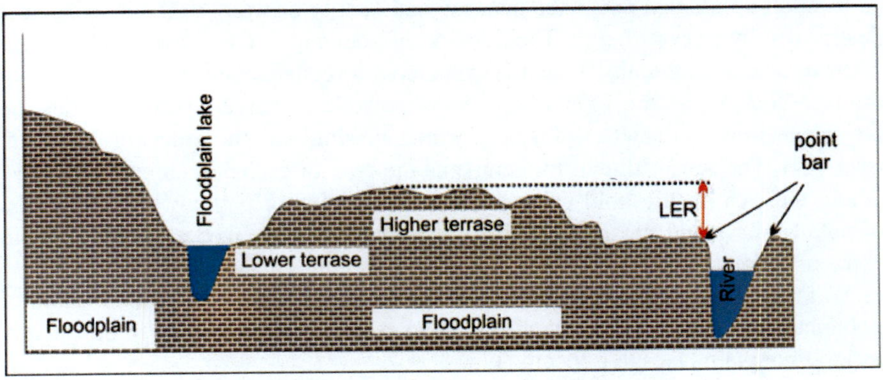

Fig. 6.5 Limnological effective rise

side). This form of basins supply occurs only when there is a difference in the value of limnological effective rise of crevasses (from the upper and lower side of the water in the river). Potamophase begins with contrafluent inflow of river waters and "old" waters pushing deep into the basin. Confluent recharge and the change of direction of basin supply starts after crossing the shore point, at the height of upper water crevasse.

Division by the manner of water exchange

The basins of the floodplain lakes are particularly dynamic ecological systems in which there is a constant exchange of matter and energy. The classic pattern of water exchange rate in the basin assumes its uninterrupted movement within the entire bowl (CSR – Continuously Stirred Reactor) [38]. The rate of water exchange in the basin results primarily from the characteristics of the basin, e.g. geological structure, relief, surface of the basin, and the bathymetry of the basin itself [39]. In the Polish scientific literature, there are two types of water exchange in relation to lakes: horizontal and vertical, and a situation where the role of these forms of exchange is so variable that none of them is dominant. In the world literature, the rate of water exchange is generally determined on the basis of the ratio of the capacity of the lake basin to the volume of the supply (in the case of nondrainage lakes). However, the capacity of the basin to the drainage (in the case of outflow lakes) or the capacity of the basin to the sum of the supply and drainage (in the case of flowing lakes) has always been implemented. The role of genetic forms of supply in terms of groundwater resources (base flow) or quick flow component of the supply is relatively rarely described in the literature.

There are three types of floodplain lakes due to the way of water exchange; evaporation-dominated, exchange-dominated and flood-dominated [40]. In the case of evaporation-dominated lakes, the vertical form of water exchange in the basin is dominant. The largest share in the revenue side of the water balance is precipitation, while the evaporation is the largest share in the expenditure side. Lakes of this type have short-lived potamophase, and for most of the year, the hydrological condition of the lake is determined by the resources of the catchment.

In flood-dominated lakes, the quantity and quality of water in the lake basin is determined by river recharge. The duration of potamophase in a hydrological year often exceeds six months, which is conducive to equalization of water levels in the reservoir during the year. Long-lasting potamic recharge also determines the physicochemical parameters of waters, which are similar to the waters of the parent river [24]. The most difficult to identify is the type of exchange-dominated lakes, due to the lack of threshold values of, e.g. evaporation or outflow, or the duration of potamophase and limnophase. The functioning of such lakes is shaped by both extrazonal potamic factors and interzonal factors of the catchment.

So far, no precise and objective criterion has been developed for the division of floodplain lakes due to water exchange. Two criteria can be adopted to objectively assign floodplain lakes to a specific type:

Direct methods

Comparison of the size of expenditure side the balance equation of the basin (evaporation and outflow), the dominance of which would indicate evaporation or flood-dominated lake type. The absence of clear dominance of one of the components may mean that it belongs to a transition type.

Determination of the lake type, based on the water residence time in the lake, regardless of how it is calculated. Lakes with a long residence time are evaporation-dominated reservoirs, whereas fast flushing is characteristic of flood-dominated lakes. The group of transitional lakes requires the determination of the limits of the range of variability.

Indirect methods

- Method of recharge. In the case of direct supplying of floodplain lake basin with potamic waters, one deals with the flow type. When potamic recharge takes place through a basin of another lake or a system of such lakes (cascade), one deals with transitional lakes, while if the recharge with river waters is episodic or does not occur, then these are evaporation-dominated lakes.
- Location of the lake basin. The flow type includes the lakes located in the modern meandering belt or within the lower floodplain terrace. Transitional lakes occur within the floodplain higher terraces, whereas the waters under the edge of the terraces are usually evaporation-dominated reservoirs.
- Bathymetric parameters. For example, the ratio of the lake surface to its maximum depth, or the ratio of the shoreline development to the mean depth. The high value of the first quotient is usually characteristic for evaporation-dominated lakes, whereas the low value for flood-dominated lakes. The remaining lakes constitute an intermediate class. Determination of the range of variability should be established individually for different river valleys. In the case of the ratio of the coefficient of shoreline expansion and the average lake depth, the values >1 are characteristic for flood-dominated type, about 1 transitional type and <1 evaporation-dominated type.
- Linking the origin of the floodplain lake with the way it is supplied. Oxbow and avulsion lakes in the vast majority belong to the group of flood-dominated lakes. The anastomotic lakes are most often transitional lakes, while the inter-levee lakes are evaporation-dominated lakes.

Division by water hydrochemistry

The quality parameters of the floodplain lake waters result from the period of their functioning cycle (potamophase, limnophase, flooding). During the potamophase, the parent rivers transform the quality of the lake waters in a short period of time as a result of intensive exchange. During overbed flows of the parent river, multiple mixing of waters in the basin takes place. Dynamic supply of allochthonous river waters determines the ionic composition of lake waters often for many months, and hydrochemical processes of catchments of the lakes are of little importance. In the period of limnophase, the physical and chemical parameters of the waters of the floodplain lakes are determined by the supplying waters of the own catchment. The quality of lake waters is then determined by atmospheric precipitation, surface and underground inflow (hypodermic and from a deeper groundwater table). In turn, the flooding is an important but only cyclical episode in flood ecosystems, during which lakes are only the recesses of submerged flood terraces [41]. General hydrochemical relationships resulting from the cycle of successive potamophases and limnophases and their effect on the quality of valley waters can be expressed with the β index. It is the ratio of Ca^{2+} cation concentration to the sum of Cl^- and $SO_4{}^{2-}$ anions concentrations [42]. Calcium, as a product of calcium carbonate dissociation processes, is treated as a geogenic ion, typical for the natural environment with a significant share of carbonate sediments. Sulfates and chlorides, being good migrants, are an indicator of the degree of anthropogenic transformation of the environment. The percentage quotient of the limnic and fluvial index β $\left(F = \frac{\beta_{\text{lakes}}}{\beta_{\text{rivers}}}\right)$ is the basis for determining the similarity or difference in the quality of these two environments. On this basis, the following groups of lakes can be distinguished:

Minor – lakes with a low F-value. The waters of these lakes have a higher share of anthropogenic components than in the waters of the parent river.
Maior – lake waters have a lower share of anthropogenic ions than that of the parent river. In this case, the quality of lake waters is shaped by processes related mainly to chemical denudation of the sediments of the reservoir's basin.
Integralis – lakes showing an apparent similarity in the size of lake and river β indices. Potamic supply is the main factor shaping the quality of lake waters. The limits of the F index ranges for the distinguished lake types are the arithmetic mean ± half of the standard deviation. When the distribution of quality parameters of the analyzed lake waters corresponds to the normal distribution, the defined limits divide the area of the variability of the index value into three ranges, the narrowest of which is the middle range (*integralis*). This is due to the need to emphasize the hydrochemical similarity of limnic and potamic waters. The value [F] below 85% means a transformation of lake waters in the direction of pollution growth and belonging to type 3 (*minor*). The opposite situation (F-value > 115%) is represented by waters of lakes of natural hydrogeochemical character, the quality of which is shaped by the supply from the own catchment. The values [F] in the range of 115–85% correspond to type 2 (*integralis*).

The way of connection between the lake and the parent river and the forms of supplying the lake basins constituted the basis for distinguishing three types of lakes within each of the lake types. Interzonal lakes (IZ) are lakes with a hydrologically dominant role of the reservoir own catchment area. When the area of lake recharge is a catchment area of the parent river, one deals with extrazonal reservoirs (EZ). The third group of mixozonal reservoirs (MZ) are those lakes in which there is no clear dominance of a particular type of catchment.

6.4.2 Anthropogenic Lakes

A separate group of lakes are anthropogenic lakes, which were created as a result of artificial regulation procedures [43]. These are the lakes located on a flood plain, which were created mainly as a result of cutting off the side arms of the river from the main river bed or closing the water space between the groynes as a result of regulatory works carried out. These lakes are often referred to as quasi-ox-bows or post-regulation reservoirs. The isolation of a lateral river bed or a fragment of the river itself results from a process aimed at narrowing the main river bed [44]. The scope of work carried out, and the type of regulation structures used to play an important role. The diagram of regulation works in the river is presented in Fig. 6.6.

River regulation includes: strengthening riverbank, works in the river bed and in the floodplain area, including flood protection. The regulation belongs to the group of hydrotechnical works, which have been carried out on a large scale and for a long time. From the very beginning, their aim was to make fuller use of the rivers and to protect the population against the threat that they may pose to the adjacent areas under extreme conditions.

The regulation of rivers according to the designed course of the river bed route is carried out with the use of appropriate regulatory structures. Such structures may be heavy or light. Heavy regulatory buildings are constructions built on a permanent basis, ensuring the durability of their construction at all water levels encountered on a given river. These structures include groynegroynes, longitudinal and transverse dams, breakages and edge bands. Light regulatory structures are those that can be easily and quickly built and, if necessary, dismantled. An example of such a structure can be different types and sizes of fences [45, 46] (Fig. 6.6).

Depending on the degree of river bed narrowing and the regulatory structures used, anthropogenic lakes may be formed in different ways. Authors [17, 47] distinguished the following methods of anthropogenic lakes formation (Fig. 6.7):

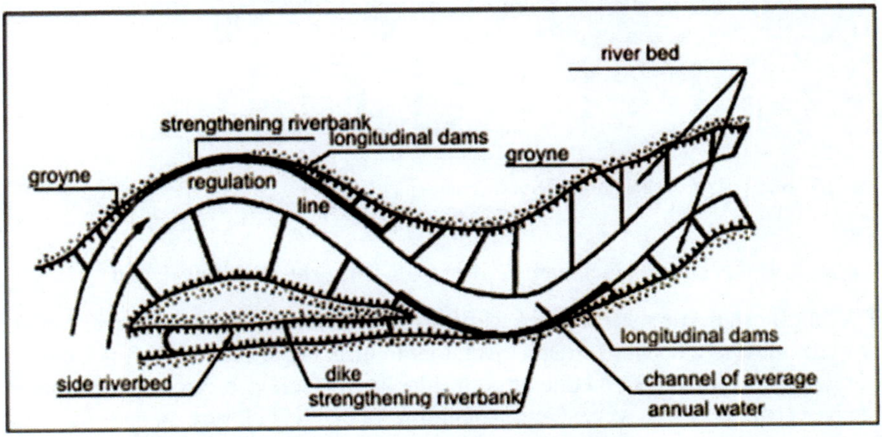

Fig. 6.6 Distribution of heavy regulation structure in a river [45]

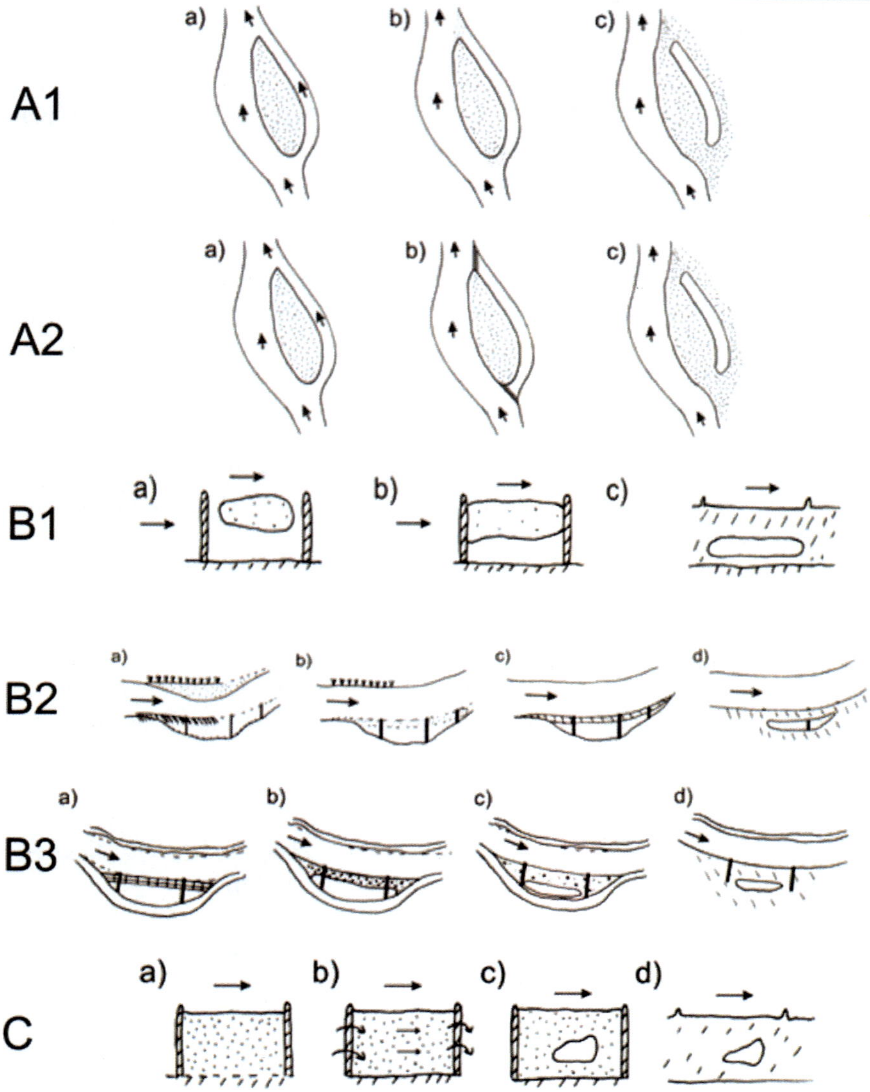

Fig. 6.7 Steps of post-regulatory reservoirs creation [17, 47]

Created as a result of river arms cutting off from the main river bed

The cutoff can occur naturally by backfilling with debris carried by the river (on one or both sides) as a result of stream diversion. In turn, the artificial cutoff is the result of closing a branch (also on one or both sides) as a result of breakage, construction of river groynes or transverse dams. A cutoff arm that had closure on one side could be used as a relief canal or work canal.

Created as a result of closing the water space between the groynes

Construction of the groynes perpendicularly to the banks of the river bed changes the existing hydrodynamic conditions as a result of a change in the course of the bed and river streams. Near the groyne heads, river streams collapse and the dragged debris is deposited in their "shade" [48]. Three mechanisms of this process can be distinguished:

- natural accumulation of material. Sandbanks formed parallel to the shore separate the water spaces cutoff from the main river bed. Gradually, the sandbanks are fixed by vegetation, while the cutoff water surfaces are continuously reduced and shallow.
- forced accumulation. It leads to a partial transformation of the spaces excluded from the active river bed into the land area. This applies especially to sections where the space between the groynes was large, and the dragged debris movement is observed in the watercourses. In the space between the groynes, at different distances from the shore, fascine mattresses or wicker curtains were usually placed at the bottom of the rivers, which slow down the speed of the flowing water and encourage the depositing of the retained material.
- artificial landing. When the space between the groynes is shallow, the land is strengthened with vegetation planting, most often with willow seedlings.

Created as a result of the floodplain erosion

This process usually occurs during floods. When the water level in the river is high, previously accumulated material is eroded and washed away. A basin is created in the place where the material is eroded by the river, and it is filled with river water.

Taking into account the nature of the inflow or outflow to lakes in regulated river valleys, a simple and complex hydrographic system can be distinguished [47].

The simple system includes single reservoirs (Fig. 6.8):

A flow lakes

A1 along the axis:

(a) with gravitational inflow and outflow,
(b) with regulated inflow and outflow,
(c) with gravitational inflow and regulated outflow,
(d) with regulated inflow and outflow.

A2 asymmetric, in close vicinity there is inflow and outflow of water:

(a) with gravitational inflow and outflow,
(b) with regulated inflow and gravitational outflow,
(c) with gravitational inflow and regulated outflow,
(d) with regulated inflow and outflow.

B lake with inflow in the form of a trench or channel from the flood plain

(a) gravitational inflow

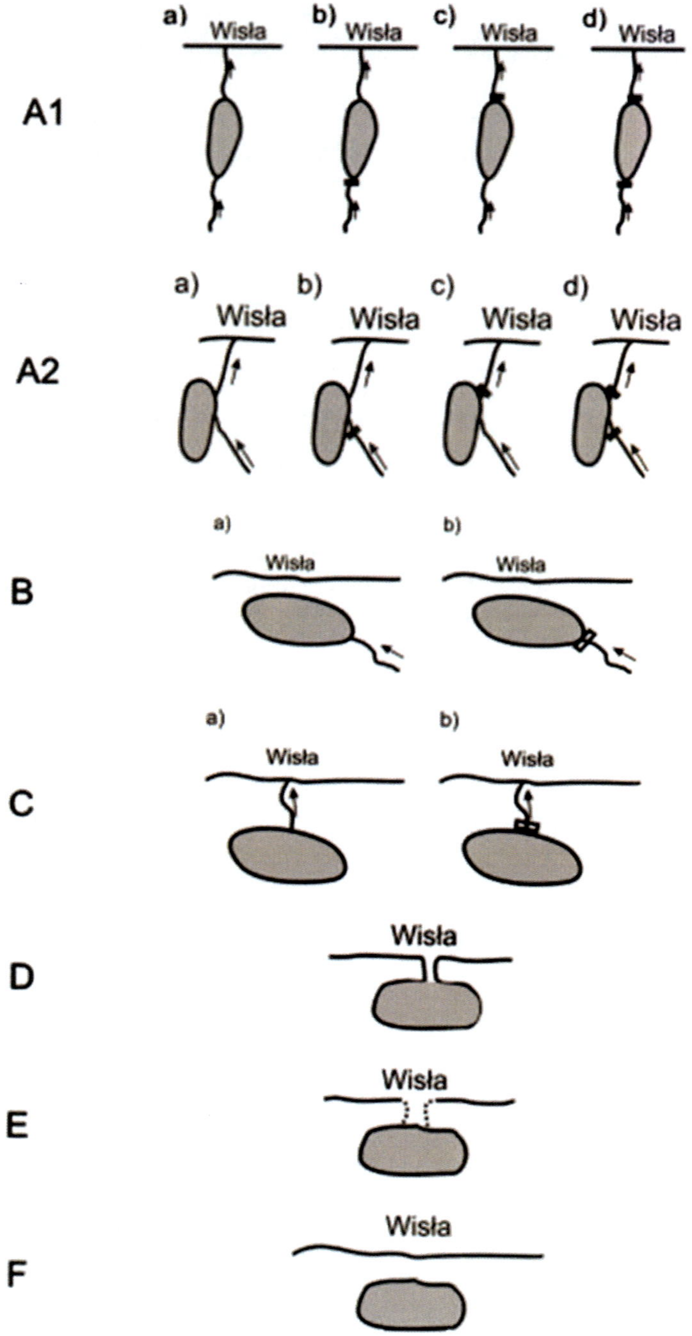

Fig. 6.8 Examples of a lake and parent river connectivity [47]

(b) inflow regulated by means of a pumping station or a flap.

C lake with outflow toward the river

(a) gravitational outflow
(b) regulated outflow.

D lake without surface outflow, isolated from the river
E lake permanently connected with the river
F lake periodically connected with the river.

The complex system is a system of several (at least 2) lake basins, which are connected by means of a narrow isthmus. As a result of dams constructions, one reservoir was divided into several smaller ones. Breaks in the dams systems occurred later, which led to the connection of some of the reservoirs. Lakes with a simple system may occur within this type.

The formation and functioning of anthropogenic floodplain lakes are determined by the type and scope of river bed regulation works. The river regulation is generally defined as hydrotechnical activities (with the use of appropriate structures) aimed at creating conditions for the formation of a uniform bed of a river with gentle meandering arches and fixed cross-sections for three characteristic stages: low water (LW), medium (MW) and high (HW). Measures to achieve this aim are: concentration of the stream, consolidation of the shore, straightening of excessively sharp curves of the river, extension of excessive narrowing, i.e., creation of conditions for the free flow of ice to avoid dangerous blockages, as well as creation and maintenance of a constant depth necessary for the proper flow [45].

The appearance of a large number of anthropogenic reservoirs in river valleys was connected with the systematic regulation of major rivers in Poland, which began in the nineteenth century. Lack of maintenance and renovation of neglected regulatory buildings caused their technical condition, as well as the whole regulatory system, was constantly deteriorating. Debris was deposited in the inter-groyne zones, and new water reservoirs were created. In places where river groynes were destroyed, blurring was created in existing, already settled inter-groyne spaces. These deformations widened as a result of the swelling wave and further erosion of the reservoir. Therefore, such waters appeared most often after the passage of large floods, which meant that these are relatively young objects.

6.5 Studies and Bathymetric Plans of Selected Floodplain Lakes in Poland

The lakes of river valleys are one of the fastest growing types of water reservoirs and at the same time the most perishable forms in the landscape. The highest disappearance rate was observed in the case of artificially cutoff anthropogenic lakes [49]. The studies on floodplain lakes in Poland were carried out to a different extent (hydrological, hydrochemical, bathymetric). They included lakes located in the valleys of

Fig. 6.9 The location of selected floodplain lakes in Poland

large rivers, e.g. the Vistula [47, 50], Odra [51], Warta [52], Obra [53], Bug [37], as well as Biebrza, Łyna, Drwęca [49].

The floodplain lakes in Poland are poorly recognized in terms of the geometry of basins. Significantly more publications concern biodiversity concerning both the composition and variability of species in the lake waters in the Słupia [54], Biebrza [55], Vistula [56], Warta [57], and Łyna [58].

Bathymetric maps of selected floodplain lakes together with their location have been presented in Figs. 6.9, 6.10, 6.11, 6.12 and 6.13.

6.6 Changes in Water Relations of Floodplain Lakes

The water relations of lakes located in river valleys are a result of the degree of transformation of the natural environment of their catchment areas. In natural and quasi-natural valleys, the water balance of the lake basins results from the way they are supplied (confluent, contrafluent, profundal, etc.) and their functional period (potamophase or limnophase). In general, it can be stated that the valleys of the

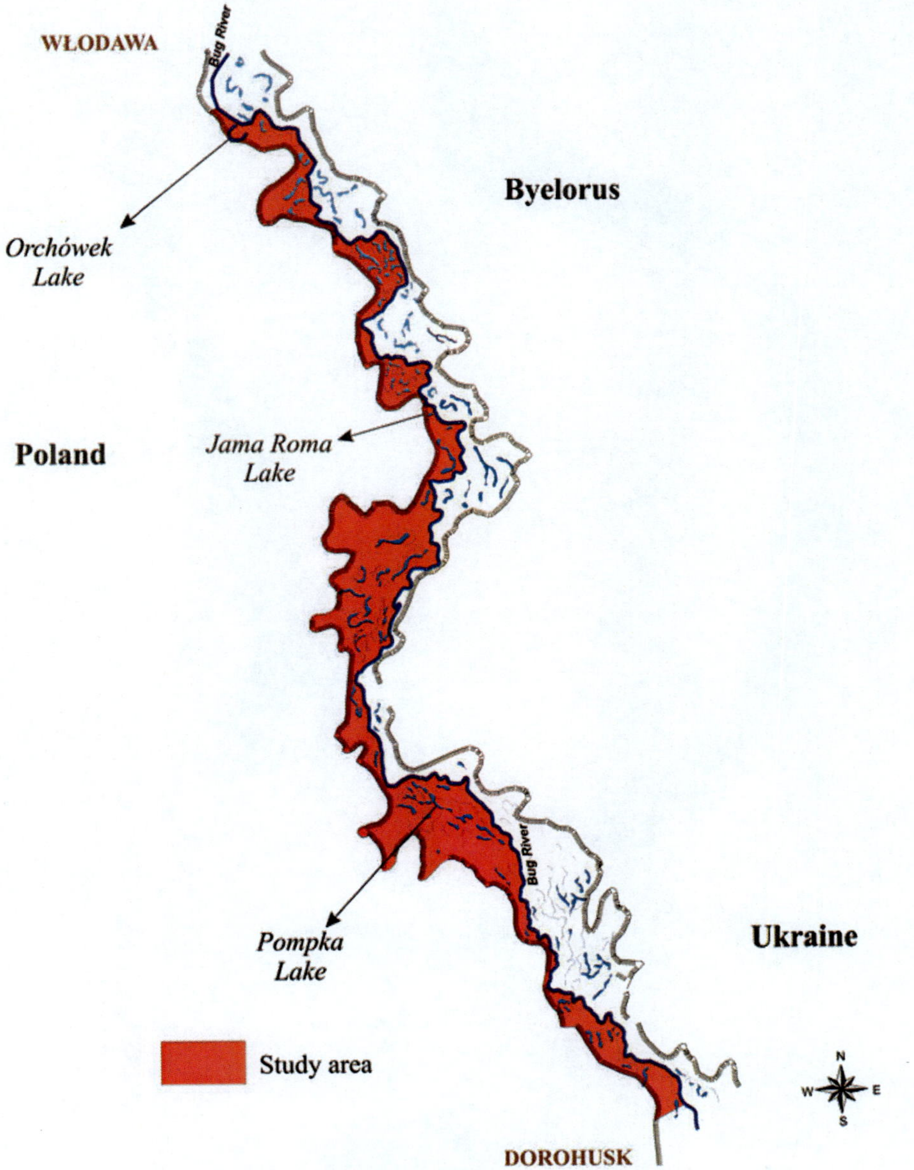

Fig. 6.10 The location of floodplain lakes in the Bug River Valley

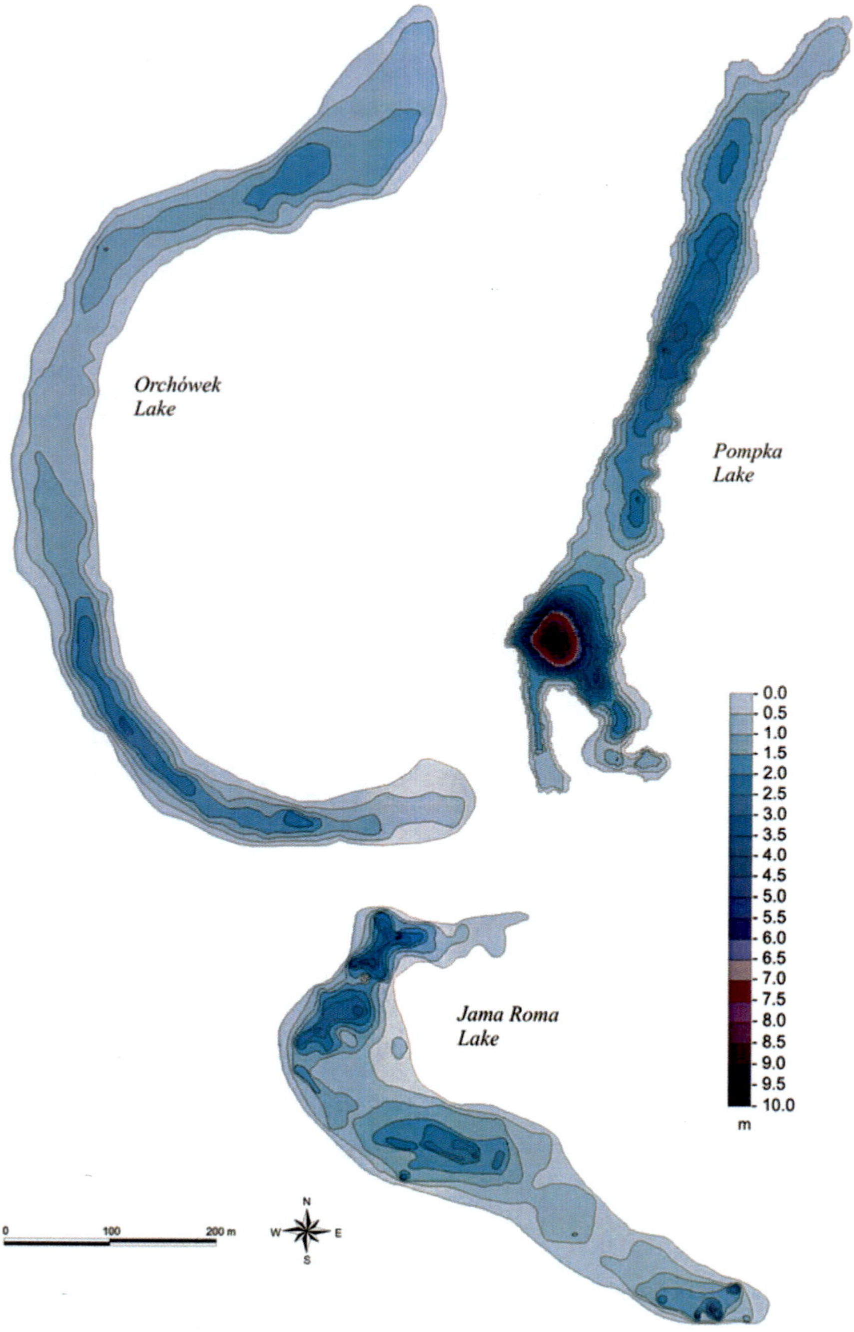

Fig. 6.11 Bathymetric scans of selected floodplain lakes in the Bug River Valley

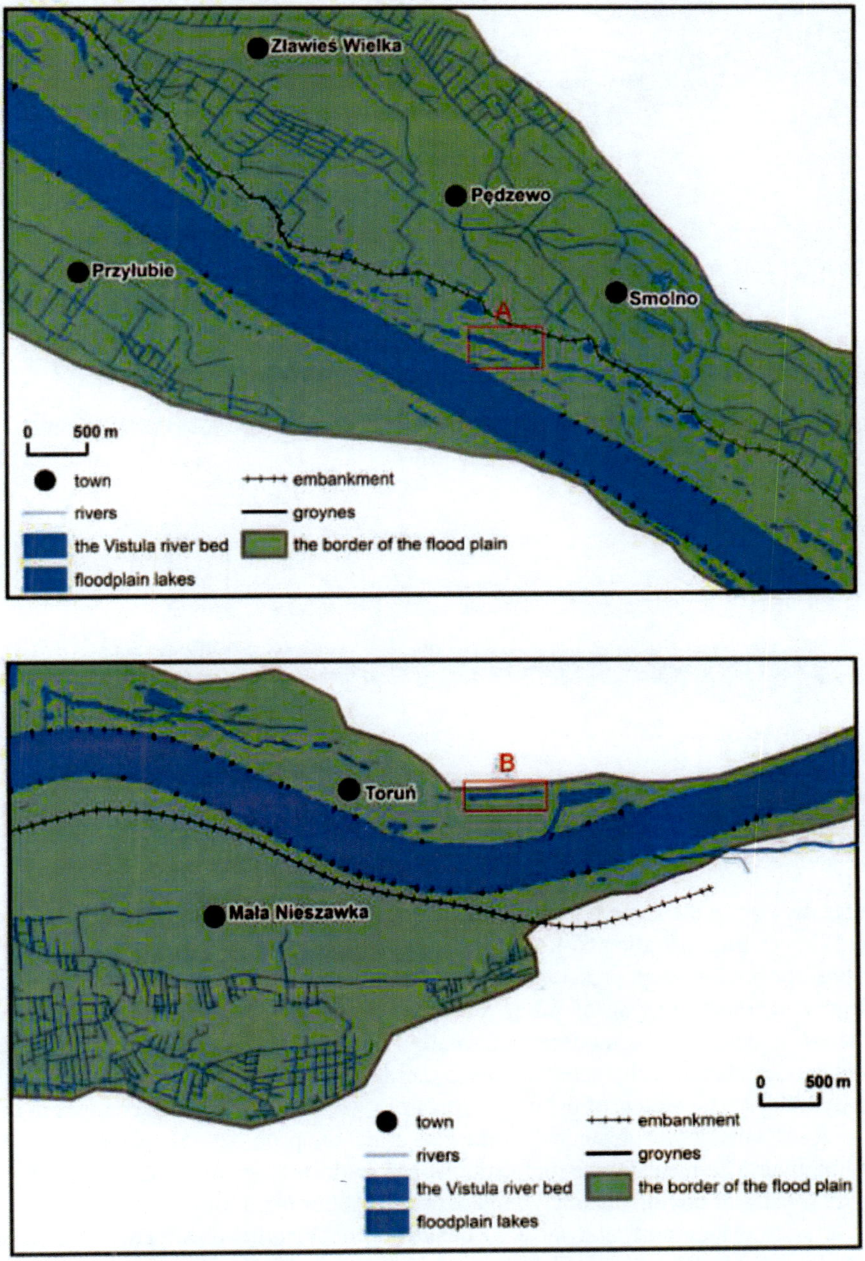

Fig. 6.12 Location of selected floodplain lakes in the Vistula River Valley (A-Smolno, B-Martówka)

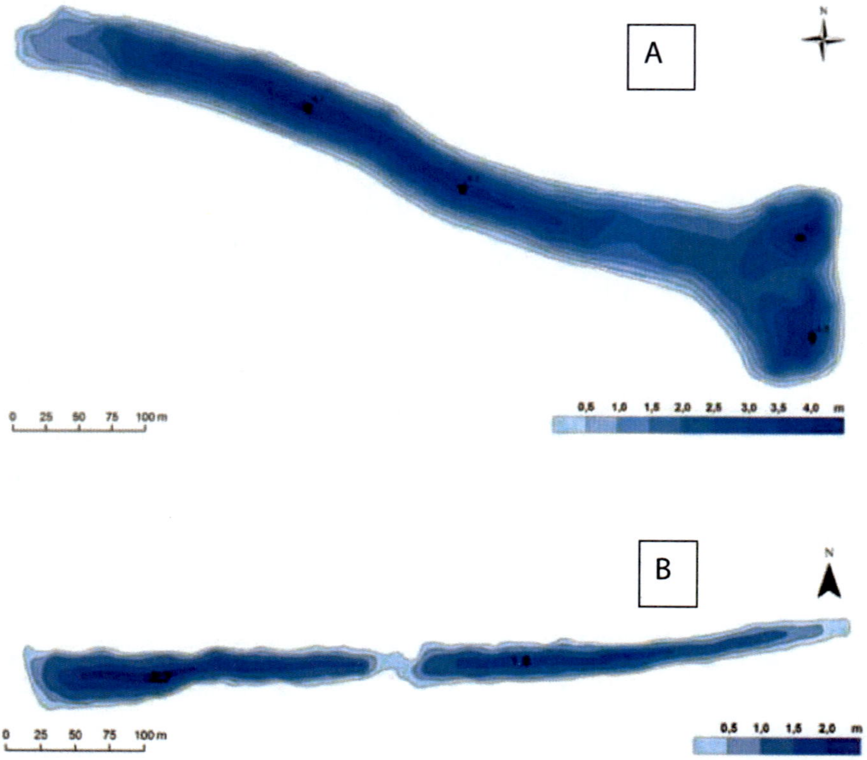

Fig. 6.13 Bathymetric scans of selected floodplain lakes in the Vistula River Valley (A-Smolno, B-Martówka)

largest rivers in the eastern part of Poland (e.g., Lower San, Lower Wieprz, Middle and Lower Bug, Middle and Lower Narew, Biebrza, Pisa) belong to the areas with relatively low anthropopression. The lakes in their valleys are numerous, and the degree of transformation of water relations in their lymnophasic drainage basins should be described as moderate or small. The western wing of the Vistula basin and the Odra basin is characterized by an increase in hydrotechnical interventions in river valleys. The effects of the main regulatory works contributed to changes in the surface of lakes. Hydrotechnical works also affect the problem of lake water quality in the hydrochemical (ionic composition) and hydrobiological sense (composition and biomass of phytoplankton, zooplankton and microplankton).

A good example of lake surface changes is the Vistula valley. There are some problems with defining the shoreline of objects, especially in the case of basins connected and forming compact complexes. However, such changes were the subject of research mainly on the regulated section of the Vistula (from Włocławek to Tczew). Studies of various details concerned both short and long sections of the valley [44, 47, 59]. They also included changes in the hydrographic network of the Vistula from the nineteenth century until 2010 (the Schroetter's map from the nineteenth century, Karte des Deutschen Reiches from the beginning of the twentieth century and a contemporary map from the beginning of the twenty-first century).

6.7 Summary

The lakes of the river valleys are very diverse reservoirs. High variability of morphometric, hydrochemical and hydrobiological as well as ecological and landscape parameters make them attractive. Moreover, they are young reservoirs subject to dynamic processes of disappearance. Although they are a very numerous group of lakes, their degree of recognition is relatively low. The small area and capacity of the basins are conducive to anthropogenic transformations. They are often backfilled, their shore zone is destroyed, and they are often the receivers of waste. Some floodplain lakes (both natural and anthropogenic ones) are used for recreation as swimming pools, fishing grounds, canoeing routes. Due to their natural values (often unique), they also have significant educational potential. The floodplain lakes allow to preserve the diversity of habitats and constitute habitats of fauna and aquatic flora. Bearing in mind the general values of floodplain lakes, active forms of lake protection and lake restoration projects can be found more often.

6.8 Recommendations

A complex mechanism of the functioning of floodplain lakes (limnophase and potamophase) and their specific environmental role (in terms of hydrochemistry and hydrobiology) require particular approach when management practices are concerned. Due to a high susceptibility of floodplain lakes on human pressure, both quantitative and qualitative protection activities are highly recommended.

In terms of quantitative issues, natural or quasi-natural type of water alimentation should be maintained. It is visible in a consistency occurrence of both periods and paths of water alimentation with a natural relation between floodplain lake and the river. In practice it means that both long-lasting potamophases or long-lasting limnophases may be observed. It often brings about high variability of water stages dynamic during the functional periods. Thus, water budget which is typical for a particular floodplain lake should be established by implicating duration of both potamophases and limnophases of the water body in question.

Qualitative approach to floodplain lakes should be focused on maintaining both typical for particular lake ionic composition and natural hydrobiological conditions of its water. It may be achieved by controlled water input from the river during the limnophase period, using drains, pipes or ditches.

Moreover a quality of the lake's catchment environment should be recognized. It is important in terms of influencing lake water quality during the limnophase period.

It is natural that water quality of floodplain lake observed during potamic supply (and afterwards) significantly differs than ionic composition formed during limnophase period.

References

1. Fisk HN (1947) Fine-grained alluvial deposits and their effects on Mississippi river activity, Vol. I. Waterways Experiment Station, War Department, Corps of Engineers, Mississippi River Commission, 82 pp
2. Camporeale C, Perucca E, Ridolfi L (2008) Significance of cutoff in meandering river dynamics. J Geophys Res 113:1–11. https://doi.org/10.1029/2006JF000694
3. Constantine JA, Dunne T (2008) Meander cutoff and the controls on the production of oxbow lakes. Geology 36(1):23–26. https://doi.org/10.1130/G24130A.1
4. Wolfe BB, Hall RI, Last WM, Edwards TWD, English MC, Karst-Riddoch TL, Paterson A, Palmini R (2006) Reconstruction of multi-century flood histories from oxbow lake sediments, Peace-Athabasca Delta, Canada. Hydrol Process 20:4131–4153. https://doi.org/10.1002/hyp.6423
5. Wren DG, Davidson GR, Walker WG, Galicki SJ (2008) The evolution of an oxbow lake in the Mississippi alluvial floodplain. J Soil Water Conserv 63(3):129–135. https://doi.org/10.2489/jswc.63.3.129
6. Martinovic-Vitanovic V, Ostojic S, Popovic N, Rakovic M, Kalafatic V (2013) Limnological study of Serbian oxbow shaped Lake Srebrno, with special emphasis on the benthic community composition and structure. Ekologia (Bratislava), 32(1), 66–86
7. Reavie ED, Robbins JA, Stoermer EF, Douglas MSV, Emmert GE, Morehead NR, Mudroch A (2005) Paleolimnology of a fluvial lake downstream of lake superior and the industrialized region of Sault Saint Marie. Can J Fish Aquat Sci 62:2586–2608. https://doi.org/10.1139/f05-170
8. Frenette JJ, Arts MT, Morin J (2003) Spectral gradients of downwelling light in a fluvial lake (Lac Saint-Pierre, St. Lawrence River). Aquat Ecol 37:7–85. https://doi.org/10.1023/A:1022133530244
9. Wojciechowska W, Pasztaleniec A, Solis M, Turczyński M, Dawidek J (2005) Phytoplankton of two river lakes in relation to flooding period (River Bug, Eastern Poland). Pol J Ecol 53(3):419–425
10. Czachorowski S, Lewandowski K, Wasilewska A (1993) The importance of aquatic insect for landscape integration in the catchment area of river Gizela (Masurian lake district, North-Eastern Poland). Acta Hydrobiol 35(1):49–63
11. Khmelewa N, Nesterowicz A, Czachorowski S (1994) The microinverbrate fauna of some Byelarussian, Karelia, and Altaian springs and its relation with certain factors. Acta Hydrobiol 36(1):75–90
12. Tockner K, Pennetzdorfer D, Reiner R, Schiemer F, Ward JV (1999) Hydrological connectivity and the exchange of organic matter and nutrients in a dynamic river-floodplain system (Danube, Austria). Freshw Biol 41:521–535
13. Kasten J (2003) Inundation and isolation: dynamics of phytoplankton communities in seasonal inundated floodplain waters of the Lower Odra Valley National Park—Northeast Germany. Limnologica 33:99–111
14. Hamilton SK, Sippel SJ, Melack JM (2004) Seasonal inundation patterns in two large savanna floodplains of South America: the Llanos de Moxos (Bolivia) and the Llanos del Orinoco (Venezuela and Colombia). Hydrol Process 18:2103–2116. https://doi.org/10.1002/hyp.5559
15. Okogwu OI (2010) Seasonal variations of species composition and abundance of zooplankton in Ehoma Lake, a floodplain lake in Nigeria. Rev Biol Trop 58(1):171–182

16. Obolewski K, Glińska-Lewczu K, Ożgo M, Astel A (2016) Connectivity restoration of floodplain lakes: an assessment based on macroinvertebrate communities. Hydrobiologia 774:23–37. https://doi.org/10.1007/s10750-015-2530-8
17. Gorączko M (2001) The hydrography of ox-bows in Fordon Valley in Bydgoszcz. Limnol. Rev. 1:109–116
18. Kubiak-Wójcicka K (2003) Changes in the area of ox-bow lakes near Toruń, based on cartographic sources. Limnol Rev 3:127–133
19. Hamilton SK, Lewis WM Jr (1990) Physical characteristics of the fringing floodplain of the Orinoco River, Venzuela. Intersciencia 15:491–500
20. García de Emiliani MO (1997) Effects of water level fluctuations on phytoplankton in a river-floodplain lake system (Paraná River, Argentina). Hydrobiologia 357:1–15
21. Dawidek J, Ferencz B (2012) Hydrological processes in the riverine systems, the origin and classifications of floodplain lakes. Ekológia (Bratislava) 31(3):331–340
22. Dawidek J (2013) Procesy limniczne w strefie aktywności fluwialnej rzeki Bug pomiędzy Dorohuskiem a Włodawą. Wydawnictwo UMCS, Lublin
23. Dembowska EA, Kubiak-Wójcicka K (2017) Influence of water level fluctuations on phytoplankton communities in an oxbow lake. Fundam Appl Limnol 190(3):221–233. https://doi.org/10.1127/fal/2017/0998
24. Hamilton SK, Lewis WM (1987) Causes of seasonality in the chemistry of a lake on the Orinoco River floodplain, Venezuela. Limnol Oceanogr 32:1277–1290
25. Castillo MM (2000) Influence of hydrological seasonality on bacterioplankton in two neotropical floodplain lakes. Hydrobiologia 437:57–69
26. Davis WM (1903) Development of River Meanders. Geol Mag 4(10):145–148
27. Brigham AP (1892) Rivers and the evolution of geographic forms. J Am Geogr Soc 24:23–43
28. Weihaupt JG (1976) Morphometric definitions and classifications of oxbow lakes, Yukon River basin. Indiana University—Purdue University, Indianapolis, Indiana, Alaska
29. Weihaupt JG (1977) Morphometric definitions and classifications of oxbow lakes, Yukon River Basin, Alaska. Water Resour Res 13(1):195–196
30. Rodrigues L, Bicudo DC (2001) Similarity among periphyton algal communities in a lentic-lotic gradient of the upper Parana river floodplain, Brazil. Rev Bras de Bot 24(3):235–248
31. Borics G, Abonyi A, Krasznai E, Várbíró G, Grigorszky I, Szabó S, Deák Cs, Tóthmérész B (2011) Small-scale patchiness of the phytoplankton in a lentic oxbow. J Plankton Res 33(6):973–981
32. Palijan G (2010) Different impact of flood dynamics on the development of culturable planktonic and biofilm bacteria in floodplain lake. Pol J Ecol 58(3):439–448
33. Koc J, Kobus Sz, Glińska-Lewczuk K (2009) The significance of oxbow lakes for the ecosystem of afforested river valleys. J Water Land Dev 13a:115–131. https://doi.org/10.2478/v10025-010-0023-8
34. Bolland JD, Nunn AD, Lucas MC, Cowx IG (2012) The importance of variable lateral connectivity Y between artificial floodplain waterbodies and river channels. River Res Appl 28:1189–1199. https://doi.org/10.1002/rra.1498
35. Glińska-Lewczuk K, Burandt P, Kujawa R, Kobus Sz, Obolewski K, Dunalska J, Grabowska M, Lew S, Chormański J (2016) Environmental factors structuring fish communities in floodplain lakes of the undisturbed system of the Biebrza River. Water 8:146. https://doi.org/10.3390/w8040146
36. Korponai J, Gyulai I, Braun M, Kövér C, Papp I, Forróet L (2016) Reconstruction of flood events in an oxbow lake (Marótzugi-Holt-Tisza, NE Hungary) by using subfossil cladoceran remains and sediments. Adv Oceanogr Limnol 7(2):131–141. https://doi.org/10.4081/aiol.2016.6168
37. Dawidek J, Turczyński M (2006) Feeding of lakes in the middle Bug River valley by river waters. Limnol. Rev. 7:65–72
38. Monsen NE, Cloern JE, Lucas LV, Monismith SG (2002) A comment on the use of flushing time, residence time, and age as transport time scales. Limnol Oceanogr 47(5):1545–1553
39. Geyer WR, Morris JT, Pahl FG, Jay DA (2000) Interaction between physical processes and ecosystem structure: a comparative approach. In: Hobbie JE (ed) Esuarine science: a synthetic approach to research and practice. Island Press, pp 177 – 206

40. Sokal MA, Hall RI, Wolfe BB (2008) Relationships between hydrological and limnological conditions in lakes of the Slave River Delta (NWT, Canada) and quantification of their roles on sedimentary diatom assemblages. J Paleolimnol 39(4):533–550
41. Hudson PF, Heitmuller FT, Leitch MB (2012) Hydrologic connectivity of oxbow lakes along the lower Guadalupe River, Texas: the influence of geomorphic and climatic controls on the "flood pulse concept". J Hydrol 414:174–183
42. Dawidek J, Ferencz B (2005) Hydrochemical classification of the river lakes situated near the Bug River. Limnol Rev 5:53–59
43. Jezierska-Madziar M (2005) Introduction. In: Jezierska-Madziar M (ed) Oxbow s as an important part of the ecosystem of the river (in Polish). Wydawnictwo Akademii Rolniczej w Poznaniu, Poznań, pp 9–10
44. Kubiak-Wójcicka K (2002) The influence of lower Vistula River's bed's regulation on creation of the ox-bows lakes. Limnol Rev 2:225–230
45. Makowski J (1997) Wały przeciwpowodziowe Dolnej Wisły historyczne kształtowanie, obecny stan, i zachowanie w czasie znacznych wezbrań. Wyd. IBW PAN, Gdańsk
46. Wołoszyn J (1974) Regulacja rzek i potoków. PWN, Warszawa
47. Kubiak-Wójcicka K (2006) Zmiany hydrograficzne i hydrologiczne w dolinie Wisły pomiędzy Włocławkiem a Bydgoszczą. Ph.D. thesis, Nicolaus Copernicus University, Toruń
48. Babiński Z (1992) Współczesne procesy korytowe dolnej Wisły. Prace Geograficzne, IGiPZ PAN, p 157
49. Glińska-Lewczuk K (2009) Influence of morphogenesis and connecting with a river on a rate of oxbow lake evolution in young glacial river valleys. In: Contemporary problems of management and environmental protection, vol 2, wetlands—their functions and protection, pp 9–28
50. Kubiak-Wójcicka K (2002) Charakterystyka starorzeczy w obszarze międzywala Wisły na odcinku Toruń-Solec Kujawski. Roczniki Akademii Rolniczej w Poznaniu, Melioracje i Inżynieria Środowiska 23:237–248
51. Kasperek R (2015) Changes in the meandering Upper Odra River as a result of flooding. Part I. Morphology and Biodiversity. Pol J Environ Stud 24 (6), 2459–2465. https://doi.org/10.15244/pjoes/59037
52. Kamiński B, Miler AT, Grajewski S, Okoński B, Schwartz K (2010) The concept for protection of flood plain forests in the Uroczysko Warta forest district. Infrastruct Ecol Rural Areas 11:31–41
53. Słowik M (2014) Holocene evolution of meander bends in lowland river valley formed in complex geological conditions (the Obra River, Poland). Geogra Ann Ser A, Phys Geogr 96:61–81. https://doi.org/10.1111/geoa.12029
54. Obolewski K, Glińska-Lewczuk K, Bąkowska M (2018) From isolation to connectivity: the effect of floodplain lake restoration on sediments as habitats for macroinvertebrate communities. Aqu Sci 80(4). https://doi.org/10.1007/s00027-017-0556-x
55. Grabowska M, Glińska-Lewczuk K, Obolewski K, Burandt P, Kobus Sz, Dunalsk J, Kujawa R, Goździejewska A, Skrzypczak A (2014) Effects of hydrological and physicochemical factors on phytoplankton communities in floodplain lakes. Pol J Environ Stud 23, 3:713–725
56. Wilk-Woźniak E, Ligęza S, Shubert E (2014) Effect of water quality on phytoplankton structure in oxbow lakes under anthropogenic and nonanthropogenic impacts. Clean Soil Air Water 42(4):421–42. https://doi.org/10.1002/clen.201200214
57. Nagengast B, Joniak T, Kuczyńska-Kippen N (2013) Water and rush plant communities in the Warta River oxbows: biodiversity and habitat requirements. Teka Kom Ochr Kszt Środ Przyr OL PAN 10:260–273
58. Glińska-Lewczuk K, Burandt P (2011) Effect of river straightening on the hydrochemical properties of floodplain lakes: observations from the Łyna and Drwęca Rivers, N Poland. Ecol Eng 37:786–795. https://doi.org/10.1016/j.ecoleng.2010.07.028
59. Kubiak-Wójcicka K, Kordowski J (2015) Uwarunkowania hydrograficzne i hydrologiczne Doliny Dolnej Wisły. In: Pająkowski J (ed) Zespół Parków Krajobrazowych Chełmińskiego i Nadwiślańskiego, vol I, pp 31–43

Chapter 7
Studies on, the Use and Protection of Springs in Poland

Paweł Jokiel and Zdzisław Michalczyk

Abstract Springs, defined as the points of a concentrated, spontaneous outflow of ground water to the land surface, are unique hydrographic objects connecting the underground and surface phases of water circulation. A symbol of purity and good quality of water, the origins of rivers and streams, in many a culture, springs are understood as a specific beginning of things and the places of enormous power to live and spiritual power. There is no doubt that the springs are phenomena of great importance to nature, science, landscape, culture, and economy. Nearly from the beginning of their existence, they have been subject to increasing human pressure, usually connected with man's inconsiderate actions in a natural environment. This study, which is based on the author's research, field observation and available literature, describes the various ways of the perception, use, and protection of springs in many locations in Poland and various cultures and social groups. Special attention has been paid to the role of springs in the natural system and culture, and their importance in local water management. These analyses and assessments were based on a wealth of crenological literature. Some of the urgently required protective or preventive actions, intended to preserve the springs of Poland in the appropriate condition, have also been indicated.

Keywords Springs · Protection · Preventive · Natural system · Poland

---

P. Jokiel (✉)
Department of Hydrology and Water Management, Faculty of Geographical Sciences, University of Łodź, Narutowicza 88, 90-131 Łódź, Poland
e-mail: pawel.jokiel@geo.uni.lodz.pl

Z. Michalczyk
Department of Hydrology, Faculty of Earth Sciences and Spatial Management, University of Marie Curie-Skłodowska, Aleja Kraśnicka 2cd, 20-718 Lublin, Poland
e-mail: zdzislaw.michalczyk@poczta.umcs.lublin.pl

M. Zeleňáková et al. (eds.), *Management of Water Resources in Poland*, Springer Water,
https://doi.org/10.1007/978-3-030-61965-7_7

7.1 The Spring Phenomenon

A spring is a form of natural, concentrated outflow of groundwater to the surface of the land. Beside dissipate objects, such as bog-springs, leakages, and seepages, it is the most spectacular manifestation and proof for water circulation in nature (Fig. 7.1). The location of the spring within the water circulation system is very specific. This makes it a kind of a connection between the underground and surface links of the water circulation system and its very existence is evidence of unity between the two links rather than their separateness. Springs, as a component of nature, are a subject of interest to researchers representing many scientific disciplines, including geography, hydrology, hydrogeology, geomorphology, hydrochemistry, hydrobiology, ecology, environmental protection, landscape architecture, as well as local history, archeology, and ethnography. Although the springs are primarily characterized by the volume and nature of the groundwater outflow as well as by the water quality and its usefulness in a wide sense (to man and the environment), its landscape- and culture-related features, including the nature and location of the spring niche, its evolution affected by the various factors, and its importance to the local community and the environment are equally important.

To a hydrologist or a geographer, a spring is the source and the starting point of a river. To find the springs of rivers, especially of those which were of key importance to the development of civilizations, has been one of the essential tasks assigned to explorers and travelers in more and less recent times. Enough to mention the explorations organized in search of the springs of the Nile, Congo, Amazon, Volga, or

Fig. 7.1 The "Strusi" springs (The Spring of Ostriches) in Imbramowice on the Kraków-Częstochowa Upland (Region VIII; cf. Fig. 7.7) – photo P. Jokiel

Danube. Even though most of the explorations failed to unambiguously locate the sources of those rivers, nearly all of them were successful in that they enabled the discovery of new places and provided more geographical information. Some geographers continue to argue about the exact location of the springs of some rivers even nowadays, thus enhancing the contemporary exploratory motif in geographical studies. Also ancient and contemporary monarchs and men of power tended to perceive the presence of a big or otherwise "important" river in their territory as an appreciated attribute, adding to the significance of the specific state or region.

On the other hand, to a geologist or a hydrogeologist, a spring is a specific exposure of groundwater, a "keyhole" enabling one to look into the underground water circulation system almost for free, identify it and even make it ready for use. There is hardly any doubt that the first water intakes (wells) were formed by digging deeper and wider (probably in the dry season) the spring niches or other points where water tended to seep out onto the land surface.

Some springs are also regarded as the "places of miracles" which provide pure water of good quality, usually satisfying drinking water standards and sometimes having either actual or legendary healing properties. Men have always willingly used the water from springs: it was freely accessible and had a kind of guarantee of natural purity. In many languages, the notion of "spring water" is a synonym of absolute purity and freshness even nowadays. In the Polish language, the words or expressions used with springs have unambiguously positive connotations: *well, water spring, fount, fountain*, and the starting sections of rivers referred to as *stream, brook river, rivulet or runnel*, are watercourses with natural water of excellent quality in nearly every case. The famous Polish ethnologist and linguist Władysław Kopaliński, in his dictionary of symbols ("Słownik symboli" – in Polish) provided the following meaning of the word spring (water spring, fount): "*A spring symbolizes the truth, spiritual awareness, wisdom, erudition, reason, justice, sensitivity, soul image, oracle, will of the people, poetic inspiration, oblivion, memory; God, repentance, joy, and bitterness; life, power to live, eternal life* ..." [1, 2].

Springs are also very interesting components of a landscape (Fig. 7.2). Whether in the form of natural spring niches or accompanied by hydroengineering facilities or architectural structures, they capably make the landscape more attractive to tourists and provide many aesthetic impressions. They may be encased in various small architectural forms, such as pools, chapels, small statues, crosses, wells), as part of old, more or less complicated tubing systems (made of wood or other materials), stonework canals or little reservoirs, many springs are the pearls of material culture and earlier technology, in addition to their original role as water drawing points. Even today, are the niches of even the smallest springs also the oases of shadow, pleasant scent, and tranquility of the softly flowing water, gladly appreciated by "*homo recreantus*."

The role of spring niches, ponds, and streams formed as the result of the functioning of springs should be mentioned as well. In the agricultural landscape, the humid enclaves of spring niches or outflow lines are the oases of biodiversity, essential components of its mosaic, and the points where the balance of matter is closed

Fig. 7.2 The „Chochołowskie" springs in the Tatras (Region XII; cf. Fig. 7.7) – photo P. Jokiel

locally. They have room both for hydrophilic and hygrophilic plants, aquatic organisms and terrestrial animals. The area around a spring has a specific microclimate; this makes a difference and accounts for the perceptible separateness of these ecosystems, developing them into the habitats of many rare or even endemic or relic plants or animals (Fig. 7.3). The spring water itself also accommodates unusual species which

Fig. 7.3 The "Ciosny" spring in the Łódź Hills (Region XIIIb; cf. Fig. 7.7) – photo P. Jokiel

do not exist anywhere else. This is not only favored by the considerable thermal inertia of spring water but also by its specific or even unique chemistry.

After having stayed among the rocks for a rather long time, groundwater can dissolve many of the substances and chemical compounds contained in the rock material. Therefore, spring water almost inevitably contains an amount of dissolved substances, so that their high or specific content may indicate either a mineral water spring (with geogenic contamination) or one with polluted groundwater (with anthropogenic pollution). Such mineralization may then either be natural or be connected with the anthropogenic degradation of a volume of groundwater or its surrounding area.

Springs have always been rather mysterious sites. The question of where the water in the spring comes from has always been appealing: it has contributed a lot to the development of life sciences and produced many a legend or fantastic theory. The water spurting from the springs has been believed to have healing properties or revered as an object of cult. Examples of this include: Massbielle in Lourdes (France), Kassotis at Delphi (Greece), Aquae Sulis in Bath (England), Boża Góra (Mount of God) in Poczajow (Ukraine), Manitou Springs in Colorado (USA), Baotu Quán in Jinan (China), Maggie Springs in Ayers Rock (Australia), Kalwaria in Wambierzyce (Poland). The so-called "miraculous" or "holy" springs have long been the places of rituals and destinations of pilgrimages in many religions and beliefs of the world. To name a few, such places include: Oyun Musa in Synai (Egypt), Narbada (India), Hot Springs (USA), Zamzam in Mecca (Saudi Arabia), Enzayimarku near Gonder (Ethiopia), Gomuhk – the holy spring of the Ganges (India), Waikoropupu Spring – the holy spring of the Maori on South Island (New Zealand), or the springs in Grabarka or Święta Woda (Poland). Many of the above sites gave rise to health resorts and baths existing now or in the past, thus confirming the "miraculous" properties of their spring waters (Bath, Baden Baden, Bormio, Lourdes, Bad Gastein, Karlove Vary, Kudowa Zdrój).

Spring water may contain dissolved gases ranging from the commonly found carbon dioxide (Fig. 7.4) or nitrogen to hydrogen sulfide and methane. There have also been very rare instances, when "noble" springs discharge a mixture of water and a noble gas, such as helium. The gas pressure in the groundwater bed may intermittently be high enough to form a kind of geysers propelled by the gas pressure. Such is the case of the quasi-geyser in Herlany in Slovakia: it ejects several thousand liters of a mixture of water and carbon dioxide a dozen meters high once in a few dozen hours; this makes it a tourist attraction of the region.

Springs which discharge thermal water (hot springs) are rather frequent though not as frequent as common springs. The existence of this type of discharge is usually connected with deep tectonic dislocation zones, the presence of young volcanoes in the area and, less frequently, with a low geothermal degree. In the extreme cases, the temperature of water in this kind of springs may be higher than 100 °C – such as in the thermal sulfur springs in the Lipari Islands (the Vulcano Island). Nearly as hot are the Hammam springs in Algieria (95 °C) and the water discharged in North Italy (Albano, Pisciarelli – 84 °C), in Canada (Rabbitkettle – 70 °C), in USA (Thermopolis – 70 °C) and Germany (Baden-Baden – 68 °C). More than 12,000

Fig. 7.4 The "Bulgotka" spring in Złockie (Region III; cf. Fig. 7.7) – photo L. Rajchel

warm and hot springs have been identified in Japan alone. Naturally discharged hot waters have been used by man since the oldest times: the use of thermal water as recreation baths and thermal treatments, in addition to culinary uses, was known to the Greeks and Romans and the Japanese more than 2000 years ago. There are no hot and no warm springs in Poland nowadays. There used to be a few small springs (for instance in Cieplice), but they were soon devastated. The operation and exploitation of thermal springs and water tend increasingly to encourage questions like these: Is it environmentally-friendly to draw water from hot springs? Is the so-called "clean energy" from geothermal systems really clean? After all, it is usually obtained by devastating the natural discharges of groundwater, and changes made in the local water circulation system cannot be overestimated.

Springs in Poland have a tradition of being valued and regarded as fascinating, unusual sites. A good example and illustration of this is the paper published, on 9 October 1938, in Głos Ziemi (Voice of the Earth) weekly issued in Vilnius. Titled *The Wonders of Nature* (Dziwy natury), it goes as follows (quote): *Then, among the many wonders of nature that exist in Poland, some of them are unique in all Europe. Let me mention some of the most interesting instances here. First of all, not known to many is the fact that there is a geyser (hot spring) in Poland. It is located near Szkło, some 50 km of Lviv. From the bottom of a sandy brook, flowing at a distance of 3 km from Szkło, there is the only spring of this type in the European continent. The locals call it "Kipiaczka." From the middle of the brook, a column of water spurts out one and a half meter high. The spring works just like a geyser: the streams of hot water spurt out intermittently, every several minutes.* Today, the "Kipiaczka" spring is situated in the Ukrainian territory and seems nothing like the former "geyser" (Fig. 7.5).

Being essential to human existence, springs have for ages been protected from destruction or contamination. Their utility and value to scientists, nature or culture long ago ceased to be a simple function of the spring yield, of the parameters of its water, or the size, origin, or nature of the spring niche. The importance and the perception of springs were and continue to be a cluster of its various physical and environmental properties and features related to culture, society or even economy. Some springs have a long and fascinating history; some other ones are regarded as the sources of "holy" water having miraculous properties [3–6]. Therefore, the ecological and social value of small springs and their didactic, cultural, scientific or environmental values are usually just as remarkable as those of the spectacular "global giants."

In the contemporary, technocratic civilization, human relationships with nature have been changed, mainly by weakening emotional states and by desacralization of human relationships with the natural environment. The relativization of values, improvement of our living conditions and, especially, easier access to water (from the tap), have changed our approach to the springs. Weaker bonds with the environment and easier availability of good water from the tap have made some of us tend to forget where its natural sources are – even though our water in the tap is drawn from a spring. The contemporary world is changing so fast, and our bonds with nature have been so reassessed that, without caring for the environment, which includes the care for water, we – intentionally or not – also stop caring for the quality of our lives. We tend to act like the fool in the poem by Adam Mickiewicz, saying [quotation is a little rephrased for the sake of the rhyme] "*who cares if the spring dries out in the mountains if we still have water for the fountains*" [original „…niech sobie źródło wyschnie w górach, byleby mi woda płynęła w rurach" (A. Mickiewicz, "Wiersze różne")]. Our resources of groundwater could be protected with the help of analyses of spring yields and distribution and evaluations of the physico-chemical and biological properties of spring water. Information about the springs and spring waters could be used as first-class geoindicators, or values which are useful for the assessment of the quantitative and qualitative changes of water in the natural environment.

Fig. 7.5 The "Kipiaczka" spring in Szkło on the Tarnogrod Plateau (Ukraine – photo T. Grabowski)

7.2 Studies on Springs in Poland

Written information on Polish springs was first published in the mid eighteenth century: first, as research notes, then in the form of specialist papers and monographs. Naturally, this does not mean than no springs had been heard of or described before that. They had been worshiped by the inhabitants of old-Slavic gords such as Biskupin ever since the neolith. Excellent spring water was commonly available, and some of the springs became very famous because their water had some healing or miraculous properties. The saint patrons of Polish springs include John the Baptist,

Fig. 7.6 The Spring of Geologists in Gołcza on the Miechów Upland (Region VIII; cf. Fig. 7.7) – photo P. Jokiel

John Nepomucene, Roch, Kinga of Poland, Florian. A number of the Polish springs have their names derived from the cult of the Virgin Mary [7, 8] (Fig. 7.6).

In the mid nineteenth century, research papers relating to the temperature of water of the springs located in the Tatras and the Prądnik valley [9, 10] and those on the thermal properties of spring waters near Kraków and Warszawa [11, 12] were published. A map of karst springs of the South-Eastern regions of Poland was also printed [13]. Other noteworthy publications include monographs on the springs of the Galicia – territory annexed by Austria [14, 15], a paper describing the locations and features of submontane springs of the North face of the Tatras [16, 17], papers describing the springs of the Przemsza and Szreniawa [18] and a description of Błękitne Źródła (Blue Springs) near Tomaszów Mazowiecki [19]. Interest in the natural discharge of groundwater was stimulated by the development of hydrographic studies and works required for the development of a *Hydrographic Map of Poland* [20]. Since the implementation of uniform methods to collect and process field data and information on springs [21], mainly in universities and centers of PAN (Polish Academy of Sciences), crenological studies have become more frequent and more information has been published: on the springs of the Beskids [22–24], of the Tatras [25, 26] and those located on the uplands [27–43] – Fig. 7.5. The first crenological characterization of the Karkonosze Mountains was published [44]. On the lowlands and in the lake districts, studies on springs were carried out at the same time by several authors [45–49].

Studies on Polish springs were intensified again – in all geographic centers – at the turn of the 20th and twenty-first century. This was accounted for by two conferences held in 1996: the hydrological conference in Łódź: *Springs and their role in the environment and their importance in water economy*, and the hydrobiological conference in Olsztyn: *The springs of Poland, status of research, monitoring, and protection.* Subsequent events dedicated to the subject were held in Łódź: *The springs of Poland – selected crenological problems* [50] and in Białystok: *The springs of Poland – the refugium of geo- and biodiversity*. Field studies over the years have resulted in published papers on springs in various regions, for instance, in the West Tatras [51, 52], the Pieniny [53–55], in several mountain chains of the Sudetes [44, 56–60], of the West and East Beskids [61–66], of central Poland [67, 68], of the Kraków-Wieluń Jura and the Miechów Jura [69–71], the Małopolska Upland [72–74], the Lublin Upland and Roztocze [75–81], the vicinity of Łódź [82], of Wielkopolska [83–87], the lake districts [88–92] (Fig. 7.7).

Springs and other ground water outflows have also been taken into account and examined, sometimes thoroughly and in many aspects, within the spatial nature protection units in Poland, including national parks [93–95], landscape parks [96, 97], as well as NATURA 2000 areas and nature reserves [89]. Springs are increasingly recognized in Poland as nature monuments which are protected by law [98–102] or as ecological sites, although the process continues to involve legal and subject-matter problems. In the recent handbook on Poland's hydrology (Hydrologia Polski – in Polish), one of the chapters is dedicated to the springs of Poland [103].

7.3 The Importance of Springs

In the natural environment of Poland, springs can be found at various topographic locations, and they have differently shaped niches and outflow points. They also differ in the lithology and tectonics of aquifers, dynamics, and forms of water outflows, as well as in the physico-chemical parameters of their water. The niches and waters of Polish springs are also inhabited by numerous and very interesting organisms. In consequence, the springs of Poland are highly diversified, crucial for maintaining bio- and geodiversity, and have many different functions and numerous environmental, cultural and social values, which overlap, thus strengthening their importance as places attracting the interest of many scientific and practical disciplines. Thus, for a long time now, the springs in Poland have been used for various purposes outline below.

Subjects of scientific research. Due to the historical, cultural, and economic importance of places where water of good quality can be found, the springs are an attractive subject of interest to specialists representing different scientific disciplines, including life sciences, engineering, and the arts, as well as many other fields. For many years now, the springs understood as natural outflows of ground waters have been the subject of geographical and geological studies. The detailed information about them is usually provided when describing environmental features of a given area, and conditions for use of mineral materials found there. It usually specifies water

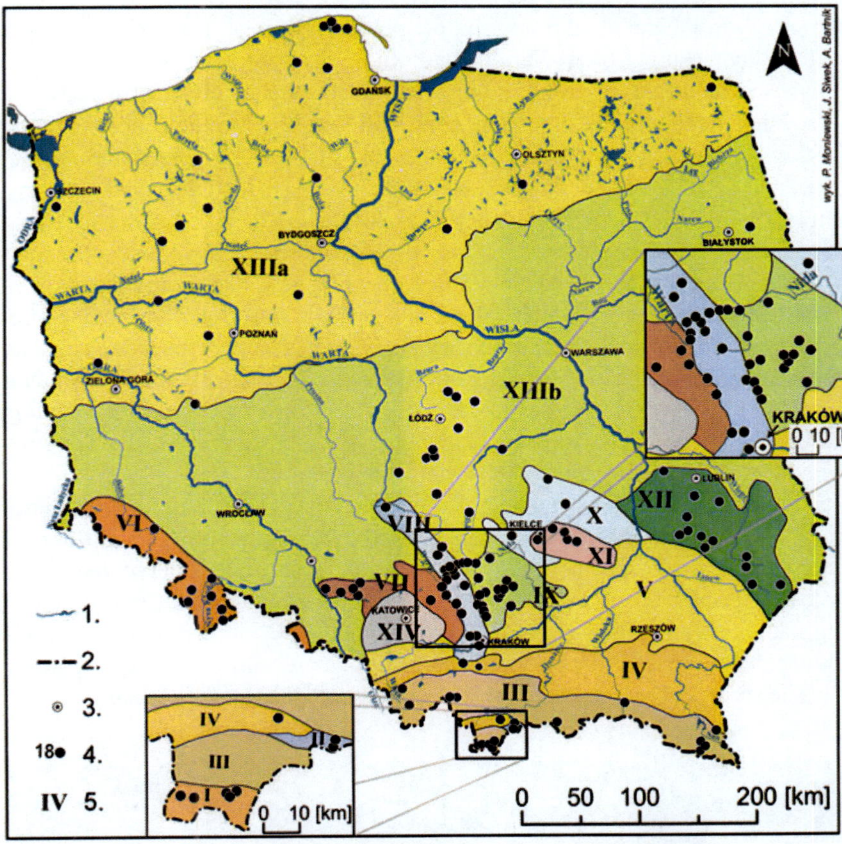

Fig. 7.7 Locations of studied springs with normal water against the background of crenological regions of Poland. Explanations: Key: 1 – main water courses; 2 – state border; 3 – voivodship capitals; 4 – important springs studied or observed; 5 – crenological regions (according to Dynowska, 1986; modified): I – Tatra Region (numerous high-discharge fissure and karst springs in limestone and dolomites, and low-discharge fissure springs in crystalline rocks); II – Pieniny Region (numerous low-discharge fissure springs in limestone, marl, and sandstone); III – Outer Carpathians Region (very numerous fissure springs in flysch and porous springs in the saprolite cover, of very low discharge); IV – Subcarpathia Region (less numerous fissure springs in flysch and porous springs in the saprolite cover, of very low discharge); V – Precarpathian Depression Region (very scarce, extremely low-discharge porous outflows in sands and in saprolite cover); VI – Sudete Region (low-discharge fissure springs in crystalline rocks and in limestone, dolomites, and sandstone, as well as porous springs in the saprolite cover); VII – Silesian-Kraków Upland (very high-discharge fissure springs, ascending in some cases, in limestone and dolomites, and porous springs in sands); VIII – Kraków-Częstochowa Upland Region (very high-discharge fissure and karst springs, ascending in some cases, in limestone); IX – Małopolska Upland Region (very high-discharge fissure and fissure and layer springs, ascending in some cases, mainly in marls); X – Świętokrzyskie Region (high-discharge fissure and karst springs, ascending in some cases, in limestone and sandstone); XI – Łysogóry Region (very low-discharge porous outflows in the saprolite cover); XII – Lublin Upland and Roztocze Region (very high-discharge fissure and fissure and layer springs, ascending in some cases, in marls and limestone); XIII – Polish Lowlands Region (scarce porous outflows in sands and gravels, of different discharge, in the young (XIIIa) and old (XIIIb) glacial areas); XIV – Upper Silesia Region (absence of springs due to human economic intervention)

discharge, hydrological conditions, and the physico-chemical parameters of water, as well as, increasingly often, various hydrobiological properties of spring waters and niches. Recently, the geomorphological aspect, seepage erosion, in particular, has also been noted with its significant contribution to the development of the relief and a fluvial system. It was documented that, in Poland, the groundwater outflows initiate the formation of new watercourses and they also extend the existing courses by stimulating headward erosion. They may also lead to the branching of a local river network. Therefore, the seepage processes result in the development (expansion) of spring niches leading to the development of a new, local hydrogeographical network and an increase in groundwater drainage [85]. The "Stoki" spring is a good example here. It is located under a steep slope, with 35 fissures grouped in ten small niches found on a stretch of 60 m. The lowest niche (Fig. 7.8) started to develop in the middle of the twentieth century. The outflow was initiated by erosion caused by a rapid surface runoff from the steep slope. In the successive years, the slope was undercut by headward erosion to a depth of five meters.

Stable thermal, humidity, and solar exposure conditions usually found in spring niches are decisive to the development of highly varied, though specific flora and fauna. As many as 447 species of vascular plants and 102 species of bryophytes were recorded in spring areas in Poland. Already a long time ago, plants associated with spring habitats were called crenophytes, with four groups distinguished amongst them, depending on their dependency on conditions prevailing in the spring niches [104]. The springs are also interesting objects of zoological studies because they are

Fig. 7.8 The "Stoki" spring in Wierzchowiska on the Lublin Upland (Region XII; cf. Fig. 7.7) – photo Ł. Chabudziński

a contact area for groundwater, surface water, terrestrial, and soil fauna. In Poland, a share of crenobionts in the spring niche fauna is usually small, when compared to crenophiles and crenoxenes. All the same, a significant individualism in fauna is observed at individual locations [105].

Components of a natural landscape. Springs, as the points of groundwater outflow, play an important role in shaping the structure of natural and transformed landscapes of Poland. When they are present, they usually have a significant share in their shaping. The groundwater outflow points and their niches can be found at different altitudes and different morphological locations. They also differ in the forms and dynamics of water outflow, as well as in water quality and temperature, and in the shape of spring niches (Fig. 7.9). Nevertheless, they are always prominent places in the landscape, with specific conditions for plant and animal life, and having a beneficial effect on humans. Therefore, they should be the enclaves of the natural landscape. In Poland, the simplest way to protect the springs is to have them turned into ecological sites. Nearly all the springs in Poland, together with their closest environment, should be covered by this type of protection.

Nowadays, even the morphology of spring niches in Poland is frequently exposed to intensive transformations. The springs change in their character due to the following

1. Ground subsidence (mining and urban areas),
2. Intense bottom erosion of river and stream beds resulting from hydroengineering works (meliorated areas,
3. The valleys of regulated rivers and streams),

Fig. 7.9 The spring of Bystrzyca Dusznicka river in the Orlické Mountains (Region VI; cf. Fig. 7.7) – photo P. Jokiel_

4. Development of local and regional depression zones in groundwater's (opencast areas, urban areas, zones affected by water intakes),
5. Ground leveling and backfilling of niche hollows, frequently with waste (suburban forests, areas of new districts, industrial and road investment areas), and
6. Many other factors related to the wide notion of urbanization and industrialization.

At the same time, concentrated outflows of leachate waters appear at the edges of many coal waste heaps in Silesia, forming specific "anthropogenic springs." In some of them, the discharge can even exceed 10 $dm^3 \times s^{-1}$. In most cases, they are descending outflows. In some cases, (e.g., the heap in Skrzyszów near Jastrzębie Zdrój), such "spring" can be of an ascending type, due to colmatation of some heap material layers [106]. Also, the construction of artificial water reservoirs supports the formation of anthropogenic springs. An example here is the dam and the reservoir on the Czarna Przemsza in Przeczyce, where the river water damming resulted in the water inflow to fissure systems located higher and what was called "vampires" by local people appeared below the dam. A horror of horrors, one of them was even found in the cellar of the dam management building [107].

Part of economic infrastructure. For many years, the springs represented a basic source of water supply for individual farms, particularly in the mountains or upland areas (Fig. 7.10). The access to spring water frequently determined the location of

Fig. 7.10 A laundry in the spring in Dzwola on Roztocze (Region XII; cf. Fig. 7.7) – photo S. Głowacki

farms and villages. However, in some cases, Polish springs became a source of local conflicts. In a small village of Rzepin (Świętokrzyskie Voivodeship), until 1962, water was drawn from a spring located 4 km away and carried by women, in buckets suspended from yokes. Suddenly, a new spring appeared in the center of the village at one of the private lots. However, this miracle of a kind would soon become a source of discord among the neighbors, deeply dividing the local community and resulting in a long-term local social conflict. The story became the basis of a widely recognized Polish documentary film Źródło (Spring) of 1962 (http://ninateka.pl/film/zrodlo-tadeusz-jaworski). Today, rural farms are turned into rural tourist boarding houses or small hotels, animal breeding for one's own needs is transformed into industrial production, and high-capacity water intakes for snow guns, indispensable on ski slopes, appear on mountain meadows and within the springs present there. In consequence, water consumptions raise drastically, and degradation follows which affects not only the groundwater outflow points but also their alimentation zones. In Poland in the past, springs were used to feed fish ponds, and they are still used for the purpose. Only the number and the scale of this type of intakes have increased. This concerns, in particular, large-scale trout ponds.

Spring waters parameters are usually characterized by good quality. Furthermore, their temperature is low in the summer. This highly facilitates the dissolving of oxygen, the high levels of which are necessary to maintain optimal conditions for trout breeding. The construction of fish ponds and, in some cases, also of reservoirs (for leisure, fire prevention, or other purposes) frequently causes an irreversible degradation of spring niches and valley spring lines, as such facilities are usually located on flood plains. Frequently, entire lines of ponds are created by damming upper sections of streams draining water from the springs. In consequence, the spring niches are flooded and the outflow points degraded [82]. One of the largest springs in the Polish lowlands – the spring in Rosanów, of an average discharge of 40 $dm^3 \times s^{-1}$ and with very clean water, drains water-rich sands of Grotnicki-Lućmierski sander (Fig. 7.11). The groundwater resources drained by that outflow, estimated for the discharge discharge mentioned above, amount to 2.27×10^6 m^3, and they are renewed in less than one year [108]. A water intake installed at this outflow would cover a water demand of over 20 thousand people. Many attempts have been made to use this spring for the production of table sparkling water. Fortunately, these plans were unsuccessful, mainly due to protests of naturalists from Łódź and of local foresters.

Tourist attractions and leisure sites. Springs are a popular attraction for tourists and "holiday-makers," mainly because their water is of good quality and they offer a mixture of values and qualities of inanimate and animate natural environment. Contact with natural water has a beneficial influence; it improves one's mood, soothes emotions, calms, and stimulates imagination. The aspects of local, historical, personal, or even emotional significance are also important. It is evident in locally used expressions and names of some outflow points: *spring of love, spring of hope, spring of cleansing, comforting water, smelly spring*. Before the World War II, a spring draining several small valleys formed in Jurassic rocks was flowing in the

Fig. 7.11 The spring of Rosanów in the Łódź Hills (Region XIIIb; cf. Fig. 7.7) – photo M. Stolarska

Prądnik valley, at the feet of the Kraków Gate. Until the end of 1960, water from it was drawn through an intake by a local table water factory, which moved the outflow point a little, for its convenience. When the factory and the intake were liquidated, a new outflow point appeared in its vicinity and was called "Spring of Love" (Fig. 7.12). A couple that drinks its water together will love each other forever. However, more inquisitive couples will notice the information on a plaque nearby, stating that it is a "secondary spring." So will this feeling truly last forever?

Balneological sites. Due to their slow flow and long contact with a rock, underground waters flowing from the spring differ highly in their chemical compositions. Natural elements dissolved in them can have healing properties or mineral water characteristics. For a long time now, these waters have been used in Polish balneotherapy (baths, inhalations, crenotherapy) at health resorts, but their uses go much further than that. The Źródło Warzelniane spring in Sola near Żywiec, due to its very high levels of dissolved salt (40 g $\times$ dm^{-3}), has been used as a brine since at least 1662 [109]. These "spring salt works" were closed only in 1933 by a so-called "Fiscal Police Flying Squad" because the state had a monopoly over salt mining at the time. The technology for obtaining carbon dioxide from carbonated waters saturated with it, which is unique in Poland, is used at two plants located in health resorts: in Duszniki-Zdrój and in Krynica-Zdrój. The plant in Duszniki-Zdrój was constructed as first, in 1924. Until the end of the 1930s, so-called dry ice was also manufactured there. Gas is separated from the old intakes of "Pieniawa Chopina" and "Jan Kazimierz," and from two more recent wells. From Pieniawa alone, 3500 kg CO_2 a day can be

Fig. 7.12 The spring of Love in the Prądnik valley, at the Kraków-Częstochowa Upland (Region VIII; cf. Fig. 7.7) – photo P. Jokiel

obtained as a very pure, liquefied gas, highly sought after by Polish manufacturers of table waters and …carbon dioxide extinguishers.

Religious cult sites. The springs are frequently associated with the symbolism of water perceived as a foundation for the development of life. In Poland, these springs which in the consciousness of generations have been perceived as locations where healing waters flow, are places of special worship (Fig. 7.13). Cases of miraculous healing are frequently attributed to the water from the springs. This concerns, in particular, outflow points located at religious cult sites. Such springs are usually perceived as miraculous, or even holy. The healing role of these sites and waters is, of course, disputable; however, chemically and bacteriologically pure spring water always has a beneficial influence on humans – it refreshes and purifies the body, brings relief in some disorders, and heals minor infections in some cases. In Łódź, or more precisely, in the Łagiewnicki Forest, thoroughly restored and well maintained

Fig. 7.13 The spring and a chapel in Krasnobród on the Lublin Upland (Region XII; cf. Fig. 7.7) – photo P. Jokiel

ancient wooden chapels dedicated to Saint Anthony and Saint Roch stand to this day. First of them was constructed probably in 1676, on a site where a church of the Franciscan Friars of Łagiewniki would be erected. The chapel existing on the site at the time was moved to its current location, called Pustelnia (Hermitage) in the vicinity of a small forest spring, and the second chapel, dedicated to Saints Roch and Sebastian, was constructed opposite it. Today only a dry hollow marks the location of the spring, though a well with allegedly healing water still exists in the older chapel. The neighboring Franciscan monastery keeps votive records with thousands of thanks for healing, and even the information that one of the pilgrims was resurrected from the dead.

Subjects of cultural research. The springs represent an important component of cultural heritage, usually associated with a history of a given site or its specific meaning for the local community (Fig. 7.14). In Poland, 106 springs (wells) are considered by local Catholic or other communities as holy or culturally important [110]. A good example here is the small spring of Grabarka, flowing out at the feet of the Holy Mountain of Grabarka in the district of Siemiatycze. An Orthodox sanctuary (a church, a monastery, and pilgrim houses) and over 10,000 Orthodox votive crosses are erected on that mountain. The small spring is enclosed in a well, and the whole structure is further enclosed in and adorned with a white chapel. According to a local legend, water from the spring saved the lives of hundreds of people who washed in it during a cholera outbreak in 1710. Since those days, thousands of Orthodox pilgrims have been visiting the place several times a year, trusting that also today the holy

Fig. 7.14 The enclosed spring in Szczebrzeszyn on Roztocze (Region XII; cf. Fig. 7.7) – photo Ł. Chabudziński

water from Grabarka will heal their diseases and improve their mood. Moreover, the citizens of the town Urzędów, located in the Voivodship of Lublin, know well that "*... there are no better cucumbers pickles in brine than those made with water from Saint Odile*". Some mischievous folk claim the real reason for this is the *E.coli* count in water from the Saint Odile spring much above the normal levels, as found by the Polish Sanitary and Epidemiological Service in 2014.

7.4 How to Save Springs

The current knowledge of crenological conditions in Poland can hardly be regarded as satisfactory, even though Poland has one of the lowest numbers of springs per km^2 in Europe and most of its springs are too small to be ranked as Meinzer's class III or IV. Any existing regional or national spring inventories are insufficient but, first of all, the understanding of the spring yield regime and its relationships with the nature and water abundance of drained aquifer horizons is far from being satisfactory. Regrettably, measurements of and studies on springs are not well coordinated, and their results are rarely published. The problem of establishing a harmonized database with information on Polish springs has been the subject of discussions for many years. However, a lot has been either proposed or declared only [111]. Only two databases, which are maintained by the Polish Geological Institute (PIG-PIB), comprise relatively harmonized information on the yields and essential parameters of common springs in Poland (data for 87 springs). Part of the information has been published in various issues of Rocznik Hydrogeologiczny (Hydrogeological Year-book) of the Polish Hydrogeological Survey (information on 37 springs). A great majority of the information relates to small springs located in mountainous regions (in the Carpathians or Sudetes). Other, even the largest, springs in the lowlands or uplands are only very rarely monitored. Even less information is available on the chemistry of spring water. This relates both to the hydrogeochemical status of a given spring but also to its changes within a single turnover of the spring-drained groundwater reservoir and anthropogenic changes of the hydrological and hydro-chemical regime of the springs. Those mineral water springs which are operated for business are an exception: their monitoring is ongoing, but the information is kept in a few different databases of the PIG-PIB. Under the circumstances, making an inventory and identifying the features of springs both in the regional and national scale in the hydrological, hydrochemical, and hydrobiological aspects seems simply indispensable.

Not much, either, is known about the disappearance of springs and their "travel" down the slopes. Opinions on the phenomenon are rather unambiguous: the drying-out of springs and lowering their altitude in Poland is a fact and is progressing: it affects both urban areas and a quasi-natural landscape. So far for the declared opinions. When it comes to validation of the opinions using detailed data, it appears that credible, reliable and comparable information on the subject remains to be scarce. Beside the spring drying-out process, the disastrous phenomenon of thoughtless devastation of springs has emerged recently. The natural form of a majority of karst springs and other large springs in Poland has been altered a lot before even though many of them are located in national parks, landscape parks or nature reserves, or are otherwise protected, for instance, as nature monuments or ecological sites. Also, far too often do we see that existing spring niches are abused as dumping sites and the spring water is polluted because of the absence of suitable protection zones. Also, the so-called "utilization" of spring niches and valley bottoms having water outflow lines by creating rearing ponds, swimming pools or garden ponds in them leads to the devastation of existing groundwater outflows and annihilation of their biotopes, many of which are unique.

Springs are an important environmental component having a high cognitive and scientific value; they are important for the landscape, culture, and economy. They are objects with unique features which ought to be protected by law as inanimate nature monuments or, at least, as ecological sites. Such protection ought to include both the groundwater outflow point (spring niche) and, as far as possible, even the entire underground catchment that supplies water to the spring. Therefore, both point-focused and area-focused protection of springs are required. Also, the historical, cultural, and social role of springs ought to be protected. It is important for every spring to keep or have its historical or local name (Fig. 7.15). Also, the ethnographic function of springs ought to be protected. Historical facts, folk stories, legends, proverbs or didactic parables of water outflows and their role in creating local non-material culture ought to be meticulously recorded, studied and popularized. Any

Fig. 7.15 The destroyed enclosing of Źródło Joanny (the Joan Spring) in Lądek Zdrój (Region XII; cf. Fig. 7.7) – photo A. Bartnik

available method that is permitted by law should be used, and any instance of illegal new projects to reconstruct or to encase springs should be restricted, whereas existing facilities of this type should be inspected, those more valuable should be provided with legal protection and some other ones – simply taken apart. There is a lot to do for various formal or informal interdisciplinary social groups and organizations, scientific circles and local societies.

The requirement to regularly study and monitor the springs is connected with the alarming depletion of the resources of the hydrosphere and with their qualitative degradation. On the other hand, it also results from the prevalence of a kind of social permission for uncontrollable, illegal transformations of selected components of the environment at the local or even regional scale. At the present stage of social awareness, it is important to provide education in the field of the aquatic environment and its links with other natural components. In children, such education should be provided at every stage of their intellectual development, starting at an early age. In adults, education must focus on fighting their prevailing habits, stereotypes in their ways of thinking and behaviors, which may result from insufficient knowledge of the environment and expose it to danger. Water has to be given back its "sacred" nature and respect as a superior good, which plays a decisive role in the health and condition of organisms and influences the physical and spiritual development of human beings.

Fig. 7.16 The protected spring "Źródło Hydrografów" (The Spring of Hydrographers) in Imbramowice on the Kraków-Częstochowa Upland (Region XII; cf. Fig. 7.7) – photo P. Jokiel

Groundwater outflow points should be protected on an ongoing basis and maintained in as a natural state as possible because these areas are highly exposed to human pressure. This should be a task of local communities. It is known that the best form of the protection of springs and specific habitats existing in the spring niches is to counteract changes in water conditions in their catchments. Establishing protection zones, similar to those which protect municipal groundwater intakes, should also be taken into account in the case of objects of special value (Fig. 7.16). Extremely important are nowadays also the system field studies intended to create an inventory of most, if not all, springs in Poland, assess their yield regime, physical and chemical properties of their water, explore the spring habitats and identify the causes of their accelerated disappearance and devastation. This requires an effort of multi-disciplinary academic centers and the development of a realistic, cooperative program to make an inventory of springs, assess their natural value, and create a concept for their protection. This shows that a lot depends on us, on our activity and effort. Rephrasing Paulo Coelho's words: *Let us be like a pulsating spring rather than a pond with stagnant water.*

7.5 Conclusions

The area of Poland is very diverse both in terms of the occurrence and efficiency of springs, as well as considering the level of knowledge about them. Springs constitute an important element of the natural environment, having scientific, landscape, cultural and economic significance. Despite many research papers, still little is known about the regime of spring output, the physical and chemical characteristics of water, and their links to groundwater resources. A separate, important problem concerns the phenomenon of the disappearance of springs, degradation of water quality and the transformation of spring niches, stemming from the increasing anthropogenic pressure. The springs, as points of groundwater outflow of unique features, should be legally protected. The idea of protecting springs, as significant parts of the natural and cultural environment, ought to be promoted for ecological awareness, especially in local communities. Springs, points of groundwater outflow to the surface should be preserved for future generations in the unaltered state.

7.6 Recommendations

The aim of the publication is to present the current knowledge about springs in Poland, not only in the historical and cognitive aspects, but also from practical and social view. The authors of this issue draw attention to the urgent need to save and preserve these objects and indicate the ways and directions of necessary actions. The results of their study should be interesting for environmental protection services, local authorities and social organizations dealing with the protection of environmental and cultural heritage.

References

1. Górnicki Z (2008) Woda w duchowych przeżyciach człowieka. Kraków, pp 1–159
2. Kopaliński W (2001) Słownik symboli. Warszawa, pp 1–521
3. Baścik M, Chełmicki W (2000) Święte źródła. In: Zemanek B (ed) Przyroda, nauka, kultura. Kraków, pp 223–233
4. Michalczyk Z (ed) (2001) Źródła Wyżyny Lubelskiej i Roztocza. Lublin. pp 1–298
5. Rajchel L (1997) Źródło świętego Jana z Kęt. Chrońmy Przyrodę Ojczystą 5:70–76
6. Wiśniewski W (1998) Lecznicze źródełka w miejscach sakralnych. Kraków, pp 1–184
7. Baścik M (2012) Wieczystym tryskam źródłem. Kraków 1–119
8. Baścik M, Chełmicki W (2004) Źródło jako obiekt badań interdyscyplinarnych. In: Izmaiłow B (ed) Przyroda-Człowiek-Bóg. Kraków, pp 149–170
9. Zejszner L (1844) O temperaturze źródeł tatrowych i pasm przyległych. Biblioteka Warszawska 2:257–281
10. Zejszner L (1860) O temperaturze źródeł w dolinie Ojcowa. Biblioteka Warszawska 1:453–461
11. Pusz JB (1844) O temperaturze źródeł w okolicy Warszawy. Biblioteka Warszawska 3:1–36
12. Pusz JB (1845) Jeszcze uwag kilka o temperaturze źródeł i powietrza w Krakowie. Biblioteka Warszawska 2:647–654
13. Pol W (1869) Obrazy z życia i natury (z jednym drzeworytem), 1. Towarzystwo Przyjaciół Oświaty, Kraków, pp 1–329
14. Szajnocha W (1891) Źródła mineralne Galicyi. Akademia Umiejętności, pp 1–111
15. Torosiewicz T (1849) Źródła mineralne w Królestwie Galicyi i na Bukowinie. Lwów, pp 1–208

16. Kowalski L (1920) Podreglowe źródła północnej strony Tatr. Sprawozdania Komisji Fizjograficznej PAU 53(54):12–27
17. Świerz L (1887) Ciepłota źródeł i stawów tatrzańskich mierzone w 1876 r. Pamiętniki Towarz Tatrzańskiego 2:107–108
18. Przesmyski P (1921) Źródła rzek Przemszy Białej i Szreniawy. Przegląd Górniczy i Hutniczy 13:452–453
19. Lewiński J (1933) Źródła Błękitne i Przepaść pod Tomaszowem Mazowieckim. Zabytki Przyrody Nieożywionej 2:69–72
20. Drwal J, Jankowski AT, Kaniecki A, Michalczyk Z (1977) Mapa hydrograficzna Polski, skala 1:50,000. In: Wytyczne techniczne K-3.4, Warszawa, pp 1–61
21. Jokiel P (1994) Wieloletnie i sezonowe zmiany wydajności wybranych źródeł Polski. Wiadomości IMGW 17(4):117–130
22. Jokiel P (1996) Wzorcowe krzywe wysychania i potencjały zasobności kilkunastu źródeł karpackich. Wiadomości IMGW 19(2):67–77
23. Pawlik-Dobrowolski J (1965) Uźródłowienie południowej Polski. Zesz Nauk Uniw Jagiellońskiego. Pr Geograficzne 12:7–42
24. Waksmundzki K (1971) Zmienność naturalnych wypływów wody podziemnej w górskich obszarach fliszowych. Zeszyty Naukowe UJ. Pr Geograficzne 29:87–94
25. Wit K, Ziemońska Z (1960) Hydrografia Tatr Zachodnich. Objaśnienia do Mapy Hydrograficznej Tatry Wysokie 1:50 000. Dokumentacja Geograficzna IG PAN 5:1–99
26. Ziemońska Z (1960) Związek temperatury źródeł morenowych z wysokością ich występowania na północnych stokach Tatr Zachodnich. Przegląd Geograficzny 32(3):321–328
27. Chełmicki W (1986) Źródła Niecki Nidziańskiej. Stud Ośrodka Dokumentacji Fizjograficznej 14:249–271
28. Chełmicki P (2006) Przestrzenna baza danych pomiarowych na przykładzie bazy danych źródeł dorzecza Dłubni. Przegląd Geol 54(8):689–692
29. Czarnecka H (1973) Rozmieszczenie źródeł na Wyżynie Małopolskiej. Pr i Stud IG UW 4:5–70
30. Czarnecka H (1975) Reżim źródeł na Wyżynie Małopolskiej. Pr IMGW 6:59–151
31. Drzał M, Dynowska I (1981) Cenne przyrodniczo źródła na Wyżynie Krakowsko-Wieluńskiej. Stud Ośrodka Dokumentacji Fizjograficznej 8:327–381
32. Drzał M, Dynowska I (1982) Cenne przyrodniczo źródła na Wyżynie Miechowskiej. Stud Ośrodka Dokumentacji Fizjograficznej 10:323–359
33. Duszyńska E (1969) Najsilniejsze źródło w dorzeczu Wieprza. Przegląd Geofizyczny 1:79–84
34. Dynowska I (1983) Źródła Wyżyny Krakowsko-Wieluńskiej i Miechowskiej. Stud Ośrodka Dokumentacji Fizjograficznej 11:1–243
35. Dynowska I (1986) Regionalne zróżnicowanie źródeł w Polsce. In: Folia geographica, geographica physica, vol 18. Kraków, pp 5–30
36. Gołąb J (1947) Krótka charakterystyka źródeł okolic Krzeszowic. Biul Państwowego Inst Geol 32:26–38
37. Janiec B (1972) Źródła południowej krawędzi Wyżyny Lubelskiej i ich związek z tektoniką. Biul Lubelskiego Towarz Naukowego, seria D, 14, Lublin, pp 63–67
38. Michalczyk Z (1979) Wydajność i mineralizacja wód wybranych źródeł Roztocza Zachodniego. Biul Lubelskiego Towarz Naukowego, seria Geogr Lublin 21, 2:57–61
39. Michalczyk Z (1982) Charakterystyka hydrogeologiczna dorzecza Łady. Biuletyn Instytutu Geologicznego. Z badań hydrogeologicznych w Polsce 339, 6:109–161
40. Michalczyk Z (1983) Źródła sanny w wierzchowiskach. Ann UMCS, sectio B 35(36):175–192
41. Michalczyk Z (1986) Warunki występowania i krążenia wód na obszarze Wyżyny Lubelskiej i Roztocza. Lublin, pp 1–195
42. Rederowa E (1965) Źródła bystrzycy lubelskiej. Ann UMCS sectio B. 18:229–244
43. Rederowa E (1971) Występowanie źródeł na Wyżynie Lubelskiej i w obszarach przyległych. Przegląd Geograficzny 43(3):355–360
44. Tomaszewski J (1977) Charakterystyka krenologiczna masywu krystalicznego na przykładzie Karkonoszy. Acta Univ Wratislaviensis 358:1–70

45. Jekatierynczuk-Rudczyk E (1999) Hydrochemia czwartorzędowych źródlisk dorzecza Supraśli. Białystok University, Białystok, Typescript
46. Kobędzina J (1949) Źródliska rzeki Łyny. Chrońmy Przyrodę Ojczystą 5(4–6):62–66
47. Kopczyński S (1982) Stosunki wodne Basenu Grudziądzkiego i jego otoczenia. Studia Societarum Torunensis, vol 9, 4, sectio C, PWN, Warszawa-Poznań-Toruń, pp 1–88
48. Maksymiuk Z (1977) Wody gruntowe i strefy ich wypływu na powierzchnię w regionie łódzkim. Zesz Nauk Uniwersytetu Łódzkiego 5:39–52
49. Nowakowski C (1976) Charakterystyka źródeł strefy czołowomorenowej Pojezierza Suwalskiego. Biul Geol Warszawa, 21:177–191
50. Jokiel P, Moniewski P, Ziułkiewicz M (eds) (2007) Źródła Polski. Wybrane problemy krenologiczne. Łódź, pp 1–352
51. Barczyk G (2008) Wywierzyska tatrzańskie. Krasowe systemy wywierzyskowe Tatr Polskich. Zakopane, pp 1–178
52. Małecka D (1997) Źródła Masywu Tatrzańskiego. Acta Univ Lodziensis Folia Geogr Phys 2:9–26
53. Humnicki W (2006) Reżim źródeł pienińskich w świetle obserwacji limnimetrycznych. Pieniny. Przyroda i Człowiek 9:29–39
54. Humnicki W (2012) Analiza recesji wydatku źródeł pienińskich w świetle obserwacji limnimetrycznych. Pieniny. Przyroda i Człowiek 12:13–31
55. Humnicki W (2015) Zmiany wydajności monitorowanych źródeł w Pienińskim Parku Narodowym w latach 2003–2014 (pieniński pas skałkowy). Przegląd Geol 63, 10, 1:750–755
56. Buczyński S, Staśko S, Modelska M, Olichwer Tarka R (2011) Charakterystyka krenologiczna masywów górskich Ziemi Kłodzkiej na podstawie bazy danych "Źródło". Biul Państwowego Inst Geol 445:17–26
57. Buczyński S, Rzonca B (2013) Wstępne wyniki badań hydrogeologicznych źródeł w Górach Orlickich (Rejon Zieleńca i Zakouti). Biul Państwowego Inst Geol 456:45–50
58. Staśko S (1984) Charakterystyka hydrogeologiczna wybranych źródeł Śląska Opolskiego. Materiały i Studia Opolskie, vol 52–53. Opole, pp 277–298
59. Tarka R (2011) Charakterystyka krenologiczna masywów górskich ziemi Kłodzkiej na podstawie bazy danych "Źródło". Biul Państwowego Inst Geol. 445, 12, 1:17–26
60. Wojtkowiak A (2000) Reżim źródeł krystalicznych Sudetów Zachodnich. Biul Państwowego Inst Geol 390:167–206
61. Jokiel P (1997) Podstawowe cechy reżimu wydajności wybranych źródeł karpackich. Acta Univ Lodziensis, Folia Geogr Phys 2:27–44
62. Łajczak A (2017) Typy źródeł i ich rozmieszczenie na Babiej Górze. Przegląd Geograficzny 89(1):111–132
63. Mocior E, Rzonca B, Siwek J, Plenzler J, Płaczkowska E, Dąbek N, Jaśkowiec B, Potoniec P, Roman S, Zdziebko D (2015) Determinants of the distribution of springs in the upper part of a flysch ridge in the Bieszczady mountains in southeastern Poland. Episodes 38(1):21–30
64. Moniewski P (2015) Seasonal variability of discharge from selected springs in central Europe. Episodes 38(3):189–196
65. Moniewski P (2016) Zmienność wydajności wybranych źródeł Polski i środkowej Europy w latach 2003–2014. Ann UMCS sectio B 71(1):29–44
66. Mostowik K, Górnik, Jaśkowiec B, Maciejczyk K, Murawska M, Płaczkowska E, Rzonca B, Siwek J (2016) High discharge springs in the outer flysch carpathians on the example of the high bieszczady mountains (Poland). Carpathian J Earth Environ Sci. Open Access Volume 11(2):395–404
67. Burchard J, Maksymiuk Z (1997) Źródła w dorzeczu Widawki. Acta Univ Lodziensis, Folia Geogr Phys 2:133–152
68. Maksymiuk Z, Mela S (1995) Źródła Polski środkowej. Acta Universitatis. Lodziensis, Folia Geogr Phys 20:109–119
69. Chełmicki W (2001) Źródła Wyżyny Krakowsko-Wieluńskiej i Miechowskiej. Kraków, pp 1–127

70. Siwek J (2004) Źródła w zlewniach Prądnika, Dłubni i Szreniawy. Naturalne i antropogeniczne uwarunkowania jakości wód. Kraków, pp 1–98
71. Siwek J, Baścik M (eds) (2013) Przyrodnicze i antropogeniczne przemiany źródeł Wyżyn Krakowsko-Wieluńskiej i Miechowskiej oraz ich rola w krajobrazie naturalnym i kulturowym. Kraków, pp 1–317
72. Baścik M, Pociask-Karteczka J (2002) Źródła Wyżyny Śląsko-Krakowskiej i Wyżyny Małopolakiej o znacznych walorach przyrodniczych. Propozycje ochrony. In: Ciupa T, Biernat S, Suligowski R (eds) Obieg wody w zmieniającym się środowisku. Prace i Studia Akademii Świętokrzyskiej, vol 7. Kielce. pp 23–41
73. Rogalińska I, Rogaliński J (1985) Przejawy występowania wód podziemnych na powierzchni i wskaźnik krenologiczny w Łysogórach. Rocz Świętokrzyski 12:47–61
74. Siwek J, Pociask-Karteczka J (2017) Springs in South-Central Poland—Changes and threats. Episodes 40(1):38–46
75. Janiec B, Michalczyk Z (1991) Wydajność i skład chemiczny wód największych źródeł Roztocza i Wyżyny Lubelskiej. In: Współczesne problemy hydrogeologii, Warszawa, pp 134–139
76. Michalczyk Z (ed) (1993) Źródła zachodniej części Wyżyny Lubelskiej. Lublin, pp 1–200
77. Michalczyk Z (ed) (1996a) Źródła Roztocza—monografia hydrograficzna. Lublin, pp 1–199
78. Michalczyk Z (ed) (1996b) Źródła województwa lubelskiego. Wydajność i parametry fizykochemiczne w 1996 roku. Lublin, pp 1–238
79. Michalczyk Z, Chabudziński Ł (2013) Zmiany wydajności źródeł w Wierzchowiskach w latach 1970–2012 (Roztocze Zachodnie). Biul Państwowego Inst Geol 456, 14, 2:405–411
80. Michalczyk Z, Chmiel S, Głowacki S, Zielińska B (2009) Changes of springs' yield of Lublin Upland and Roztocze Region in 1998–2008. J Water Land Dev 12:113–125
81. Michalczyk Z, Chmiel S, Głowacki S, Zielińska B (2015) Monitoringowe badania źródeł wyżyny lubelskiej i roztocza. Przegląd Geol 63, 10, 2:935–939
82. Moniewski P (2004) Źródła okolic Łodzi. Acta Geogr Lodziensia Łódź 57:1–140
83. Choiński A, Ptak M (2009) Inwentaryzacja obiektów krenologicznych dorzecza Odry. Pr Fizjograficzne nad Polską Zachodnią, seria A 60:95–106
84. Girjatowicz JP (2013) Atrakcyjne turystycznie źródła na niżu polskim. Turyzm 23(1):49–57
85. Mazurek M (2010) Hydrogeomorfologia obszarów źródliskowych (dorzecze Parsęty, Polska NW), Poznań, pp 1–304
86. Szczucińska AM (2009) Wypływy wód podziemnych w Rynnie Gryżyńsko-Grabińskiej. Poznań. pp 1–117
87. Szczucińska A (2017) Hydrochemical background based on spring waters (Lubuskie Lakeland, Western Poland). Geochem Int 55:1–7
88. Cieśliński R, Leśniowski P (2013) Wypływy wód podziemnych w województwie pomorskim. Gospodarka Wodna 8:306–311
89. Fac-Beneda J (2008) Naturalne wypływy wód podziemnych na obszarach chronionych w północnej Polsce. In: Partyka J, Pociask-Karteczka J (eds) Wody na obszarach chronionych. Kraków, pp 133–141
90. Górniak A, Pietryczuk A (2015) Źródła okolic Lęborka (północna Polska). Gospodarka Wodna 5:135–139
91. Jaworska-Szulc B, Pruszkowska-Caceres M, Przewłócka M (2015) Zmiany wydajności wypływów wód podziemnych młodoglacjalnego obszaru morenowego na Pojezierzu Kaszubskim. Przegląd Geol 63, 10, 1:774–779
92. Jekatierynczuk-Rudczyk E, Zieliński P, Puczko K (2017) Czy ochrona źródlisk w Puszczy Knyszyńskiej jest skuteczna i zadowalająca? Chrońmy Przyrodę Ojczystą 73(2):135–147
93. Bogdanowicz R, Jokiel P, Pociask-Karteczka J (eds) (2012) Wody w parkach narodowych Polski. Kraków, pp 1–402
94. Grabowski T, Harasimiuk M, Kaszewski BM, Kravchuk Y, Lorens B, Michalczyk Z, Shabliy O (eds) 2015. Roztocze, przyroda i człowiek. Zwierzyniec, pp 1–527
95. Łajczak A (2016) Wody Babiej Góry. Monografia babiogórskie. Maków podchalański, pp 1–423

96. Pius B, Kruszewska P, Glazik R (2016) Występowanie wypływów wód podziemnych na obszarze Welskiego Parku Krajobrazowego. Stud Limnol i Telmatologica 10(1):2–13
97. Pukacz A, Pełechaty M (eds) 2015. Łagowsko-Sulęciński Park Krajobrazowy. Różnorodność ekologiczna i gatunkowa. Gorzów Wielkopolski, pp 1–256
98. Baścik M (2010) Podstawy prawne i problemy z ochroną źródeł w Polsce. Prace i Materiały Muzeum im. W. Szafera. 20, Prądnik, 115–134
99. Baścik M, Urban J, (2007) Konserwatorska ochrona źródeł w Polsce. In: Jokiel P, Moniewski P, Ziułkiewicz M (eds) Źródła Polski. Wybrane problemy krenologiczne, Łódź, pp 320–329
100. Chełmicki W, Jokiel P, Michalczyk Z, Moniewski P (2011) Distribution, discharge and regional characteristics of springs in Poland. Episodes 34(4):244–256
101. Janiec B (1992) Ochrona źródeł na Roztoczu. Ann UMCS, sectio B 47:198–213
102. Michalczyk Z, Turczyński M (1999) Charakterystyka hydrologiczna źródeł w wąwolnicy na wyżynie lubelskiej. Ann UMCS sectio B 54:217–234
103. Moniewski P, Siwek J (2017) Źródła i inne wypływy wód podziemnych. In: Jokiel P, Marszelewski W, Pociask-Karteczka J (eds) Hydrologia Polski. Warszawa, pp 89–96
104. Kucharski L (2007) Flora źródlisk—skład i gatunki wskaźnikowe. In: Jokiel P, Moniewski P, Ziułkiewicz M (eds) Źródła Polski. Wybrane problemy krenologiczne. Łódź. pp 62–68
105. Czachorowski S (2007) Fauna źródeł—stan poznania i perspektywy badań. In: Jokiel P, Moniewski P, Ziułkiewicz M (eds) Źródła Polski. Wybrane problemy krenologiczne. Łódź, pp 55–61
106. Molenda T (2007) Antropogeniczne wypływy wód podziemnych—charakterystyka hydrograficzna. In: Jokiel P, Moniewski P, Ziułkiewicz M (eds), Źródła Polski. Wybrane problemy krenologiczne. Łódź, pp 69–77
107. Pulina M (1999) Kras. Formy i procesy. Katowice, pp 1–375
108. Jokiel P (2002) Woda na zapleczu wielkiego miasta (Możliwości wykorzystania i problemy ochrony zasobów i obiektów wodnych w małej zlewni strefy podmiejskiej Łodzi). Warszawa, pp 1–148
109. Rajchel L (2007) Fascynujące źródła karpackie. In: Jokiel P, Moniewski P, Ziułkiewicz M (eds) Źródła Polski. Wybrane problemy krenologiczne. Łódź, pp 285–290
110. Pytka P (2006) "Święte źródła" w polskim krajobrazie kulturowym. In: Wołoszyn W (ed). Krajobraz Kulturowy—cechy, walory, ochrona. Problemy Ekologii Krajobrazu, vol 18. Lublin, pp 451–456
111. Bartnik A, Moniewski P (2018) Zawartość informacyjna baz danych PIG-PIB w aspekcie badań wydajności źródeł Polski. Przegląd Geol 66(5):284–293

Chapter 8
Groundwater Resources of Poland

Arkadiusz Krawiec and Andrzej Sadurski

Abstract More than 70 percent of water supply in Poland is cover at present from the groundwater resources. Total exploitation of the groundwater intakes reaching 1,5 km^3 a year in the country and still growing up due to good and stable quality of water, as well stable amount of exploitation resources and also low cost of purification technology in contradiction to surface water from the rivers. The most abundant in water are porous Quaternary aquifers in the northern and central parts of Poland, that means along the lake lands and Polish - German lowland. In the southern part, along the Sudetes and Carpathians and Holly Cross Mts., and their forelands dominate the fissure aquifers, with low groundwater capacity. Droughts and scarcity of water resources overlapping this belt comprises central lowlands from Belarus to Brandenburg in Germany. This situation occur during the years of low precipitations (e.g. two month without rainfall) in summer. In the beginning of XXI century great influence of water management in Poland had implementation of Water Framework Directive to the polish low system. New approach to the groundwater resources was defined and ecological aspects of water demand in the management plans for river basins were established. Available groundwater resources of the river basins means that significant volume of water for flora and fauna has to remain in the area. Total groundwater resources of the country is 36.4 million m3 a day, including almost 15 million m^3/day of estimated prospective resources. Mineral and thermal waters have been used in Poland for therapeutic purposes since the early Middle Ages. Dominate mineral waters of the Polish Lowlands is Cl-Na type, whereas in waters of the HCO_3-Na-Mg or HCO_3-Mg-Na types also over saturated with CO_2 occur in the Sudetes and Carpathians. Unique mineralized, of oversaturated with CO2, water name Zuber, occur in the Carpathian Province. The highest concentrations of radon in healing waters have been identified in the Sudetes Province (e.g. Świeradów Spa and Lądek Spa). Different types of mineral and thermal waters are exploited in more than 40 of Spas for therapeutic purposes. Thermal waters are exploited for recreation and heating purposes in twenty places and health resorts.

A. Krawiec (✉) · A. Sadurski
Department of Geology and Hydrogeology, Faculty of Earth Sciences and Spatial Management, Nicolaus Copernicus University, Lwowska 1, 87-100 Toruń, Poland
e-mail: arkadiusz.krawiec@umk.pl

M. Zeleňáková et al. (eds.), *Management of Water Resources in Poland*, Springer Water,
https://doi.org/10.1007/978-3-030-61965-7_8

Keywords Groundwater resources of poland · Disposable groundwater resources · Available resources · Therapeutic · Mineral and thermal waters

8.1 Introduction

The scientific term "groundwater resources" was properly introduced into hydrogeology over a century ago due to the analogy of solid mineral resources applied in economic geology. It was originally used for the sake of spatial planning and investments in groundwater intake constructions. The accession of Poland to European Union required the implementation of the Water Framework Directive (2000/60 WE), which took account of hydro-environmental needs – inviolable river flows, water levels of great ecological significance, which are legally protected areas. Since 2004 groundwater resources available for development have been designated and water management plans, whose aim is to preserve good water status or take measures to improve poor groundwater quality in water economic regions, have been drawn up. The assumption accepted in new water policies expressed in the Water Framework Directive concerns a full realisation of catchment-based approach to water management according to the requirements and standards of European Union as well as the adopted principle of Integrated Water Resources Management.

About 99.7% of the territory of Poland is located within the Baltic Sea catchment area (Fig. 8.1). Only 0.2% of the country's area belongs to the Black Sea catchment area whereas approximately 0.1% to that of the North Sea. This is why the country's water balance is mainly based on precipitation, evaporation, and flows of the rivers situated within the area of Poland. Only about 5% of water volumes participating in the water recharge in Poland comes from foreign tributaries. Over 70% of groundwater resources is used for human water consumption.

Water supply for the inhabitants of small towns and villages comes primarily from groundwater intakes, whereas municipal water intakes of big cities and industrial plants take up water also from surface bodies such as rivers. The use of water for agricultural irrigations is rather insubstantial, and even during dry years did not exceed 1 km^3. Open-cast and underground coal mines have a higher share of the exploitation of water resources, and they supply about 1.2 km^3 of water through their drainage systems, which frequently greatly impairs the quality of surface waters. The greatest water abstraction is recorded in power plants, particularly those generating coal-based energy, and exceeds 8 km^3 a year.

During the process of planning water management policy of the country, including water licenses for groundwater intakes as well as water status assessments according to the Water Framework Directive, which has been implemented in the Polish Water Law, groundwater resources available for development play a significant part. Such groundwater resources are determined taking into account natural river flows and the preservation of good groundwater status in dependent ecosystems, which are legally protected.

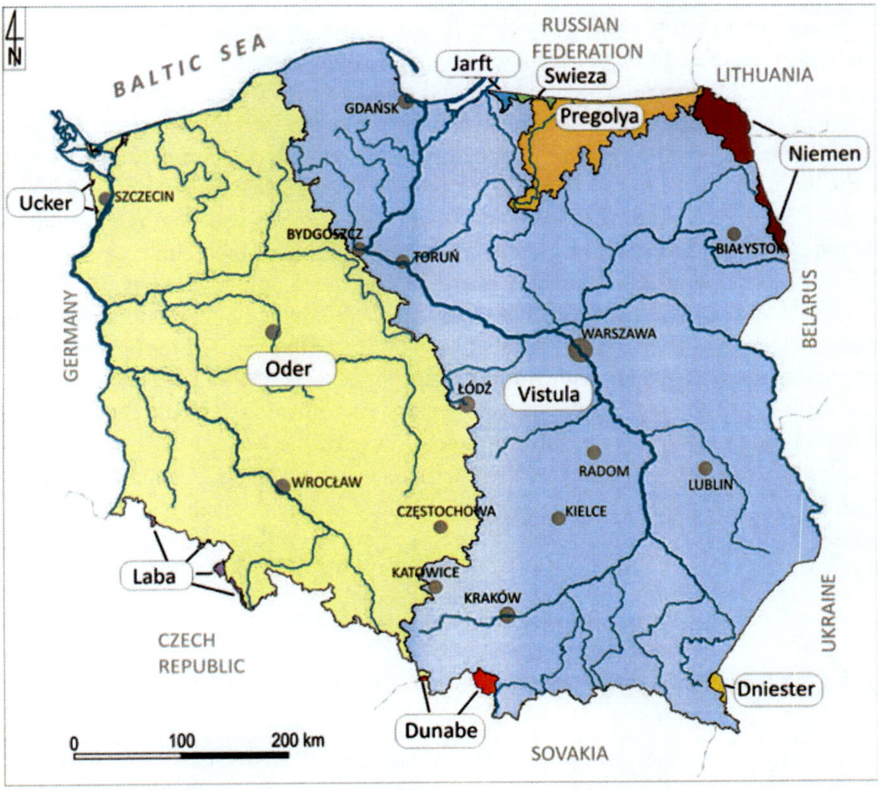

Fig. 8.1 River basins in the territory of Poland

8.2 The Topography of the Land

In the area of Poland, a band-shaped system of parallelly running geomorphological forms can be distinguished. It has a considerable impact on the fresh groundwater conditions in the whole country [1–5].

The range of mountains and foothills of the Sudetes stretches along the southern border of Poland in the west, whereas the mountains and foothills belonging to the range of the Carpathians are located in the east. The genesis and geological evolution of these mountains vary. The Sudetes were formed during the Variscan orogeny and belong to the Palaeozoic tectonic platform of Western Europe. The Sudetes largely consist of volcanic and metamorphic rocks, while the Carpathians are relatively young mountains, belonging to the Alpines, which is the reason for the prevailing occurrence of flysch in the massive layers of the sheets from the Jurassic period to the Neogene. Only in the highest parts of the Tatras has the occurrence of hard crystalline rocks been recorded.

From the centrally located Silesian Upland in the Carpathian Foreland, a range of highlands stretches through the Świętokrzyskie Mountains towards the east. The area is predominantly corrugated and hilly, and several rivers which later become tributaries of the Vistula and the Oder, have their sources in the highland area. Through the central part of the country, from Belarus in the east to Germany in the west, stretches a gently rolling flat landscape of the elevation levels between 50 and 200 m amsl. Classified as part of the Polish Lowlands or Polish and German Lowlands. In the north, the Baltic Coast is separated from the rest of the country by a belt of Lakelands characterised by diverse landscapes of the young glacial type.

The central and northern parts of the country are covered by the sediments of the youngest ice age – the Vistulian Glaciation. The contemporary hydrographic network as well as the Baltic Sea result from the morphogenesis of the Vistulian Glaciation in the period of the past 45,000–12,000 years. The inland area of the Baltic Seais a relatively young reservoir in the geological sense of the word. Its history does not exceed 12,000 years. It is a brackish (saline) water reservoir. As it is intensively recharged by rivers, the overall average tributary of which reaches 14,500 m^3/s, it would become a freshwater lake within 50 years if the inflows of saline water from the North Sea through the Danish straits were cut off.

8.2.1 *Outline of Hydrogeological Conditions in Poland*

In the range of highlands, foothills and mountains, groundwater occurs mainly in fractured aquifers as well as in pore and fractured aquifers and karstic and fractured aquifers. The systems of cracks, tectonic dislocations, and discontinuities form complicated and heterogeneous aquifers. The deepest aquifers, which have the broadest range and carry substantial amounts of groundwater, are connected with dislocation zones and faults. Only in river valleys, groundwaters are connected with sandy and gravelly alluvial deposits. In the greatest morphological depressions – in drainage areas – groundwaters are carried beyond the range of elevations and mountains. They simultaneously function as buffers between surface waters in the beds of streams, creeks and rivers, and rock aquifers. In the periods of high water levels and high precipitation, the groundwater table rises by a few metres, whereas during the droughts water stored in alluvial soil are drained by flows. The main groundwater reservoirs – MGR [2], in the foothill and mountain zones are connected mainly with the valleys of major rivers or buried valleys of former rivers. The thickness of alluvial soils in these reservoirs reaches even 100 m and are characterised by high hydraulic conductivity (transmissivity).

The zone of Mid-Polish uplands, which also includes the Holy Cross Mountains is characterised by a low thickness of the Quaternary layers of the glacial, fluvioglacial, fluvial and lake origin. Inland abatements, they frequently form the first aquifer level below ground. At greater depths, there occur fractured aquifers as well as pore and fractured aquifers in the Mesozoic layers, particularly in the Cretaceous and Jurassic aquifer levels. In the areas of the highest ground ordinates, the Mesozoic strata appear directly on the surface, forming monadnocks and aquifer supply areas. In the Świętokrzyskie Mountains, Sudetes and in Upper Silesia there occur locally on the surface aquifer levels from older geological periods, including the Palaeozoic aquifers. Large parts of rock formations are made of limestone and dolomites, which contain tunnels, channels, passages and karst craters, which in turn form fractured and karst water-bearing systems. By virtue of their dominant position in the land morphology – being surrounded by a belt of foreland depressions (for example, the Carpathian Foredeep) in the south and the zone of Mid-Polish lowlands in the north, vast uplands constitute aquifer supply areas. In the past twenty years, buried valleys whose depths reach approximately 140 m have been detected in the upland zone by means of research boreholes and geophysical surveys 140 m. They eroded in solid rocks of much older surface and were filled with Quaternary sediments [3]. These valleys can contain substantial reserves of groundwater in the case of being filed in with sandy and gravelly sediments.

The Mid-Polish Lowlands, which belong to the region of Polish and German Lowlands are distinguished by their extensive groundwater reservoirs in a deep basin or trough structures, in which aquifer levels belong to the Quaternary, Neogene, Paleogene, Cretaceous and Jurassic sediments. In the Mesozoic layers, saline and salt waters occur, but they are not suitable for the supply of potable water for local communities. In the aforementioned layers, pore ranges have paramount importance for groundwater resources, and only locally in the Upper Cretaceous and Jurassic periods can one find fractured aquifers as well as karstic and fractured ones. In the area of Mid-Polish Lowlands, a range of Major Underground Water Reservoirs have been identified, covering the total area of several thousand km^2 [2, 5]. These reservoirs provide the greatest volumes of groundwater extracted at large municipal intakes. Substantial reserves of groundwater can be located in wide river valleys whose sandy alluvial deposits on average do not exceed 20 m of thickness, but in the event of their interference with buried valleys, the thickness of sandy water-bearing levels can reach even 80 m.

Considerable groundwater resources in regional reservoirs also occur in the Neogene and Paleogene aquifers. In the areas where Quaternary aquifers do not occur, these resources constitute the major usable water-bearing horizon. The reservoirs are perfectly insulated and are impervious to surface pollution.

The upper roof of the Mesozoic aquifers in the Mid-Polish Lowland area occurs on average at a depth of 100 m. It is composed of the Upper Cretaceous and Jurassic layers of marls, limestones, and siliceous limestones. Only in the area of the Mid-Polish anticlinorium (from Cuiavia to the Holy Cross Mountains) is the upper roof of the Mesozoic covered with a series of the Kenozoic sediments of a rather low thickness (30–50 m). In this range layers of Jurassic limestones and sandstones prevail. Occasionally brackish and salt waters occur in the Mesozoic water-bearing levels.

The major utility significance in the area of central and northern Poland have the Quaternary aquifers, whereas the water-bearing levels of the Neogene and Paleogene are of secondary importance [6]. Locally potable water is extracted from the Upper Cretaceous aquifers, and in the area of the Mid-Polish anticlinorium, the Jurassic aquifers are also exploited. In the Quaternary range, aquifers occur in the forms of intermoraine water-bearing levels, lenses, buried valleys, or outwash ranges (see Fig. 8.2). On average the thickness of the Quaternary levels locally amounts to 50–100 m, but can reach even up to 300 m in the buried river valleys. The majority of Neogene and Paleogene layers have the thickness of 30–60 m, but they occur sporadically in the northern part of Poland due to erosion. The upper roof of the Mesozoic is located in the Lakeland zone at the elevation of 70–110 m amsl.

The groundwater in Quaternary aquifers is predominantly of the HCO_3–Ca type, whereas water in the Neogene, Paleogene, and Cretaceous aquifers is mostly of the HCO_3–Na type [7]. Locally, the Mesozoic aquifer contains abnormally high concentrations of fluorine compounds released to water from fluorine-bearing phosphate nodules occurring in the Paleogene and in the roof of the Upper Cretaceous [8, 9].

The coastal zone constitutes the extension of the Lakelands zone in terms of its geological structure. However, it is distinguished by the boundary of two different hydrogeochemical environments formed by the area separating fresh groundwater and brackish water of the marine origin, which has been diffused due to the spreading of sediments. The saltwater wedge is located at the bottom of the aquifer in the range of coastal lowlands, but in the estuary parts of rivers, it expands inland causing the degradation of groundwater resources. This process, commonly called saltwater ingression, can be attributed to the excessive extraction of potable water at municipal intakes in the coastal zone. An additional hazard acutely affecting the coastal regions is the ascension of saline waters (brines of Na–Cl type) from the deeper Mesozoic basement. The salinization of groundwater can occur here as a result of both ascensions of brines from the Mesozoic strata and local ingression of seawater along the Baltic Coast [9–12].

In the western part of the Polish Baltic coast, the Quaternary aquifers have reduced thickness which does not exceed 80 m, except buried river valleys. On the eastern coast, the upper roof of the Mesozoic occurs on average at the elevation of -100 m amsl. The aquifer of the Quaternary and Upper Cretaceous sediments are those that abound in groundwater. The Neogene and Paleogene layers occur locally and thus are of little utility significance. In the Quaternary sediments porous groundwater aquifers prevail, and only in the upper roof of the Cretaceous and Jurassic fracture aquifers can be found in hydraulic connection with sandy sediments situated above.

8.2.2 *Thermal and Therapeutic (Healing) Groundwater in Poland*

Therapeutic water is a specific type of groundwater used in hydrotherapy of people. In Poland healing waters are precisely defined by the Regulation of Minister of Health of 2006 (Journal of Law, No. 80, Item 565) as well as by the Geological and Mining Act of 2011 (Journal of Law, No. 163, Item 981, as amended). In light

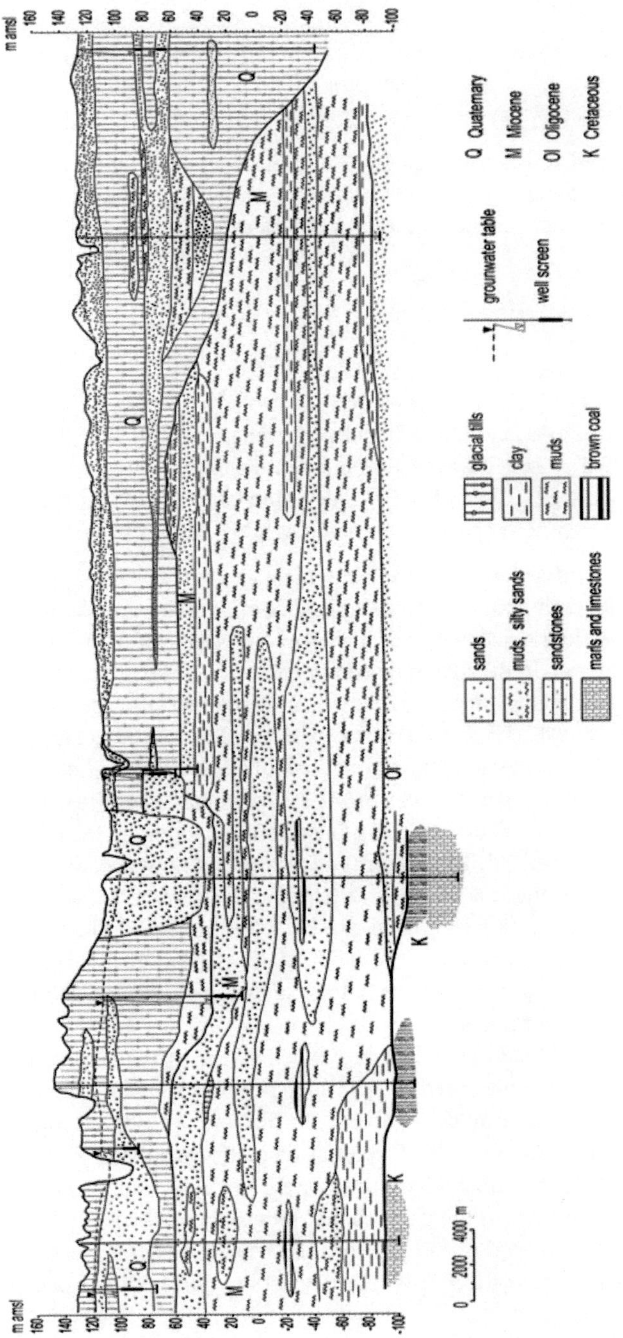

Fig. 8.2 Schematic hydrogeological cross-section of the Polish Lowlands

of these legal regulations, groundwater is considered to be therapeutic when it is uncontaminated chemically or microbiologically, and it is characterised by natural variability of physical and chemical properties. Additionally, healing mineral water must meet at least one of the following criteria:

- total dissolved solids – at least 1000 mg/dm^3
- the contents of one or potentially more of the so-called specific components listed below in the minimum concentration of at least:
  - 10.0 mg Fe(II)/dm^3 (ferruginous water);
  - 2.0 mg F^-/dm^3 (fluoride water);
  - 1.0 mg I^-/dm^3 (iodide water);
  - 1.0 mg S(II)/dm^3 (sulphide water);
  - 70.0 mg H_2SiO_3/dm^3 (silica water);
  - 74 Bq Rn/dm^3 (radon or radioactive water);
  - 250–999 mg free CO_2/dm^3 (carbonated water)
  - >1000 mg free CO_2/dm^3 (CO_2-rich water, carbonated water); and
  - the outlet temperature of water reach at least 20 °C (thermal water).

The earliest information concerning the use of mineral waters in Poland is connected with salt extraction and production. This information comes from Ciechocinek, Kołobrzeg (seventh century AC), as well as Rabka-Zdroj and Szczawa (thirteenth century). In the first and second centuries AD the first wooden images of carbonated water intakes in Szczawno-Zdrój were made and the first written record included in the so-called Book of Henryków from 1221 mentions these healing waters [13]. A significant event was carrying out the first research of healing waters in Lądek-Zdrój by Conrad of Berg as early as in 1498. His research confirmed the presence of "sulphur, alum, salt, and copper." This fact allows us to regard his research as one of the earliest in Europe [14]. The majority of currently known healing water resources were discovered in the nineteenth and twentieth centuries. Detailed specifications concerning healing waters in Poland can be found in the studies of Paczyński and Płochniewski [15], Ciężkowski et al. [13], Rajchel [16], Krawiec and Ciężkowski [17] as well as on the websites of the Polish Geological Institute – National Research Institute (https://mineralne.pgi.gov.pl/) [18].

Poland can be divided into four provinces as far as the occurrence of mineral and healing water is concerned [15]. These provinces include (see Fig. 8.3): the Precambrian platform province, Palaeozoic platform province, Sudetes province, and the Carpathian province. The Precambrian platform and the Palaeozoic platform occupy the area of the Polish Lowlands, where healing waters have been identified mainly in the sediments of the Cretaceous, Jurassic, and Triassic. The oldest health resorts, which are located in the area of the Polish Lowlands, were created on the basis of the occurrence of Cl-Na type brines from which salt has been produced for over a millennium.

Province A – the Precambrian platform occupies the north-eastern area of Poland. This province is the least abundant in mineral and thermal waters. Healing waters are extracted mainly in Ustka (the Cl–Na type I; mineralisation 34.3 g/dm^3),

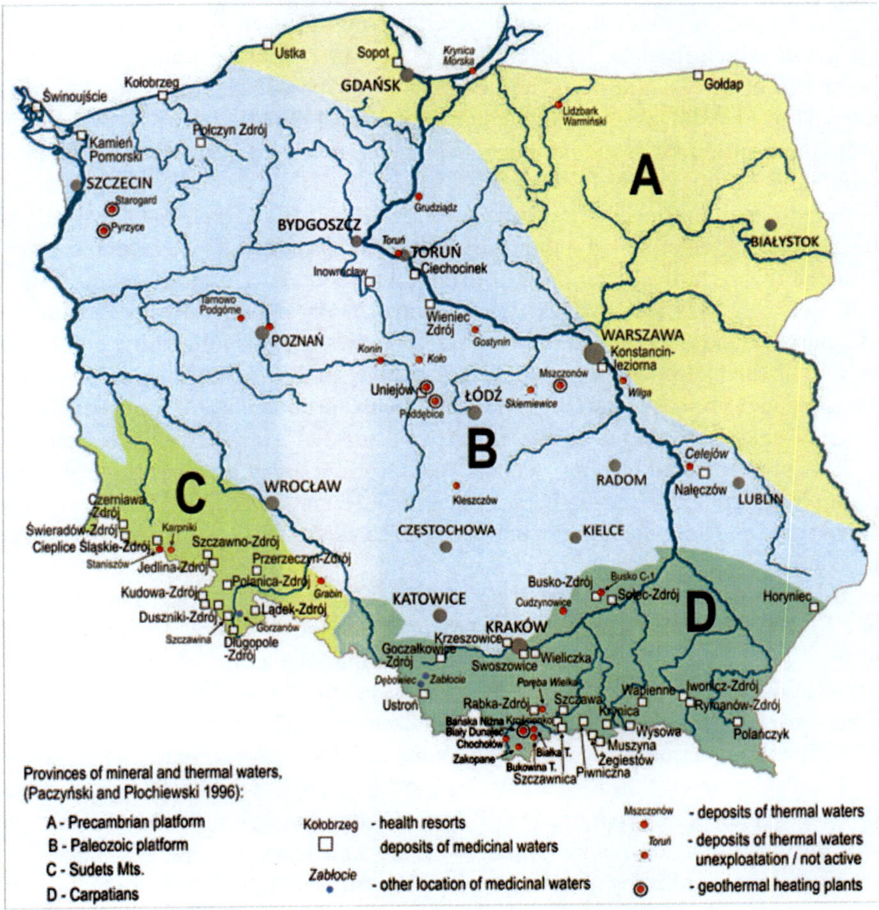

Fig. 8.3 Thermal and therapeutic water in Poland

Sopot (the Cl–Na type, I; mineralisation 42 g/dm^3) and Gołdap (the Cl–Na, F, T types; mineralisation 6 g/dm^3). Mineral waters of the Cl–Na type have also been identified in the lias sediments in the vicinity of Gdańsk, Krynica Morska and Lidzbark Warmiński (the temperature of approximately 18–20 °C).

Province B – the Paleozoic platform occupies the north-western and central parts of Poland. Salt-bearing formations of Zechstein had a profound impact on the creation of mineral waters. Here occur almost exclusively brines of the Cl–Na type. Healing waters occurring in the vicinity of Kamień Pomorski, Świnoujście, Kołobrzeg, Połczyn-Zdrój, Ciechocinek, Inowrocław, Wieniec-Zdrój and Konstancin are exploited for medicinal purposes.

In Świnoujście healing waters of the Cl–Na, I, Fe, the mineralisation of which reaches 43–45 g/dm^3, are extracted from the Early Cretaceous aquifers. The health resort of Kamień Pomorski is supplied with water of the Na–Cl type, the mineralisation of which reaches the level of 34 g/dm^3. In Kołobrzeg healing water from boreholes of the Cl–Na, I type and Cl–Na, I, Fe type with mineralisation ranging from 51 to 61 g/dm^3 is used for medicinal purposes [12, 19].

In the central part of the province, healing water is extracted in one of the most prominent health resorts in Poland, Ciechocinek, as well as in Inowrocław and Wieniec-Zdrój. In Ciechocinek thermal brines (28–32 °C) of the Na–Cl, I type with the mineralisation of 43–54 g/dm^3 are used in therapies [20]. This health resort was founded owing to the occurrence of brines, which were used in the production of salt. Between 1824 and 1828 brine graduation towers and saltworks were built for this purpose and they have been in operation ever since, manufacturing edible salt as well as the Ciechocinek hydroxide and sludge. In the vicinity of Grudziądz and Konstancin the brines of the Cl–Na, I, Fe type with the mineralisation between 75 and 79 g/dm^3 and the temperature reaching 40.5 °C and 29 °C respectively are exploited.

This province also belongs to one of the most important regions whose thermal waters have properties and temperatures justifying their potential use for heating. The Cretaceous and Jurassic aquifers are recognized as the most prospective in energy production.

In Pyrzyce and Stargard, the brine of the Cl–Na, I type with the mineralisation of 120–140 g/dm^3 and temperature between 61 and 84 °C is used. In Poznań water of the Cl–Na type with the mineralisation reaching 20.8 g/dm^3 and temperature of 41 °C at the outflow is used to supply a leisure centre "Malta Thermas" whereas in Tarnowo Podgórne the brine with the mineralisation of 80 g/dm^3 and temperature of 44 °C is extracted from the depth of approximately 1.2 km for recreational purposes.

The reservoir of Late Cretaceous in the Mogilno and Łódź Trough also has favourable temperature parameters. In Uniejów thermal waters of the Cl–Na type, with the mineralization of approximately 7 g/dm^3 and the temperature reaching 67 °C are used in balneotherapy, recreation as well as for heating purposes. In the nearby village of Poddębice, water of such high temperature as 71 °C, and low mineralization amounting to only 0.4 g/dm^3 was extracted from the depth of 2.1 km. In Mszczonów water from the Late Cretaceous formations reaching the temperature of 40.5 °C, and mineralisation of 0.5 g/dm^3 is exploited. Thermal brines have also been identified in the area of Konin, where the water of the outflow temperature reaching 92 °C and mineralisation at the level of 150 g/dm^3 has been obtained. Similarly, in Toruń, the mineralization of the thermal brine reaches 120 g/dm^3, and its temperature at the outflow can amount to approximately 60 °C. Thermal water has also been extracted in Kleszczów (Cl–Na type of the temperature reaching 52 °C), and thermal wells have been identified in the vicinity of Wilga and Celejów (temp. 30 °C).

In the area of the Sudetes Province, there occur three major types of healing waters: carbonated water, thermal water, and radon water. The most common type of water in this province is carbonated water – mineralized bicarbonate water containing carbon dioxide. Carbonated waters extracted in the region of Czerniawa-Zdrój and Świeradów-Zdrój are springs and wells located at the depth of 360 m. Their mineralization reaches 3 g/dm^3, and the content of carbon dioxide exceeds 2000 mg/dm^3.

These waters are of the HCO_3–Ca–Mg, (Fe), (F), (Rn) types. The highest concentrations of radon in healing waters have been identified in Świeradów-Zdrój, where they can reach the value of 3000 Bq/dm^3.

The impact of carbonated waters can also be found in the vicinity of Wałbrzych (Stare Bogaczowice, Szczawno-Zdrój, Jedlina-Zdrój) and the of Kłodzko area (Kudowa, Polanica-Zdrój, Stary Wielisław, Duszniki-Zdrój, Szczawina or Długopole-Zdrój). These are mainly waters of the HCO_3–Ca, Mg or HCO_3–Ca and HCO_3–Na–Ca types [13, 17]. These are primarily ferrous waters and occasionally radon waters. Moreover, in Duszniki-Zdrój carbonated water, the temperature of which reaches 36 °C has been extracted.

Thermal waters have been recognised as having healing properties in Cieplice Śląskie-Zdrój and in Lądek-Zdrój, in which the origins of using water for treatment date back to the twelfth and thirteenth centuries. The waters of Cieplice flow out from within the Karkonosze granites. Their mineralization reaches approximately 0.6 g/dm^3, and they represent the SO_4–HCO_3–Na type and contain around 13 mg/dm^3 of fluorides and up to 106 mg/dm^3 of metasilicic acid [21]. The temperature of groundwater 87.6 °C at the outflow has been extracted. The waters of Lądek-Zdrój are characterised by low mineralization (approximately 0.2 g/dm^3), and they are mostly of the HCO_3–F–Na type. The fluoride concentration reaches 13 g/dm^3, the content of the hydrogen sulphide amounts to 3.6 mg/dm^3, whereas radon concentration exceeds 1350 Bq/dm^3 [22]. The temperature of the springs is between 20 and 28 °C.

The Carpathian Province – includes the Inner Carpathians, Outer Carpathians, and Carpathian Foredeep. This area is rich in mineral waters. In this region, fresh water commonly coexists with mineral and thermal waters [13, 16, 23].

In the area of Podhale Trough (Inner Carpathians), thermal waters have been documented and exploited present. The temperature of these waters reaches 90 °C deeper in the Podhale Trough with the mineralisation ranging from 0.3 to 3 g/dm^3. These thermal waters are used both for recreational and heating purposes in Zakopane, Bukowina, Białka, Biały Dunajec, and Chochołów [24].

In the Outer Carpathian flysch, carbonated waters, saline waters and brines commonly occur. Carbonated waters prevail along the Poprad river valley (e.g., Krynica Zdrój). These carbonated waters are of the HCO_3–Ca types, and their mineralization does not exceed 3 g/dm^3, whereas earlier waters of the HCO_3–Na–Mg or HCO_3–Mg–Na types reveal a much higher level of mineralization reaching even 12 g/dm^3 [16]. The healing water Zuber, occurring near Krynica Górska constitutes a unique type of carbonated water. This is water of the HCO_3–Na + I or HCO_3–Na–Mg + I, with the mineralization reaching approximately 30 g/dm^3. Chloride carbonated waters are used mainly for healing purposes [13, 16, 25].

In the area of the Carpathians, the occurrence of 125 springs of sulphurous waters has been identified. They contain hydrogen sulphide in concentrations from 1 to 160 mg/dm^3, with the water mineralization ranging from 0.3 to 3.6 g/dm^3 [26].

In the south-western part of the Carpathian Foredeep, there occur brines of the Cl–Na, Br, I type with the mineralization ranging from 30 to 185 g/dm^3, which are used for the production of curative salt (Zabłocie, Dębowiec, Łapczyca). In the north-central part of the Foredeep, in the vicinity of Busko and Solec-Zdrój, healing sulphurous waters and brines of the Cl–Na, S, Cl–SO_4–Na, H_2S, Cl–Na, I types and mineralization ranging from several to 40 g/dm^3 can be found [27, 28]. In Busko-Zdrój, from the C-1 well, water of the Cl–Na–Ca, H_2S type with the temperature of 24 °C was extracted. Healing sulphurous waters also occur in the region of Cracow (Swoszowice, Krzeszowice). In the eastern part of the Foredeep, healing waters occur in Horyniec (HCO_3–Ca–Na, H_2S, type with mineralization between 0.5 and 0.7 g/dm^3) and Latoszyn (SO_4–Ca, H_2S, type with mineralization reaching 2.8 g/dm^3).

The interest in all kinds of undertakings connected with the management of groundwater classified in the category of minerals in Poland has been rising steadily over several years. It is a consequence of more documentation of new resources as well as a greater interest in geothermal waters, which are more commonly used in modern geothermal power stations and leisure facilities.

8.3 Groundwater Resources

Fresh groundwater is in constant circulation, and therefore, one of its characteristics is sustainability. For the calculation of groundwater resources, different methods are applied from those which are used in geological documentation of non-regenerative resources of solid or liquid fossils such as thermal waters or crude oil. The term fixed resources (stored) of groundwater means the quantity of groundwater, most frequently expressed in volumetric units per unit time, while in the case of the dynamic groundwater reserves, it means the quantity in the groundwater reservoir, drainage basin or another hydrological unit [29]. The quantitative and qualitative assessment of these resources is conducted for a specific period on the basis of the research data collected over a multiannual period. Under Polish legislation, the groundwater resources are treated as disposable resources of a GWB catchment area or water economic region. Appropriate assessment of disposable resources is a starting point for any remaining types of hydrological resource assessments, not only for ordinary waters, but also for renewable resources of healing and thermal waters.

Under Polish legislation, there are currently two kinds of groundwater resources: disposable resources, which are determined in the course of ongoing regional research and exploitational resources calculated for groundwater well field intakes. In the assessments of the state of bodies of groundwater reserves (GWB) for reporting in compliance with the European Union regulations, the term disposable groundwater resources is also used. These are disposable resources reduced by the quantity of water essential for the environmental river flow and maintenance of aquatic ecosystems dependent on legally protected groundwater resources.

European Union's Water Framework Directive is commonly referred to also as the community policy regarding water management, the main aim of which is maintenance and potential improvement of the status of water in the case of groundwater bodies (GWB). In the Water Framework Directive (WFD) of 2000, it was adopted as a guiding principle that water resource management is implemented within the borders of river basins and water catchment areas. It could be assumed that a large water catchment area is approximately proportional to the underground basin. Consequently, multiannual meteorological and hydrogeological data can be found for river drainage areas, and additionally, for the majority of catchment areas and water economic regions in Poland, there is abundant hydrogeological documentation containing disposable groundwater resources assessment calculated using the numerical flow models.

The most important feature of groundwater reservoirs, from the perspective of disposable resource assessments, is the sustainability of their resources understood as regular replenishment of groundwater resources of a particular reservoir using natural infiltration of meteoric water in order to complement the loss of groundwater due to natural drainage or exploitation.

According to Krajewski [30], disposable resources of aquifer levels in hydraulic connection with surface waters should be understood as renewable resources reduced by underground run-off ensuring the natural river flow, which protects aquatic life and other environmental resources and qualities. The author also claims that the definition above does not concern Artesian aquifers and other forms of deep groundwater occurrence, where there is no direct link between groundwater and surface waters. Paczyński [31] emphasized a close connection between groundwater renewable and disposable resources, maintaining that the former guarantee the credibility and sustainability of disposable resources, their position in the water balance, and indirectly, environmental consequences of water extraction. According to him, the quantitative characteristics of groundwater recharge form an integral part of the model simulation of disposable and exploitable resources.

The first stage in the process of identifying disposable resources in the balance area using this method is the assessment of groundwater resource sustainability. The size of groundwater sustainability is determined by the average value of the multiannual groundwater run-off to rivers, which is most frequently accepted to be equal to the multiannual mean low flow (SNQ) in rivers [32–34]. The SNQ value, which is the result of hydrological measurements, is the starting point in the process of disposable groundwater resources assessment using the method in question. By identifying sustainability with efficient infiltration recharge, it is possible to assess the distribution of the water recharge within the borders of a given water balance area using the constant volume conversion method [35]. The weighting factors controling the distribution of water recharge is here the lithology of surface formations and the distribution of mean long-term precipitation rates.

The second stage of disposable resources identification concerns the development of the mathematical model for groundwater flow in which the estimated distribution of water recharge, equal to the groundwater run-off to rivers, is a boundary condition. At this stage, the criteria of disposable resources assessment as well as weighting factors controlling their distribution are defined. Using the method of successive approximations, it is possible to determine the participation of disposable resources in the overall renewable resources as well as the proportion of disposable resources division between the modelled aquifers [36]. The distribution of the disposable resources in particular aquifers of the model is calculated using the constant volume conversion algorithm.

The aggregate value of manageable standard groundwater resources understood as the sum total of disposable and prospective resources in Poland amounts to approximately 36.4 mln m^3/d (as at 31.12.2015), including almost 21.4 mln m^3/d established as disposable and 15 mln m^3/d estimated as prospective resources (only in the areas where the numerical modelling of groundwater flows has not been carried out yet) [37, 38] (Fig. 8.4; Table 8.1).

For groundwater resources management, with due consideration given to their interaction with surface waters within the country, groundwater management regions,

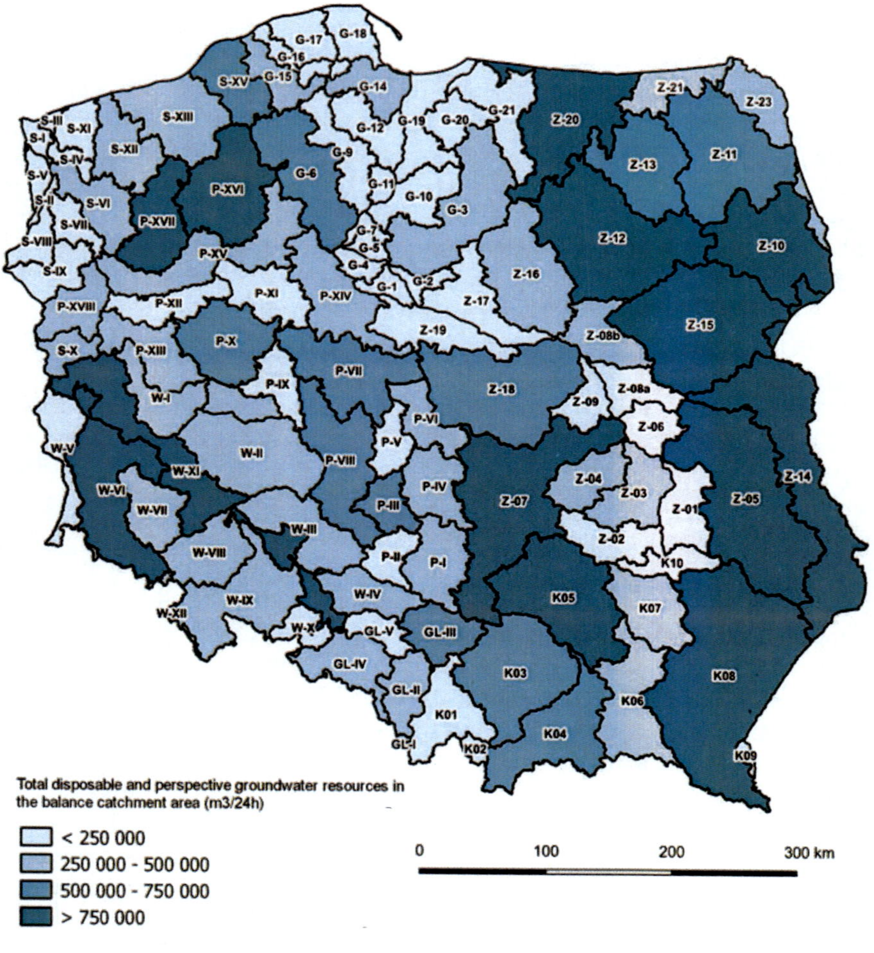

Fig. 8.4 Available groundwater resources (disposable and/or prospective resources), after [38]

Table 8.1 Available groundwater resources – ZDG (*ZD* – disposable, *ZP* – prospective) divided according to river basins and water regions (as of 31.12.2015); after [37, modified by authors]

River basin	Water region	Area	Disposable groundwater resources	Estimeted prospective groundwater resources	Available groundwater resources
			ZD	*ZP*	*ZDG (ZD + ZP)*
		[km^2]	[m^3/d]	[m^3/d]	[m^3/d]
Vistula	Lower Vistula	35 084.10	3,107,412	423,585	3.530,997
	Upper Vistula	43 110.31	1,521,075	3,033,720	4,554,795
	Little Vistula	3 942.47	961,879	7912	969,791
	Middle Vistula	101,039.94	6,626,366	4,054,454	10,680,820
Oder	Lower Oder and Western Baltic Area	20,405.78	2,682,855	18,755	2,701,610
	Upper Oder	3,829.79	518,962	–	518,962
	Middle Oder	39,299.68	2,497,191	2,070,783	4,567,974
	Warta	54,480.00	3,453,214	3,701,696	7,154,910
Pregoła	Łyna and Węgorapa	7,521.69	72,511	1,226,065	1,298,576
Neman	Neman	2,515.14	14,694	299,998	314,692
Dniester	Dniester	233.06	–	27,000	27,000
Danube	Czadeczka	24.59	655	–	655
	Black Orava	359.67	–	41,000	41,000
	Morava	0.71	–	86	86
Jarft	Jarft	210.08	–	–	36,155
Elbe	Jizera	47.12	–	5713	5713
	Elbe and Upa	19.42	–	2355	2355
	Metuje	99.38	–	12,051	12,051
	Orlica	72.52	–	8795	8795
Świeżej	Świeżej	161.41	–	27,779	27,779
Ücker	Ücker	14.71	2945	–	2945
Country's overall area		**312,471.57**	**21,459,759**	**14,997,902**	**36,457,661**

which simultaneously constitute basic groundwater balance units, have been distinguished. In the predominant number of **668** water management regions, which collectively occupy 90% of the country's area, there occur sizeable reserves of available resources (utilisation degree of groundwater resources does not exceed 30% there). The average groundwater reserves occur in water management regions covering 5.7% of the country's area – (approximately 30–60% of all resources are used there). Low and very low reserves (the utilisation level between 60 and 90% of all resources) can be observed in water economic regions occupying only 1.9% of the country's area. The risk of groundwater shortage (the utilisation level exceeding 90%) concerns only 1.4% of the country's area. These are water economic regions where dewatering systems for brown coal open-cast mines and a large concentration of potable water intakes in large urban and industrial agglomerations can be found [4, 39] – see Fig. 8.5.

The groundwater resources in Poland include the estimated aquifer reserves in particular balance regions, available for utilisation provided due consideration is

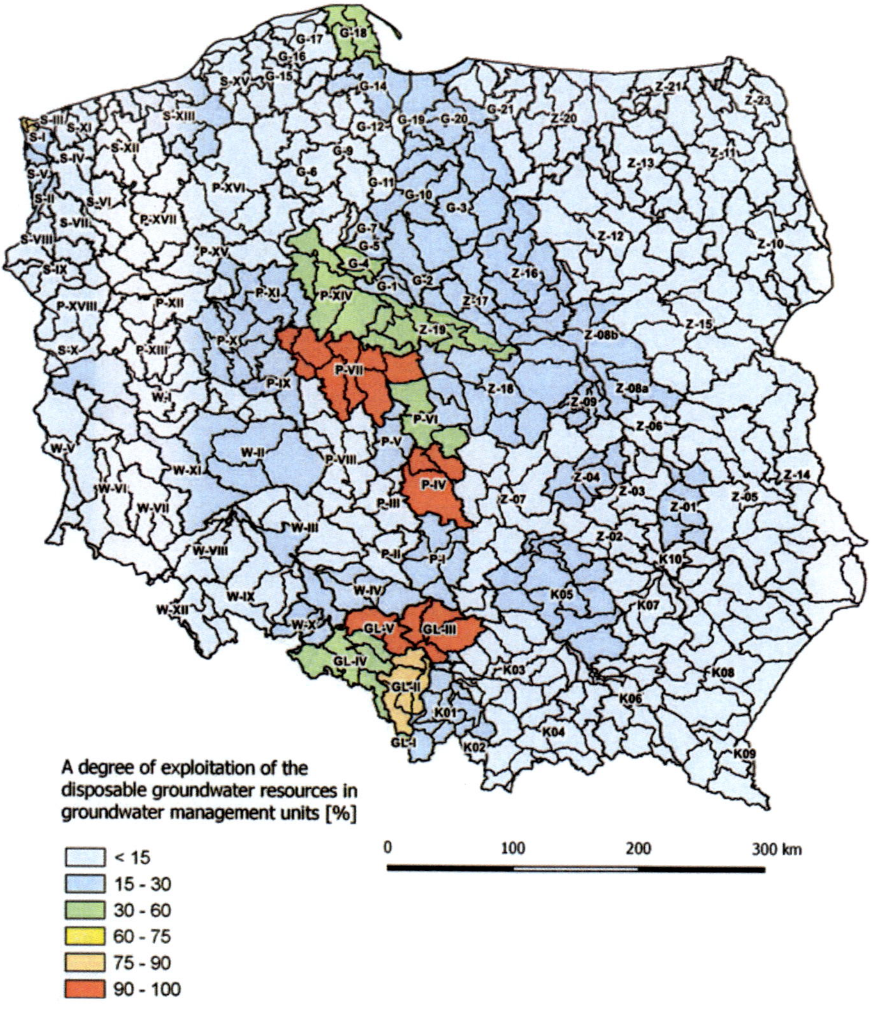

Fig. 8.5 A degree of exploitation available groundwater resources in Poland [4, modified by authors]

given to the conservation of the already existing ecosystems which depend on these water management regions. The estimation of groundwater resources is understood as hydrogeological research and calculations whereby approximation methods of assessing renewable groundwater resources are applied, using model-based verification of the permissible level of the hydrodynamic field conversion and groundwater circulation balance. The water needs of aquatic and terrestrial ecosystems dependent on groundwater resources are taken into account in a simplified manner, without numerical modelling of groundwater flows. However, they are duly considered by preserving the resources of groundwater outflow to rivers on the level of hydrobiological flows inviolable in gauge profiles and by accepting the groundwater run-off to rivers as the basis for determining disposable resources. It has been accepted that the difference between renewable aquifer resources and the groundwater run-off to rivers constitutes a reserve to cover the evapotranspiration groundwater drainage in meadow and swamp ecosystems typical of river valleys.

The fieldwork above and model research work constitute elements which identify renewable resources and determine disposable groundwater resources. Conducting this research work has made it possible to attain a high degree of accuracy during resource assessment, which in turn has enabled making authoritative administrative decisions concerning water management. This applies in particular to reviews and approvals of projects involving groundwater intakes, groundwater exploitation resources intakes, reviews of of applications regarding groundwater intake permits and formulations of hydrogeological recommendations concerning groundwater exploitation for the whole catchment basin (GWB).

The comparison of values of available groundwater resources with their current extraction figures constitutes the basis for sustainable management of groundwater in river basins groundwater bodies (GWB) and water management plans for regions.

8.4 Conclusions

1. Poland's accession to the European Union took place on May 1, 2004. The intensive preparatory process for the accession had been conducted since 1999, shortly before Water Framework Directive was enforced. Taking into consideration the necessity to implement considerable changes in the area of water management, water resource conservation, water status reports as well as taking coordinated action in all these fields, the idea was conceived to organize properly the institution of Polish hydrology within the framework of national service.
2. There is rising pressure on groundwater resources exploitation according to the expanding needs of water supply at present. Water-works enlarge the volume of potable water exploitation from groundwater resources. At present more than 70% of the total community is supplied by groundwater. The important factor for groundwater protection is the phenomena of droughts and high water occurrence in the last 25 years and also proper recognition of old dumping sites and hidden (wild) storage yards.

3. In recent years water management groundwater balance reports have been compiled in relations to surface waters, which provides the basis for a holistic approach to water resource management and effective action aiming at the conservation of water resources.
4. The public interest in healing water, thermal water, and highly mineralised water (brines) has been growing considerably for some time. All these factors make scientists and specialists show greater interest in deep water resources, also in the form of a mathematical description of phenomena occurring in the stagnant water zone, which is virtually outside water circulation.
5. The application of mathematical modelling methods for groundwater filtration processes creates the opportunity for assessment of disposable groundwater resources along with hydrogeological and environmental results of its extraction, which has been visible with increasing intensity in Poland in recent years. The most frequently applied software worldwide for hydrogeological modelling is an American programme called MODFLOW with its numerous versions such as for instance Visual Modflow, Groundwater Vistas, Processing Modflow or Groundwater Modelling System. The method of constant volume conversion allows direct combination in the calculation process of disposable resources assessment and renewable resources assessment [40, 41].

8.5 Recommendations

Water supply for the inhabitants of small towns and villages comes primarily from groundwater intakes, water intakes of big cities and industrial plants take up water also from rivers. Over 70% of groundwater resources is used for human water consumption. Care should be taken to ensure water quality to maintain access to groundwater resources in the coming decades. Rational distribution of available groundwater resources in the country and comparison with the groundwater exploitation is the background of proper, sustainable management of its resources.

Water resources are constantly exposed to negative changes and transformations under the influence of various sectors of the economy and anthropopressure. Attention should be paid to the protection of groundwater resources through appropriate provisions and restrictions in spatial development plans. Particular attention should be paid to control and monitoring of the quality and quantity of groundwater in the areas of mining activity related to mine drainage and areas intensively used for agriculture.

In Poland has been rising steadily over several years greater interest in geothermal waters, which are more commonly used in modern geothermal power stations and leisure facilities. Their resources should also be controlled.

References

1. Kleczkowski AS (1988) Regionalizacja słodkich wód podziemnych Polski w zmodyfikowanym ujęciu. Mat. III Ogólnopolskiego Symp. Aktualne problemy hydrogeologii. Cz. 3. Wyd. Inst. Morski. Gdańsk
2. Kleczkowski AS (edit. 1990) Mapa obszarów głównych zbiorników wód podziemnych (GZWP) w Polsce wymagających szczególnej ochrony, 1:500,000. Wyd. IHiGI AGH, Kraków
3. Paczyński B, Sadurski A (edit. 2006) Hydrogeologia regionalna Polski, tom I. Państ. Inst. Geol., Warszawa
4. Herbich P, Mordzonek G, Przytuła E (2017) Stopień wykorzystania dostępnych do zagospodarowania zasobów wód podziemnych w Polsce, PIG Warszawa 2017. https://www.pgi.gov.pl/psh/materialy-informacyjne-psh/stan-srodowiskowy-wod-podziemnych.html)
5. Mikołajków J, Sadurski A (edit. 2017) Major groundwater reservoirs in Poland. Polish Geological Institute, Warsaw, 413 p
6. Jamorska I (2015) Hydrogeological conditions of Southern Kujawy region (in Polish). Prz Geol 63(nr 10 cz. 1):756–761
7. Jamorska I, Krawiec A (2017) Zmiany chemizmu wód podziemnych w województwie kujawsko-pomorskim w latach 2006–2015. Prz Geol 65(nr 11/2):1270–1275
8. Kozerski B, Maciaszczyk A, Pazdro Z, Sadurski A (1987) Fluorine anomaly in the groundwater in the Gdańsk Region [in Polish, Eng. Sum.]. Ann Soc Geol Pol 57(3–4):349–374
9. Krawiec A (2013) The origin of chloride anomalies in the groundwaters of the Polish Baltic coast (in Polish, Eng. Sum). Nicholas Copernicus Univ. Publ. Toruń, pp 1–143
10. Kozerski B (1981) Salt water intrusions in to coastal aquifers of Gdansk region. In: Proceedings of 7th Salt Water Intrusion Meeting, Uppsala, vol 27, pp 83–89
11. Burzyński K, Sadurski A (1989) Influence of near slop drainage of Lakeland plateau on the Cretaceous aquifer on the Vistula delta plane [in Polish, Eng. Sum]. Kwart Geol t. 33(2):301–312
12. Krawiec A, Rübel A, Sadurski A, Weise SM, Zuber A (2000) Preliminary hydrochemical, isotope, and noble gas investigations on the origin of salinity in coastal aquifers of western pomerania, Poland. In: Proceedings of 16th Salt Water Intrussion Meeting, Toruń, pp 87–94
13. Ciężkowski W, Chowaniec J, Górecki W, Krawiec A, Rajchel L, Zuber A (2010) Mineral and thermal waters of Poland. Prz Geol 58(nr 9/1):762–773
14. Ciężkowski W (1998) Lądek-Zdrój. Dolnośląskie Wydawnictwo Edukacyjne
15. Paczyński B, Płochniewski Z (1996) Wody mineralne i lecznicze Polski. Państw. Inst. Geol., Warszawa
16. Rajchel L (2012) Szczawy i wody kwasowęglowe Karpat polskich. Wyd. Nauk. AGH, Kraków, pp 1–144
17. Krawiec A, Ciężkowski W (2017) Geologia i górnictwo wód leczniczych i peloidów w Polsce. s. 103–130. In: Ponikowska I, Kochański JW (edit.) Wielka Księga Balneologii, Medycyny Fizykalnej i Uzdrowiskowej. Tom I. Część ogólna, pp 1–836. Wyd. Aluna
18. https://mineralne.pgi.gov.pl
19. Kalwasińska A, Deja-Sikora E, Szabó A, Krawiec A, Felföldi T, Swiontek-Brzezinska M, Walczak M (2019) Microbial communities of low temperature, saline groundwater used for therapeutical purposes in North Poland. Geomicrobiol J. https://doi.org/10.1080/01490451.2018.153563
20. Krawiec A (1999) New results of the isotope and hydrochemical investigations of therapeutical waters of Ciechocinek Spa (in Polish). Prz Geol 47:255–260
21. Ciężkowski W, Gröning M, Leśniak PM, Weise SM, Zuber A (1992) Origin and age of thermal waters in Cieplice Spa, Sudeten, Poland, inferred from isotope, chemical and noble gas data. J Hydrol 140:89–117
22. Zuber A, Weise SM, Osenbrück K, Grabczak J, Ciężkowski W (1995) Age and recharge area of thermal waters in Lądek Spa (Sudeten, Poland) deduced from environmental isotope and noble gas data. J Hydrol 167:327–349

23. Chowaniec J (2009) Studium hydrogeologii zachodniej części Karpat polskich. Biul. PIG-PIB, 434, Hydrogeologia, VIII
24. Kępińska B (2018) A review of geothermal energy uses in Poland in 2016–2018 (in Polish). Technika Poszukiwań Geologicznych, Geotermia, Zrównoważony Rozwój t. 57 z. 1:11–26
25. Zuber A, Chowaniec J (2009) Diagenetic and other highly mineralized waters in the Polish Carpathians. Appl Geochem 24:1899–1900
26. Rajchel L (2000) Źródła wód siarczkowych w Karpatach polskich. Kwartalnik Geologia, AGH 26:309–373
27. Zuber A, Weise SM, Osenbrück K, Mateńko T (1997) Origin and age of saline waters in Busko Spa (Southern Poland) determined by isotope, noble gas and hydrochemical methods: evidence of interglacial and pre-Quaternary warm climate recharges. Appl Geochem 12:643–660
28. Deja-Sikora E, Gołębiewski M, Kalwasińska A, Krawiec A, Kosobucki P, Walczak M (2018) Comamonadaceae OTU as a remnant of an ancient microbial community in sulfidic waters. Microbial Ecol:1–17. https://doi.org/10.1007/s00248-018-1270-5
29. Dowgiałło J, Kleczkowski AS, Macioszczyk T, Różkowski A (edit. 2002) Słownik hydrogeologiczny. Wyd. II. Min. Środ. Wyd. PIG-PIB. Warszawa
30. Krajewski S (1980) Odnawialność a dyspozycyjność zasobów wód podziemnych kredy lubelskiej. Symp. "Współczesne problemy hydrogeologii regionalnej" w Jachrance, Warszawa
31. Paczyński B (edit. 1995) Atlas hydrogeologiczny Polski, 1:500,000. Cz. II, Zasoby, jakość i ochrona zwykłych wód podziemnych. Wyd. PIG. Warszawa
32. Jokiel P (1994) Zasoby, odnawialność i odpływ wód podziemnych strefy aktywnej wymiany w Polsce. Acta Geographica Lodziensia, Łódź
33. Herbich P (2005) Zasoby perspektywiczne wód podziemnych—cel ustalenia i metodyka obliczeń dla zlewniowych systemów wodonośnych. W: Współczesne problemy hydrogeologii. 12:261–268. Wyd. UMK Toruń
34. Herbich P, Nowicki Z, Sadurski A (2004) Groundwater resources, management and protection in Poland. In: Common pool groundwater resources. In: Brentwood M, Robar SF (Edit) An international perspective. Paeger Publ. USA, pp 155–163
35. Sadurski A, Śmieański L (2015) Problem zasobów wód podziemnych. Przegląd Geologiczny T.63(10/2):1047–1052
36. Herbich P, Kapusciński J, Nowicki K, Rodzoch A (2013) Metodyka określania zasobów dyspozycyjnych wód podziemnych w obszarach bilansowych z uwzględnieniem potrzeb jednolitych bilansów wodnogospodarczych. Poradnik metodyczny, HYDROEKO, Warszawa
37. Filar S, Mordzonek G, Przytuła E, Węglarz D (2015) Ustalenie możliwych do zagospodarowania zasobów wód podziemnych i przeprowadzenie bilansu wodnogospodarczego z uwzględnieniem oddziaływań z wodami powierzchniowymi w polskiej części dorzeczy: Dnierstu, Dunaju, Jarft, Łaby, Niemna, Pregoły. Świeżej i Ucker. Informator państwowej służby hydrogeologicznej, PIG-PIB, Warszawa
38. Sadurski A, Przytuła E (2016) Zasoby dyspozycyjne wód podziemnych dorzeczy w Polsce w świetle zrównoważonego gospodarowania wodami. Biul Państw Inst Geol 466:261–270
39. Frankowski Z, Gałkowski P, Mitręga J (2009) Struktura poboru wód podziemnych w Polsce. Wyd. PIG, Warszawa
40. Śmieański L (2012) Zastosowanie przekształcenia stałoobjętościowego do oceny odnawialności zasobów wód podziemnych wschodniej części Pojezierza Pomorskiego. Biul Państ Inst Geol 451:227–234
41. Śmieański L (2010) The quantitative evaluation of the catchment available groundwater resources—the case study. Biul Państw Inst Geol 441:1

Part III
Change of Flow in Polish Rivers

Chapter 9
Flow Regime Patterns and Their Changes

Dariusz Wrzesiński

Abstract Poland is characterized by relatively significantly diversified environmental conditions, reflected in various water supply conditions and seasonality of the river flow, which determines the flow regime. Based on the Pardé coefficient, five types of river regime: three nival (snowy) – poorly, moderately and well-formed, nival-pluvial (snowy-rainy) and pluvial-nival (rainy-snowy), respectively, can be distinguished. While the nival regime is represented mainly by lowland rivers in the central and northern parts of the country, rivers in the southern, upland and mountainous parts have the nival-pluvial and pluvial-nival regimes. Rivers representing the nival poorly formed type are characterized by the most even flows and the highest share of the groundwater flow in the total flow in the annual cycle. In contrast, rivers with the nival well-formed regime are distinguished by the most contrasting types of periods, from deep low-water to very high high-water. Climate changes and the human impact on water relations make the features of the river flow regime change. Rivers with similar changes in flow conditions, except from those with transformation of water relations caused by human activity, also represent similar geographical regions. This indicates the importance of climatic conditions in the modification of the characteristics of the flow regime. Certain evidence of the influence of climatic conditions on changes in the flow regime in the winter-spring period may be a significant decrease in winter flows observed in many rivers, and a delayed increase in the spring thaw in the 1950s and 1960s. In turn, in the 1970s and 1980s in these rivers, there was a disappearance of the winter low-water stages associated with a remarkable increase in the winter flow. These regularities indicate the noticeable impact of changes in the intensity of the North Atlantic Oscillation on the transformation of the characteristics of the flow regime of rivers in Poland. Research confirms that there is a temporarily and spatially differentiated impact of the North Atlantic Oscillation on the level of river flows. This impact is not strong but noticeable. It is observed with varying intensity in rivers in many regions of the country, mainly in winter, spring and summer.

D. Wrzesiński (✉)
Department of Hydrology and Water Management, Institute of Physical Geography and Environmental Planning, Adam Mickiewicz University, Bogumiła Krygowskiego 10, 61-680 Poznań, Poland
e-mail: darwrze@amu.edu.pl

M. Zeleňáková et al. (eds.), *Management of Water Resources in Poland*, Springer Water,
https://doi.org/10.1007/978-3-030-61965-7_9

Keywords River regime · Flow regime typology · River flow regime change · North Atlantic Oscillation · Poland

9.1 Introduction

The country area, even at the regional scale, is characterized by relatively significantly diversified environmental conditions, both climatic, as evidenced by Woś [1], who distinguished 28 climatic regions, and hydrological, reflected in various water supply conditions and seasonality of the river flow, which determines the flow regime. It defines the state and responses of the river system in relation to the climatic system and physical-geographical characteristics of the basin [2]. Therefore, the river regime is the regularity of variability of all phenomena occurring in the river in an average in a multi-year period annual cycle, depending on the properties of the natural environment, mainly climate and the catchment structure. Consequently, the flow regime, the thermal, ice, chemical regime, etc. may be considered. In this chapter, the analysis refers only to the flow regime of rivers in Poland. The analysis of the temporal diversity of hydrological phenomena in the annual cycle may refer to the most characteristic and distinctive periods of the hydrological cycle – the high and low water levels and discharges [3, 4] or it may cover the variability of phenomena in the yearly cycle. In the latter case, the contractual division of the year into monthly periods or seasons is applied, or the variability of phenomena is analysed in hydrological seasons, i.e., periods in the annual cycle, which are characterized by a similar course of climatic and hydrological phenomena and processes.

In order to determine the hydrological regime of rivers, various methods are applied. The recognition of the regularity of the flow variability in the annual cycle is based on both the supervised and unsupervised approaches. In the supervised approach, a few indicators (types of regime) are first defined, and then regularities in the multi-year monthly or seasonal discharges are sought according to the predefined indicators. Such an approach can be exemplified by the typologies of the regime proposed by Pardé [5], Lwowicz [6] or, for rivers in Poland, by Dynowska [7]. In the unsupervised approach, based mainly on the grouping of variables selected for the analysis, there are no such indicators, and determination of the type of regime is made on the basis of data structure so that objects within a single type are as similar as possible. Such an approach was proposed by Gottschalk [8] in the hydrological regionalization of Sweden, and by Rotnicka [9, 10] in Poland. While in the first approach, determining a priori an indicator that decides about the type of regime, exhibits its certain subjectivity, in the second, objectified, unsupervised approach the result of the analysis is affected by the so-called 'stop rule' and the acceptance of a certain degree of compliance within the identified groups.

The earliest, more extensive works on the regime of rivers in Poland were published in the 1970s. Dębski [11] classified our rivers as of a nival-pluvial type in four variants: complex mountain (e.g. the Odra River at gauge Połęck), complex Tatra Mountain (e.g. the Vistula River at gauge Karsy), oceanic (the Słupia River at gauge

Słupsk) and lowland (e.g. the Bug River at gauge Wyszków and the Vistula River at gauge Korzeniewo). However, the most detailed typology of the river regimes in Poland was presented by Dynowska [7]. When determining the type of river regime, she took into account the following factors: the type and share of water supply (groundwater, rain, snow), the nature of the flow variability based on the variability of daily discharges and irregularity of monthly and annual average discharges, and the time of occurrence of the highest discharge. She identified three main types of hydrological regime: balanced, moderate and unbalanced, within which she distinguished 13 subtypes, depending on the number and periods of the occurrence of the largest discharges and the prevailing type of water supply. A cartographic synthesis of the river regime in Poland Dynowska [12] presented in the "Atlas of the Republic of Poland." Based on monthly flow rates, she distinguished five types of river regime: three nival (snowy) – poorly, moderately, and well-formed, nival-pluvial (snowy–rainy) and pluvial-nival (rainy-snowy). An interesting approach to the study on water regime was proposed by Rotnicka [9] on the example of rivers in the Odra drainage basin and the Przymorze catchments. In that approach, different from the previous ones, river regime is understood as the type and time structure of river flows in the normal hydrological cycle. Elements of this structure are the so-called hydrological periods, which are a tool for studying the regime, and the basis for its characteristics. On the basis of the number, types, and sequences of hydrological periods, Rotnicka distinguished six main types of regimes of the studied rivers:

- the five-period regime, contrasting, with deep low-water stage in summer-autumn and pronounced high-water stage in spring (in two variants, e.g. the Odra River at gauge Miedonia, the Warta River at gauge Poznań),
- the four-period regime, with average low-water stage in summer-autumn and pronounced high-water stage in early spring (in two variants, e.g. the Odra River at gauge Gozdowice, the Ner River at gauge Dąbie),
- the three-period lowland regime, with average low-water stage in summer-autumn and low high-water stage in late winter or early spring (e.g. the Gwda River at gauge Piła), the three-period mountain and sub-montane with low low-water stage in summer-autumn and pronounced high-water stage in spring (e.g. the Nysa Kłodzka River at gauge Bystrzyca),
- the two-period regime, with low low-water stage in summer and low high-water stage in winter-spring (e.g. the Parsęta River at gauge Bardy),
- the single-period regime (e.g. the Kłodnica River at gauge Łany Małe).

9.2 Types of Regime

A detailed spatial analysis of the typology of features of the river flow regime in Poland in accordance with the criteria proposed by Dynowska [12] in the "Atlas of the Republic of Poland" was presented by Wrzesiński [13], based on the course and amount of monthly discharge rates (Pardé coefficients). In that work, the author used a rich hydrometric material collected at 516 water gauge stations, located on 295 rivers from the period 1971–2010.

Type 1 – the nival poorly formed regime. It is characteristic of rivers whose average flow of the spring month does not exceed 130% of the average annual flow. This refers mainly to the rivers of Przymorze (east of the Parsęta River) and the Pomeranian Lake District (the upper Drawa River, and the Gwda and Brda rivers). This type of regime is also represented by some rivers in the Masurian Lake District, the Lubusz Land, and the Silesian-Kraków Upland (Fig. 9.1). In the annual cycle, these rivers are characterized by the most even flows and the highest share of the groundwater flow in the total flow. In the case of the coastal and lake-district rivers, this is mainly due to a high retention capacity of the catchments (favorable infiltration

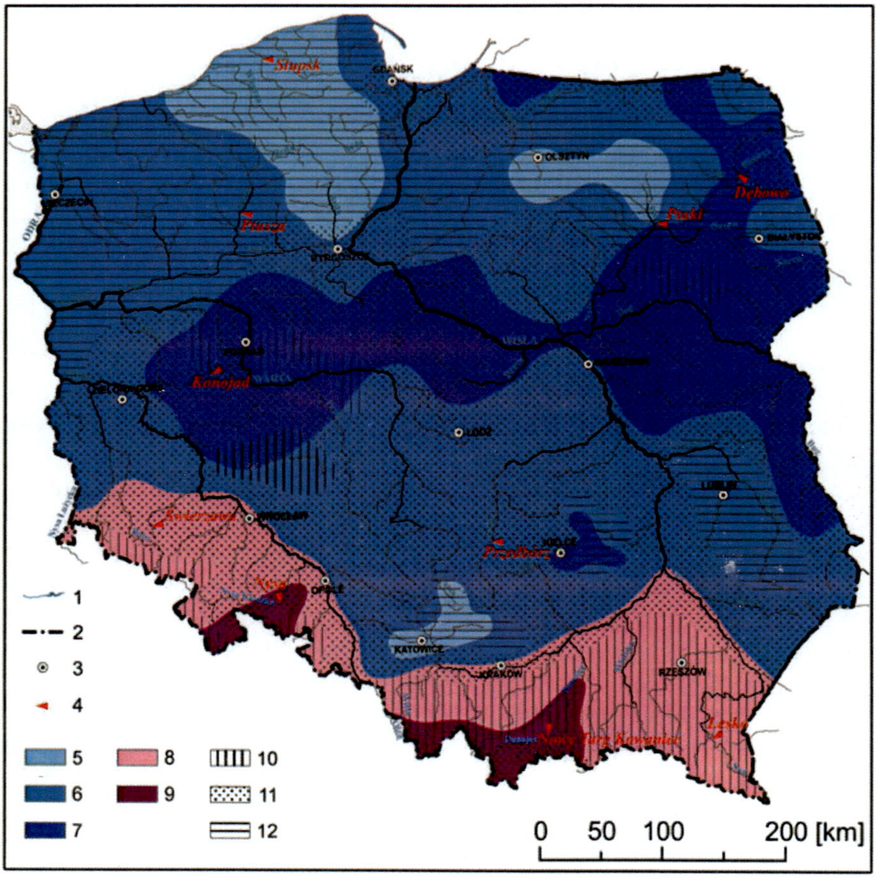

Fig. 9.1 Types of flow regimes of rivers in Poland (after [13]). 1 – Rivers and reservoirs; 2 – state borders; 3 – voivodship cities; 4 – gauges; Types of regime: 5 – nival poorly formed; 6 – nival moderately formed; 7 – nival well-formed; 8 – nival-pluvial; 9 – pluvial-nival; Share of groundwater supply: 10 – below 40%; 11 – 40–60%; 12 – above 60%

conditions), a high lake density and a large number of closed depressions. Upland rivers have balanced flows due to significant retention capabilities of heavily fissured and karstified carbonate rocks, while those in industrial areas due to the human interference in the water cycle. The Przymorze and lake-district rivers are characterized by the smallest variability of flows as well as their extreme irregularity. These rivers are also distinguished by high total flow, over 200 mm, and in the case of the Przymorze rivers even over 300 mm, and the largest share of the groundwater supply in the total flow, which in many rivers is higher than 80% (Table 9.1). The date of occurrence of high water in the rivers of Przymorze and on individual rivers of the Masurian Lake District is relatively stable and occurs in the winter–spring season. In the majority of rivers, the first minimum of monthly flows has a relatively stable or stable date of occurrence, usually in the summer or the summer–autumn period. A strongly stable date of the appearance of a period of low flows is characteristic only of the Brda River (June–July). On the other hand, relatively unstable minima occur in individual rivers, mainly transformed by the human activity, for example in Silesia, and their date falls in the summer–autumn (the Bytomka River) or the autumn–winter (the Brynica River) period. Due to the low variability of flows, these rivers are characterized by the low-contrasting hydrological periods, with the following sequence in the annual cycle: normal period → low high-water → normal period → low low-water (Fig. 9.2). In extreme cases, in rivers with changes in flow conditions caused by human activity, only one normal hydrological period is observed with very even flows (the Brynica River).

Type 2 – the nival moderately formed regime. This is represented by rivers whose average flow of the spring month is from 130 to 180% of the average annual flow. Rivers with this regime are located in:

- the northern part of the country, e.g. in the western part of Przymorze (the Rega and Parsęta rivers), in the Masurian Lake District (the Łyna and Omulew, Czarna Hańcza and Supraśl rivers),
- in the central part – the transit rivers, e.g. the Vistula, Odra and Warta rivers with the Noteć River,
- in the upland belt – rivers located between the Warta and Vistula rivers, the Lublin Upland rivers in the Wieprz catchment.

These rivers are characterized by average flows (100–200 mm), only the riparian and coastal rivers are distinguished by the flow of over 200 mm or even 300 mm. Rivers in the northern part and in the Wieprz catchment are also characterized by a high share of the groundwater supply (60–80%), which in the other rivers of this type of regime is 40–60%. The date of the high-water phase is usually relatively stable and stable and falls in the winter–spring or the spring period. In most rivers, the low-water phase has a relatively stable or a stable date of occurrence, usually in the summer–autumn or the summer period. In comparison with the previous type rivers with the nival moderately formed regime are characterized by greater variability of flows, but the types of hydrological periods and their sequence are similar. The following ones can be distinguished in the yearly cycle: normal period → high-water (from low to high) → normal period → low low-water.

Table 9.1 Types and characteristics of river flow regime in Poland (1971–2010) (after [13])

River	Gauge	Catchment area [km^2]	H [mm]	H_{ground} [%]	Flow variability		High-water phase		Low-water phase		Type of regime
					Daily *Cv*	Extreme irregularity Q_{max}/Q_{min}	SC	Period	SC	Period	
Carpathian mountains											
Soła	Oświęcim	1386	447	22.6	1.963	908	0.316	March–July	0.400	September–November	5
Raba	Proszówki	1470	365	31.8	1.953	2652	0.328	March–April	0.388	September–November	4
Dunajec	Nw. Targ-Kowaniec	681	675	42.8	1.171	239	0.467	April–July	0.513	January–February	5
Wisłoka	Żółków	581	419	21.4	1.978	3300	0.413	March–April	0.412	August–October	4
San	Lesko	1614	581	33.5	1.063	323	0.470	April–May	0.424	September–November	4
Sudety mountains											
Nysa Kłodzka	Nysa	3276	281	39.4	1.089	475	0.372	April–July	0.353	October–December	5
Bystrzyca	Krasków	683	211	30.0	1.714	668	0.324	March–June	0.447	September–November	4
Kaczawa	Świerzawa	134	280	46.2	1.777	927	0.347	March–May	0.486	August–October	4
Kwisa	Nowogrodziec	736	311	48.2	1.217	382	0.318	March–May	0.400	September–November	4
Uplands (built of carbonate rocks)											
Warta	Kręciwilk	66.2	384	68.2	0.610	44.2	0.389	February–April	0.353	July–August	1
Wieprz	Krasnystaw	3001	133	66.9	0.553	36.3	0.504	March–April	0.359	July–September	2
Bystrzyca	Sobianowice	1265	128	65.9	0.624	57.5	0.463	February–April	0.281	July–October	2
Kielce-Sandomierz Upland											
Czarna Nida	Tokarnia	1216	173	50.3	1.193	160	0.486	February –April	0.394	August–October	2
Koprzywianka	Koprzywnica	502	118	44.8	1.815	1072	0.544	February –April	0.463	July–September	3
Lowlands											
Mała Panew	Staniszcze Wlk	1107	203	51.8	1.053	144	0.417	February–April	0.441	August–November	2

(continued)

Table 9.1 (continued)

River	Gauge	Catchment area [km^2]	H [mm]	H_{ground} [%]	Flow variability		High-water phase		Low-water phase		Type of regime
					Daily *Cv*	Extreme irregularity Q_{max}/Q_{min}	SC	Period	SC	Period	
Mogilnica	Konojad	663	79	33.1	1.440	1325	0.556	February–April	0.441	July–October	3
Liwiec	Łochów	2466	137	47.1	1.148	240	0.509	February–April	0.544	July–September	3
Wkra	Trzciniec	1928	168	53.0	0.746	94.0	0.525	February–April	0.478	July–September	2
Lake districts											
Drawa	Drawsko Pom	609	226	71.1	0.554	22.2	0.516	January–April	0.422	July–October	2
Rega	Łobez	609	239	72.5	0.462	12.2	0.458	January–March	0.486	July–September	2
Słupia	Słupsk	1450	348	72.1	0.293	8.3	0.382	January–March	0.486	June–August	1
Pisa	Ptaki	3562	186	81.8	0.383	12.7	0.417	February–April	0.366	July–October	1
Łyna	Sępopol	3647	215	62.9	0.630	28.0	0.509	February–April	0.475	June–August	2
Guber	Prosna	1568	170	36.5	1.149	223	0.486	February–April	0.431	August–November	3

Types of regime: 1 – nival poorly formed, 2 – nival moderately formed, 3 – nival well-formed, 4 – nival-pluvial, 5 – pluvial-nival. Regime character (stability coefficient): SC < 0.20 – unstable; 0.20 < SC < 0.31 – relatively unstable; 0.31 < SC < 0.45 – relatively stable; 0.45 < SC < 0.62 – stable; SC > 0.62 – strongly stable

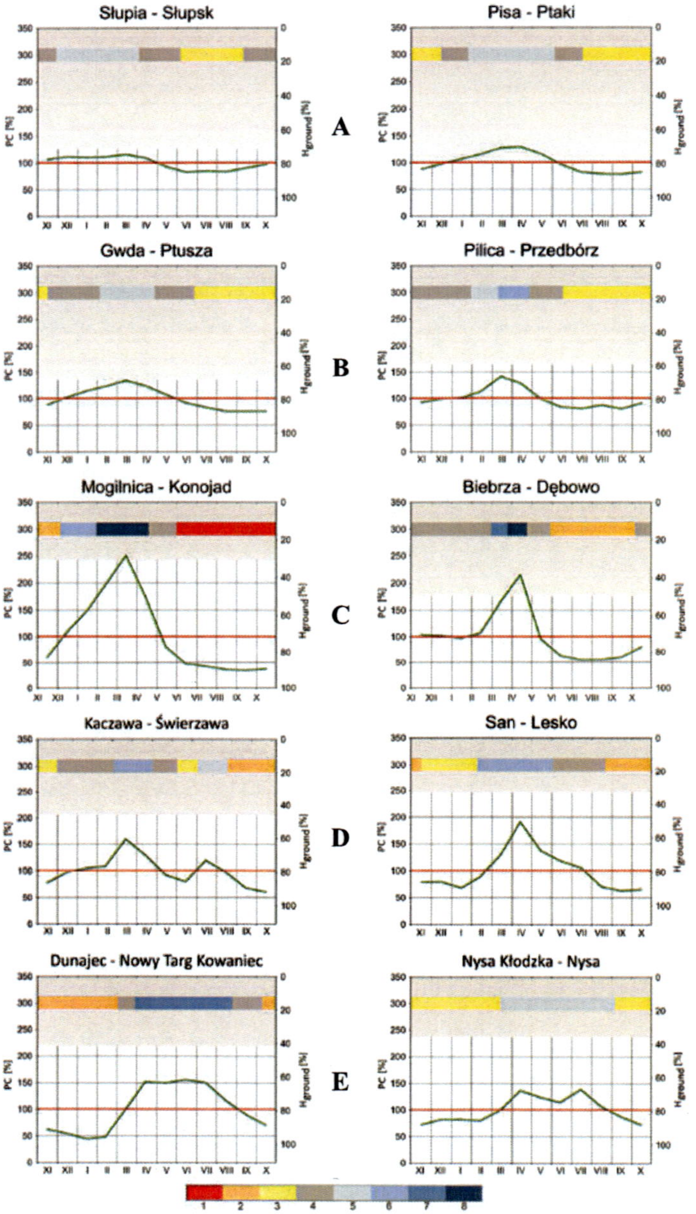

Fig. 9.2 Monthly flow coefficients – Pardé coefficients (PC), share of groundwater supply (H_{ground}) and sequences of hydrological periods of selected rivers of particular types of regime: A – nival poorly formed, B – nival moderately formed, C – nival well-formed, D – nival-pluvial, E – pluvial-nival; Types of periods: 1 – deep low-water, 2 – average low-water, 3 – low low-water, 4 – normal period, 5 – low high-water, 6 – average high-water, 7 – high high-water, 8 – very high high-water (after [13])

Type 3 – the nival well-formed regime. This type of regime is characteristic of rivers with an average flow of the spring month higher than 180% of the average annual flow. These rivers are distinguished by the largest changes in flow in the annual cycle. They are located in the lowland part of the country, from the Greater Poland Lake District and the South-Silesian Lowland through the Central Mazovian Lowland, to the majority of rivers in the eastern part of the country, in the Narew and Bug rivers catchments. Rivers of this type of regime still exist in the Kielce-Sandomierz Upland (in the Kamienna River catchment) and locally in the Myślibórz Lake District (the Myśla River) and on the Szczecinski Coastland (in the Ina River catchment). These rivers lie in the belt characterized by the lowest flows, which usually do not exceed 150 mm. A higher flow, over 200 mm, is observed only in the north-east and in the Kielce Upland. Extremely low flows, below 80 mm, are observed in the Mogilnica and Flinta rivers in the Greater Poland Lake District and on the Tążyna River, the left tributary of the Vistula River in Kujawy (60 mm). The volume of the groundwater supply and the flow variability are also diversified. High variability of daily flows is characteristic of some rivers of the Kielce Upland (the Koprzywianka and Czarna rivers), and on the lowlands the Brok River, which is the right tributary of the Bug River, and the Tążyna River. The latter, like many other rivers of the Great Poland-Kujawy Lake District (the Sama and Mogilnica rivers), is also distinguished by high variability of annual flows ($Cv > 0.600$). At the same time, these rivers usually have a very low groundwater supply (below 40%). There is a short, intense period of meltwater feeding in spring and formation of high-water stage on these rivers. Then, there is a rapid recession of the flood flow and a transition to the low-water stage in the summer-autumn period. A strongly stable period (March–April) of the occurrence of floods is observed in many rivers of north-eastern Poland. In the other rivers, the date of occurrence of the high-water period is stable and earlier (January–March, February–April). Stable, summer–autumn (July–September) is also the date of appearance of the low-water period. Rivers with the nival well-formed regime are distinguished by the most contrasting types of periods with the following sequence: average low-water (or normal period) → low (or average) high-water → very high high-water → normal period → deep (or average) low-water.

Type 4 – the nival-pluvial regime. It is characteristic of rivers whose average flow of the spring month is generally 130–180% of the average annual flow and the flow increase in the summer months is marked, with at least 100% of the average annual flow. This type is represented by rivers of the Sudety Mountains and most of the Carpathian rivers, and by the transit Vistula River up to gauge Puławy, whose regime in this section is shaped by its Carpathian tributaries. These rivers are characterized by a large range of annual flows from 100 mm (the left tributaries of the upper Odra River) to over 800 mm (the upper Vistula River, tributaries of the San River in the Western Bieszczady Mountains). These rivers are characterized by a small share of the groundwater supply in the total flow (20–40%) and a large range of variability of daily flows, from $Cv < 1.0$ in the lower reaches of the mountain rivers to $Cv > 2.0$ in the upper reaches. The date of the high-water phase is usually relatively stable and stable and falls in the spring or the spring–summer period. The least regular date of appearance of the maxima (relatively unstable type) is usually associated with a strong human impact on the water cycle. It is observed, among others, in the Vistula River between gauges Goczałkowice and Nowy Bieruń, in rivers of the Silesian Foothills, in the Skawa River, and in the Odra River catchment in its tributaries – the Bystrzyca River with the Piława River, and the Ślęza River. In most rivers, the low-water phase has a relatively stable or stable date of occurrence, usually falling for the autumn, autumn–winter, and winter periods. A stable low-season period is characteristic of many rivers in the Dunajec River catchment and the tributaries of the Odra – the Kaczawa River and the Bystrzyca River with the Strzegomka River in the Sudety Mountains. Relatively unstable minima occur in individual rivers in the Sudety Mountains and the San River catchment in the summer-autumn period. In the annual cycle, the sequence of hydrological periods is represented by low low-water (or normal period) → average high-water → normal period (or low low-water) → low high-water → average low-water. Sometimes the high-water period in spring is prolonged and connected with the high-water period in summer (the San River).

Type 5 – the pluvial-nival regime. This regime is represented by rivers whose average flow of the summer month is higher or almost equal to the average flow of the spring month, and in both cases, the flow is generally 130–180% of the average annual flow. Such regularity in the Sudety Mountains is exhibited by the Nysa Kłodzka River, and in the Carpathian Mountains by the upper Vistula River, the Soła and Uszwica rivers, and rivers in the Dunajec catchment. At the country level these rivers are distinguished by the highest flows, which on average are about 500 mm, and in the case of mountain streams are higher than 1000 mm, or even 1500 mm (the Potok Kościeliski River, the Białka River at gauge Łysa Polana). That flow consists mainly of surface flow, while the share of the groundwater supply is usually lower than 40%. As a rule, rivers with that regime are also distinguished by high variability of daily flows ($Cv > 1.5$). The date of flooding in these rivers is relatively stable and stable. Such a feature of the flow regime is characteristic of rivers in the Nysa Kłodzka, Soła and Dunajec catchments, in which the stable date of monthly maxima falls in the spring or the spring–summer period. In turn, strongly stable maxima are recorded only in the Białka and Potok Kościeliski rivers (May–July). In most rivers, the minimum monthly flows have a relatively stable date of occurrence, usually falling in the autumn, autumn and winter periods, and in the Tatra Mountains rivers in the winter period. Many rivers in the Dunajec River catchment are characterized by the stable low-water period. The most stable date of the appearance of monthly minima is characteristic only of the Tatra Mountains streams: the Białka and Potok Kościeliski. In the sequence of hydrological periods, the regime of these rivers is characterized by the occurrence of the average (low) low-water → normal period → long, high high-water in spring–summer → normal period → average low-water.

9.3 Changes of Flow Regime Patterns

Contemporary climate changes and the human impact on water relations make the features of the river flow regime change. Studies on the regularity of their variability in the multi-year period (1951–2010) and the hydrological cycle allow identifying groups of rivers having different scales and causes of modification of the hydrological regime [14]. Rivers with the largest changes in the flow regime caused by human activity are found in different regions of the country. These include: the Kłodnica, Brynica and Czarna Przemsza rivers in the Silesian region, the Oława River in Lower Silesia, the Ner River flowing through the Łódź agglomeration, rivers in the hydrographic system of the Drwęca, Ina and Drawa river catchments in the Pomeranian Lake District, and rivers of Przymorze – the Wieprz and the Łupawa. These rivers are distinguished by a decrease in flow in the years 1951–1970 and an increase in 1971–1990. Observed changes usually occur in the winter period (from mid-December to mid-February) and in the summer-autumn season (from mid-June to the end of the hydrological year). In the summer and autumn seasons, a higher flow was also observed in 1981–2000 (from mid-June to the end of the hydrological year). The different nature of changes in the flow regime is presented by other rivers transformed by the human activity in Silesia: the Brynica and Biała Przemsza, but also the lower Nysa Kłodzka River, the upper Warta with the Oleśnica, Barycz, Pilica, Łasica, Rawka rivers, the upper Noteć River with the Gąsawka and Wierzyca rivers, the Wda and Brda rivers in the lower Vistula basin. Until 1985, higher, and after that year, lower flows than the average were observed. The biggest changes usually occur at the beginning of the hydrological cycle, that is in November and December. In the years 1961–1980, these rivers were characterized by an increase, and in the years 1983–2002 a decrease in flow. In 1961–1980 a higher flow was also observed from mid-April to early July, and in some rivers even up to the end of the hydrological year. The more stable flow regime is represented by rivers located in the north-eastern part of the country – in the Narew River catchment (except for the Bug River), some rivers of Przymorze (the Rega, Parsęta and Słupia rivers) and the lake-district Gwda River. However, in 1951–1970 usually lower than average flows were observed in these rivers, especially in the winter season (in the Narew River from January to March). In turn, in the years 1970–1989 higher flows occurred in the summer-autumn period, and in the years 1975–2000 also in winter. Other regularity can be observed in the case of the majority of rivers in the middle and lower Warta River catchment, in the middle and lower reaches of the Odra basin and in rivers in the Pilica catchment. In the years 1961–1990, these rivers were characterized by a clear increase in flows from December to the beginning of February, as well as from May to July. At the scale of the whole country, the most stable features of the flow regime are represented by rivers of the upper and middle Vistula basin together with the Bug River and the Sudety Mountains tributaries of the Odra River. Exceptions include the Dunajec River at gauge Czorsztyn, the Nysa Kłodzka River, and the Wieprz River at gauge Zwierzyniec.

Rivers with similar changes in flow conditions, except from those with transformation of water relations caused by the human activity, also represent similar physical and geographical regions. This indicates the importance of climatic conditions in the modification of the characteristics of the flow regime, and the spatial diversity of these changes results from the local variability of the flow caused by the environmental conditions of the catchment. Certain evidence of the influence of climatic conditions on changes in the flow regime in the winter-spring period may be a significant decrease in winter flows observed in many rivers, and a delayed increase in the spring thaw in the 1950s and 1960s. In turn, in the 1970s and 1980s in these rivers there was a disappearance of the winter low-water stages associated with a remarkable increase in the winter flow, sometimes prolonging the less pronounced spring high-water period. These regularities indicate the noticeable impact of changes in the macroscale intensity of the air circulation type, which is the North Atlantic Oscillation, on the transformation of the characteristics of the flow regime of rivers in Poland.

The North Atlantic Oscillation (NAO) is attributed to a very important, climactic role in Poland [15–18]. There are also more and more papers documenting the impact of the North Atlantic Oscillation on the flow of rivers in Poland. Kaczmarek [19] pointed to the impact of the NAO on the magnitude of high-water stages caused by thaw in the Central European rivers. In the positive phase of the NAO, lower spring high-water stages are usually observed than in the negative phase. Other studies also confirmed the impact of the NAO on the Warta River flows [20] and the existence of asynchronous relationships between the winter NAO indices and flows of some Carpathian rivers and the Vistula River [21, 22]. Analyses of changes in hydrological periods and features of the river flow regime in various phases of the NAO_{DJFM} were made by Wrzesiński [23–26], and Wrzesiński and Paluszkiewicz [27]. The height of the flow in various NAO_{DJFM} phases is temporally and spatially differentiated. In the positive NAO_{DJFM} phase in the winter months (January–February), higher flows are observed in rivers in the north-eastern part of the country and the mountains. In the spring months, higher flows in rivers almost all over the country are in the negative phase of the NAO_{DJFM}. Significantly lower than the average flows in the positive phase of the NAO are observed in the summer months. As a consequence, at this stage of the NAO_{DJFM} the annual flows in most rivers are lower than average, and in rivers between the Odra and Vistula by as much as 20–40%. This is also confirmed on the maps of average flows in rivers in both NAO_{DJFM} phases (see Figs. 9.3 and 9.4).

Wrzesiński [28] presented the directions of the transformation of the types of the flow regime of rivers in Poland in the years 1971–2010 in various phases of the NAO_{DJFM}. In that paper, values of the NAO_{DJFM} winter index were used. That index is the normalized mean difference in the atmospheric pressure from December to March, between Lisbon and Stykkisholmur and Reykjavik in Iceland [29]. The type of regime was established in accordance with the criteria proposed by Dynowska [12] for the entire period 1971–2010 and for 10 years with high ($NAO_{DJFM} > 2.20$) and 10 years with low ($NAO_{DJFM} < -0.23$) values of the NAO_{DJFM} winter index. These numbers correspond to the first and third quartiles from the entire NAO_{DJFM} index set from the years 1971–2010. In the examined variants, the rivers in the distinguished types of the regime are characterized by a similar distribution and range of changes in monthly flows in the average annual cycle. The distinguished groups of rivers represent the same types of regime. However, they differ in numbers, which has consequences in a different picture of their spatial distribution (Fig. 9.3). The analysis shows that in the studied NAO_{DJFM} phases,, there is a frequent transformation of the flow regime of many rivers. Compared to the average conditions (1971–2010), in

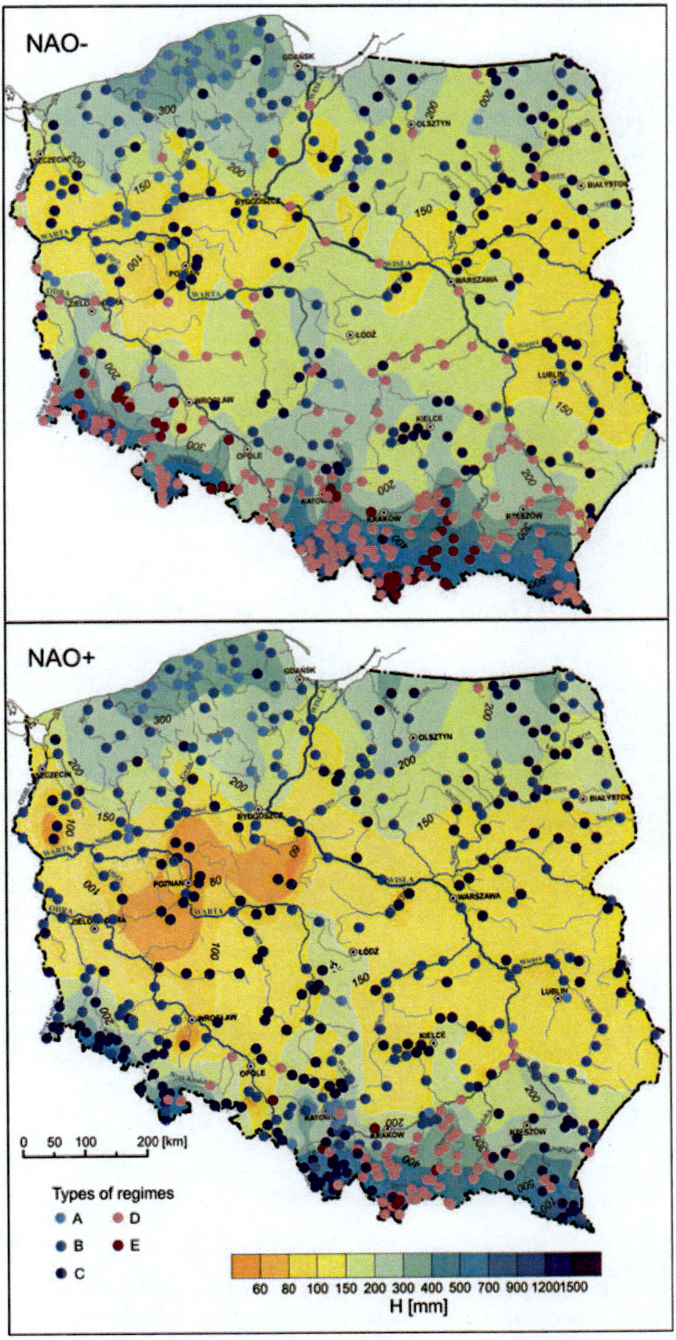

Fig. 9.3 Spatial distribution of types of regime against the average annual flow in the negative (NAO–) and positive (NAO+) phases of the North Atlantic Oscillation. Types of regime: A – nival poorly formed; B – nival moderately formed; C – nival well-formed; D – nival-pluvial; E – pluvial-nival

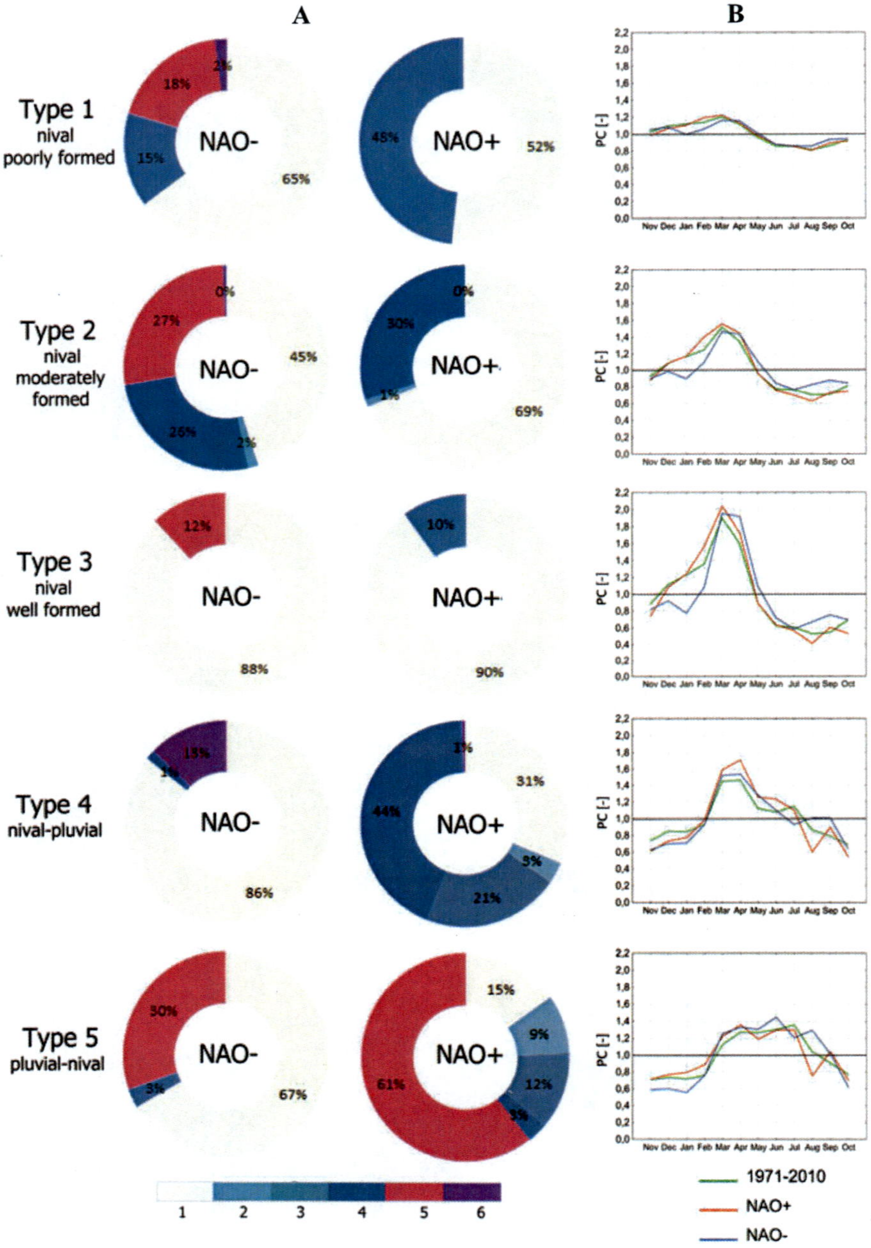

Fig. 9.4 Directions of transformation of regime types (A) and the monthly flow coefficients – Pardé coefficients (PC) (B) in different NAO phases. 1 – no change; 2 – change into type 1; 3 – change into type 2; 4 – change into type 3; 5 – change into type 4; 6 – change into type 5

the negative phase of the NAO_{DJFM} the most stable types of regime are type 3 (the nival well-formed regime) and type 4 (the nival-pluvial regime), which in 85% of river profiles have not changed. Few changes of the regime consist of transformation from type 3 into type 4, and from type 4 into type 5, which proves an increase in flows in the summer months. In this phase, a less stable type of regime is type 1 – the nival poorly formed regime, and type 5 – the pluvial-nival regime. In both cases, in approximately 35% of river profiles representing these types of regime a transformation of type occurs. In the case of type 1 of regime, there is a change into the nival moderately formed regime or the nival-pluvial regime, while the pluvial-nival regime is transformed into the nival-pluvial one. Most often, in more than 55% of cases, the nival moderately formed regime is transformed. This regime is changed into the nival well-formed or the nival-pluvial regime. The observed direction of transformation indicates that in the negative phase of the NAO_{DJFM} in many rivers, there is an increase of the spring and summer flows. In the positive phase of the NAO_{DJFM}, type 3 remains the most stable type of regime – the nival well-formed regime. It represents 90% of all profiles, at which it was identified for the average conditions. On a small number of lake-district rivers, this type of regime is changed into the nival moderately formed one. Compared with the average conditions in the positive phase of the NAO_{DJFM}, more frequent transformations are subject to type 2 – the nival moderately formed regime (mainly changed into the nival well-formed one – in some lake-district and upland rivers, and in the Prosna River catchment) and type 1 – the nival poorly formed regime, which in rivers, mainly of the eastern part of the Pomeranian Lake District, is transformed into the nival moderately formed regime. Compared to the average conditions, in the positive NAODJFM phase, types 4 and 5 are of the least stability. The nival-pluvial regime is transformed at almost 70% of profiles into the nival moderately formed type (the middle and lower reaches of some rivers in the Sudety Mountains and the San River), but mostly into the nival well-formed type (most of the upper reaches of the Sudety Mountains rivers, the Vistula and San rivers). The most frequent transformations are observed in the case of rivers, which in average conditions represent type 5 – the pluvial-nival regime. In the positive phase of the NAO_{DJFM} they account for only 15% of cases, in the other cases the regime is changed, mainly into type 4 – the nival-pluvial regime. This direction of transformation is observed in most rivers in the Dunajec catchment. Research confirms that there is a temporarily and spatially differentiated impact of changes in the intensity of the North Atlantic Oscillation on the level of river flows in Poland. This impact is not strong but noticeable. It is observed with varying intensity in rivers in many regions of the country, mainly in winter, spring and summer. In the negative phase of the NAO_{DJFM}, it was observed in 31% of the examined rivers, and in the positive one – in 43%. In particular, changes in flows of the spring and summer seasons are crucial due to the adopted typology criteria of the river regime. Thus, the observed transformations of the regime type in the two NAO_{DJFM} phases against the background of average conditions are understandable and indicate the possible destabilization of the characteristics of the flow regime of many rivers in Poland in the changing climatic conditions caused by the varying intensity of the macroscale type of circulation, which is the North Atlantic Oscillation.

9.4 Conclusions

Due to different environmental conditions, rivers in Poland are characterized by a relatively large diversity of water supply conditions and seasonality of the river flow, enhanced by its variability. This results in the distinction of five types of flow regime. The lowland areas are dominated by three simple nival regimes: poorly, moderately and well-formed, while in the highlands and in the mountain areas two complex regimes, namely nival-pluvial and pluvial-nival can be distinguished. Determination of the features of the river flow regime based on different approaches: supervised and unsupervised, allowed a better recognition of the regularity of the river flow variability in the annual cycle. Key elements determining water resources, such as duration, time of appearance and stability of the high and low water periods were included. The paper also describes how contemporary changes in climatic conditions and human impact on water conditions affect the characteristics of the river flow regime. Multiannual and seasonal flow changes were found, which are of particular importance in the spring and summer seasons due to the adopted criteria of the typology of the river regime. Studies confirmed that the destabilization of the flow regime characteristics of many rivers in Poland may be triggered by the changing climatic conditions. These are caused by the varied intensity of macro-scale air circulation types, such as the North Atlantic Oscillation, whose impact is not strong but noticeable and statistically significant in the winter season. Recent research revealed that the climatic conditions and flow regime of rivers in Poland may also be influenced by the North Atlantic Thermohaline Circulation (NA THC) [30].

9.5 Recommendations

The obtained results of this study indicate that further research is needed to detect changes in the water cycle, the amount of the river flow, and its regime at various spatial scales in order to identify the hydrological consequences of contemporary climate change and human impact.

Acknowledgements The author is grateful to the Institute of Meteorology and Water Management in Warsaw for providing the data used in this paper.

References

1. Woś A (2010) Climate of Poland in the second half of the 20th century. Wyd. Naukowe UAM, Poznań, 489 p (In Polish)
2. Gutry-Korycka M (2001) Geographical conditions of river regimes. WGSR UW, Warszawa, 192 p (In Polish)
3. Dębski K (1961) Hydrological characteristics of Poland. PWN, Łódź-Warszawa (in Polish)

4. Mikulski Z (1963) Outline of hydrography of Poland. PWN, Warszawa ((In Polish))
5. Pardé M (1957) Rivers. Warszawa, PWN, 234 p (in Polish)
6. Lvovich MI (1938) Experience from classification of the USSR's rivers. Trudy GGI, 6, Leningrad (in Russian)
7. Dynowska I (1971) Types of river regimes in Poland. Zeszyty Naukowe UJ, CCLXVIII, Prace Geogr, vol 28, 150 p (in Polish)
8. Gottschalk L (1985) Hydrological regionalisation of Sweden. Hydrol Sci J 30:65–83
9. Rotnicka J (1988) Taxonomic foundations of the classification of river regime; the example of the catchments of the Odra and Polish coastal rivers. Wyd. UAM, Seria Geografia, vol 40, 130 p (in Polish)
10. Rotnicka J (1993) A typology of hydrological periods for use in river regime studies. Quaestiones Geographicae 15(16):77–95
11. Dębski K (1970) Hydrology. Arkady, Warszawa ((In Polish))
12. Dynowska I (1997) The river flow regime. In: Atlas of the Republic of Poland, Główny Geodeta Kraju, Warszawa
13. Wrzesiński D (2017) River regimes. In: Jokiel P, Marszelewski W, Pociask-Karteczka J (eds) Hydrology of Poland, PWN, Warszawa, pp 215–221 (in Polish)
14. Wrzesiński D, Sobkowiak L (2018) Detection of changes in flow regime of rivers in Poland. J Hydrol Hydromech 66(1):55–64
15. Marsz A, Żmudzka E (1999) The North Atlantic oscillation and the lenght of a vegetative period in Poland. Przegl Geofiz 44(4):199–210 ((In Polish))
16. Marsz A (1999) The North Atlantic oscillation and the thermal regime in the area of north-west Poland and the Polish coast of the Baltic Sea. Przegl Geogr 71(3):225–245 ((In Polish))
17. Marsz A, Styszyńska A (2001) North Atlantic oscillation and air temperature in Poland. Wyższa Szkoła Morska, Gdynia, 107 pp (in Polish)
18. Styszyńska A (2001) North Atlantic oscillation and precipitation on the territory of Poland. Prace Stud Geogr 29:232–241 ((in Polish))
19. Kaczmarek Z (2003) The impact climate variability on flood risk in Poland. Risk Anal 23:559–566
20. Styszyńska A, Tamulewicz J (2004) Warta river discharges in Poznań and atmospheric circulation in the North Atlantic region. Quaestiones Geogr 23:63–81
21. Limanówka D, Nieckarz Z, Pociask-Karteczka J (2002) The North Atlantic Oscillation impact on hydrological regime in Polish Carpathians, interdisciplinary approaches in small catchment hydrology: monitoring and research FRIEND international conference, Demanovska Dolina, 132–213
22. Pociask-Karteczka J, Limanówka D, Nieckarz Z (2002–2003) The North Atlantic Oscillation impact on hydrological regime in Polish Carpathians (1951–2000). Folia Geographica, series GeographicaPhysica, 33–34, 89–104 (in Polish)
23. Wrzesiński D (2005) Changes of the hydrological regime of rivers of Northern and Central Europe in various circulation periods of the North Atlantic Oscillation. Quaestiones Geogr 24:97–109
24. Wrzesiński D (2008) Typology of spatial patterns seasonality in European rivers flow regime. Quaestiones Geogr 27A(1):87–98
25. Wrzesiński D (2011) Regional differences in the influence of the North Atlantic Oscillation on seasonal river runoff in Poland. Quaestiones Geogr 30(3):127–136
26. Wrzesiński D (2013) Entropy of river flows in Poland. Studia i Prace z Geografii i Geologii 33. Bogucki Wydawnictwo Naukowe, Poznań, 204 p (in Polish)
27. Wrzesiński D, Paluszkiewicz R (2011) Spatial differences in the impact of the North Atlantic Oscillation on the flow of rivers in Europe. Hydrol Res 42(1):30–39
28. Wrzesiński D (2018) Typology of the river flow regime in Poland in different phases of the North Atlantic Oscillation. Badania Fizjograficzne R. IX, Seria A—Geografia Fizyczna (A69):265–278

29. Hurrell J (1995) The climate data guide: Hurrell North Atlantic Oscillation (NAO) Index, National Center for Atmospheric Research, Boulder, USA. https://climatedataguide.ucar.edu/climate-data/hurrell-north-atlanticoscillation-nao-index-station-based. Last modified 20 Oct 2015
30. Wrzesiński D, Marsz AA, Styszyńska A, Sobkowiak L (2019) Effect of the North Atlantic Thermohaline circulation on changes in climatic conditions and river flow in Poland. Water 11(8):1622

Chapter 10
Flow Seasonality in Two Big Polish Rivers – The Vistula and the Oder

Paweł Jokiel and Przemysław Tomalski

Abstract The aim of the chapter is to identify the flow seasonality of Vistula river (14 water gauges) and the Oder river (11 water gauges) located along the river's course. For this purpose seasonality index (*IS*), concentration date (*PK*) half-flow date (*TPO*), and seasonality coefficient (*GMO*) were used. IMGW-PIB shared hydrometric data for the period from 1951-2016. Average concentration dates and half-flow dates occurred in April or May in analysed cross-sections. Along Vistula and Oder rivers mean *PK* and *TPO* regress towards winter (are earlier and earlier). Average seasonality index comprises in the range 20–32%, while seasonality coefficient falls into the range 9.5–14.5. The highest *IS* characteristic are for the upper part of the rivers. Next *IS* rapidly decreases, due to the summary of the flows from the rivers' tributaries of which are characterized by various conditions of outflow. A clear correlation was found between the concentration date (*PK*) and half-flow date (*TPO*) for both rivers. These parameters demonstrate a linear trend and *PK* grows in line with *TPO*. However, their regression lines have different inclinations. The rise is slower for the Vistula but this difference ceases as *TPO* and *PK* decrease with a growing area of both basins. Covariation between average seasonality indices *IS* and flow seasonality coefficient *GMO* is less pronounced in each of the rivers than in the case of *PK* and *TPO*. Multiannual changes in the concentration date (*PK*) for the Vistula and the Oder irregularly oscillate around average values, and average *PK* values for the upper, middle and lower parts of the catchments are nearly identical. The similarity of *PK* indices in individual years for the Vistula and the Oder in the analyzed multi-year period occurred in the lower and middle course of the rivers. For the upper sections, the differences between rivers are larger, and for the last 30 years there is a growing tendency for earlier concentration date for the Oder. Changes in the flow seasonality index *IS* showed huge variability and depended on the catchment size. Contrary to *PK*, relatively the greatest similarities between the Vistula and the Oder were noticed for their upper sections ($R = 0.72$). As the catchment area increased,

P. Jokiel (✉) · P. Tomalski
Department of Hydrology and Water Management, Faculty of Geographical Sciences, University of Łódź, Narutowicza 88, 90-139 Łódź, Poland
e-mail: pawel.jokiel@geo.uni.lodz.pl

P. Tomalski
e-mail: przemyslaw.tomalski@geo.uni.lodz.pl

M. Zeleňáková et al. (eds.), *Management of Water Resources in Poland*, Springer Water,
https://doi.org/10.1007/978-3-030-61965-7_10

the seasonality of the Vistula flows became considerably higher than of the Oder, and the differences were the most pronounced in the 1960s and 1970s. In recent years, their significance has been declining.

Keywords Flow seasonality · Multiannual changes · Spatial diversity · Big Polish rivers

10.1 Introduction

The catchment area of the Baltic Sea drains 99.7% of Poland's territory. Polish borders are largely hydrographic boundaries, as the country comprises nearly 87% of the Vistula basin and over 89% of the Oder basin. These are favorable conditions for rational water management, as the vast majority of Polish water resources origin within the same country. The basins of the Vistula, the Oder, rivers of the Baltic Sea coast and the catchments of the Vistula Lagoon and the Szczecin Lagoon drain nearly 98.9% of Poland's area. Only 1.1% of the country territory is found within the catchments of the Niemen (the Świsłocz, the Czarna Hańcza, the Szeszupa rivers), the Dniester (the Strwiąż river), the Danube (the Czarna Orawa and the Metuje river) and the Łaba (the Izera river). The springs of two out of four largest Polish rivers are located outside Poland, i.e. the Oder in the Czech Republic and the Bug in the Republic of Ukraine. This is one of the reasons why total river flow from Poland is formed within the area by 12.3% larger than the country's area and covers 351,028 km^2. At the same time, only 87% of the country's river water resources originate from the territory of Poland.

The Quaternary evolution of the Polish river network resulted in a considerable asymmetry of left and right basins of major Polish rivers (the Vistula: 72.8% ÷ 27.2%; the Oder: 70.5% ÷ 29.5%). The same processes caused the main right tributaries of the Vistula and the Oder (the Narew with the Bug and the Warta) to be longer and have larger or slightly smaller basins than the main rivers. Another characteristic feature of the Polish river network, particularly within the Polish Lowlands, are relatively low drainage divides running across wide ice marginal valleys. One of these is the main drainage divide separating two largest basins of the Vistula and the Oder.

Due to human pressure and local corrections of the river network, the number of Polish rivers is difficult to estimate and varies over time. According to IMGW, it amounts to 4,656 including:

- 2,913 watercourses with catchment area smaller than 50 km^2 – (62.6% of all rivers),
- 1,553 watercourses with catchment area between 50 km^2 and 500 km^2 – (33% of all rivers),
- 182 rivers with catchment area between 500 km^2 and 10 000 km^2 – (4% of all rivers), and
- 8 rivers with catchment area larger than 10,000 km^2 – (0.4% of all rivers) [3].

Poland harbors 488 medium and large rivers with catchment area exceeding 200 km^2, of which 250 (over 51%) belong to the Vistula basin and 177 (36.3%) to the Oder basin. Only 61 rivers collect water from other basins of the Łaba, the Dniester and the Niemen. The total length of those 488 rivers amounts to 33,456 km, of which 55% are 20–50 km long, and 30% are 50–100 km long. Only four Polish rivers are longer than 500 km (the Vistula, the Oder, the Narew with the Bug, and the Warta), and only the Vistula is longer than 1000 km.

10.2 Physiogeographic Features of the Studied Rivers

The Vistula is considered to originate at the springs of the Czarna Wisełka, and the Biała Wisełka streams at the northern slopes of Barania Góra in the Silesian Beskids at an elevation of 1116 m amsl The mouth of the river is curbed with a 7.1 km long canal with embankments constructed in 1895 that directs the river water into the Baltic Sea near Świbno. The Vistula stretches for about 1022 km, and its basin area is 194,424 km^2. Of these, 169,000 km^2 belong to Poland and comprise 54% of the country territory.

The first, longest section of the river that runs for nearly 400 km from the springs to the mouth of the San is called the Upper Vistula. This section collects water from the most water-abundant area in Poland – the Upper Vistula catchment (c.a. 30% of Polish water resources). The basin area belonging to Poland covers 45,900 km^2. Its key regions include the Western and Eastern Beskids drained by a few dozens of large mountain rivers (e.g. the Soła, the Skawa, the Raba, the Dunajec, the San and many other), with pluvial-nival and nival-pluvial regimes, and a large fragment of Central Poland Highlands drained by a few dozens of highland rivers (e.g. the Przemsza, the Szreniawa, the Nida, the Tanew and other), with poorly or moderately developed nival regime and considerable share of underground water supply [11, 25]. As the Vistula is considered a navigable river only from the mouth of the Przemsza, this spot is treated as the starting point of its chainage. The section of so-called the Small Vistula is roughly closed with a gauge in Nowy Bieruń.

The central, 256 km long section of the Vistula begins below the mouth of the San (a gauge in Zawichost) and ends below the mouth of the Narew (slightly above a gauge in Kępa Polska). This part of the Vistula basin (51%) covers about 98,900 km^2, of which 88,800 km^2 belong to Poland. This section of the river is mainly supplied from the right, lowland part of its catchment (83.4% of the entire catchment area) drained by the Wieprz, the Bug and the Narew and the Biebrza, i.e. rivers with the continental regime and moderately or strongly developed nival supply. Left tributaries are less important from a hydrological perspective, and the largest ones comprising the Kamienna, the Radomka and the Pilica are typical highland rivers with usually moderately or sometimes strongly developed nival regime. Interestingly, if the Narew is treated as a tributary of the Bug, then that later is longer than the Vistula at the point of their junction [13].

The Lower Vistula begins below the mouth of the Narew and runs for 391 km until the Baltic Sea. This part of the basin is the smallest and covers only 34,100 km^2 (23% of the entire basin), but, unlike the other two, it belongs entirely to Poland. The catchment of this part of the Vistula basin is almost symmetrical, and the draining rivers are either of lowland character (e.g. the Bzura, the Skrwa), with minimal water resources and strongly or moderately developed nival regime or of lakeland character (e.g. the Drwęca, the Brda, the Wda, the Wieżyca) with huge water resources, high retention, and underground supply, and poorly or moderately developed nival regime. Mean multiannual (1951–2016) flow at the most downstream gauge of the Vistula at Tczew was SSQ (multiannual mean flow) = 1059 $m^3 \cdot s^{-1}$, and extreme flows recorded for the same period reached WWQ (multiannual high flow) = 7750 $m^3 \cdot s^{-1}$ (maximum, June 1962) and NNQ (multiannual low flow) = 264 $m^3 \cdot s^{-1}$ (minimum, January 1954).

The springs of the Oder are located in the Czech Republic at 632 m amsl on the slope of Fidlův Kopeč (the Oderské vrchy). The river enters the Baltic Sea via a shallow part the Szczecin Lagoon (the Oder Bay) at 840 km of its chainage (726 km in Poland). The Oder is navigable for 736 km, and almost 162 km of the river constitutes a border between Poland and Germany.

Mountain features can be spotted for only the first 50 km of the Oder. As soon as it enters the Moravian Gate, its character changes to submontane river, and its slope drops down to 0.7–0.9‰. Along the Moravian-Racibórz section, the Oder is supplied by three tributaries (the Opawa, the Ostrawica, and the Olza) that increase its water supply by several times. The Opawa mouth is also a starting point of the Oder chainage. Below the Ostrawica mouth, the Oder enters a flat, marshy Racibórz Basin where it meanders and forms numerous ponds. The river crosses Czech-Polish border near the gauge in Chałupki. Between the border and the mouth of the Gliwice Canal, the Oder is supplied by its left tributary the Psina and right tributaries the Ruda and the Bierawka. The Kędzierzyn-Koźle gauge marks the upper, 202 km long section of the Oder. The river catchment area at this gauge is 9,173.6 km^2. Water resources of this area are not large, but they vary substantially over time [1]. Right tributaries have moderately developed nival regime, and left ones have nival-pluvial character [25].

The longest, middle section of the Oder begins in the Racibórz Basin. Further on the river crosses two vast lowlands – Silesian Lowland and Great Poland Lowland. The end of the Oder middle section is marked by the mouth of the Warta at the border of the South Pomeranian Lakeland (at 618 km). The Warta's mouth is located nearly exactly in between the gauges in Słubice and Gozdowice. Physical geography of the middle section of the Oder basin is highly diverse. The south-western (left) part is drained by water abundant rivers originating from the Sudetes and submountain regions (e.g. the Nysa Kłodzka, the Bystrzyca, the Kaczawa, the Bóbr, the Nysa Łużycka, and several others). Right tributaries are usually of lowland character (lakeland in the north) and their water resources are typically low (rarely moderate). The largest right tributaries of this section of the Oder include the Mała Panew, the Widawa, and the Barycz. The Warta, the largest tributary of the Oder, is 795 km long and its catchment area reaches 53,666 km^2. Water resources of the Warta catchment show high spatial diversity – they are moderate in the upper (upland) section, small in the middle and lower (lowland) section, to very small locally [18]. Together with the Warta catchment, the central part of the Oder basin covers an area of 99,018 km^2. The rivers in this part of the basin have diverse regimes. The left tributaries fed from Sudeten springs have nival-pluvial or locally pluvial-nival regimes, while lowland right and some left tributaries have moderately or locally strongly developed nival regime [25].

The Lower Oder is a lowland, engineered river widely used for navigation. Its lower section is 144.3 km long, and its basin covers 10,796 km^2 (7,221 km^2 in Poland). The watercourses entering the Oder from Poland include the Myśla, the Rurzyca, the Płonia (via Dąbie Lake) and the Ina, and from Germany the Finowkanal and the Weise. These watercourses have moderately developed nival regime. Polish rivers typically feature a high share of underground water supply, while the flows in the watercourses entering the Oder from Germany depend on the needs for draining some sections of old Oder beds and requirements of navigation within E70 waterway. Mean multiannual flow of the Oder slightly above its mouth (Widuchowa water gauge) is SSQ (multiannual mean flow) = 534.6 $m^3 \cdot s^{-1}$ (1974–2013), while the maximum and minimum flows reached WWQ (multiannual high flow) = 2980 $m^3 \cdot s^{-1}$ (August 1997) and NNQ (multiannual low flow) = 153 $m^3 \cdot s^{-1}$ (August 1992), respectively. Interestingly, both extreme flows occurred in the same month.

10.3 Physical and Geographical Conditions of Flow Seasonality in Polish Rivers

The hydrological effects triggered by increased area and physiographic diversity of a catchment are well known among hydrologists. They generally affect the flow regime in qualitative and quantitative terms and change the character of the river and its valley. As the catchment area increases, medium and maximum unit flow is usually reduced, and minimum flow rises. Therefore, flow variability drops and its inertia grows as a consequence of elevated retention. In larger catchments, extreme events (floods, low water) develop and cease more slowly and last longer than in small catchments. Also, extreme flows less closely correlate with the onset of the triggering factor. On top of all these, there are also effects caused by changes in stationary and non-stationary conditions and factors of flow formation.

One of the features shaping the catchment flow regime is its seasonality. Information on seasonality of river flow is the most basic and most important in hydrological practice and water management. The temporal regime of the river flow obviously depends on seasonal diversity in abundance and type of precipitation and intensity of surface evaporation. However, apart from external (e.g. climatic) factors, the flow is affected by a number of phenomena and processes associated with the river and its catchment. The catchment and the river system form a set of filters in which the external impulses (weather, climatic or anthropogenic) are so strongly transformed that their final form and spatial distribution differ strongly from the input. For example, different types of retention create a specific hydrological memory that increases flow inertia, while reshaping and slowing down precipitation events advancing downstream the catchment [11].

The annual, more or less clear repeatability of hydrological events in Poland is an obvious consequence of climate-dependent changes in seasons and hence repeated, seasonal changes in the size and type of precipitation and its redistribution. Spring is expected to bring snowmelt, rainfall and high flows in rivers. Summer features low water and occasional rainfall-related floods. Low water levels caused by limited or lack of precipitation in the summer may further drop in the autumn or revert due to increased moisture and reduced evaporation. In the winter, river flows usually build up and are later transformed into mid-winter and then spring thaws. It seems therefore that the annual cycle is closed and its phases clearly outlined. However, its repeatability over years is only relative. This is because winters can be frosty and long or mild and short. They can also be snowy or nearly snowless. These six parameters are enough to generate 12 types of winter with features triggering a highly variable course of hydrological events. They would affect not only winter, but also spring and even summer events, as Poland has large retention related flow inertia, and water resources in the country are formed mainly in the winter half-year (October-April).

Other seasons also show high multiannual variability of climatic conditions and features. As a result, the times of floods and low waters change rather freely over the year timeline. These changes in timeline position bring about variations in the event parameters and causes. If we combine the already noticeable hydrological effects of permanent or temporary climatic changes and human pressure, it is clear that the term "flow seasonality" is increasingly imprecise and requires clarification or even periodic redefinition [9].

Large physical and geographical variations (relief, geological structure, groundwater, climate, land cover) of the Vistula and the Oder basins affect not only the conditions and factors determining the river flow but also change the course of water concentration, drainage, and transport along the river beds. In consequence, water regimes of the Vistula and the Oder vary along their length, as they become resultants of an increasing number of factors and conditions occurring within a growing and more diversified area. The aim of this work is therefore to identify the level and the character of flows seasonality in the Vistula and the Oder and their fluctuations with the increasing length of the rivers, their catchment areas and physical and geographical diversity (Fig. 10.1). We will also attempt to assess changes in flow seasonality of the rivers over a multiannual period to identify various spatial and temporal regularities. The study will be based on four seasonality metrics: seasonality index *IS*, concentration date *PK*, seasonality coefficient *GMO,* and half-flow date *TPO*.

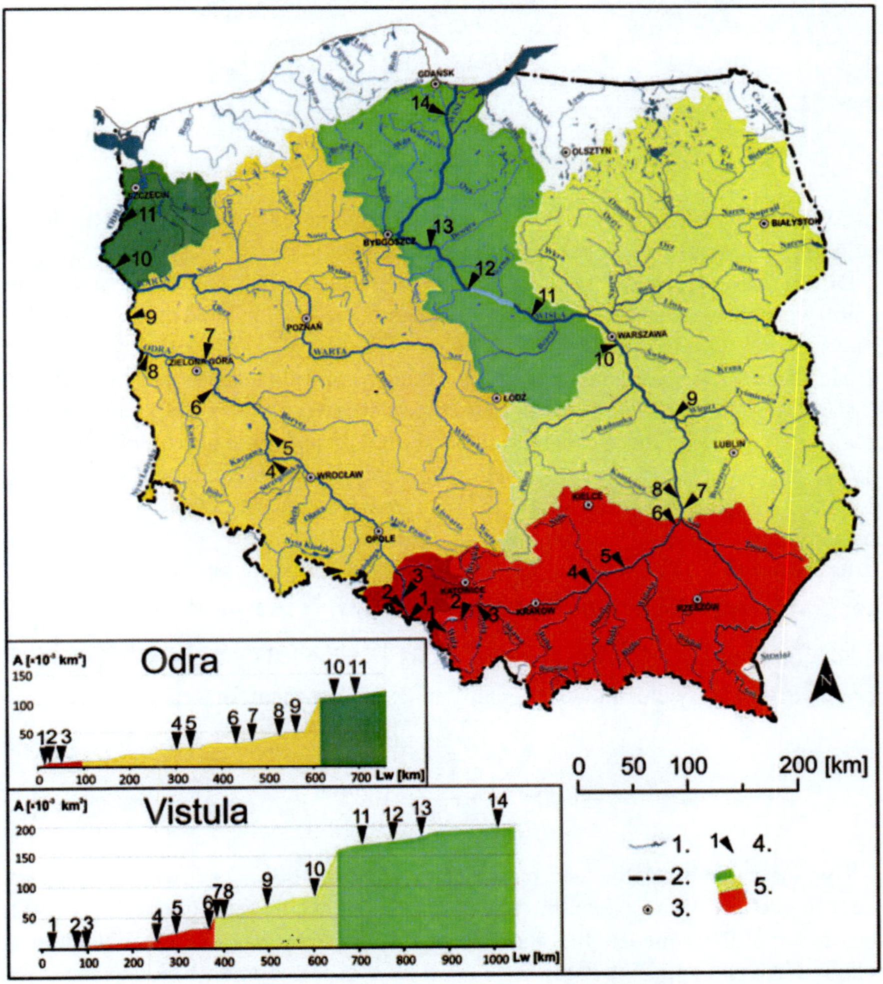

Fig. 10.1 Selected water gauges along the Vistula and the Oder. Explanations: 1 – rivers, lakes, 2 – state borders, 3 – cities, 4 – water gauges (numbering according to Tables 10.2 and 10.3), 5 – the Vistula and the Oder basins; upper basins are marked in red, middle in yellow and lower in green, Lw – river chainage, A – catchment area

10.4 Materials and Methods

The analysis included a series of daily flows from 14 water gauges located along the Vistula and 11 gauges along the Oder (Table 10.1, Fig. 10.1). The series were usually recorded over the years 1951–2016. Shorter series (40–61 years) were analyzed for two gauges on the Vistula (Kępa Polska and Włocławek) and four gauges on the Oder (Krzyżanowice, Malczyce, Nowa Sól and Widuchowa). The data were obtained from

Table 10.1 Calendar of times and dates for *PK* and *TPO*

Day	93	107	122	136	152	166	182	196	213	228	242	257	275
Date	1 Feb	15 Feb	1 Mar	15 Mar	1 Apr	15 Apr	1 May	15 May	1 Jun	15 Jun	1 Jul	15 Jul	1 Aug

IMGW-PIB database. They were used to determine necessary characteristic flows, including matrices of mean monthly flows and flow curves.

Different aspects of flow seasonality at the investigated gauges over the multi-year period were assessed with four metrics of seasonality. Two absolute metrics included concentration date (*PK*) and half-flow date (*TPO*) [4, 6, 15, 17] and two relative metrics were flow seasonality index (*IS)* (Ibidem) and flow seasonality coefficient (*GMO*) [12, 19, 21]. The first and third metrics were suggested by Markham for daily precipitation [15]. Jokiel and Bartnik [5] adjusted them for mean monthly flows.

One may consider a series of average monthly flows each element of which is identified with a vector $\vec{r}(1, 2, 3, \ldots 12)$ with length directly proportional to its value and direction expressed with an angle α_i. The angle is directly proportional to the period between a day in the middle of a month and the beginning of a hydrological year. A resultant of 12 vectors determined this way $\vec{r}_i$ is a vector $\vec{R}$ with module $\left|\vec{R}\right|$ and direction ω. Length of the resultant vector $\left|\vec{R}\right|$ divided by the total length of all partial vectors $|\vec{r}_i|$ yields flow seasonality index (IS_j) typical for year *j*.

$$IS_j = \frac{\left|\vec{R}_j\right|}{\sum |\vec{r}_i|} \cdot 100\% \tag{10.1}$$

This indicator is relative and may range from 0 to 100%. Its lowest values may occur in two situations: when the flow is distributed evenly throughout the year (all vectors are of the same length), and there is no flow seasonality, or when the partial vectors form two opposing resultants (e.g. flow concentration on 1 November and 1 May). For *IS* close to 100%, monthly flows for a given year are extremely seasonal, e.g. the river is a periodic one.

The other parameter suggested by Markham for seasonality investigation is the flow concentration date (*PK*). It is based on the same assumptions as the seasonality index, but it is an absolute value. The angle of the vector $\vec{R}$ is the measure here. It indicates (angle measure) weighted time of flow concentration in the year expressed as ω. In hydrological analysis, it is more convenient to use a specific date instead of ω. It may be obtained from the following formula: $360^\circ = 365$ (366) days. The angle ω for average monthly flows is calculated with the following formula:

$$\omega = arctg\left(\frac{\sum_{i=1}^{12} |\vec{r}_i| \cos \alpha_i}{\sum_{i=1}^{12} |\vec{r}_i| \sin \alpha_i}\right) \quad (10.2)$$

Another absolute measure of seasonality, and more precisely a measure of flow distribution throughout the year was defined almost simultaneously by Bartnik and Jokiel [6] and McCabe and Clark [17]. It is called half-flow date *TPO* or central of mass data *CMD*. The measure is easy to construct and interpret. It is created following an analysis of the relative summary (accumulated) curve of annual flow and indicates the day in the year on which the accumulated flows (from the beginning of the hydrological year) reach 50% of the annual sum. For the year *j*, *TPO* will be determined according to the formula:

$$TPO_j = \{i : V_i = V_{365(12)}/2\} \quad (10.3)$$

where:

V_i total flow [m^3] since 1 November (beginning of the hydrological year),
$V_{365(12)}$ annual total flow [m^3],
TPO_j day (month) when $V_i = V_{365(12)}/2$

The result obtained for *TPO* and *PK* calculation is a day from the beginning of the hydrological year. Traditionally, a hydrological year in Poland begins on 1 November and ends on 31 October. Therefore, all measures and diagrams presented below should be analyzed with this information in mind. To make the analysis easier, Table 10.1 contains a simplified calendar of times and dates.

The team led by Oliver, already mentioned before, used a measure for investigating precipitation distribution over the year (so-called precipitation regime) to work out flow seasonality coefficient (*GMO*). *GMO* is a relative measure based on average monthly flows and therefore time intervals that may also constitute a basis for determination of two previously mentioned Markham's measures. Flow seasonality coefficient for year *j* is calculated using the following formula:

$$GMO_j = \frac{\sum_{i=1}^{12} (SQM_i)^2}{(\sum_{i=1}^{12} SQM_i)^2} \cdot 100 \quad (10.4)$$

where:

GMO_j flow seasonality coefficient for year *j*,
SQM_i average monthly flow for month *i*

The method for *GMO* construction indicates that in the absence of flow seasonality (identical mean flow for all months of the year), its value is 8.3. When the flow happens for only one month (maximum seasonality) the coefficient equals 100.

It should be emphasized, however, that all the above mentioned characteristics have some specific disadvantages. While defining them, we assume that a hydrological year is a "closed" time unit in which daily or monthly flows are independent of the previous ones and would not affect those occurring after them. Considering large flow inertia that is due to, e.g. retention, this assumption is not true, and it should be constantly kept in mind [4].

To assess different types of changes in the vectors of resulting seasonality measures, we used basic tools of statistical data analysis, including time series analysis, correlation, and regression, as well as simple tests and statistical diagrams. The calculations were performed with EXCEL and STATISTICA packages.

10.5 Changes in Flow Seasonality Along Both Rivers

Average multiannual flow in the Oder along its nearly 680 km^2 length ranges from 42 $m^3 \cdot s^{-1}$ to over 535 $m^3 \cdot s^{-1}$. In the Vistula, the flow increase along the river course reaches over 1000 $m^3 \cdot s^{-1}$ (Tables 10.2 and 10.3). Half of the annual flow is released from the upper Oder catchment (*TPO*) between 21 and 25 April, and in the lower section, this time is two weeks earlier and falls between 11 and 12 April. Average flow concentration date (*PK*) for the Oder gauges follows a similar pattern. In the upper section, it falls between 27 and 30 April and in the lower between 3 and 6 April – three weeks earlier. *TPO* changes along the Vistula course are of an analogous character. In its upper section half of the annual flow is released from 28 to 30 April and in the lower section two weeks earlier, i.e. from 15 to 16 April. Importantly, half of the annual water resources of the Vistula, both in its middle and lower section, is released by 4–5 days later than half of the annual water resources of the Oder. The difference is small but stable along both rivers (except for their upper sections). Concentration date coefficients (*PK*) do not show similar regularities. Their dates in both rivers fell from the end of April (upper sections) until the beginning of April (lower sections). The difference between the rivers is about 30 days.

In the Vistula and the Oder, there is a clear relationship between the concentration date (*PK*) and half-flow date (*TPO*). It is of a linear character, and the rise in *PK* accompanying the rise in *TPO* is markedly slower for the Vistula (red points in Fig. 10.2) than for the Oder (blue points in Fig. 10.2). At the same time, the strength of the relationship between the two parameters is greater and the spread of points is smaller for the Oder than for the Vistula (Fig. 10.2). For the same concentration dates (*PK*), the half-flow dates (*TPO*) on the Vistula gauges fall later than on the Oder gauges. This difference disappears rapidly as *TPO* and *PK* decrease with a growing area of both basins.

Comprehensive measures of flow seasonality (*GMO* and *IS*) change significantly as the area of both river basins increases. For the Oder, mean *GMO* and *IS* drop

Table 10.2 Average measures of flow seasonality in the Oder

No	Water gauge	Basin area A [km^2]	River length L [km]	Mean flow [m^3·s^{-1}]	Range of data	Mean TPO		Mean GMO	Mean IS	Mean PK	
						Day	Date			Day	Date
1	Chałupki	4,662.2	20.7	42.02	1951–2016	173	22 Apr	12.4	29.5	180	29 Apr
2	Krzyżanowice	5,874.8	33.6	57.26	1956–2016	173	22 Apr	11.6	25.6	181	30 Apr
3	Racibórz-Miedonia	6,744.0	55.5	66.24	1951–2016	172	21 Apr	11.7	26.6	178	27 Apr
4	Malczyce	26,812.4	304.8	158.35	1971–2016	173	22 Apr	10.2	19.7	181	30 Apr
5	Ścinawa	29,583.8	331.9	180.40	1951–2016	175	24 Apr	10.2	21.2	185	4 May
6	Nowa Sól	36,780.3	429.8	204.64	1971–2016	170	19 Apr	10.0	20.3	173	22 Apr
7	Cigacice	39,888.2	471.3	221.17	1951–2016	170	19 Apr	10.1	21.4	174	23 Apr
8	Połęcko	47,152.0	530.3	255.73	1951–2016	169	18 Apr	10.0	21.1	172	21 Apr
9	Słubice	53,382.0	584.1	302.17	1951–2016	167	16 Apr	9.8	20.8	169	18 Apr
10	Gozdowice	109,729.1	645.3	519.78	1951–2016	163	12 Apr	9.6	20.5	157	6 Apr
11	Widuchowa	110,524.3	701.8	535.36	1974–2016	162	11 Apr	9.5	20.3	154	3 Apr

Explanations: TPO – half-flow date; GMO – seasonality coefficient; IS – seasonality index; PK – concentration date

Table 10.3 Average measures of flow seasonality in the Vistula

No	Water gauge	Basin area A [km^2]	River length L [km]	Mean flow [m^3·s^{-1}]	Range of data	Mean TPO		Mean GMO	Mean IS	Mean PK	
						Day	Date			Day	Date
1	Skoczów	296.7	35.1	6.0	1951–2016	179	28 Apr	13.7	30.7	175	24 Apr
2	Goczałkowice	738.1	68.4	8.8	1951–2016	194	13 May	14.5	31.8	182	1 May
3	Nowy Bieruń	1747.7	109.8	21.1	1951–2016	180	29 Apr	11.6	22.5	170	19 Apr
4	Jagodniki	12,058.2	259.3	129.0	1951–2016	181	30 Apr	10.8	22.4	178	27 Apr
5	Szczucin	23,900.6	300.3	234.2	1951–2016	186	5 May	10.8	24.7	184	3 May
6	Sandomierz	31,846.5	374.6	293.7	1971–2016	182	1 May	10.9	25.4	180	29 Apr
7	Zawichost	50,731.8	393.8	425.6	1951–2016	175	24 Apr	11.1	27.0	165	14 Apr
8	Annopol	51,518.1	404.6	431.1	1951–2016	175	24 Apr	11.0	26.3	165	14 Apr
9	Dęblin	68,234.3	499.2	499.3	1951–2016	175	24 Apr	10.6	24.7	163	12 Apr
10	Warszawa Nad.	84,540.5	610.3	547.1	1951–2016	173	22 Apr	10.4	23.7	167	16 Apr
11	Kępa Polska	168,956.1	712.7	927.0	1969–2016	167	16 Apr	10.0	23.0	153	2 Apr
12	Włocławek	172,389.2	785.6	914.8	1961–2016	166	15 Apr	10.2	24.0	153	2 Apr
13	Toruń	181,033.4	840.2	972.3	1951–2016	166	15 Apr	10.2	23.9	154	3 Apr
14	Tczew	194,376.0	1,014.8	1,058.8	1951–2016	167	16 Apr	10.1	23.4	156	5 Apr

Explanations: TPO – half-flow date; GMO – seasonality coefficient; IS – seasonality index; PK – concentration date

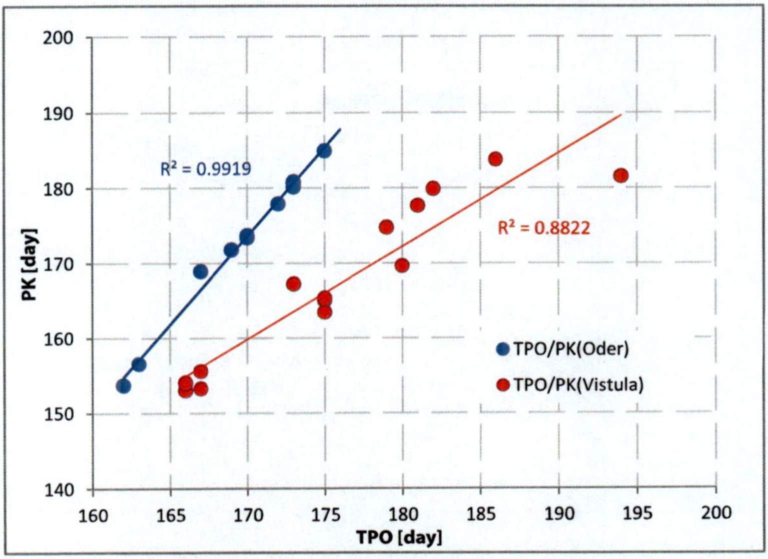

Fig. 10.2 *TPO* and *PK* relationships for the Vistula and the Oder gauges

from 12.4 to 9.5 and from 29.5 to 20.3, respectively. A similar decrease is observed for the Vistula, i.e. from about 14 to 10.1 for *GMO*, and from 31.8 to 23.4 for *IS* (Tables 10.2 and 10.3). However, due to changes in flow regime, these drops do not follow a regular pattern as the area of the basins increases. Exemplary irregularities include a local decrease in seasonality (*IS* and *GMO*) at Nowa Sól gauge for the Oder and local increase in seasonality (*IS* and *GMO*) at Zawichost gauge for the Vistula. These phenomena will be discussed in more detail later in this chapter.

The correlation between both seasonality parameters in the gauge data of both rivers is slightly weaker than that described above for *PK* and *TPO*. However, if all water gauges from both basins are pooled together, the relationship between *IS* and *GMO* is clear and has a curvilinear character (Fig. 10.3). The form of the equation shows that an increase in *GMO* is accompanied by a clear albeit diminishing growth in *IS*. Therefore, the increase in the catchment area significantly reduces flow seasonality.

Changes in all four parameters of flow seasonality depending on the length of the Vistula and its basin area are illustrated in the diagrams in Fig. 10.4. The diagrams clearly show some of the regularities indicated before. *IS* curve collapses along the Vistula course correspond perfectly with regime changes and seasonal structure of the flow caused by entering of two large, water-rich mountain rivers: the Dunajec (Szczucin gauge) and the San (Zawichost gauge), as well as the impact of Włocławek Reservoir. A different seasonal structure of the flow from the catchments of these two rivers causes abrupt fluctuations in the Vistula flow seasonality in the middle and lower sections. The Włocławek Reservoir affects the Vistula in a similar manner, while the entrance of the Narew (Kępa Polska gauge) does not evoke this type of

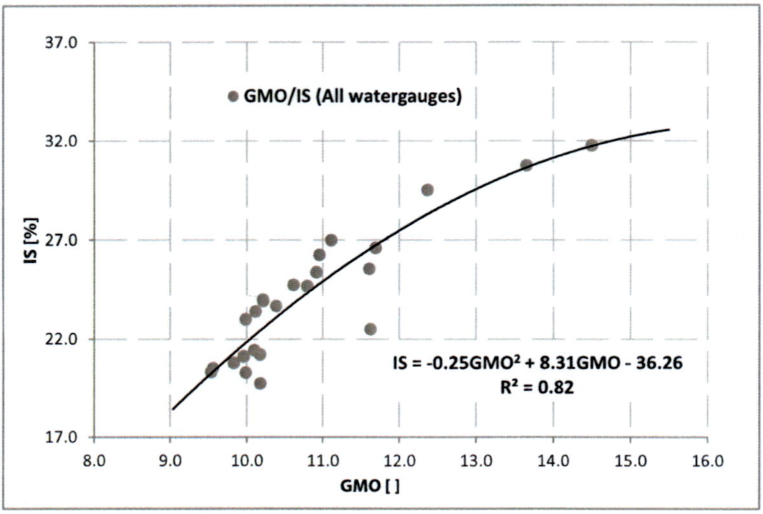

Fig. 10.3 The relationship between the seasonality index *IS* and seasonality coefficient *GMO* for all water gauges of the Oder and the Vistula

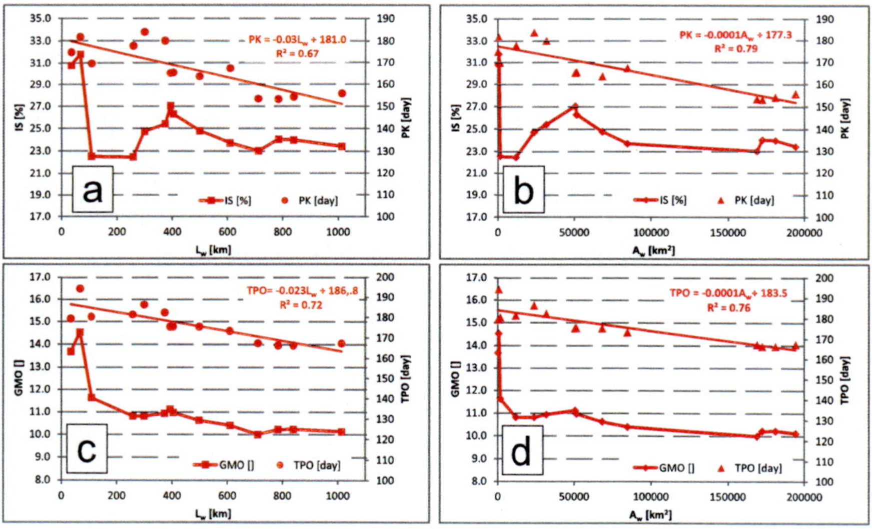

Fig. 10.4 Changes in seasonality parameters depending on increasing catchment area along the course of the Vistula. Explanations: TPO – half-flow date; GMO – seasonality coefficient; IS – seasonality index; PK – concentration date; L_W – the Vistula length; A_W – the Vistula catchment area; R^2 – determination coefficient

changes. Huge increase in the Vistula flow and catchment area after inclusion of the Narew and the Bug does not enhance the river flow seasonality. On the contrary, *IS* of the Vistula below the mouth of the Narew is reduced, proving that the seasonal structure of the Narew and the Bug flow is similar to that of the Vistula at the gauges located above the mouth of the Narew. Similar patterns, although less visible, are also observed for *GMO* changes along the Vistula course.

Concentration date (*PK*) and half-flow date (*TPO*) gradually decrease with the increasing length of the Vistula and the area of its basin. Therefore, the river flow seasonality is significantly reduced. These trends can even by described by statistically significant ($\alpha = 1\%$; F-Snedecor test) regression equations as functions of the river chainage (Fig. 10.4a and c) and catchment area (Fig. 10.4b and d). Statistical errors of the equations are large, but the determination coefficients (R^2) in all cases indicate a good or very good degree of explanation. A slightly better match of the equations was obtained for *TPO* and its correlation with the catchment area (A_w) than for *PK* and the Vistula length (L_w).

A decrease in *IS* and *GMO* along with the river course and increasing basin area is also clearly visible for the Oder (Fig. 10.5). Contrary to the Vistula, the changes are not so abrupt, even if the entrance of the Olza at 20 km of the Oder course results in considerable growth in *IS* and *GMO* despite a small increase in area. Downstream the river, a systematic drop in the Oder flow seasonality is observed with small fluctuations between the gauges in Malczyce (305 km) and Cigacice (472 km). Local peaks and valleys in the Oder flow seasonality are due to alternate entrance of water

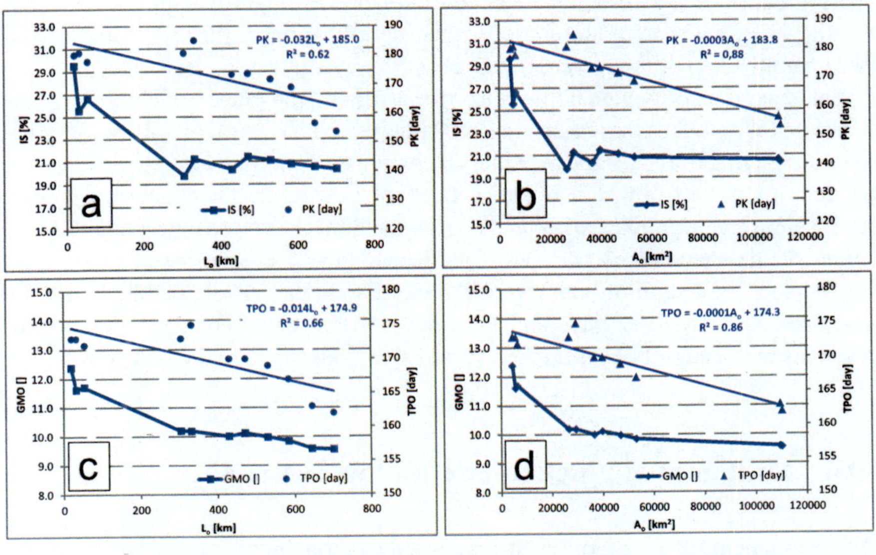

Fig. 10.5 Changes in seasonality parameters depending on increasing catchment area along the course of the Oder. Explanations: TPO – half-flow date; GMO – seasonality coefficient; IS – seasonality index; PK – concentration date; L_O – the Oder length; A_O – the Oder catchment area; R^2 – determination coefficient

from large mountain rivers including the Kaczawa (between Malczyce and Ścinawa), which slightly enhance the seasonality, and lowland rivers including the Barycz (between Ścinawa and Cigacice), which exert contrary effects. Interestingly, the entrances of the Bóbr (between Cigacice and Połęka) and the Nysa Łużycka (between Połęcko and Słubice) do not cause major changes in the seasonal nature of the Oder flow. No significant changes in the Oder seasonality are observed downstream of the Warta entrance, similarly as for the Narew entrance into the Vistula. The index and coefficient of seasonality (*IS* and *GMO*) in Gozdowice and Widuchowa are only slightly lower than in Słubice although the Oder basin area more than doubles along this section (Table 10.2, Fig. 10.5). This clearly shows that the Oder flow regime, and particularly seasonality at Słubice are similar to that at the mouth of the Warta.

The longer the Oder and the larger its basin, the earlier the concentration date and half-flow date fall. In the upper section of the river *PK* and *TPO* reach 180–185 and 170–175 days, respectively. In the lower section, they drop to 150–155 and 162–163 days. This fall is not smooth, and the spread of its values is slightly larger (especially in the middle section) than for the Vistula. Between Racibórz-Miedonia and Ścinawa both parameters indicate relatively late *PK* and *TPO*, and downstream both dates fall rather early in the year (Fig. 10.5). This delay in *PK* and *TPO* is clearly due to a significant supply of water from water-rich mountain rivers (the Nysa Kłodzka, the Kaczawa and a few others), while the acceleration downstream results from the entrance of many lowland rivers, mainly the Barycz. Regression models of the relationships between *TPO* and *PK* and A_o and L_o are statistically significant but burdened with large regression errors, similarly as for the Vistula. The relationships between the concentration date (*PK*) and half-flow date (*TPO*) with the Oder basin are, like for the Vistula, slightly stronger than the correlation of these parameters with the river length.

The equations presented above allow for assessing the rate of *PK* and *TPO* drop associated with increasing basin area and length of both rivers (Figs. 10.4 and 10.5). An increase in the Vistula and the Oder catchment area by 1000 km^2 causes the same acceleration of half-flow date by on average one day. As for the concentration date, the acceleration for the Oder is clearly greater and reaches three days per 1000 km^2, while for the Vistula it is one day. This dependency is even more complex when we consider the length of both rivers. An increase in the Vistula length by 100 km accelerates *PK* by three days and *TPO* by over two days. For the Oder, average gradients reach slightly over three days and about one and a half day, respectively.

10.6 Multiannual Fluctuations in Flow Seasonality

Our assessment of long-term changes and fluctuations in the scale, directions and covariability of various aspects of flow seasonality in the two largest Polish rivers (the Vistula and the Oder) is based on the data obtained for the following water gauges: Jagodniki, Warsaw and Tczew (the Vistula) and Ścinawa, Połęcko and Gozdowice (the Oder). The analysis of various seasonality parameters along the course of both rivers indicates significant changes at those sites (Figs. 10.4 and 10.5). Moreover, the first two gauges (Jagodniki and Ścinawa) roughly close the parts of the basins featuring mountain river regimes (Western Beskids – the Vistula; Sudetes – the Oder). The flow regime at the next two gauges (Warsaw and Połęcko) is considerably modified by waters supplied from the areas where many rivers show regimes typical of central Poland uplands. The last two gauges (Tczew and Gozdowice) close nearly the entire basins of the largest Polish rivers. The observed changes and oscillations in the seasonal structure of flow may reflect fluctuations and large-scale changes in flow regimes in Poland. Data obtained at the selected water gauges cover the same time interval of 1951–2016, which allows for reliable comparisons of the recorded changes and regularities.

Multiannual changes in two different seasonality aspects in six selected water gauges are analyzed based on the concentration date (*PK*) and seasonality index (*IS*). The other two investigated parameters (*TPO* and *GMO*) closely correlate with the previous two (Figs. 10.2 and 10.3).

Analysis of variability and covariability of average flows in the Vistula and the Oder at the selected water gauges (Fig. 10.6) seems to be a good starting point for assessing multiannual changes in their seasonality. The correlation coefficients between the pairs of the water gauges are $R(SQs \div SQj) = 0.74$; $R(SQp \div SQw) = 0.69$; $R(SQg \div SQt) = 0.81$. All are significant for $\alpha = 1\%$ (Student's t-test). The greatest convergence of multiannual changes in average flows occurred for the entire basins of the Oder and the Vistula, and the smallest for the gauges closing the mountain and upland sections of both basins. All gauges showed multiannual, sinusoidal fluctuations with phases and relative amplitudes for individual periods not identical but close regarding the dates of maxima and minima. The amplitude of changes in average flow at the gauge closing the upper part of the Oder basin (Ścinawa) was greater than at the corresponding gauge for the upper Vistula in Jagodniki. A reverse situation was observed at downstream water gauges. Relative dynamics of changes in the Vistula flows was larger than that for the Oder. This was observed for both year to year and multiannual fluctuations.

Multiannual changes in the concentration date (*PK*) at the six water gauges at the Oder and the Vistula irregularly oscillated around average values, and average *PK* for the three pairs of gauges were nearly identical (Fig. 10.7). Changes in the concentration dates for the Oder and the Vistula flow at the gauges closing the upper sections of their basins (Ścinawa and Jagodniki) were too irregular to identify any correlation ($R = 0.14$). *PK* extrema (local minima and maxima) concerned the same years (1974, 1983, 1997), or occurred in different years (Fig. 10.7a). There were years when the concentration date was very late in the upper Vistula and very early in the upper Oder or the other way round (1958, 1978, 2009). In the years 1951–1983 *PK* oscillation rate in both rivers was more rapid than later, and a declining trend in the upper Oder that began in the 1980s caused the concentration date to fall increasingly early.

For the other two pairs of water gauges (Połęcko÷Warszawa and Gozdowice÷Tczew), correlation of *PK* parameters over time was higher and significant ($\alpha = 1\%$), although in some years late concentration date in the Vistula was

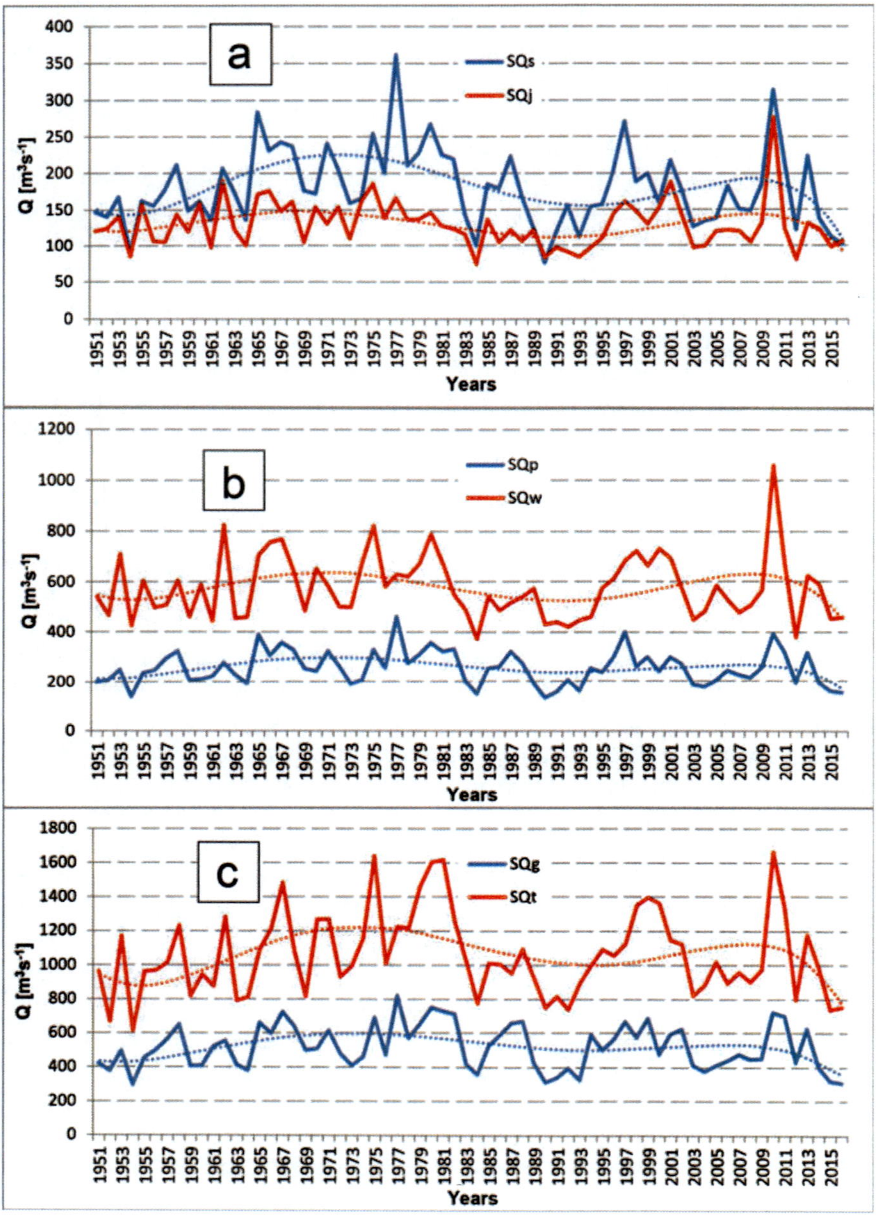

Fig. 10.6 Multiannual changes in average flow at selected water gauges of the Oder and the Vistula. Explanations: **a** – water gauges Ścinawa (*SQs*) and Jagodniki (*SQj*); **b** – water gauges Połęcko (*SQp*) and Warszawa (*SQw*); **c** – water gauges Gozdowice (*SQg*) and Tczew (*SQt*); the changes were smoothed with a quintic polynomial

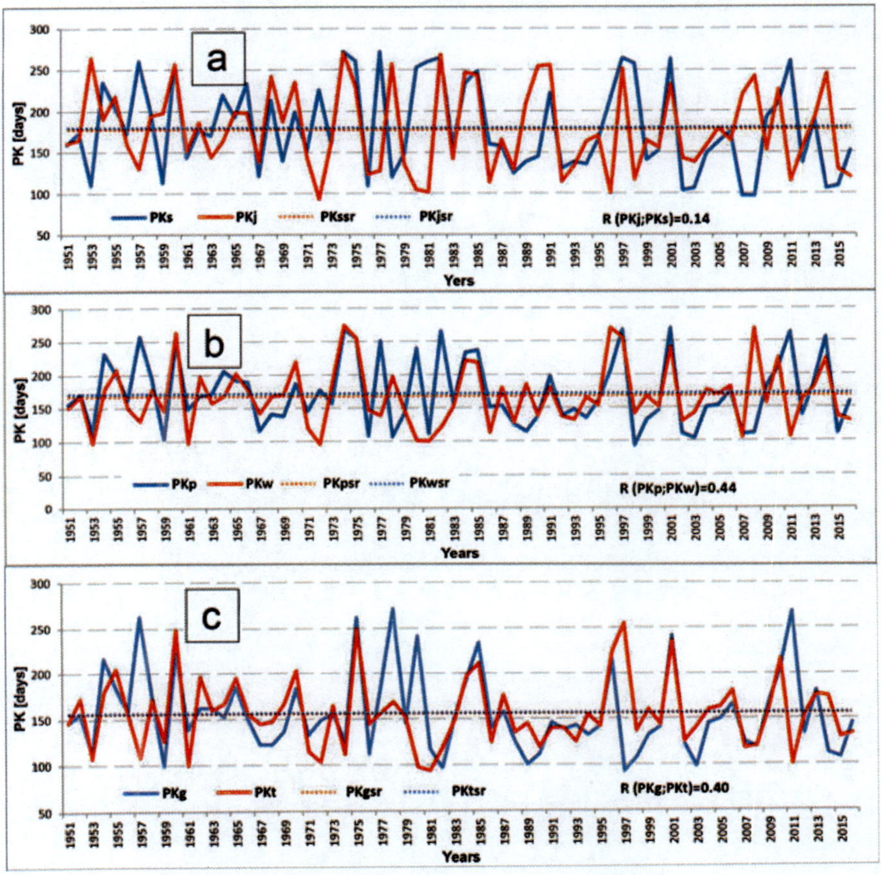

Fig. 10.7 Multiannual changes in concentration date (PK) for six selected water gauges of the Vistula and the Oder. Explanations: **a** – water gauges Ścinawa (*PKs*) and Jagodniki (*PKj*); **b** – water gauges Połęcko (*PKp*) and Warszawa (*PKw*); **c** – water gauges Gozdowice (*PKg*) and Tczew (*PKt*); *PKsr* – mean concentration date; *R* – correlation coefficient

accompanied by early concentration date in the Oder (Fig. 10.7b and c). Reverse situations happened as well. A characteristic feature of the fluctuations observed for these water gauges are two phases of relatively early concentration date (1961–1972 and 1987–1995) in both rivers, preceded with and separated by periods of highly variable and occasionally relatively late concentration date. In the last 15 years, flow concentration date varied greatly in both rivers. Moreover, the trend for earlier concentration date observed for the upper Oder catchment and indicated above, was not noticed.

Changes and fluctuations of the seasonality index *IS* at the gauges closing three the Oder and the Vistula catchments of different sizes demonstrate great variability (Fig. 10.8). At the gauges closing the upper parts of the basins (Fig. 10.8a), the fluctuations were comparable ($R = 0.72$); similarly as the length of the multiannual

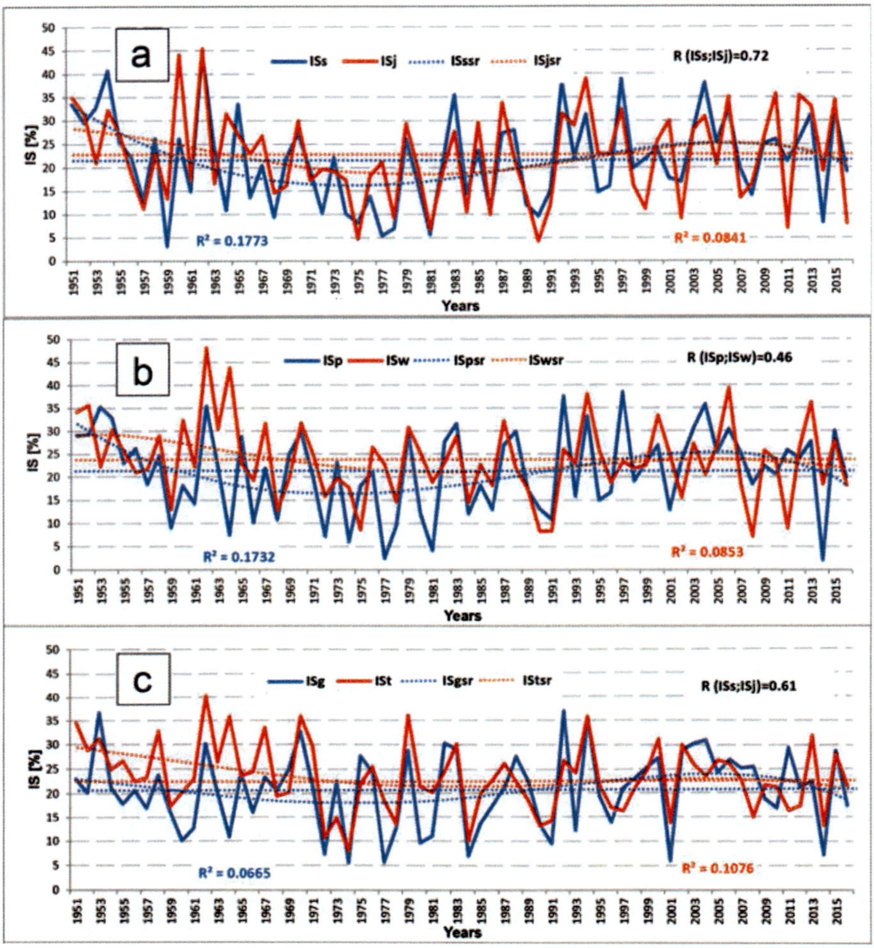

Fig. 10.8 Multiannual changes in seasonality index (*IS*) for six selected water gauges of the Vistula and the Oder. Explanations: **a** – water gauges Ścinawa (*ISs*) and Jagodniki (*ISj*); **b** – water gauges Połęcko (*ISp*) and Warszawa (*ISw*); **c** – water gauges Gozdowice (*ISg*) and Tczew (*ISt*); *ISsr* – average concentration date; *R* – correlation coefficient; R^2 – determination coefficient

fluctuation phases and year to year scale changes. A small difference was found for the amplitudes of polynomial trend curves. The difference was considerably bigger at Ścinawa gauge with a significant trend. The average multiannual flow seasonality indices at both water gauges are comparable, similarly as the scale of long-term *IS* variability.

Multiannual *IS* differed significantly at the gauges closing middle parts of both basins (Fig. 10.8b). The Vistula flow showed *s*lightly greater seasonality than that of the Oder, while the dynamics of multiannual changes in the Oder was, similarly as in the upper section, slightly greater than in the Vistula. This was confirmed by, e.g. greater amplitude of the curve of significant polynomial trend of multiannual changes.

In both rivers, the difference in flow seasonality was particularly visible in the 1960s and 1970s. *IS* indices for both Oder gauges were then clearly smaller than those for both Vistula gauges. The difference ceased to be perceptible later on. Signs of this regularity can be found when analyzing *IS* dynamics for the gauges closing the entire basins (Fig. 10.8c). Again, the index for the Vistula was slightly higher than for the Oder, but the difference was the most pronounced for the first 30 years. Then the average flow seasonality indices were similar in both rivers and demonstrated a similar nature of multiannual changes in relation to years ($R = 0.61$), year to year changes and multiannual fluctuations.

10.7 Conclusions

Physico-geographical regions of Poland are divided roughly along latitudinal lines. Mountains in the south are gradually replaced towards north with uplands, lowlands and finally lakelands [14]. Two largest Polish rivers follow basically the same course from south to north as they drain the surrounding regions. It is therefore clear that their flow seasonality structure shows spatial differences. However, as the courses of both rivers are similar, greater differences are observed along the rivers than between their basins. This is confirmed by long-term studies on the variability of their flows that indicate strong and significant correlation of the Vistula and the Oder SSQ in their lower sections [16]. The analyses presented above allow for drawing a few basic conclusions regarding changes in flow seasonality with increasing size of the river basin.

- Half-flow date (*TPO*) in the upper sections of the rivers falls roughly two weeks later than in the lower ones. The average *TPOs* recorded for these watercourses occur between 162 and 194 days of a hydrological year and are similar to analogous measures calculated for 12 Carpathian rivers [10, 12]. The difference between the Vistula and the Oder was insignificant. *TPO* for the Oder is usually by 4–5 days earlier than that for the Vistula. The value of *TPO* is relatively stable along the course of both rivers, even though it is the highest in the upper sections.
- Flow concentration dates (*PK*) in the Vistula and the Oder are getting earlier along their course. The similarity of *PK* in both rivers is even greater than of *TPO*. No major shifts were observed for flow concentration date, which typically occurred between 153 and 185 days of the hydrological year. This indicated a gradual transformation of the flow regime downstream from a complex to a simple one. Average concentration dates for the gauges closing the upper sections of the Vistula and the Oder were comparable to those computed for the rivers of the Western Beskids, and those inferred for the lower sections were similar to times obtained for the rivers of the Eastern Beskids [12].

- Average seasonality indices *IS* changed with the course of both rivers in a very similar way from 32% to 22% and were comparable to those obtained for large Carpathian, upland and lowland rivers [4, 12] but clearly larger (by about 10%) than the values calculated for large lake rivers [8]. Similar changes were observed for flow seasonality coefficient *GMO* as it describes basically the same aspect of flow seasonality. *GMO* declined from about 15 to 9.5. These results are comparable to those obtained for the Carpathian rivers [12, 21].
- A clear correlation was found between the concentration date (*PK*) and half-flow date (*TPO*) for both rivers. These parameters demonstrate a linear trend and *PK* grows in line with *TPO*. However, their regression lines have different inclinations. The rise is slower for the Vistula, but this difference ceases as *TPO* and *PK* decrease with a growing area of both basins.
- Covariation between average seasonality indices *IS* and flow seasonality coefficient *GMO* is less pronounced in each of the rivers than in the case of *PK* and *TPO*. However, when the water gauges from both river basins are considered, the relationship between average *IS*, and *GMO* is clear and its curvilinear model indicates *IS* growth along with *GMO* increase. Thus, the increase in the catchment area significantly reduces flow seasonality decline.

The river flow shows long-term oscillations due to changes in the water balance components. Genetic diversity of multiannual fluctuations in flow time series has been advocated by many authors [2, 7, 24]. This analysis also shows about 40 years long cycle of changes in average flow (*SQ*) for the gauges closing the upper, middle and lower sections of the Vistula and the Oder. A similar, though slightly longer cycle was identified for an aggregated flow from both those basins [16]. These fluctuations are often associated with North Atlantic Oscillation *NAO* [20, 22, 23], or even North Atlantic Thermahaline Circulation *NA THC* [16]. In most cases, a negative correlation of various *NAO* indices with the flows of rivers from southern and even central Poland can be confirmed. This analysis also showed a significant ($\alpha = 1\%$, Student's t-test) correlation between the gauges closing the catchments of the upper and middle sections of the Vistula and the Oder (*R*: −0.55 and −0.43 for the Vistula and −0.46 and −0.42 for the Oder) with *NAO* index (Hurrell). Despite the confirmed influence of the North Atlantic Oscillation *NAO* on flow size, flow seasonality measures show only insignificant correlation coefficients with these indices at $\alpha = 1\%$ (Student's *t*-test). This may suggest that *NAO* related circulation changes over Poland affect the amount of water discharged with the rivers but not seasonality of the flow. Our analysis enabled drawing some basic conclusions on changes in flow seasonality over time:

- Multiannual changes in the concentration date (*PK*) for the Vistula and the Oder irregularly oscillate around average values, and average *PK* values for the upper, middle and lower parts of the catchments are nearly identical (Fig. 10.7).
- The similarity of *PK* indices in individual years for the Vistula and the Oder in the analyzed multi-year period occurred in the lower and middle course of the rivers. For the upper sections, the differences between rivers are larger, and for the last 30 years, there is a growing tendency for earlier concentration date for the Oder.

– Changes in the flow seasonality index *IS* showed huge variability and depended on the catchment size. Contrary to *PK*, relatively the greatest similarities between the Vistula and the Oder were noticed for their upper sections ($R = 0.72$). As the catchment area increased, the seasonality of the Vistula flows became considerably higher than of the Oder, and the differences were the most pronounced in the 1960s and 1970s. In recent years, their significance has been declining.

10.8 Recommendation

In 2016, the Polish government presented a strategy for the development of inland waterway transport in Poland. It assumes the adaptation of waterways in our country to the fourth class of navigability. It will require major changes in the river valleys. Therefore, further research should be carried out on the flow seasonality. It will allow to check whether the introduced changes do not disturb the natural regime of these rivers. These studies will be helpful for decision makers in the Ministry of Maritime Economy and Inland Navigation as well as for ecological organizations monitoring changes in rivers.

References

1. Czaja SW (2011) *Powodzie w dorzeczu górnej Odry.* Katowice. pp 1–211
2. Gutry-Korycka M, Boryczka J (1990) Długookresowe zmiany elementów bilansu wodnego w Polsce i zlewisku Bałtyku. *Przegląd Geofizyczny* 35(3–4):19–32
3. Jokiel P (2004) *Zasoby wodne środkowej Polski na progu XXI wieku.* Łódź. pp 1–114
4. Jokiel P (2009) O sezonowym rozmieszczeniu odpływu w wybranych rzekach środkowej Polski. *Wiadomości Instytutu Hydrologii i Gospodarki Wodnej* 2–3:15–29
5. Jokiel P, Bartnik A (2000) Zmiany w sezonowym rozkładzie odpływu w środkowej Polsce w wieloleciu 1951–1998.*Wiadomości Instytutu Hydrologii i Gospodarki Wodnej* 24 (2):3–17
6. Jokiel P, Bartnik A (2005) Niektóre problemy zmian i zmienności rocznego hydrogramu przepływu rzeki na podstawie Pilicy w Przedborzu. *Wiadomości Instytutu Hydrologii i Gospodarki Wodnej* 2:5–27
7. Jokiel P, Kożuchowski K (1989) *Zmiany wybranych charakterystyk hydroklimatycznych Polski w bieżącym stuleciu*, Dokumentacja Geograficzna, 6. Warszawa. pp 1–94
8. Jokiel P, Stanisławczyk B (2016) Zmiany i wieloletnia zmienność sezonowości przepływu wybranych rzek Polski. *Prace Geograficzne UJ*. Kraków, 144:10–33
9. Jokiel P, Tomalski P (2015) Identyfikacja i analiza sezonów hydrologicznych na przykładzie dwóch rzek z obszaru środkowej Polski. In: Jokiel P (ed) *Metody statystyczne w analizach hydrologicznych środkowej Polski.* Łódź, pp 201–213
10. Jokiel P, Tomalski P (2016) Zmiany i zmienność sezonowej struktury odpływu rzecznego w świetle terminu połowy odpływu.*Gospodarka Wodna* 805(1):12–18
11. Jokiel P, Tomalski P (2017) Formy odpływu rzecznego i ich zróżnicowanie przestrzenne. In: Jokiel P, Pociask-Karteczka J, Marszelewski W (eds) *Hydrologia Polski.* Warszawa, pp 160–167
12. Jokiel P, Tomalski P (2017) Sezonowość odpływu w wybranych zlewniach karpackich. *Przegląd Geograficzny* 89(1):29–44
13. Jokiel P, Tomalski P (2018) Zróżnicowanie i zmienność wieloletnia sezonowości przepływu w wybranych przekrojach wodowskazowych Wisły. *Prace Geograficzne* UJ, Kraków, 155:27-45

14. Kondracki J (1998) *Geografia regionalna Polski*, Warszawa. pp 1–460
15. Markham Ch. G (1970) Seasonality in the precipitation in The United States. Ann Assoc Am Geogr 593–597
16. Marsz AA, Styszyńska A, Krawczyk WE (2016) Długookresowe wahania przepływów rocznych głównych rzek w Polsce i ich związek z cyrkulacją termohalinową Atlantyku Północnego. *Przegląd Geograficzny* 88(3):295–316
17. McCabe GJ, Clark MP (2005) Trend and variability in snowmelt runoff in the Western United States. J Hydrometeorol, 6, Boston. pp 476–482
18. Michalczyk Z (2017) Odpływ średni, zmienność w czasie i zróżnicowanie przestrzenne. In: Jokiel P, Marszelewski W, Pociask-Karteczka J (eds) *Hydrologia Polski*, Warszawa. pp 153–160
19. Oliver JE (1980) Monthly precipitation distribution: a comparative index. Prof Geogr 32(3):300–309
20. Pociask-Karteczka J, Limanówka D, Nieckarz Z (2002–2003) Wpływ Oscylacji Północnoatlantyckiej na przepływy rzek karpackich (1951–2000), *Folia Geographica – Series Geographica Physica*, 33–34, 89–104
21. Soja R (2002) *Hydrologiczne aspekty antropopresji w polskich Karpatach*. Prace Geograficzne, Warszawa, pp 1–186
22. Styszyńska A (2002) Związki między przepływem warty w Poznaniu a zimowymi wskaźnikami NAO w okresie 1865–2000. In: Marsz A, Styszyńska A (eds) *Oscylacja Północnego Atlantyku i jej rola w kształtowaniu zmienności warunków klimatycznych i hydrologicznych Polski*, Gdynia, pp 173–180
23. Stanisławczyk B (2016) Wieloletnia dynamika odpływów charakterystycznych z wybranych zlewni Polski w świetle zmian indeksu NAO. *Przegląd Geograficzny* 89(3):413–428
24. Wrzesiński D (2009) Tendencje zmian przepływu rzek Polski w drugiej połowie XX w., *Badania Fizjograficzne nad Polską Zachodnią, Ser. A.*, 60:147–162
25. Wrzesiński D (2017) Reżimy rzek Polski. In: Jokiel P, Marszelewski W, Pociask-Karteczka J (eds) *Hydrologia Polski*, Warszawa. pp 215–221

Chapter 11
Low-Flows in Polish Rivers

Edmund Tomaszewski and Katarzyna Kubiak-Wójcicka

Abstract The chapter discusses spatial, seasonal, and multiannual variability of low-flows in Poland as well as their influence on river regime, and also availability, dynamics, and management of water resources in hydrological drought condition. The study covered 17 gauging cross-sections closing small and middle catchments as well as located in big, transit rivers. Basic data on daily discharge series was collected in the years 1951–2015. Low-flow periods were identified with reference to threshold level method matching 70 and 95 percentile at a flow duration curve as constant, multiannual truncation level ($Q_{70\,\%}$, $Q_{95\,\%}$). The research covered temporal and spatial variability of minimum flows, low-flow duration, inter-low-flow spacing, and rate of low-flows recession and rise. The maximum low-flow duration at different levels of non-exceedance probability is analysed as well. As a result, there were indicate areas with a high-risk rating of water shortage hazardous for water ecosystems and water management operations. Analysis of selected features of extreme low-flow events is important. It allowed us to identify a group of determinants which should be taken into consideration in low-flow river regime analysis. Also, it could help in local water resources exploitation planning or in long-term strategies of water management in the scale of the whole country.

Keywords Low-flows · Threshold level method · Hydrological drought · Flow duration probability · Poland

E. Tomaszewski (✉)
Department of Hydrology and Water Management, Faculty of Geographical Sciences, University of Łódź, Narutowicza 88, 90-131 Łódź, Poland
e-mail: edmund.tomaszewski@geo.uni.lodz.pl

K. Kubiak-Wójcicka
Department of Hydrology and Water Management, Faculty of Earth Sciences and Spatial Management, Nicolaus Copernicus University, Lwowska 1, 87-100 Toruń, Poland
e-mail: kubiak@umk.pl

M. Zeleňáková et al. (eds.), *Management of Water Resources in Poland*, Springer Water,
https://doi.org/10.1007/978-3-030-61965-7_11

11.1 Introduction

Hydrological extremes are of great significance for river flow structure. They define the range of river regime dynamics as well as indicate directions of water management operations concerning mitigation of their negative effects. Low-flows, in contrast with floods, develop very slowly. On the one hand, it is relatively simple to estimate drought streamflow deficit volume because of high process inertia, on the other, it is difficult to predict when the low-flow episodes terminate because of high inertia as well. Therefore the occurrence of low-flow periods is determined not only by stochastic factors connected with restricted alimentation but also depends on the spatial pattern of water balance structure which allows identifying areas with different risk levels of drought appearance. Many projections and scenarios of climate changes in Poland prove that further rise in air temperature is expected. It will result in higher evapotranspiration and adverse structure of water balance. Moreover, there is a predicted higher density and intensity of hydrometeorological extremes, especially in interior zones [1–3]. Wide and multidirectional analyse of the low-flow regime, in the context of presented facts, seems to be very important for research and practical purposes. Results should improve low-flow prediction methods and support the implementation of water shortage effect mitigation strategies, especially in the field of hydropower, water supply, agriculture, inland water transport, etc.

The river low-flow is commonly defined as a period of low flows (water levels) in a river or flows during prolonged dry weather [4, 5]. This process is usually initiated by rainfall shortage—meteorological drought (see chapter *Dynamics, Range, and Severity of Hydrological Drought in Poland*). Prolonged lack of precipitation combined with intense evapotranspiration results in gradual loss of soil moisture within the vadose zone—agricultural drought. It may lead to significant depletion of groundwater resources in the hydrologically active zone (groundwater low-flow) and in hydraulically connected river beds (surface water low-flow)—hydrological drought. Thus, the streamflow drought that occurs at the end of the chain of events (adverse in terms of water management) can be considered as a reliable indicator of hydrological drought development [5, 6]. Winter low-flows in rivers follow a different course. Limited runoff is then due to temporary water retention in the snow cover often combined with riverbed freezing during severe frosts that stop all forms of drainage. It is worth noticing that the low-flow episode has a highly seasonal nature in this case because all of the water entrapped in snow cover will break the drought period during spring snowmelt alimentation. Low-flow appearance may temporary disorder river and valley ecosystems existence. Low-flow episodes can seriously hinder realization of water management tasks and during severe hydrological drought may lead to serious consequences of social, economic, and environmental matter—socioeconomic drought [7].

Observations of river low-flows have been conducted from the down of history. The oldest notes about dramatically low water stages on Polish lands concern years: 988, 1121, 1332, 1473 [8]. However, the contemporary methodology of this phenomenon assessment not until 50. of XX century appeared. Its origin was an exceptionally long period of severe drought depleting seriously Polish water resources in the 1940s and 1950s. New methodical approaches and extending data set of discharge series allowed to do research on the scale of the whole country [9–15]. Presented authors analysed the spatial pattern and temporal dynamics of low-flows in different time scales, identifying hydrometeorological and physiographical determinants of this process. Very interesting studies concerning regional scale, e.g. [6, 16–22], discussed widely relations between factors and processes important for the low-flow formation and very often gave the base for local modelling of water circulation under drought condition. Much attention was given to case studies of simple or group of rivers. On their base, many methodological and applicable solutions were presented. Significant progress was made in a matter of probability assessment of low-flow maximum duration and deficit [6, 23–26]. As a result, a very interesting algorithm and software were made by Jakubowski & Radczuk [27]. It allows to estimate automatically all important parameters of low-flow distributions, applying the best fitted theoretical functions. In recent times, two-dimensional distributions to extreme low-flows assessment are discussed and tested—copula function and generalized Pareto distribution. Many interesting conclusions gave studies on the impact of dammed and natural lakes on low-flow [28–30], analyses of low-flow formation in transit rivers [31, 32] or multidirectional studies on anthropogenic impact [33–35].

Low-flows were also a component of investigations on river regime typology. In analyses presented by Dynowska [36] or Wrzesiński [37], low-flow periods constitute a few seasonal phases of flow. In recent times, research in the field of drought risk management operational systems supporting is developed. It includes characteristics and models of river low-flows [38]. This approach, based on strategical management in pursuit of sustainable assurance of water safety for social and natural systems, ensures the durability of freshwater ecosystems services responsible for maintaining biodiversity, life processes, and environment reclamation as well as providing people with economic benefits.

11.2 Methods and Data

Periods of drought streamflow deficits are usually defined on the base of the threshold where deviations of daily runoff from truncation level are estimated. One of the most common methods for the low-flow episode delimitation is the threshold level method, also known as a peak over the threshold. In this approach, a period during which discharge attains values below an established limit is defined as a streamflow deficit period. Its two basic parameters are the low-flow duration and deficit volume (Fig. 11.1). Threshold values are estimated based on second degree main flows (maximum, mean or median value of annual minimum flows), periodic flows from the flow duration curve such as the percentage of exceedance from the range of between 70 and 95% or analysis of annual minima distribution. There also conventional flows can be applied. They refer to specific challenges of water management

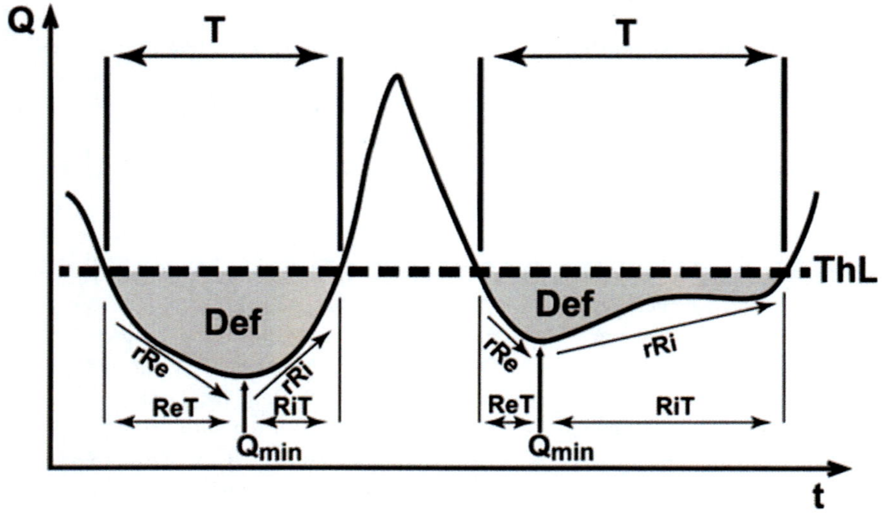

Fig. 11.1 Basic parameters of the low-flow episode (after [6], modified). Q—discharge, t—time, T—low-flow duration, Def—drought streamflow deficit, Q_{min}—minimum low-flow discharge, ReT—low-flow recession time, RiT—low-flow rise time, rRe—rate of low-flow recession, rRi—rate of low-flow rise, ThL—threshold level

and environmental management such as environmental flows, minimum flow for proper water intake management, etc. [5, 39–42].

In the present chapter low-flow period was identified with reference to the threshold at the level of the 70th percentile, derived from the flow duration curve ($Q_{70\,\%}$). Only those low-flows, during which the flow was below the threshold level for at least 7 days were subject to identification on a time scale. In terms of separating individual episodes, it was arbitrarily assumed that the low-flows separated by an interruption lasting no longer than 3 days should be treated as inherently homogeneous events, combining their duration and volume. There also severe low-flow episodes were identified. Their appearance is determined by depletion of seasonally renewable water resources in the active exchange zone. It results in river alimentation from groundwater reservoirs characterized by multiannual rhythm only. For this sort of episodes, the truncation level was established at 95% of the flow duration curve [6].

The dynamics of low-flow episodes progression was assessed based on the low-flow discharge recession rate [6]:

$$r\mathrm{Re} = \frac{q_{ThL} - q_{\min}}{\mathrm{Re}T} \tag{11.1}$$

where:

rRe low-flow discharge recession rate [$\mathrm{dm^3 \cdot s^{-1} \cdot km^{-2} \cdot d^{-1}}$],
q_{ThL} threshold level, [$\mathrm{dm^3 \cdot s^{-1} \cdot km^{-2}}$],

$q_{\min}$ minimum discharge during low-flow episode [$dm^3 \cdot s^{-1} \cdot km^{-2}$],
ReT low-flow discharge recession time [days].

The dynamics of receding of the low-flow episode was estimated based on the low-flow discharge rise rate:

$$rRi = \frac{q_{ThL} - q_{\min}}{RiT} \tag{11.2}$$

where:

rRi low-flow discharge rise rate [$dm^3 \cdot s^{-1} \cdot km^{-2} \cdot d^{-1}$],
q_{ThL} threshold level, [$dm^3 \cdot s^{-1} \cdot km^{-2}$],
$q_{\min}$ minimum discharge during low-flow episode [$dm^3 \cdot s^{-1} \cdot km^{-2}$],
RiT low-flow discharge rise time [days].

Both defined characteristics were recalculated to specific values in $dm^3 s^{-1} km^{-2}$ per day. Thank to this, the comparison of results for differently sized catchments was possible (Fig. 11.1).

The study covered 17 gauging cross-sections in Polish rivers (Table 11.1). Their location reflects a full spectrum of possible river regimes, especially low flow regimes occurring in Poland, as well as the variety of physico-geographical conditions determining the process of river channel alimentation. Catchments selected for investigation are differently sized, from 681 km^2 up to 194376 km^2 which allows for study on the influence of water resources capacity and flow formation inertia on low-flows features. Moreover, some of the gauging stations are located in reaches of rivers crucially important for water management—channel regulations and damming water table for water intakes, hydropower, irrigation, and inland water transport purposes. Among the analysed water-gauges, there were 7 cross-sections located on the Vistula and Oder River, thanks to which it was possible to take into consideration the specificity of low-flow episode progression along the course of the transit river.

The basic calculations were based on the daily discharges series measured in studied cross-sections and made available by the Polish Institute of Meteorology and Water Management—National Research Institute (IMGW-PIB). The analysis covered the observation period 1951-2015. The presented multiannual period is characterized by the occurrence of seasons with various moisture conditions and various structures of the water balance, and thus also the appearance of hydrological droughts and river low-flows of various severity, extent, and duration. As a result, the study included periods of deep and long-lasting low-flows in the 1950s and 90 s, moderate shortage periods at the beginning of the XXI century or shallow and short low-flows in the 1970s which enabled the assessment of the almost full spectrum of the conditions that determining low-flows.

Table 11.1 Selected characteristics of low flows (1951–2015)

No.	River	Gauging cross-section	A	ALq	LLq	$q_{70\,\%}$	$q_{95\,\%}$	a	R^2
			km^2	dm^3s^{-1} km^{-2}				–	–
1	Oder	Racibórz-Miedonia	6744.0	2.49	0.99	4.52	2.28		
2	Oder	Ścinawa	29583.8	2.24	0.79	3.85	2.06		
3	Oder	Gozdowice	109729.1	2.26	1.13	3.27	2.04		
4	Nysa Kłodzka	Kłodzko	1084.0	3.20	0.87	5.82	3.07		
5	Warta	Sieradz	8139.6	2.66	1.25	3.71	2.48	0.015	0.157
6	Noteć	Pakość	2356.2	0.75	0.22	1.20	0.52	0.005	0.065
7	Wieprza	Stary Kraków	1518.7	6.10	3.69	8.10	6.20	0.025	0.194
8	Vistula	Nowy Bieruń	1747.7	3.11	0.86	5.49	3.02	0.028	0.229
9	Vistula	Sandomierz	31846.5	3.08	1.79	4.93	3.14	0.009	0.074
10	Vistula	Warszawa	84540.0	2.58	1.32	4.12	2.61	0.012	0.127
11	Vistula	Tczew	194376.0	2.16	1.03	3.50	2.21	0.007	0.074
12	Dunajec	Nowy Targ (Kowaniec)	681.1	5.04	2.35	9.93	5.24	0.023	0.077
13	San	Przemyśl	3686.5	2.83	0.54	6.05	2.85	0.030	0.420
14	Wieprz	Krasnystaw	3001.0	2.06	1.16	2.78	1.86	0.018	0.287
15	Pilica	Przedbórz	2535.9	2.28	1.01	3.75	2.17		
16	Narew	Suraż	3376.5	1.21	0.50	2.03	1.05	0.016	0.435
17	Drwęca	Elgiszewo	4959.4	2.88	1.16	4.03	2.62		

A—catchment area, ALq—specific average minimum flow, LLq—specific lowest minimum flow, q70%, q95%—specific low-flow threshold (percentile of the flow duration curve); parameters of statistically significant ($\alpha = 0.05$) linear trend of multiannual course of minimum flows: a—slope coefficient, R^2—determination coefficient

11.3 Minimum Flows

The minimum value of runoff is an integral part of a low flow regime. It indicates limitations during recession of flow on the dry weather curve and very often, the regime of a basic groundwater aquifer drained by river channels during a drought. Minimum value may also be modified by such factors as lakes, wetlands, or forest retention. It may depend on some water management operations, especially downstream of dammed lakes because of the guaranteed flow.

Distribution of annual flow minima is characterized by strong variation (Fig. 11.2). The highest minimum specific flow exceeded 10 dm^3s^{-1} km^{-2} in Dunajec river whereas the lowest one reached two orders of magnitude less—0.2 dm^3s^{-1} km^{-2} in Noteć river. A similar distribution of the Lq at the high average level is observed in high mountainous and seaside catchments—Dunajec river—Nowy Targ and Wieprza river—Stary Kraków which indicates that high annual precipitation determines low flow regime much more than catchment size and its quantity of water resources.

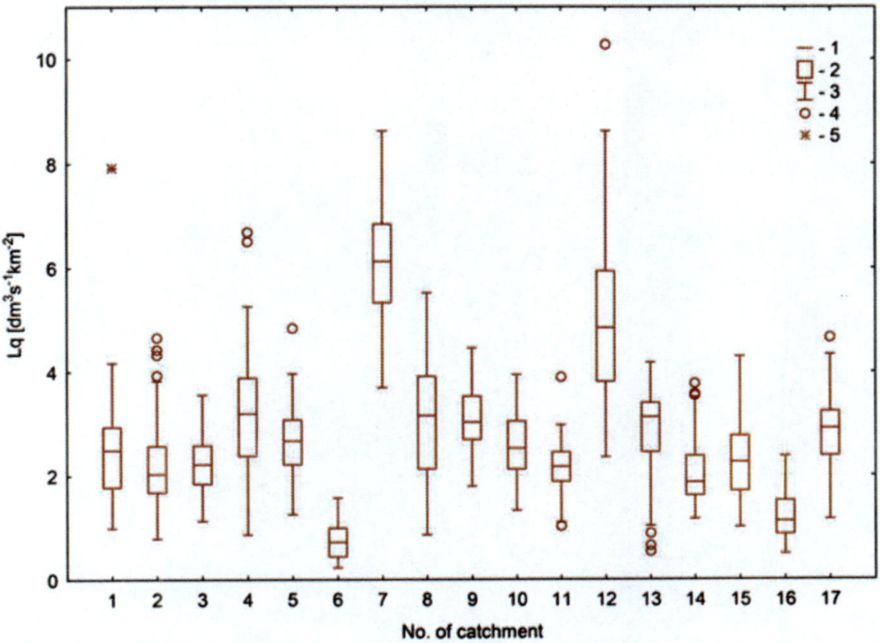

Fig. 11.2 Distribution of annual minimum specific flow (1951–2015). 1—median, 2—range between first and third quartile, 3—range limited by 1 quartile deviation, 4—outliers under 1.5 quartile deviation, 5—extremes over 1.5 quartile deviation; number of catchment—see Table 11.1

Exceptionally low values of Lq occurred in lowland catchments of Noteć and Narew river where low groundwater resources cannot be supplemented significantly with lake or wetland retention. It is also determined by water management, especially by open cast mine dewatering and discharging mine water that lead to groundwater depression cone development as well as some irrigational purposes. It is worth noticing that wide interquartile range of Lq with its interquartile deviation reach indicates predominance of natural or quasi-natural factors crucial for low-flow forming in the catchment. Very narrow range is mostly determined by water management systems and units which are manifested by stable or systematically occurring operations. In some catchments, statistical outliers and extremes were observed (Fig. 11.2). Their appearance might signify lower stability of low flow regime because in some years minimum flows significantly depart from typical changeability range. Distributions of Lq particularly adverse for water management will be characterized by asymmetry skewed to the left with very low extremes of course.

Minimum flows also show spatial differentiation (Fig. 11.3). Catchments with high average minimum specific flow occupy the north and south part of the country because of high precipitation alimentation in seaside and mountainous areas. The mean value of ALq occurs in upland rivers which is determined by high and stable groundwater flow from capacious fissure-karst aquifers. In lowland areas, average minimum flow differs and seems to be dependent on local factors such as groundwater retention level,

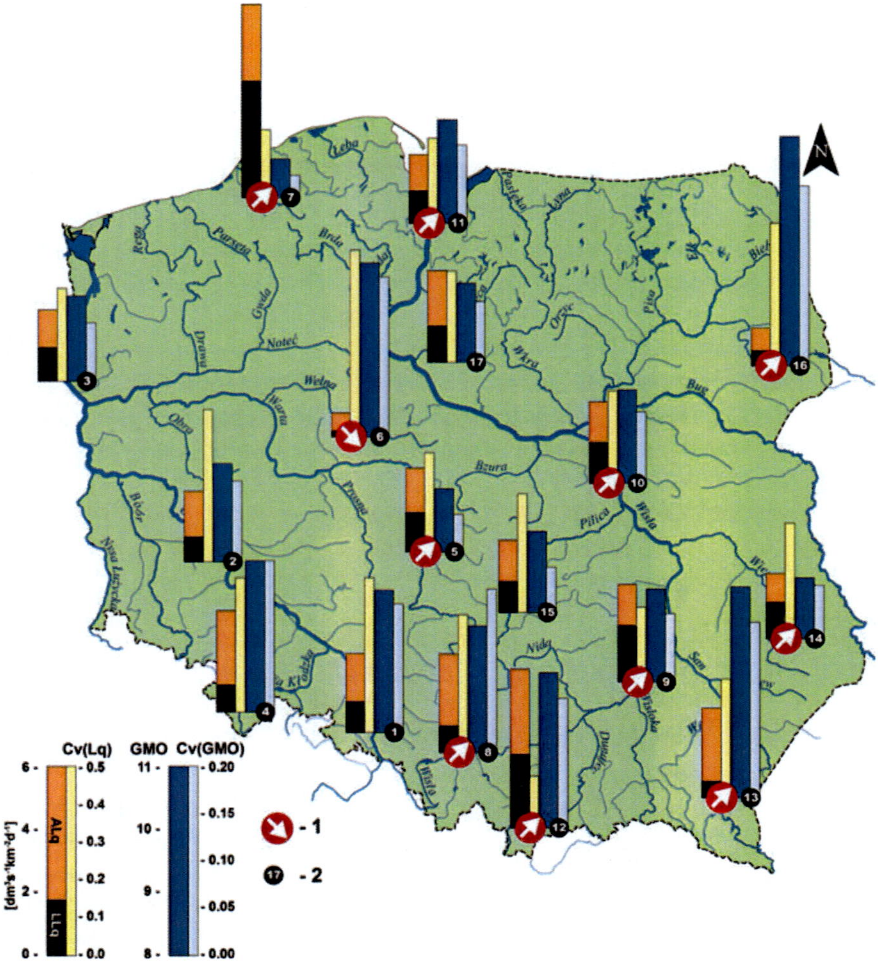

Fig. 11.3 Spatial differentiation of selected characteristics of minimum flows (1951–2015). ALq—average annual minimum specific flow, LLq—the lowest annual minimum specific flow, Cv(Lq)—variation coefficient of annual minimum specific flow, GMO—index of minimum flow seasonal concentration, Cv (GMO)—variation coefficient of minimum flow seasonal concentration index, 1—statistically significant ($\alpha = 0.05$) linear trend of multiannual course of minimum flows, 2—number of catchment—see Table 11.1

river-lakes system occurrence, and variety of water management impacts. The spatial pattern of the lowest specific minimum flow confirms conclusions from previous paragraph (Fig. 11.3). However, interesting findings might be made on the base of analysis of the relationship between average and the lowest minimum flow. In rivers where LLq contributes about 50% of ALq or more, there is a good background for water management planning and low risk of water ecosystems degradation during hydrological drought. To this group belong some catchments with good retention conditions and also transit reach of Vistula river where water resources are discharged downstream on a high level of inertia making water shortage periods less arduous for water management. In rivers with the contribution of LLq below 30%, such as San river, Nysa Kłodzka river or upper Vistula river, severe hydrological droughts may be a considerable threat to basic water resources and proper structure of water balance. The variation coefficient of an annual minimum specific flow, calculated for the whole multi-year period, increases in inverse proportion to the LLq contribution, presented above (Fig. 11.3). It proves in multiannual scale, the relation between low flow dynamics and susceptibility to severe drought streamflow deficit occurrence.

In the hydrological system, deficit periods, similarly to any other phenomena, are forming dynamically. Already Dębski [43] noticed that particularly oppressive low-flows occur when water deficits happen in typical resource-feed periods. As a result, low flow analysis should refer to those features of the flow regime which will facilitate the identification and valorisation of periods with significant disturbances in the water balance structure. Large irregularities in the occurrence of low-flows crucially impede the seasonal analysis based on time series in a monthly step. The solution of this problem may be the characteristic describing the seasonal variability comprehensively. Level of low flows seasonality may by estimated by application a complex index used by Olivier [44] to seasonal precipitation concentration analysis. This characteristic was successfully adopted to total river flow analysis in Carpathian rivers by Soja [34]. After transformation the formula to low flow purposes, the modified equation of GMO index is as shown in Eq. 11.3 which reads:

$$GMO_j = \frac{\sum_{i=1}^{12} (LqM_i)^2}{\left(\sum_{i=1}^{12} LqM_i\right)^2} \cdot 100\% \tag{11.3}$$

where:

GMO_j index of low flow seasonality in year j
LqM_i minimum flow in moth i

Results of GMO index are within the range of 8.3–100%. Its hydrological interpretation indicates that absolutely lack of low flow seasonality which means that all of 12 values of monthly minimum flows are equal will result in 8.3%. Theoretically, if the flow (and low flow) will occur in one month only, index value reaches 100% which means absolutely concentration and total seasonality of low flow.

Indices of seasonality estimated for studied cross-sections varied in the range between 8.61 and 11.6% (Fig. 11.3). For comparison, Jokiel and Tomalski [45] who calculated GMO index for average flow in 14 cross-sections in the Vistula river in a similar period, received results from the range 10.0–14.5%. It indicates a lower level of low flow seasonality. However, relatively small differences with the average flow may prove the high similarity of factors determining these components of river regime. Spatial order of GMO follows the degree of precipitation continentalism and catchment resources capacity. The exception is the transit rivers Vistula and Oder where GMO is almost stable along the whole course. Multiannual variability of GMO index was estimated on the base of variation coefficient (Fig. 11.3). In general, there is a proportional relationship between low flow seasonality and its variability. Therefore in rivers with high GMO such as Dunajec, San, Nysa Kłodzka, Narew or Noteć serious seasonal perturbations of water resources managing may be expected. This conclusion does not have to concern natural hydrological systems because very often they have resistance mechanisms let them survive seasonal water shortage periods.

In the analysis of the long-term variability of low flow, the existence of a linear trend at the significance level of $\alpha = 0.05$ was tested. The studies on systematic components in the investigated time series of Lq led to the identification of linear trends, statistically crucial for the majority of the catchments. Significance was verified using the Student's t-test and the Mann-Kendall trend test [46, 47]. Lack of statistically crucial trends was observed in the west part of the country, along the whole course of the Oder River and its tributary—Nysa Kłodzka river (Fig. 11.3, Table 11.1). Such a spatial pattern may suggest the influence of climate continentalism. However, the Oder channel is much more improved then Vistula which may have an important impact as well. Exceptions are Pilica and Drwęca river where huge water resources stored in lake basins or aquifers make low flow regime resistant to multiannual systematic changes. The other investigated gauging cross-sections reflected positive linear trends of minimum flow, and in one case only (Noteć) the slope coefficient was negative (Fig. 11.4). This last case is determined by dewatering and interbasin transfer of water for mining and irrigational purposes. All of the positive trends in Lq series seems to be caused by multiannual progress in water management objects and operations, especially by damming and controlling water level which is connected with a commitment to ensuring the guaranteed flow and environmental flows. It is worth adding that there were no clear positive multi-year tendency in precipitation in Poland during the investigated period, but in some areas, seasonal tendencies in both directions occurred [48].

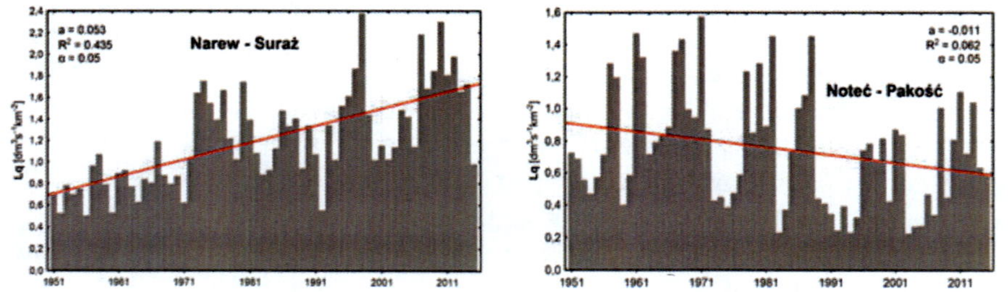

Fig. 11.4 Examples of linear trends of multiannual course of minimum flows. a—slope coefficient, R^2—determination coefficient, α—significance level

11.4 Duration and Dynamics of Low-Flow Episodes

Average low-flow duration, estimated as median, varied between 17 and 67 days (Fig. 11.5). The maximum value of this characteristic in most cases did not exceed 250 days. However, in few rivers, this time was much longer and reached up to 500 days. In the distribution of low-flow duration, 4 catchments are characterized by a relatively wider interquartile range which means that almost all occurred low-flow episodes possess features typical for the regime of these rivers. In the others, numerous group of outliers and extremes should be connected with events extended far beyond standard water shortage period.

Severe low-flow duration, in median position varied between 10 and 35 days (Fig. 11.6). In the majority of investigated catchments, the third quartile of T_S did not exceed 30 days which may indicate a real limitation of severe streamflow deficits duration in Poland. Of course in a few rivers range of interquartile and non-outliers is much wider and caused by local factors. It is worth noticing that outliers appeared in almost every catchment (two or more), but their differentiation looks very stochastic which may suggest that hydrogeological determinants are more important than hydrometeorological.

The shortest low-flows periods occurred in mountainous and seaside catchments (Fig. 11.7). In lowland areas, low-flow duration depends on basin water resources and

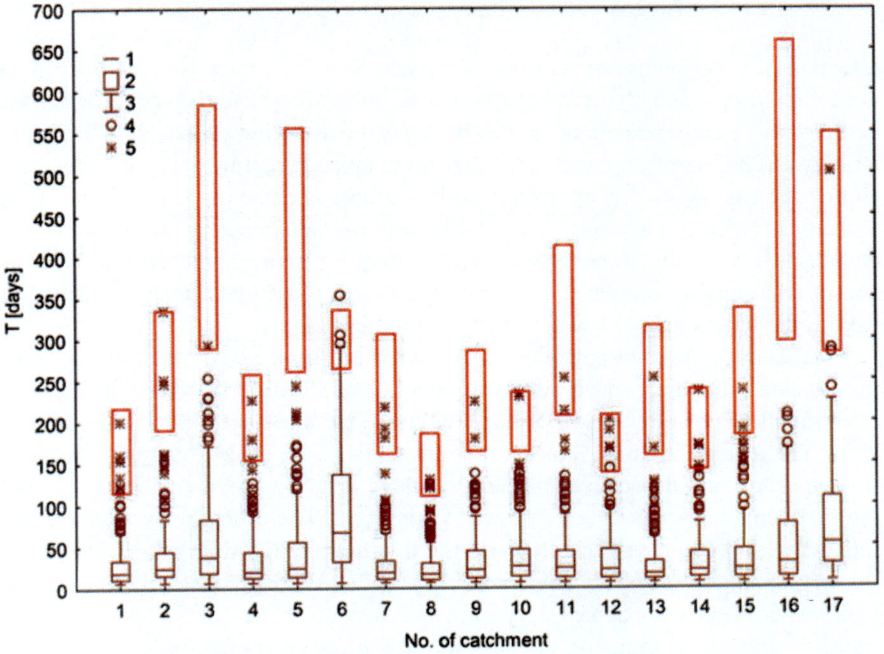

Fig. 11.5 Distribution of low-flow duration (1951–2015). Rectangle marks probability of maximum duration non-exceedance range 95–99%, remaining sign.—see Fig. 11.2

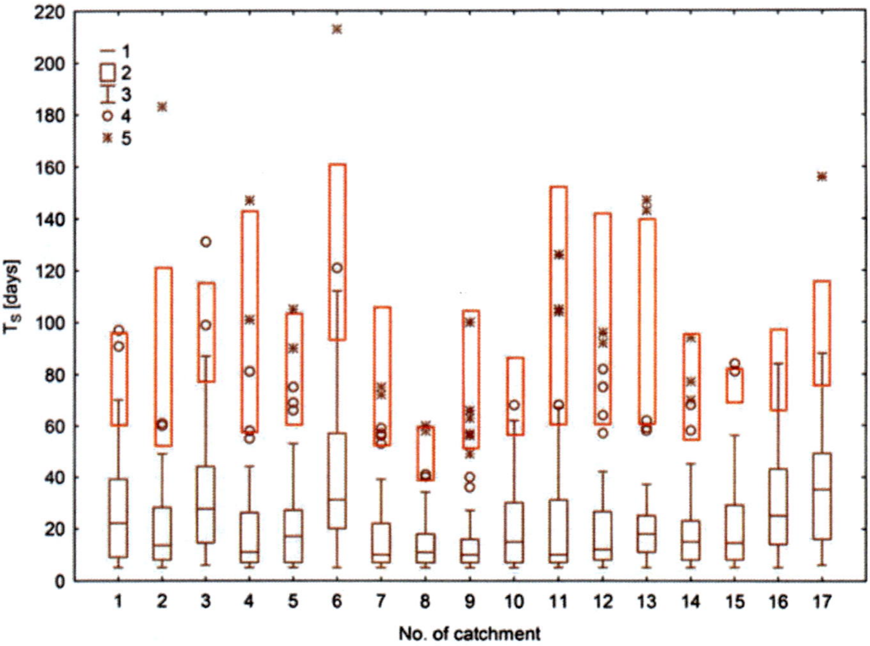

Fig. 11.6 Distribution of severe low-flow duration (1951–2015). Explanations—see Fig. 11.5

direction of water management operational activity. Severe low-flow duration does not follow proportionally to the previous one. In San river and the upper Oder river parameters T_S and T are almost equal. It means that low-flows after beginning, very quickly reach the severe phase and at the end of episode rapidly disappear. The high contribution of T_S in T is characterized by mountainous rives, downstream of the Vistula and Oder river as well as Wieprza. In the other catchments, both characteristics change proportionally. It seems that such kind of information about severe low-flows contribution might improve strategies of hydrological drought effects mitigation as well as the functionality of water management objects.

Frequency of low-flow episodes occurrence was assessed on the base of inter-low-flow spacing. Such approach allows to avoid generalization which appears during counting number of low-flow episodes within a year and does not lose information about the real position of shortage period on the time scale. Average inter-low-flow spacing varied between 68 and 254 days (Fig. 11.8a). Distribution of intT (av) is strongly asymmetric and skewed to the right. Therefore the average value is relatively low (100 days) and closes to the first quartile (88 days). It is reflected in the spatial pattern of this characteristic (Fig. 11.7). Probably it is safe to conclude that low-flow episodes appear very rarely or very often. To the first group belong transit rivers (for example the Vistula river at Warsaw cross-section—average every 254 days) and catchments with lakes (Noteć river—every 230 days, Drwęca river—every 217 days). In the other catchments, inter-low-flow spacing is much shorter

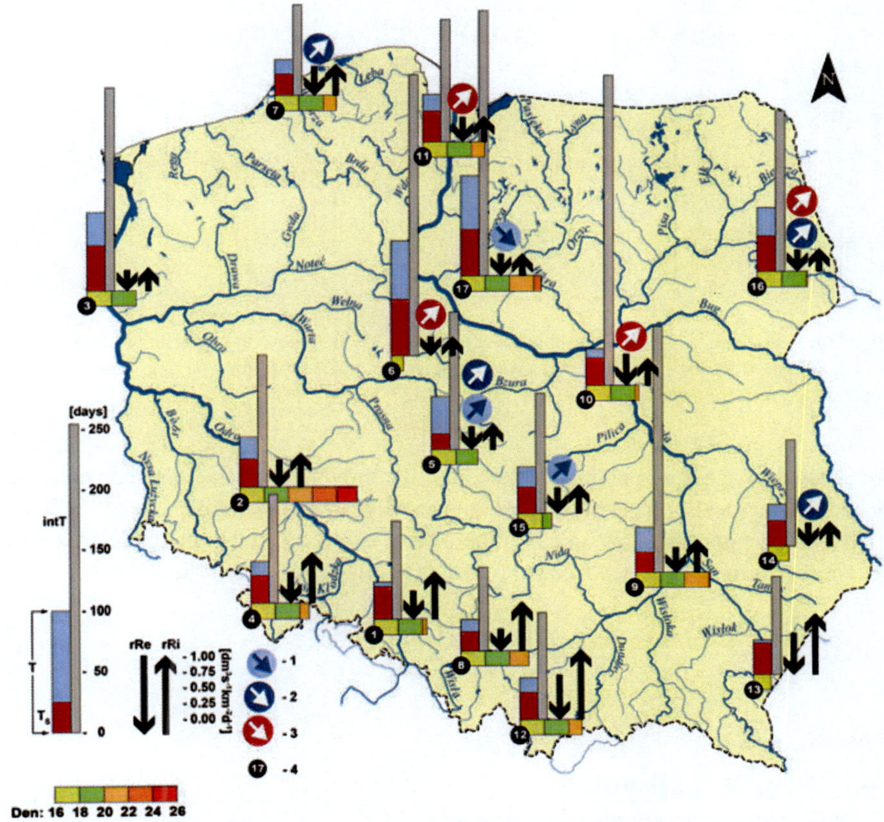

Fig. 11.7 Spatial differentiation of selected characteristics of low-flow episodes duration and dynamics (1951–2015). T—average low-flow duration, T_S—average severe low-flow duration, intT—average inter-low-flow spacing, rRe—average rate of low-flow recession, rRi—average rate of low-flow rise, Den—low-flow density index; statistically significant ($\alpha = 0.05$) linear trend of multiannual course of: 1—low-flow duration, 2—inter-low-flow spacing, 3—severe inter-low-flow spacing; 4—number of catchment—see Table 11.1

and does not exceed 100 days. Distribution of maximum observed period between low-flows is very close to symmetrical (Fig. 11.8b). Its median value reaches a bit more than 1000 days. However, outliers and extremes were placed very far from the interquartile range in this case. The longest waiting time for low-flow was noticed in the Oder river at Ścinawa cross-section (more than 8 years) and in the upland catchment of Wieprz river (almost 6 years). There are also rivers were low-flow episodes occur extremely often because the longest inter-low-flow spacing did not exceed 620 days (San, Dunajec and the upper Oder River).

Inter-low-flow spacing for severe episodes is obviously longer (Fig. 11.8c). Its average value varied between 340 and 850 days with a median on the level of 460 days. Severe low-flow frequency reached about 1.5 years should be very useful information

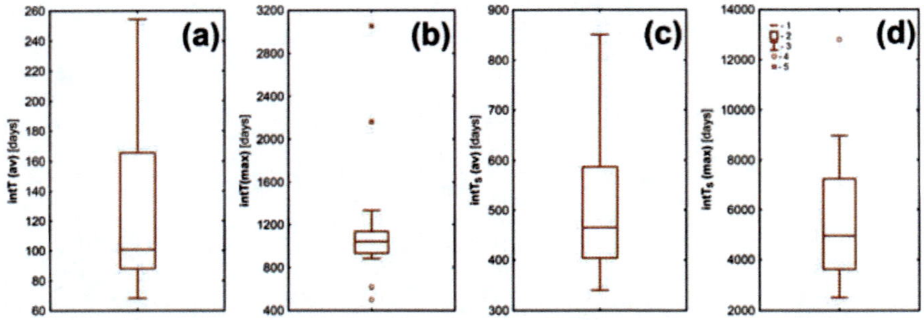

Fig. 11.8 Distribution of inter-low-flow spacing (1951–2015). a—average inter-low-flow spacing, b—maximum inter-low-flow spacing, c—average severe inter-low-flow spacing, d—maximum severe inter-low-flow spacing, 1—5—see. Figure 11.2

for water management planning, especially that the longest space between extreme water shortage periods may last from 7 up to 35 years (Fig. 11.8d).

Based on inter-low-flow spacing, there were estimated relative characteristic of severe low-flow occurrence frequency which is low-flows density index [6]:

$$Den = \frac{\text{int}Ts_{AV}}{Ts_{AV}} \tag{11.4}$$

where:

Den	low-flow density index
int Ts_{AV}	average severe inter-low-flow spacing
Ts_{AV}	average severe low-flow duration

The presented index indicates how much longer (Den > 1) or shorter (Den < 1) than average low-flows are average breaks between them. Relative matter of this characteristic allows receiving a few pieces of new information about the temporal structure of low-flows appearance which may be useful for water management planning. In the group of studied rivers, severe inter-low-flow spaces were on average 20.5 times longer than drought episodes (Fig. 11.7). The interquartile range of severe low-flow density index was very narrow and closed between 20 and 21.5 which proves high stability of this feature of the low-flow regime in Poland. Severe low-flow episodes frequency higher than average value is characterized by mountainous catchments and the almost whole course of the Vistula river. An exception is San river where besides Noteć and Wieprz river, index of density reached the lower value on the level of 17. It is worth noticing that information about severe low-flows density may be useful for hydro-technical installations work cycle planning and damming constructions density along the rivers in order to effective reactions to undesirable fluctuations of low-flows.

Recession phase belongs to this part of low-flow episode in which runoff recession is determined by retention level of groundwater reservoirs drained by river channels.

At the beginning of low-flow episode, rivers may be alimented in mixed way by quick and base flow. At the end of the low-flow event, the period of runoff rise is usually determined by direct runoff as an effect of effective precipitation or snowmelt alimentation. A very wide variety of intensity of separate form of alimentation modified by current hydrometeorological conditions and seasonal factors result in different reactions to feeding. Average low-flow recession rate was highly varied (Fig. 11.7). Low-flow episodes develop very slowly in lake catchments (buffering function of lake basins) an in uplands (good retention in groundwater reservoirs). In mountainous catchments low-low recession rate is relatively high and similar; the highest value is characterized by Dunajec river—0.452 $dm^3km^{-2}s^{-1}d^{-1}$. The average rate of low-flow rise is not fully proportional to the recession. Its value varied in a very wide range from 0.023 $dm^3km^{-2}s^{-1}d^{-1}$ (Noteć river), up to 0.882 $dm^3km^{-2}s^{-1}d^{-1}$ (Dunajec river). Mountainous catchments much quicker terminate low-flow episodes than the others (rRi = 0.498-0.882 $dm^3km^{-2}s^{-1}d^{-1}$) because of rapid reaction for precipitation or snowmelt intensified by quick surface flow over steep slopes. It is worth noticing that along transit rivers rate of low-flow recession and the rise decreases gradually which should be connected with a continuous increase of basin water resources as well as lack of synchronicity between low-flows occurrence in the main river and its tributaries.

The multiannual course of low-flow duration is characterized by wet and dry group of years (Fig. 11.9). Periods with severe hydrological droughts, common for whole country, covered the first half of the 1950s, the first half of 1960s, the middle part of 1980s, beginning of 1990s, the middle part of 2000s and beginning of 2010s. It is easy to calculate that serious water shortage periods occur in Poland every 10-15 years on average. However, long low-flows with huge streamflow deficit appeared in all rivers at the same time very rarely. For example in the lower course of transit rivers, longer low-flows occurred in the Vistula river for first 20 years of the investigated period, and then the Oder river was characterized by longer duration of these episodes (Fig. 11.9a).

Moreover, low-flows in the Vistula river are generally shorter and occur more often than in the Oder river. It is worth noticing that big rivers are deprived of low-flows in the late 1970s and early 1980s because of the flood years. However, very high floods which covered the majority of Poland territory in 1997 and 2010 very often accompany low-flow episodes in the same year.

In smaller catchments, low-flow periods are more evenly distributed on the time scale (Fig. 11.9b). Hydrological droughts in the 1960s and the 1970s were more severe for mountainous rivers whereas 1980s and 1990s impacted more seriously lowland catchments. In every case, the multiannual course of low-flow duration looks more like fluctuation or oscillation than a linear trend or chaotic distribution. Severe low-flow duration distributions also confirm the above observations (Fig. 11.9c, d). Common long water shortage periods are connected with macro hydrometeorological conditions. However, the others seem to be under local factors influence, especially hydrogeological determinants and some water management installations.

In the series of low-flow duration, statistically significant trends exist very rarely (Fig. 11.10a). Observed tendency very often was driven by the separate group of long low-low episodes. Estimated determination coefficients varied between 0.022 and 0.049 which means that from 2% to 5% of observed multiannual low-flow duration variability might be explained by systematic linear tendency. Therefore presented results should be interpreted very carefully. Positive trends of low-flow duration occurred only in lowland catchments where seasonally precipitation shortage and agricultural drought occurs (Fig. 11.7). Restricted water resources, adverse structure of water balance, and tendency to lengthening of low-flow periods should be taken into consideration in further plans of water management development in this part of the country. There were also statistically significant tendencies in inter-low-flow spacing (Fig. 11.10b). All of them have positive direction and appeared in the north and east part of Poland (Fig. 11.7). It means that low-flow occurrence frequency in these catchments systematically decreases.

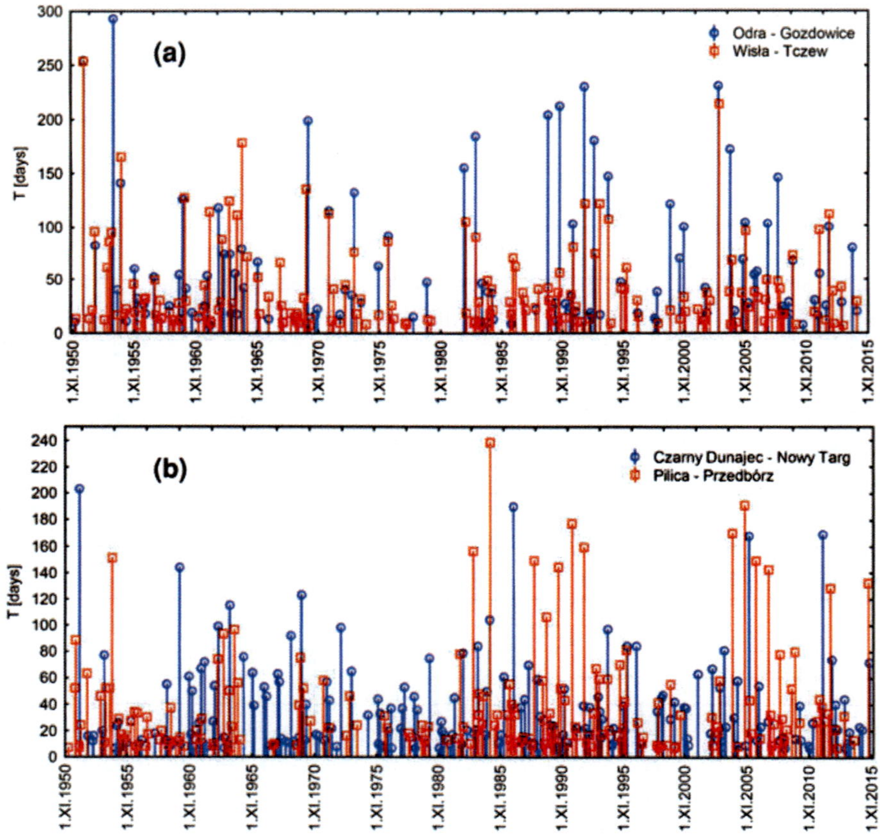

Fig. 11.9 Examples of multiannual course of low-flow duration (1951–2015). T—low-flow duration, T_S—severe low-flow duration

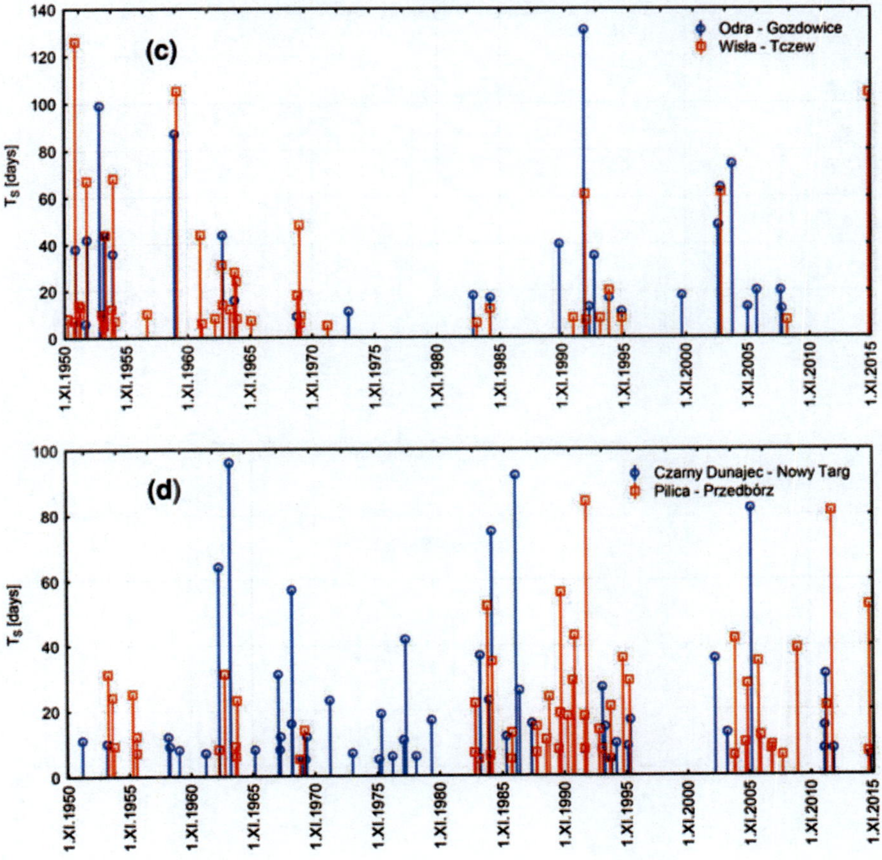

Fig. 11.9 (continued)

11.5 Low-Flow Extremes

Low-flows occurring as a result of hydrological drought can lead to a situation in which separate branches of the national economy will record losses caused by an excess of water needs over the disposal water resources at a specific time and space. Therefore assessment of low-flow appearance probability and risk of potential losses caused by water shortage is a very important component of contemporary water management analyses.

Assessment of maximum probable low-flow duration was executed on the base of "Niżówka 2003" software [27]. Distribution of discrete variable, connected with the low-flow occurrence, was most well-fitting to Poisson's function. Continuous variable, which is an estimator of low-flow duration, was described by log-normal distribution (Fig. 11.11). It is worth noticing that maximum low-flow duration on

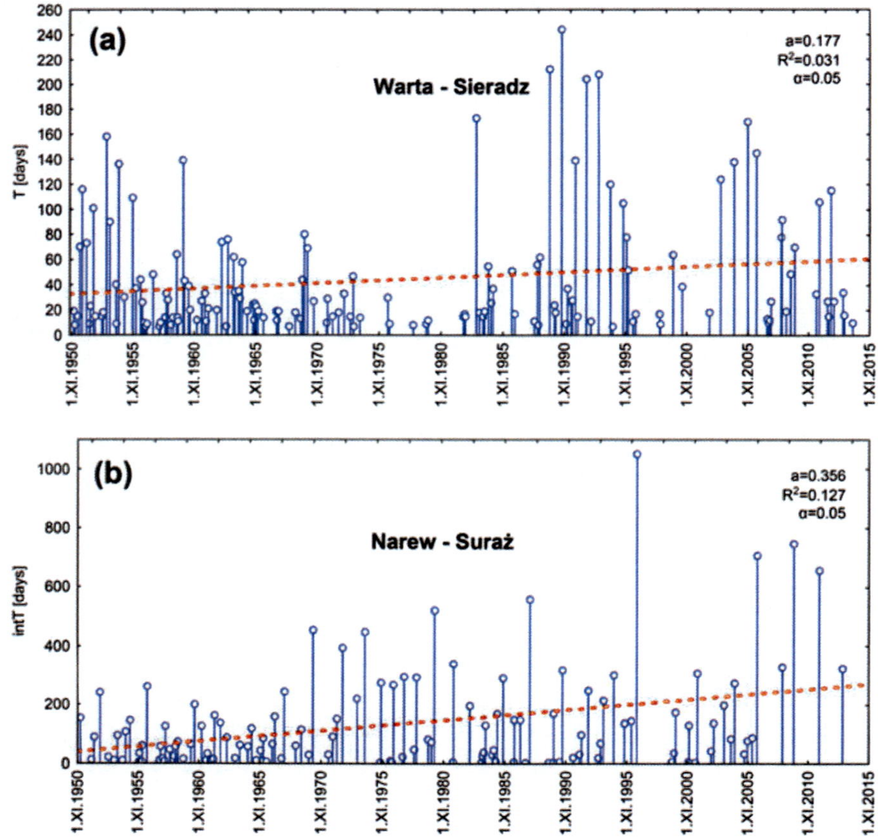

Fig. 11.10 Examples of multiannual trends of low-flow duration **a** and inter-low-flow spacing **b** (1951–2015). Explanations—see Fig. 11.4

95% probability of non-exceedance is mentioned in literature as maximum credible (or likelihood) low-flow [5].

Average low-flow duration on 95% probability of non-exceedance in Polish rivers is equal 162 days (Fig. 11.12a). In half of the investigated gauging cross-sections, this parameter varied between 154 and 254 days. Extremes were observed in Narew river (296 days) and the upper course of the Vistula river (114 days). Spatial differentiation of this characteristic generally reflects the availability of water resources (Fig. 11.13). The average severe credible low-flow duration was much shorter—60 days. Dispersion of distribution is very small, and extremes are characterized by Noteć river (92 days) and the upper Vistula river again (39 days) (Fig. 11.12b). The spatial pattern of this characteristic is very similar to the previous one (Fig. 11.13). Differences between catchments are determined mostly by the abundance of water resources and the pace of their replenishment. As a result, maximum probable (95%)

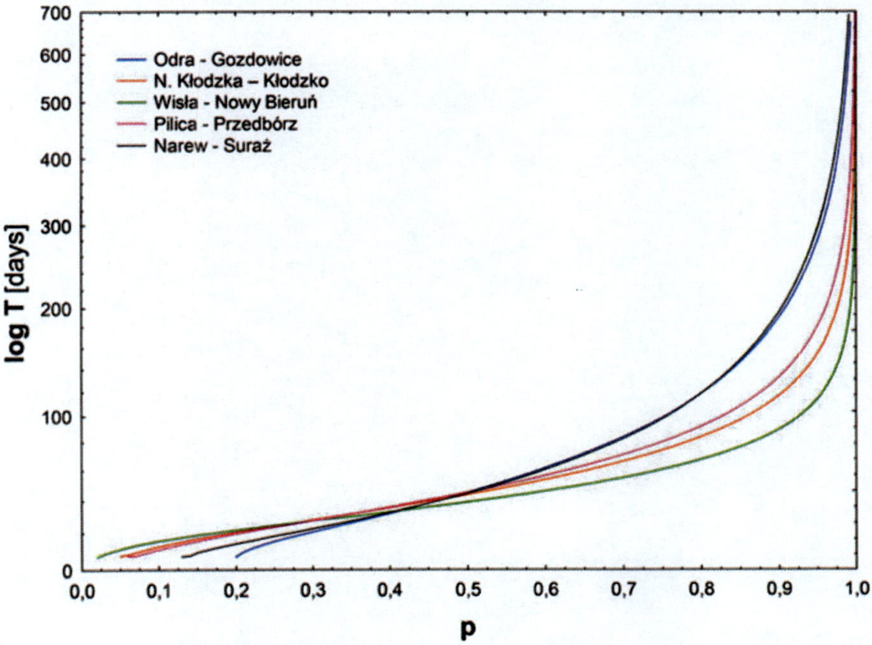

Fig. 11.11 Examples of non-exceedance probability distribution of maximum low-flow duration. p—probability, T low-flow duration

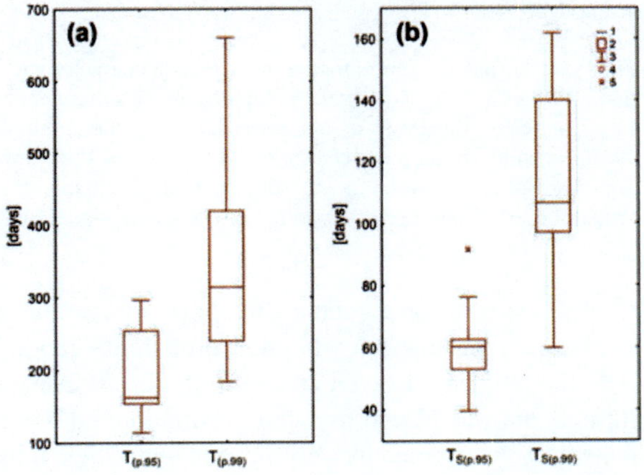

Fig. 11.12 Distribution of maximum probable low-flow duration (1951–2015). T—low-flow duration, T_S—severe low-flow duration, p. 95, 99—probability of non-exceedance, 1—5—see. Figure 11.2

severe low-flow duration in big transit rivers is beginning to be similar to this in smaller streams.

Toughening the probability level up to 99% of non-exceedance generates much longer low-flow duration (Fig. 11.12). On an average level, low-flows reach 314 and 60 days but maximally this time was lengthened 660 and 92 days. Such a probability

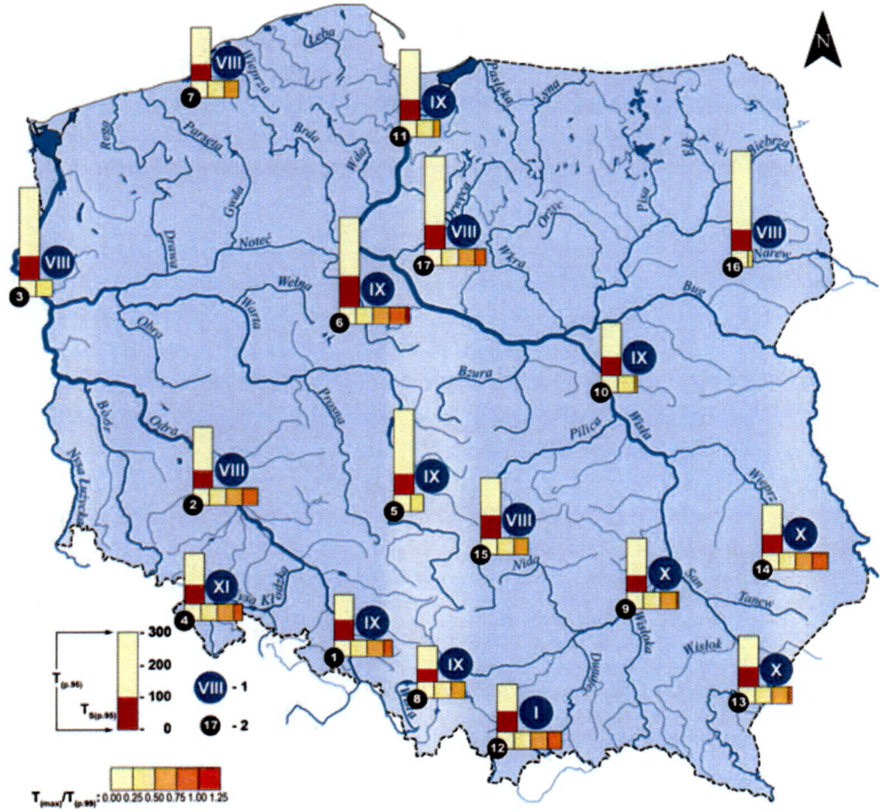

Fig. 11.13 Spatial differentiation of selected characteristics of low-flow extremes and their risk (1951–2015). $T_{(p.95)}$—95% probability of non-exceedance of maximum low-flow duration, $Ts_{(p.95)}$—95% probability of non-exceedance of maximum severe low-flow duration, $T_{(p.99)}$—99% probability of non-exceedance of maximum low-flow duration, $T_{(max)}$—maximum observed low-flow duration, 1—the month with the highest number of low-flow episodes exceeding 95% probability of its duration, 2—number of catchment—see Table 11.1

of 99% seems to be a good estimator of an extreme drought event. It is interesting to compare the low-flow duration probability range between 95% and 99% with real, the longest observed episodes. In almost every catchment, two or more low-flow events was placed over the maximum credible duration (Fig. 11.5). They are, generally, determined by very serious hydrological droughts in the 1950s and 90 s and might be identified as estimators of severe water deficiencies in basins and should be considered in water balance predictions and water management planning. In one case the longest observed low-flow exceeded the limit of 99% probability (Noteć river). However, its origin should be explained by the anthropogenic impact caused by dewatering groundwater resources for mining purposes which were enhanced by natural hydrological drought conditions and caused extraordinary long low-flow event.

Analysis of the relationship between maximum duration of observed and predicted (99%) low-flow event allows assessing the potential risk of very extremely long drought period occurrence. This parameter was estimated as a quotation of these two characteristics (T_{max} / $T_{p.99}$). The highest potential risk if the further low-flow episode would be much longer than observed, appears in rivers with a low value of index: lower course of the Vistula and Oder river and some lowland catchments with restricted water resources (Fig. 11.13). Relationships between the low risk of crucial low-flow lengthening and available water resources are characterized by upland (significant groundwater alimentation) and mountainous (high and dynamic precipitation alimentation) catchments.

Severe low-flow episodes exceeding maximum credible duration appeared very often (Fig. 11.6). In 10 cases, the limit of 99% probability was exceeded as well. Moreover, the probability range 95–99% in many catchments involved the interquartile deviation reach. Such image of distribution indicates that hydrogeological limitations, as well as physiographical features of the catchment, are more important than macroscale hydrometeorological conditions. In rivers where maximum observed severe low-flow duration exceeded level of 99% probability, there is a very small chance for significantly longer event appearance. Seasonal assessment of this phenomenon might be important for prediction of water ecosystem condition and improvement of water management effectiveness. Therefore, for every catchment, the month with the highest number of low-flow episodes exceeding 95% probability of its duration was identified (Fig. 11.13). Selected months, which are estimators of extremely long severe low-flows appearance time, occurred in summer and autumn season. It proves that precipitation shortage and evapotranspiration process are the main determinants of severe hydrological drought in the scale of the whole country. The only excepts are high mountainous areas (Dunajec) where winter low-flows are basic low-flow regime determinant.

11.6 Conclusions and Recommendations

Poland belongs to territories which are put at risk of occurrence of hydrological droughts at serious degree of severity. It results in periodic water shortage, a decrease of groundwater table, and occurrence of low-flows in rivers. Multidirectional analysis of low-flows allowed to assess low-flow development dynamics as well as identify multiannual and seasonal features of drought streamflow deficits regime. Good knowledge of these processes and phenomena may effectively support the tools and strategies of optimal reduction of drought results, its prevention, and prediction. It is worth noticing that knowledge of low-flow regime is very important for water balance structure which may be crucial for the identification of quantity and dynamics of available water resources.

Dynamics of low-flow episodes progression is characterized by high seasonal and multiannual variability. Its basis is determined by strong differentiation of hydrometeorological conditions which generate particular episodes as well as modifying the role of anthropogenic and physiographical factors (mainly hydrogeological conditions and lakes). Multiannual tendencies of selected characteristics of low-flows, such as minimum flows, low-flows duration, and inter-low-flows spacing, covered only a part of investigated catchments and seem to be caused by natural as well as anthropogenic factors. Analysis of the probability of maximum low-flow episodes occurrence allowed to indicate areas with a high-risk rating of water shortage hazardous for water ecosystems and water management operations. In the matter of probability and risk of maximum low-flows duration there was possible to identify a full spectrum of determinants which should be taken into consideration in local water resources exploitation planning as well as in long-term strategies of water management in the scale of the entire country.

References

1. Kundzewicz ZW (2008) Hydrological extremes in the changing world. Folia Geogr. Ser Geogr Phys 39:37–52
2. Kundzewicz ZW, Jania JA (2007) Extreme hydro-meteorological events and their impacts. From the global down to the regional scale. Geogr Pol 80(2):9–23
3. McMahon TA, Finlayson BL (2003) Droughts and anti-droughts: the low flow hydrology of Australian rivers. Freshw Biol 48:1147–1160
4. Smakhtin VU (2001) Low flow hydrology: a review. J Hydrol 240:147–186
5. Tokarczyk T (2013) Classification of low flow and hydrological drought for a river basin. Acta Geophys 61(2):404–421
6. Tomaszewski E (2012) Wieloletnia i sezonowa dynamika niżówek w rzekach środkowej Polski (Multiannual and seasonal dynamics of low-flows in rivers of central Poland). Wyd. Uniwersytetu Łódzkiego, Łódź
7. Sene K (2010) Hydrometeorology. Forecasting and Applications, Springer, Dordrecht-Heidelberg-London-New York
8. Girguś R, Strupczewski W (1965) Wyjątki ze źródeł historycznych o nadzwyczajnych zjawiskach hydrologiczno-meteorologicznych na ziemiach polskich w wiekach od X do XVI. Instr. i podręcz. PIHM, 87, WKiŁ, Warszawa
9. Bartnik A (2005) Odpływ niski w Polsce (Low flow in Poland). Acta Geogr Lodz 91:1–95
10. Biernat B (1987) Typowe okresy występowania niżówek (Typical periods of low-flow occurrence). In: Stachý J (ed) Atlas hydrologiczny Polski (Hydrological atlas of Poland). Wyd. Geol, Warszawa, p 64
11. Farat R, Kępińska-Kasprzak M, Kowalczak P, Mager P (1995) Susze na obszarze Polski w latach 1951–1990 (Droughts on the area of Poland in the years 1951-1990). Mat. Bad. IMGW, Seria: Gosp. Wodn. i Ochr. Wód, 16
12. Kępińska-Kasprzak M (2014) Zagrożenie wystąpieniem niżówki w Polsce (Hazard of a low-flow occurrence in Poland). Monogr. Kom. Gosp. Wodn. PAN, XX, Warszawa, pp 163–172
13. Mikulski Z (1963) Zarys hydrografii Polski (Hydrography of Poland—an overview). PWN, Warszawa
14. Mikulski Z (1998) Gospodarka wodna (Water management). Wyd. Nauk, PWN, Warszawa
15. Tomaszewski E (2017) Niżówki i susze (Low-flows and droughts). In: Jokiel P, Marszelewski W, Pociask-Karteczka J (eds) Hydrologia Polski (Hydrology of Poland). Wydawnictwo Naukowe PWN, Warszawa, pp 175–182
16. Dubicki A (ed) (2002) Zasoby wodne w dorzeczu górnej i środkowej Odry w warunkach suszy (Water resources of the upper and middle Oder river basin in drought condition). Wyd. IMGW, Warszawa

17. Jokiel P (2010) Niżówki i odpływy niżówkowe w małych rzekach środkowej Polski w drugiej połowie XX wieku, In: Wrzesiński D (ed) Odpływ rzeczny i jego regionalne uwarunkowania, Studia i Prace z Geogr. i Geol., Bogucki Wyd. Nauk., Poznań
18. Kasprzyk A (2004) Występowanie susz hydrologicznych w Europie Środkowej (Occurrence of hydrological droughts in Central Europe). Prace Inst Akad Święt w Kielcach 13:181–201
19. Kubiak-Wójcicka K, Bąk B (2018) Monitoring of meteorological and hydrological droughts in the Vistula basin (Poland). Environ Monit Assess 190:691
20. Strzebońska-Ratomska B (1994) Metodyka oceny intensywności i zasięgu suszy hydrologicznej, cz. I: Susza hydrologiczna na Podkarpaciu w latach 1961–1990. Wiad. IMGW 4:15–42
21. Tlałka A (1982) Przestrzenne zróżnicowanie niżówek letnich w dorzeczu górnej Wisły (Spatial differentiation of low-flows in the upper Vistula river basin). Rozpr. Hab, UJ, p 63
22. Tokarczyk T (2001) Zmienność przepływów niskich na obszarze Kotliny Kłodzkiej (Variability of low flows in the Kłodzka Valley). Zesz. Nauk. Akad. Roln. we Wroc. Inż Środ 413:105–127
23. Jakubowski W (2006) An application of the bivariate generalized pareto distribution for the probabilities of low flow extremes estimation. Hydrol Earth Syst Sci Discuss 3:859–893
24. Jakubowski W (2008) The low flow extremes in a small Beskydy Mts. experimental watershed, [w:] Chełmicki W, Siwek J (eds) Book of abstracts of Biennal International Conference ERB "Hydrological extremes in small basins", Jagiellonian University, Kraków, pp 157–160
25. Kasprzyk A (2010) Prawdopodobieństwo wystąpienia niżówek w zlewniach rzek województwa świętokrzyskiego (Low flows probability of occurrence in the Świętokrzyskie Province river catchment). In: Ciupa T, Suligowski R (eds) Woda w badaniach geograficznych. Instytut Geografii Uniwersytet J. Kochanowskiego, Kielce, pp 273–280
26. Tomaszewski E (2008) Maksymalny czas trwania głębokich niżówek letnich w środkowej Polsce i jego uwarunkowania (Maximum severe summer low-flow duration in Central Poland and its determinants). Acta Univ Lodz Folia Geogr Phys 8:89–98
27. Jakubowski W, Radczuk L (2004) NIZOWKA 2003 software, In: Tallaksen LM, van Lanen HAJ (eds) Hydrological drought. Processes and estimation methods for streamflow and groundwater. Developments in Water Science, vol 48, Elsevier, Amsterdam
28. Chełmicki W, Bieńkowski T (2005) Przepływy niżówkowe w dorzeczu górnego Dunajca w 2003 roku na tle wielolecia 1951–2003 (Low flows in the upper part of the Dunajec drainage basin in 2003 compared with the period of 1951–2003). Folia Geogr Ser Geogr Phys 35–36:65–75
29. Jańczak J, Kowalik A, Sziwa R (1994) Reakcja stanów wody jezior dorzecza Odry na suszę lat 1989–1992 (Reaction of water levels of lakes in the Oder river basin to drought in 1989-1992). Wiad. IMGW 2:117–124
30. Tomaszewski E (2016) Impact of Lake Gopło on low-flow regime of the upper Noteć river. Limnol Rev 16(2):95–103
31. Kubiak-Wójcicka K (2012) Charakterystyka niżówek na Wiśle w Toruniu (The characteristics of low water levels on the Vistula river in Toruń). In: Marszelewski W (ed) Gospodarowanie wodą warunkach zmieniającego się środowiska. Monografie Komisji Hydrologicznej PTG, Toruń, pp 85–94
32. Tomaszewski E (2018) Low-flow discharge deficits assessment, applying constant and variable low-flow threshold levels, as illustrated with the example of selected catchments in the Vistula river basin, Acta Sci Pol, Formatio Circumiectus, 17(3):205–216
33. Kasprzyk A (2002) Ocena surowości suszy hydrologicznej w zlewniach o różnym użytkowaniu powierzchni. Prace Inst Akad Święt w Kielcach, Kielce, pp 147–154
34. Soja R (2002) Hydrologiczne aspekty antropopresji w polskich Karpatach (Hydrological aspects of anthropopression in the Polish Carpathians. Prace Geograficzne IGiPZ PAN, vol 186, Warszawa
35. Tomaszewski E (2014) Dynamika niedoborów odpływu niżówkowego w rzekach poddanych silnej antropopresji (Dynamics of drought streamflow deficits in rivers affected by strong anthropopression). In: Ciupa T, Suligowski R (eds) Woda w mieście (Water in the city). Monografie Komisji Hydrologicznej PTG, vol 2, Kielce, 289–300

36. Dynowska I (1971) Typy reżimów rzecznych w Polsce (Types of river regimes in Poland). Nauk. UJ, Prace Grogr, Zesz, p 28
37. Wrzesiński D (2017) Reżimy rzeki Polski (Regimes of Polish rivers). In: Jokiel P, Marszelewski W, Pociask-Karteczka J (eds) Hydrologia Polski (Hydrology of Poland). Wydawnictwo Naukowe PWN, Warszawa, pp 215–221
38. Tokarczyk T, Szalińska W (2018) Drought hazard assessment in the process of drought risk management. Acta Sci Pol, Formatio Circumiectus, 18(3):217–229
39. Hisdal H, Tallaksen LM, Clausen B, Peters E, Gustard A (2004) Hydrological drought characteristics. In: Tallaksen LM, van Lanen HAJ (eds) Hydrological drought. Processes and Estimation Methods for Streamflow and Groundwater. Developments in Water Science, vol 48, Elsevier, Amsterdam, pp 139–198
40. Ozga-Zielińska M (1990) Niżówki i wezbrania—ich definiowanie i modelowanie (Droughts and Floods—Their Definition and Modelling). Przegl. Geof. 1–2:33–44
41. Tomaszewski E (2011) Defining the threshold level of hydrological drought in lake catchments. Limnol Rev 11(2):81–89
42. Yevjevich V (1967) An objective approach to definitions and investigations of continental hydrologic drought, Hydrology Paper, No. 23, Colorado State Univ., Fort Collins, Colorado
43. Dębski K (1970) Hydrologia (Hydrology). Arkady, Warszawa
44. Oliver JE (1980) Monthly precipitation distribution: a comparative index. Prof Geogr 32(3):300–309
45. Jokiel P, Tomalski P (2018) Zróżnicowanie i zmienność wieloletnia sezonowości przepływu w wybranych przekrojach wodowskazowych Wisły (Differentiation of river flow seasonality and its multiannual changeability in selected cross-sections on the Vistula river). Prace Geogr 155:27–45
46. Tomalski P, Tomaszewski E (2015) Metody, formuły i wzory obliczeniowe zastosowane w pracy (Methods, formulas and computational equations). In: Jokiel P (ed) Metody statystyczne w analizach hydrologicznych środkowej Polski (Statistical methods in hydrological analyses of Central Poland). Wyd. Uniwersytetu Łódzkiego, pp 215–272
47. Yule G, Kendall MG (1966) Wstęp do teorii statystyki (An introduction to the theory of statistics). Wyd. Nauk, PWN, Warszawa
48. Czarnecka M, Nidzgorska-Lencewicz J (2012) Wieloletnia zmienność sezonowych opadów w Polsce (Multiannual variability of seasonal precipitation in Poland). Woda Środ Obsz Wiej 2(38):45–60

Chapter 12
Dynamics, Range, and Severity of Hydrological Drought in Poland

Edmund Tomaszewski and Malwina Kozek

Abstract The chapter discusses a phenomenon of hydrological drought based on the assumption that river low flow serves as a good estimator of the drought development. The study analyzed 87 catchments of Polish rivers with a total surface exceeding 317 thousand km^2, and covering almost the entire area of the country. Basic data on daily series of discharges at the gauging cross-sections closing the catchments were collected in the years 1985–2014. Low flows were identified with reference to threshold level method matching 70 and 95 percentile at a flow duration curve as constant, multiannual truncation level ($Q_{70\%}$, $Q_{95\%}$). The identification and separation criteria allowed for identification and analysis of the course of mild and severe hydrological droughts in Poland. The research covered parameters describing duration, severity, range, and identification of periods with different patterns of hydrological drought development. An analysis of multiannual and seasonal variability of the phenomenon and selected genetic relationships enabled identification and evaluation of the factors determining the development of hydrological drought in Poland. Seasonal properties of the drought were additionally assessed with a two-parameter analysis of seasonality degree and concentration date involving angular measures. The study findings and conclusions are of cognitive as well as practical nature and can be applied to improve the effectiveness of water management aimed at mitigating the effects of drought.

Keywords Hydrological drought · Low flows · Threshold level method · Flow seasonality · Poland

E. Tomaszewski (✉) · M. Kozek
Department of Hydrology and Water Management, Faculty of Geographical Sciences, University of Łódź, 88 Narutowicza Str, 90-139 Łódź, Poland
e-mail: edmund.tomaszewski@geo.uni.lodz.pl

M. Kozek
e-mail: malwina.kozek@geo.uni.lodz.pl

M. Zeleňáková et al. (eds.), *Management of Water Resources in Poland*, Springer Water,
https://doi.org/10.1007/978-3-030-61965-7_12

12.1 Introduction

Drought is one of the most unfavourable effects of weather and climate variability. It is usually defined as an exceptionally dry (rainless) period long enough so that water deficit seriously disturbs the structure of water balance within a specific area [1, 2]. Its basic feature is limited access to water resources. Drought is usually initiated by rainfall shortage (meteorological drought) (see Fig. 12.1). However, the rainfall shortage initiates this stage of drought only when it occurs in a typical period of groundwater alimentation with precipitation or when, due to various factors, groundwater retention is very low. In practical terms, this stage is identified based on relative deviations of rainfall characteristics from standard values or multiannual means [3, 4]. Apart from the amount of total precipitation, its time distribution and intensity also play important roles. Limited rainfall supply to groundwater reservoirs is usually exacerbated by evapotranspiration. In Poland, the areas with the highest risk of atmospheric drought include Central Poland Lowlands and western part of the Pomeranian Lakeland [5, 6].

Prolonged lack of precipitation combined with intense evapotranspiration result in gradual loss of soil moisture within the vadose zone. At first, infiltrating gravitational water and suspended vadose water disappear, then capillary water, and in extreme cases even adhesive water. This process is accompanied by an increase in soil capillary potential, the value of which directly depends on the time necessary to restore field capacity so that effective infiltration could begin. As the infiltration is very slow, single precipitation events during drought do not supply groundwater, because properties of the dried soil are similar to those of the impermeable layer.

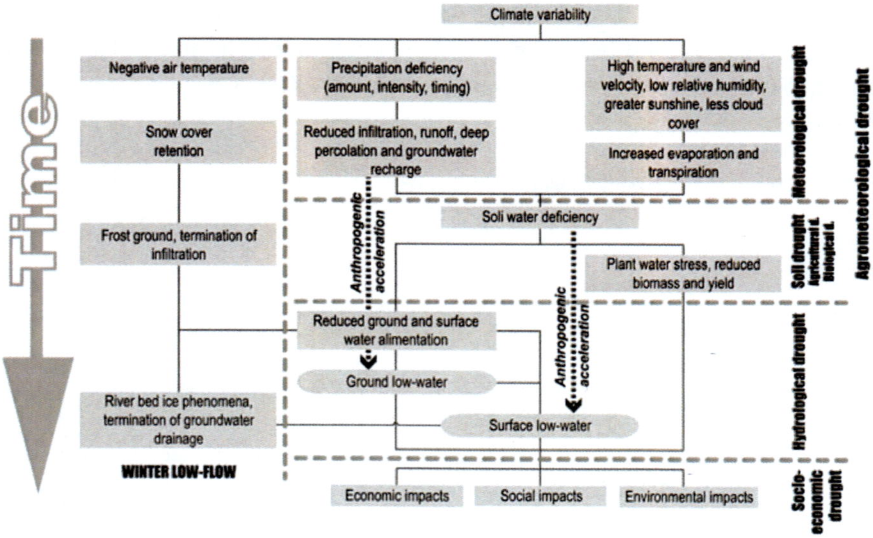

Fig. 12.1 Diagram of drought development (after [7], modified)

This may sometimes intensify surface runoff and evaporation, i.e., the processes conducive to soil drought. If water shortage is translated into measurable losses manifested by plant degradation and restricted growth or if drought periods overlap with intense field works, agricultural drought may occur [8]. Individual plant species show different tolerance to soil water deficiency, which is why the concept of agrometeorological drought involves relationships between precipitation and plant growth conditions. An official indicator of agricultural drought in Poland is the climatic water balance (CWB). The metric is used by the Agricultural Drought Monitoring System[1] that evaluates drought severity between 1st April and 30th September in 13 six-decade periods for various plant species. In communes where CWB threshold values are exceeded, the monitoring data are the basis for compensation claims.

Further increase of water deficit results in hydrological drought [2, 9–12]. The non-supplied aquifers are continually drained by watercourses and springs. As a result, the elevation of their water table declines and low groundwater period occurs. This is accompanied by a systematic decrease in surface water resources, which are usually in a hydraulic connection with groundwater. These mechanisms facilitate the appearance of summer and autumn low flows. Recession rate of the groundwater resources of the active exchange zone in this period, and thus the rate of low flow development, depend nearly exclusively on the retention level of groundwater reservoirs [13–15].

Winter low flows of surface waters follow a different course. Limited runoff is then due to temporary water retention in the snow cover often combined with riverbed freezing during severe frosts that stops all forms of drainage (Fig. 12.1). This is not a global problem, but in the countries with a harsh climate, it may severely disturb water management [16–18]. It is worth noticing that winter drought does not consist in water shortage but its entrapment in the form of snow and ice. This means it is highly seasonal and water resources are retained "on the spot".

Severe hydrological droughts bring about serious losses to water consumers, and this is why a concept of a socio-economic drought has been coined. The effects of such a drought may be perceptible at the national scale [10]. Disturbed functioning of water-power engineering or agricultural production affects economy. Water shortage impacting municipal services management is classified as social effects of drought, while the degradation of aquatic ecosystems, especially during the blooms of toxic algae causing significant decline of water quality, is the manifestation of its environmental effects.

Human activities currently interfere with all stages and links of drought development (see Fig. 12.1), and in some cases, may accelerate its advancement [7]. Agrotechnical treatments aimed at ensuring optimal conditions of plant growth need to be initiated as soon as rainfall shortage brings about meteorological drought. Irrigation is then intensified by using local groundwater resources and providing water to plants via sprinkling machines or other forms of watering. Water collected from the saturation zone does not infiltrate back to the groundwater reservoirs but is removed during evapotranspiration. This results in simultaneous development of atmospheric

[1] https://www.susza.iung.pulawy.pl/en/.

drought and low flows of low groundwater levels, often without the soil drought stage. At the next stage, groundwater resources are exhausted, and surface water reserves need to be mobilized to mitigate the soil drought. However, by this time the drought severity is so strong that water redistribution quickly leaps from local to regional scale and results in low flow of rivers. As a result, when the hydrological drought should only naturally begin, low flows of underground and surface waters are already highly advanced.

Many regions of the world, where the effects of drought are particularly harsh or dangerous, introduced early warning systems. They improve and implement methods for short-term and seasonal forecasting of drought and develop an integrated information exchange system between networks. They include various scenarios of drought development and emergency plans at the state and regional levels. The systems also assess the risk of individual events and available insurance options [19]. Drought risk assessment in Poland involves not only the strategies of mitigating its effects but also the process of drought risk management [20].

12.2 Methodology and Data

River low flow, i.e. the last stage of the response to insufficient supply, is considered a good indicator of hydrological drought [21–23]. It is usually defined as a period of low flows (water levels) in a river or flows maintained during dry weather [14, 24]. Precise definition of the low flow depends on a specific research approach. One of them is a threshold level method in which an analysis of a flow hydrograph with respect to a threshold value determined based on a selected characteristic flow enables identification of the low flows. Limit values are estimated based on second degree main flows, periodic flows from the flow duration curve, analysis of annual minima distribution or conventional flows adapted to specific challenges of water management and environmental management. In this research approach, the low flow is a period with flows lower than the established threshold flow [25–27]. As a consequence, the basic parameters of the identified phenomenon include the volume of discharge deficit in the period when it is lower than the threshold flow and the duration of the low flow event (Fig. 12.2a).

The calculated deficits of drought streamflow were transformed into a relative deficit (RVn), which made it possible to compare results from catchments of variable surface area [7]. The presented characteristic is calculated as per the following formula:

$$RVn = \frac{Vn}{Vmax} \cdot 100\% \tag{12.1}$$

where:

RVn relative drought streamflow deficit (%),

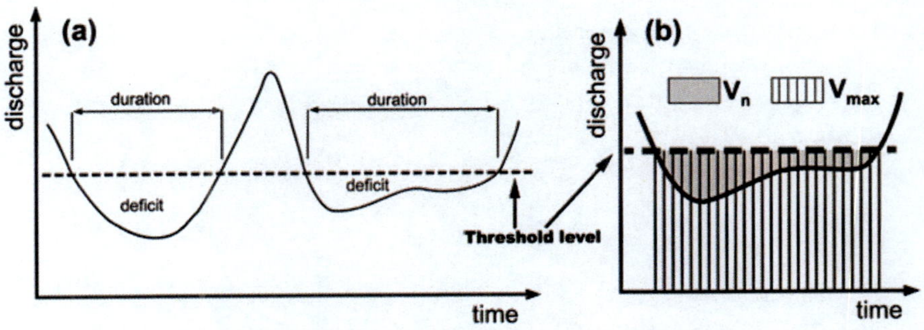

Fig. 12.2 Basic parameters of low flow episodes (**a**) and graphic analysis of the calculation of relative low flow discharge deficit (**b**) (after [7], modified). V_n – volume of drought streamflow deficit (m^3), V_{max} – maximum volume of possible drought streamflow deficit for a given period, i.e. when the river discharge value equals 0 (m^3)

Vn volume of drought streamflow deficit (m^3),
Vmax volume of maximum possible drought streamflow deficit for a given period, i.e., when the river flow value equals 0 (m^3).

This measure not only evaluates the intensity of the deficit but also indicates the degree of the catchment resources drainage that shows a hydraulic connection with the low flow. When the metric equals 100%, no flow in the riverbed should occur (Fig. 12.2b), so it can serve as an estimator of the hydrological drought severity. Moreover, it ensures full comparability of results in catchments of various sizes and is useful in the analysis of low flows occurring along transit rivers, as it is based on observations from a specific gauging section only.

The study covered 87 water gauges from Poland (see Fig. 12.3). Total catchment area closed with these gauging sections exceeds 317,000 km^2. Their location reflects a full spectrum of possible river regimes occurring in Poland, as well as the variety of physico-geographical conditions that affect the shaping process of low flows and their deficits. The analysis covered the observation period 1985–2014. It is long enough to meet the reliability threshold for hydrological analyses advocated in the literature to be at least 30 years, and it reflects the current status of the investigated phenomenon.

The calculations were based on the series of daily discharges collected and shared by the Polish Institute of Meteorology and Water Management – National Research Institute (IMGW-PIB) in Warsaw. The flow corresponding to 70th ($Q_{70\%}$) percentile at the flow duration curve was assumed as a truncation level for low flows. The study assessed the number of days with the low flow for individual months and years of the analyzed period, and the streamflow deficit volume in absolute values that was converted into the relative deficit (RVn). It also identified severe low flows that occur when the seasonally renewable resources in the hydrologically active zone are depleted, and supply is provided exclusively by the aquifers characterized by a multiannual rhythm. The truncation level for severe low flows was set at 95 percentile [7].

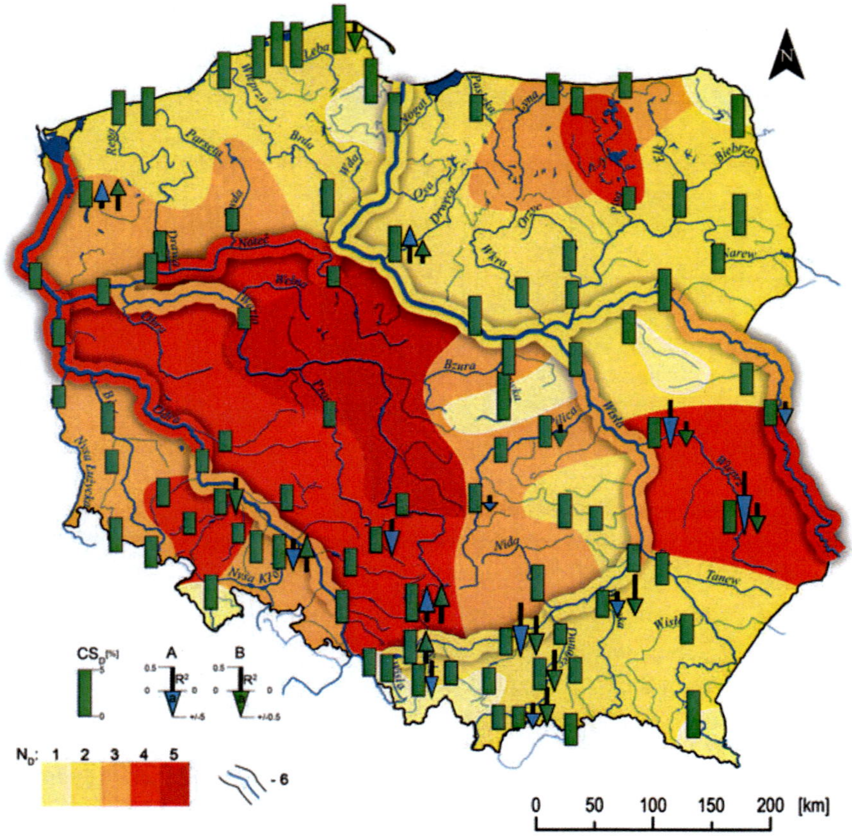

Fig. 12.3 Spatial differentiation of selected hydrological drought estimators in Poland (1984–2015). Mean annual number of days with hydrological drought (N_D) (days): 1 – 121–135, 2 – 136–150, 3 – 151–165, 4 – 166–180, 5 – 181–195; 6 – N_D changes in larger transit rivers (outside the map scale), CS_D – mean annual coefficient of severe drought contribution (%); statistically significant ($\alpha = 0.05$) linear trend in multiannual course of annual number of days with hydrological drought (A) and the coefficient of severe drought contribution (B), a – trend line slope, R^2 – coefficient of determination, arrow direction denotes positive (up) or negative (down) sign of the slope

As mentioned in the statement opening this chapter, it was assumed that river low flow occurrence indicates the development of the hydrological drought in the catchment. Drought identification criteria were as follows:

- *Events of hydrological drought per year*, if the number of days with low flow ($Q_{70\%}$) exceeded 90 in a given hydrological year,
- *Events of hydrological drought per month,* if the number of days with low flow ($Q_{70\%}$) exceeded 10 in a given month.

Finally, for each of 87 investigated catchments the study identified the years and months with hydrological drought described by: the number of days with hydrological drought (N_D), the volume of drought streamflow deficit during drought (V_D), and a coefficient of severe hydrological drought contribution calculated as a quotient of severe streamflow deficit and total streamflow deficit during drought:

$$CS_D = \frac{VS_D}{V_D} \cdot 100\% \tag{12.2}$$

where:

CS_D coefficient of severe hydrological drought contribution (%),
VS_D volume of severe streamflow deficit during drought (m^3),
V_D volume of total streamflow deficit during drought (m^3).

A comprehensive assessment of the hydrological drought at the country-wide scale required a definition of spatial measures that enabled a comparative analysis on a multiannual scale. The first measure is a range of hydrological drought that indicates the part of the country affected by hydrological drought in a given year or month [21]:

$$R_D = \frac{\sum A_D}{\sum A} \cdot 100\% \tag{12.3}$$

where:

R_D a range of hydrological drought (%),
$\sum A_D$ total area of the catchments affected by the drought (km^2),
$\sum A$ total area of all investigated catchments (km^2).

Relative discharge deficits for individual low flow periods in a catchment (RVn) may be used to work out this deficit associated with drought (RV_D), which also serves as an estimator of the drought severity. The weighted average can then be used to calculate the hydrological drought severity index for entire Poland in monthly or yearly intervals [21]:

$$S_D = \frac{\sum_{i=1}^{N} (RV_{Di} \cdot A_i)}{\sum_{i=1}^{N} A_i} \tag{12.4}$$

where:

S_D hydrological drought severity index (%),
RV_{Di} a relative deficit of streamflow during a drought in catchment *i* (%),
A_i catchment *i* area (km^2),
N number of catchments affected by hydrological drought.

12.3 Spatial Distribution of Hydrological Drought Estimators

A crucial element for the spatial analysis of hydrological drought structure in Poland is a distribution of the average annual number of days with drought (Fig. 12.4a). As per assumption made earlier in this chapter, the drought statistics were only calculated for the years in which the number of days with low flow exceeded 90. In the investigated group of catchments, average N_D, expressed as a median, equalled about 150 days. In half of the cases examined, the annual number of days with drought ranged between 140 and 162, while in extreme cases there were only 126 days or slightly more than half a year. The analyzed distribution was relatively symmetrical and close to normal. This proves a dominant role of hydroclimatic factors that determine the hydrological drought at the country level. Differences in river basin retention levels and specific factors shaping the water cycle become visible only at a regional or local scale.

The small average number of days with hydrological drought per year is typical of southern Poland (Fig. 12.3). The catchments of the Carpathian rivers feature high water resources, the prevalence of precipitation over evaporation in the water balance structure, and high rate of water exchange within the active exchange zone. As a result, hydrological droughts are rare in this region and are usually a consequence of a series of dry years. Hydrological droughts in the Sudetes last a bit longer. The average annual number of days with drought may reach up to half a year. The main reason for this is a change in the water balance structure manifested by an increased contribution of evaporation and declined supply from precipitation.

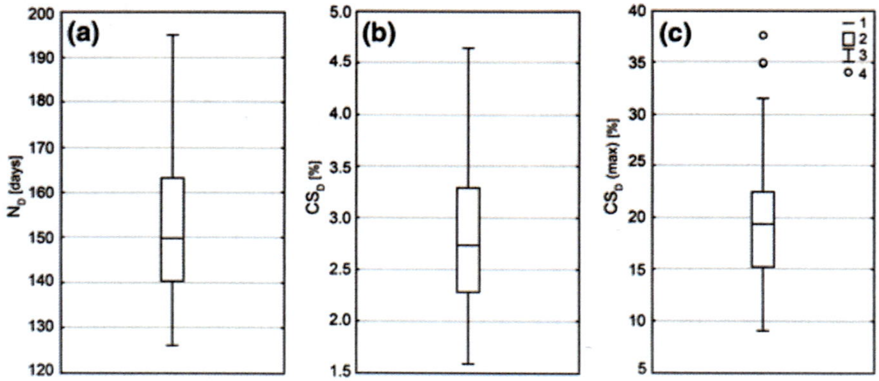

Fig. 12.4 Distribution of selected hydrological drought parameters in Poland (1984–2015). **a** – mean annual number of days with hydrological drought, **b** – mean annual coefficient of severe drought contribution, **c** – maximum annual coefficient of severe drought contribution, 1 – median, 2 – variability range limited by the first and third quartile, 3 – non-outliers within the first interquartile range, 4 – outliers exceeding the first interquartile range

In the belt of highlands and lowlands, south of the Noteć and latitudinal section of the Vistula and the Bug, the average duration of hydrological drought enhances markedly. This zone is additionally separated by a longitudinal transect along the first-order watershed divide isolating the Vistula and the Oder basins (Fig. 12.3). West of this line, the duration of low flows in the years with hydrological drought reached five to six months, and its spatial variability was negligible. This indicates a similar water balance structure, hardly beneficial for water management, and poor water resources that are due to low retention capability of hydrogeological structures. The eastern part of the analyzed area is much more diverse. It harbours catchments with the small annual number of days with drought in the Świętokrzyskie Mountains as well as the basin of the Wieprz, where droughts last up to 178 days. This is associated with a slow rate of groundwater recharge in well-fissured carbonate rocks of the Lublin Upland. Within the lakeland regions, the average annual number of days with hydrological drought is shortened and usually does not exceed 150. This is obviously due to increased rainwater supply in the moraine elevation zone and the inflow of wet polar-marine air masses from the north-west sector. Interestingly, in the catchments with flow-through lakes of considerable size (e.g. of the Pisa or the Łyna), the average number of days with hydrological drought increases up to 180. This is probably due to enhanced evaporation from the water surface and not too large retention capacity of such lake harbouring catchments.

Severe hydrological drought markedly disturbs the functioning of facilities and water management systems. It is caused by a loss of seasonally renewable water resources and often results in the degradation of water-dependent ecosystems. Riverbeds then contain limited amounts of water supplied by groundwater reservoirs renewed in the multiannual cycle. Prolonged severe hydrological drought causes gradual drying of ever-larger watercourses. This phenomenon was assessed based on an annual number of days with low flow below $Q_{95\%}$, assumed as the severe drought estimator. The contribution of streamflow deficit during a severe drought in the total volume of the deficit during drought (CS_D) provided data necessary for the analysis of the structural properties of these events.

Average annual coefficient of severe droughts contribution in Poland equals about 2.7% (Fig. 12.4b). Variability of the parameter is small, as half of the values oscillating around the median fall within a relatively narrow range between 2.3 and 3.3%. The empirical distribution of the parameter is close to normal, which proves a stationary character of severe drought formation at the country-wide level. In a significant number of cases, the coefficient of severe droughts was inversely proportional to an annual number of days with hydrological drought (Fig. 12.3). This indicates some limitations in the development of this drought phase determined mostly by hydrogeological conditions. The coefficient of severe droughts contribution in the Carpathian catchments is close to the average for Poland but also more diversified. These catchments show a lower importance of severe droughts in the rivers below dam reservoirs (guaranteed flow), and variable CS_D determined by the severity of winter hydrological droughts and variable retention capacity of the catchments that depends on geostructural and geofiltration factors. The above is true also for the Sudeten catchments, except for lower importance of winter droughts. The catchments of the western part of the lowland belt show a small and spatially variable contribution of severe droughts. In the rest of the country, severe hydrological droughts play a vital role in catchments that discharge water directly into the sea. However, considering a relatively efficient precipitation supply and favourable structure of the water balance in the coastal zone, water management should not be seriously affected by severe hydrological droughts.

The considerations presented above refer to a typical year and may differ from a situation when after a series of dry years (see Sect. 12.4) a particularly severe hydrological drought appears and causes damage to the environment and serious losses in the water management. This problem was assessed based on the maximum observed coefficients of severe drought contribution (Fig. 12.4c). The average maximum CS_D in relation to a typical year increased by seven times (19.7%). The distribution of the analyzed variable shows outliers that indicate particularly strong effects of severe droughts on hydrological systems (above 35%: the San, the Rawka, the Łeba). As these catchments represent different geographical regions (mountains, lowlands, coastal areas), it may be hypothesized that extremely severe hydrological droughts strongly depend not only on hydrometeorological conditions but also local catchment-related factors and result from both natural phenomena and human activities.

The process of hydrological drought development follows a slightly different pattern along larger transit rivers (Fig. 12.3). The average annual number of days with hydrological drought in the upper course of the Vistula is mainly determined by its Carpathian tributaries. Below the entry of the San, the number of such days increases to 165 per year, but this does not reflect the features of drought in autochthonous catchments of direct tributaries of the middle Vistula section. Only after the entry of the Narew and the Bug, the number of days with hydrological drought drops back to 150 and remains at this level until the Vistula mouth thanks to gradual increment in water resources along with the growing basin area. Despite the relatively low value of N_D estimated for the Vistula, the contribution of hydrological droughts is pretty high and can periodically disturb water management. A reverse situation occurs for the Oder. An average number of days with hydrological drought rises along the river course and is determined by its right bank lowland tributaries and droughts developing in the catchment of the Warta. The effects of these factors are so strong that the increase in water resources of the basin seems only a secondary factor determining the duration of hydrological droughts along the Oder.

The analysis of the hydrological drought duration and the contribution of severe droughts was supplemented by verification of the multiannual variability of these parameters against a systematic component. To this end, the linear trends verified with Student's t-test were identified at the significance level $\alpha = 0.05$ (Table 12.1, Fig. 12.3). Interestingly, 11 out of 14 significant, multiannual trends of an annual number of days with hydrological drought had a negative sign. These trends occurred in the selected mountain and upland catchments and indicated gradual mitigation of periods with water deficits within the investigated period. The trends with the best fit (R^2: 0.28–0.44) appeared in the rivers where large dam reservoirs were built.

Table 12.1 Parameters of significant linear trends in the annual series of selected hydrological drought estimators

River	Water-gauge	Annual number of days with hydrological drought (N_D)		Coefficient of severe drought contribution (CS_D)	
		a	R^2	a	R^2
Nysa Kłodzka	Skorogoszcz	−3.465	0.173	0.382	0.253
Bystrzyca	Jarnołtów	−4.146	0.191	–	–
Liswarta	Kule	−4.211	0.195	–	–
Ina	Goleniów	3.366	0.131	0.245	0.248
Reda	Wejherowo	–	–	−0.303	0.149
Vistula	NowyBieruń	–	–	0.327	0.218
Przemsza	Jeleń	4.149	0.168	0.378	0.280
Soła	Oświęcim	−3.504	0.275	–	–
Raba	Proszówki	−5.622	0.404	−0.360	0.326
Dunajec	Krościenko	−2.278	0.165	−0.351	0.293
Dunajec	NowySącz	–	–	−0.361	0.376
Wisłoka	Mielec	−2.472	0.140	−0.276	0.410
Wieprz	Krasnystaw	−7.049	0.441	−0.265	0.211
Wieprz	Kośmin	−6.068	0.362	−0.303	0.200
Pilica	Przedbórz	–	–	−0.163	0.150
Pilica	Białobrzegi	−2.900	0.144	–	–
Bug	Włodawa	−3.296	0.151	–	–
Drwęca	Elgiszewo	3.921	0.343	0.302	0.150

Significance level: $\alpha = 0.05$, a – slope coefficient of the trend line, R^2 – determination coefficient

Water management at these facilities allows for alleviating the effects of hydrological drought, as manifested by a significant tendency to the reduced number of days with drought. A similar observation is true for the coefficient of severe droughts contribution, which systematically decreases as a result of guaranteed flow generated by these reservoirs. However, in the rivers downstream of the reservoirs, in which municipal functions predominate (e.g., the Vistula – Nowy Bieruń, the Nysa Kłodzka – Skorogoszcz), where water resources are stored and transported over natural catchments, significant positive trends were noticed for the coefficient of severe droughts contribution with relatively high fit ($R^2 = 0.3$). Spatial analyses do not reveal dense or ordered zones with significant multiannual trends for hydrological drought estimators. It can, therefore, be assumed that the observed trends do not have a hydroclimatic background and are triggered by local factors, mainly related to water management.

12.4 Multiannual Variations

In multiannual perspective, the average annual number of days with hydrological drought in Poland ranged from 104 to 183 (Fig. 12.5). The longest droughts occurred in the years 1990–1994 when long-term and severe low flows in rivers were observed [7], caused by a series of dry years [28]. The lowest number of days with hydrological drought was noticed in 1998 and 2010 when water levels in Polish rivers were exceptionally high, resulting in catastrophic floods in some catchments. An extreme year in this respect was 1997, when droughts were followed by particularly high flood waves with peak flows exceeding 0.001 probability of exceedance (so-called 1,000-year flood) at some water gauges.

The groups of dry and wet years are also visible along the course of hydrological drought range index (Fig. 12.5). The periods with the large annual number of days with hydrological drought are usually characterized by an extensive range of the phenomenon. However, drought has never affected 100% of the analyzed area in one year for the entire period of the study. Droughts with the range exceeding 95% of the country area occurred in the years 1990–1994, 2003–2004, 2006 and 2012. The hydrological drought of the smallest range of 15% occurred in 2010.

Average annual index of drought severity in Poland ranged in the investigated multi-year period between 16.6 and 32.2% (Fig. 12.5). Years with high drought

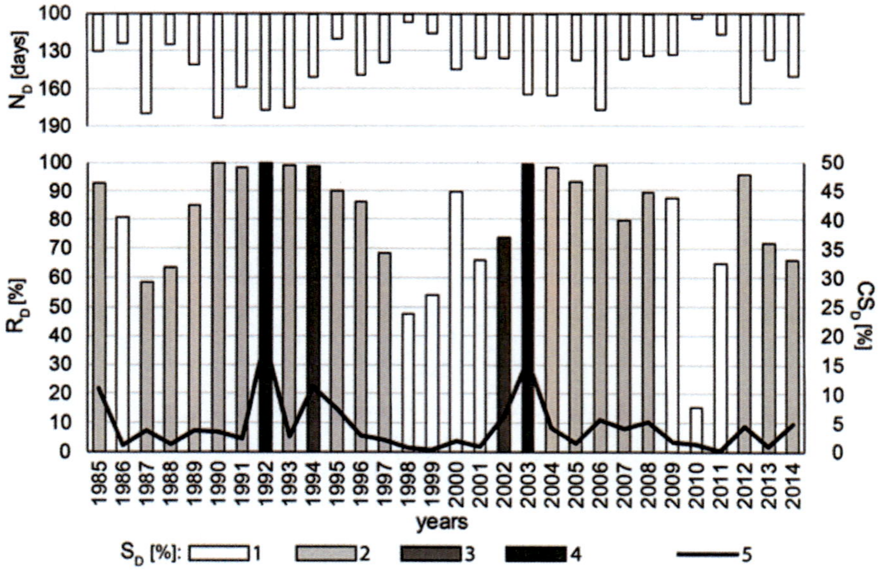

Fig. 12.5 Multiannual course of selected hydrological drought parameters in Poland. N_D – mean annual number of days with hydrological drought, R_D – range of hydrological drought, S_D – hydrological drought severity index (%): 1 – 0–4, 2 – 5–10, 11–16, 17–21, 5 – CS_D – coefficient of severe drought contribution; parameters N_D, S_D and CS_D were calculated based on weighted average where the weight was the catchment area (see Sect. 12.2)

severity index, in which the degree of drainage of catchment water resources available during low flow periods exceeded 30% usually constituted a core of a few-year long drought periods, e.g., 1990–1994, 2002–2006. Mild hydrological droughts ($S_D <$ 20%) occurred in 1986, 1998–2001 and 2009–2011. They were usually characterized by small range and lower than the average annual number of days with hydrological drought.

The index of severe drought contribution was a source of many interesting observations (Fig. 12.5). Its values in the multiannual period of investigation varied from 0.15 to 18.95% and were obviously the highest in the years with severe hydrological droughts (1992, 2003). However, in the remaining years, this relationship was not so clear and linear (Fig. 12.6a). The fit of the exponential function indicates that for mild droughts, a small increase in the number of days with severe drought triggers a rapid increase in the drought severity index. For severe droughts, a much larger increase in severe drought contribution is necessary for a similar spike in this metric. The course of the analyzed function may be of vital importance for water management as it demonstrates that mean annual drought severity index in Poland should not exceed 50%.

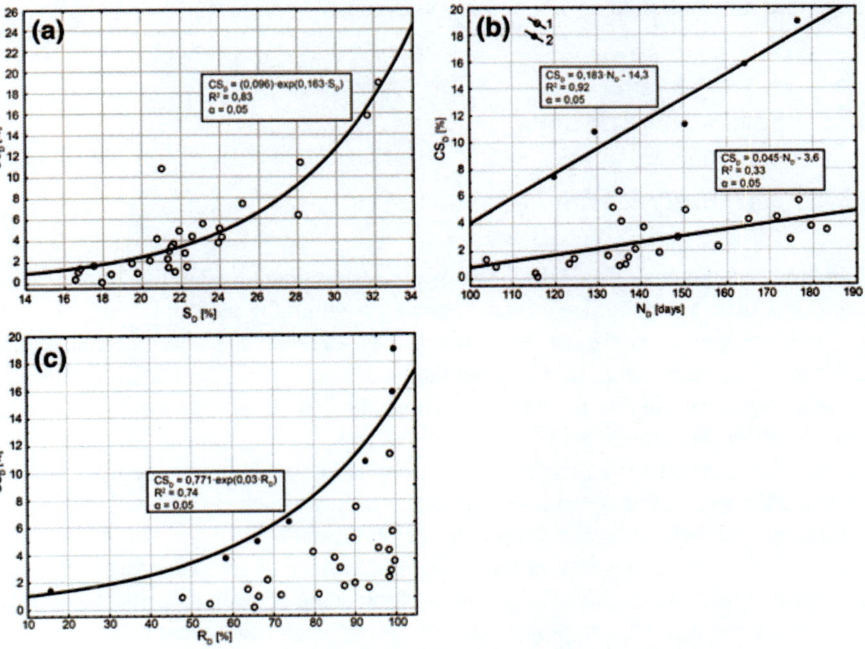

Fig. 12.6 Regression dependencies between selected parameters of hydrological drought. CS_D – coefficient of severe drought contribution; S_D – hydrological drought severity index, N_D – number of days with hydrological drought, R_D – range of hydrological drought, α – significance level of an established regression equation; R^2 – coefficient of determination, 1 – function established for points without maximum values CS_D, 2 – envelope function of maximum CS_D values

The contribution of severe droughts strongly correlates with total drought duration during the year (Fig. 12.6b). The resulting linear dependence does not show a good fit ($R^2 = 0.33$), which indicates a large impact of stochastic hydrometeorological conditions that affect the relationship. However, a strong linear dependence ($R^2 = 0.92$) could be observed for the envelope following the maximum points of severe drought contribution. This function indicates regional limitations of severe drought development in Poland and may be helpful in water management planning that considers the predicted annual number of days with hydrological drought. The severe drought contribution coefficient partly correlated with drought range index (Fig. 12.6c). Although graphic representation of the dependence shows a cloud of points impossible to approximate with a linear function, the envelope composed of points with a maximum contribution of severe droughts can be described with the exponential equation of high degree of fit ($R^2 = 0.74$). In practice, it is possible to determine regional limits of the maximum coefficient of severe droughts contribution in relation to the area affected by hydrological drought.

The multiannual course of the analyzed parameters did not reveal any linear trends statistically significant at $\alpha = 0.05$. It can, therefore, be concluded that in Poland, the long-term variability of hydrological drought parameters depends primarily on the natural fluctuation of hydroclimatic conditions.

12.5 Seasonal Variations

12.5.1 Monthly Variability

Dynamics of the hydrological system reveals significant variability of the water balance structure and the size of water resources within the seasonal cycle. Monthly intervals are basic units for determining disposable resources for water management and assessing water needs at a regional and national scale. This section provides a statistical analysis only for the months in which the number of days with hydrological drought exceeded 10 (see Sect. 12.2).

Monthly distribution of the average number of days with hydrological drought reflects a simple, seasonal variability (Fig. 12.7). Between November and March, this parameter systematically decreases until the minimum of 18 days. This is the effect of a gradually weakening evapotranspiration determined by temperature changes and disappearance of vegetation. The gradually decreasing drought severity index (from 8.6 to 2.8%) confirms the hypothesis that in recent years, catchment water resources are recharged in winter. This is mainly due to milder weather conditions during this season, resulting in high rainfall retention and relatively small evapotranspiration [7]. Winter hydrological droughts demonstrate relatively stable range (20–25%), and the local maximum of the severe drought contribution coefficient falls in January. This indicates stability of winter droughts in selected mountain catchments in which regular and periodic storage of water in the snow cover is often intensified by frost

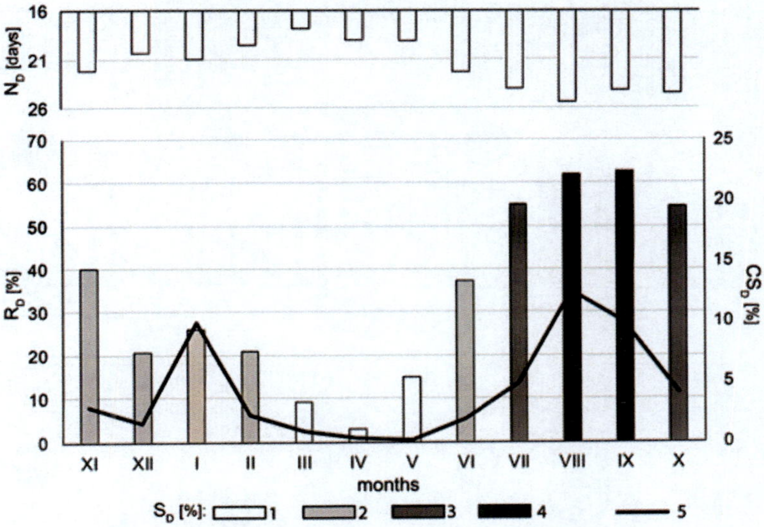

Fig. 12.7 Seasonal course of selected hydrological drought parameters in Poland. N_D – mean monthly number of days with hydrological drought, R_D – mean monthly range of hydrological drought, S_D – mean monthly hydrological drought severity index (%): 1 – 0–4, 2 – 5–10, 11–16, 17–21, 5 – CS_D – mean monthly coefficient of severe drought contribution; parameters N_D, S_D and CS_D were calculated based on a weighted average where the weight was the catchment area (see Sect. 12.2)

penetration into the riverbeds. Observations of multiannual variability of drought estimators for the autumn months on the example of November revealed a tendency for grouping the years in drought periods typical of the entire multi-year period (Fig. 12.8a). The average degree of drought severity and its duration indicates that the majority of these episodes occur as the continuation of prolonged summer hydrological droughts (see Fig. 12.8d). In dry periods, the range of November droughts falls between 40 and 95%, while in wet ones it is marginal to complete cessation as in, e.g. 1999. In winter months (January), the range of more extensive droughts varies between 50 and 70%, while smaller droughts cover below 20% of the country area (Fig. 12.8b). Droughts of longer duration and considerable coefficient of severe drought contribution occurred fairly regularly as a result of long, snowy and frosty winters. In the years following such a winter, Poland experienced high water levels in rivers that often transformed into catastrophic floods, e.g., in 1994, 1997, 2010.

In the spring (March–May), hydrological droughts are rare and of low intensity due to snow melt and accompanying floods (Fig. 12.7). Moreover, vegetation that only begins to develop does not generate intense evapotranspiration. A good illustration of this situation is a multiannual course of hydrological drought parameters for April (Fig. 12.8c). In this month, droughts appeared on average every two or three years, although there were also a few year periods without drought. The number of days

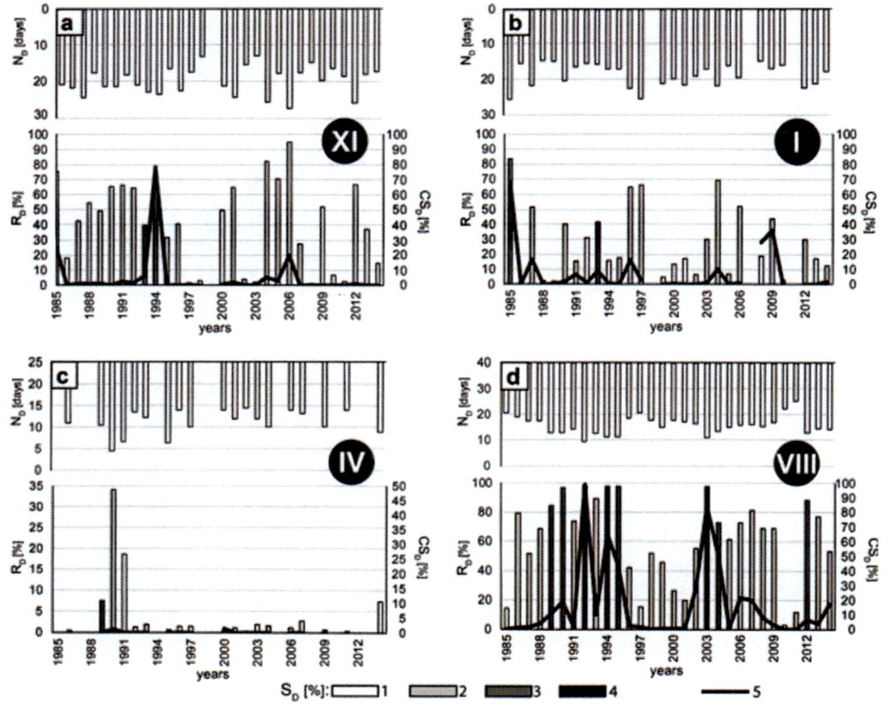

Fig. 12.8 Multiannual course of hydrological drought parameters in selected months. **a** – November, **b** – January, **c** – April, **d** – August, other symbols as in Fig. 12.7

with drought rarely exceeded 15, and usually included only a few percent fraction of the investigated area.

Severe and extensive hydrological droughts usually develop in the summer and early autumn (July–October), while June is a transitional month that may significantly prolong drought duration in the years with dry spring and summer (Fig. 12.7). A maximum number of days with drought and the highest coefficient of severe drought contribution occurs in August ($N_D = 25.5$ days, $CS_D = 12.5\%$), which indicates high stability of evapotranspiration conditions that determine water deficits in this period. Summer and autumn droughts also belong to the most extensive ones (R_D: 54.5–62.5%) and are characterized by high or very high severity that is due to the high homogeneity of factors that determine drought properties in the summer half of the year. In the month with the most intense hydrological drought throughout a year (August), the number of days with drought only occasionally falls below 20 (Fig. 12.8d). Droughts of extensive range (60–100%) and high severity occurred in the years 1986–1995, 2003–2009 and 2012–2013, and accounted for about two-thirds of the investigated multi-year period. The years 1996–2002 were a period with droughts moderate in terms of range and intensity. A clear grouping of dry and wet years depends not only on precipitation but also results from the ongoing recession of groundwater resources in the catchment. Depletion and recharge of these water reserves show huge inertia that causes droughts of similar properties to group in a few year-long periods.

12.5.2 Degree of Seasonality and Concentration Date of Hydrological Drought

The degree of hydrological drought severity depends mainly on hydrometeorological conditions in the period preceding its occurrence. Deficit of water resources may be determined by seasonal anomalies associated with a lack of supply in typical time of alimentation or multiannual disturbances evoked, e.g. by a series of dry years. Both groups of factors overlap in a random and practically inseparable way, making droughts and low flows highly variable events with unpredictable frequency and intensity.

This strong irregularity in the appearance of drought streamflow deficits considerably hampers seasonal analyses based on time series analyzed in monthly steps (see Sect. 5.1). A solution to this problem may be applying characteristics that describe the seasonal variability comprehensively. The variability of the annual course of drought streamflow deficits and their concentration date may be identified using angular measures introduced to the literature by Markham [29]. Originally, they served to analyze seasonal variability of precipitation in the US, but following a few methodological transformations, two seasonality measures of hydrological drought were proposed: seasonality index (SI) and seasonal concentration date index (CI) [7].

Both measures assume that monthly volume of drought streamflow deficit is represented by a vector (r_i) of a length proportional to the volume of this deficit and an angle of inclination (α_i) depending on the position of the middle of a given month relative to the beginning of the hydrological year:

$$\alpha_i = \frac{360 \cdot S}{365} \tag{12.5}$$

where:

S number of days between the beginning of the hydrological year and the middle of the month.

As a result, 12 vectors are obtained for which a resultant vector R of the module |R| and direction ω is determined (Fig. 12.9). The quotient of the resultant vector |R| and the total length of the partial vectors $|r_i|$ allows for the calculation of the seasonality index SI:

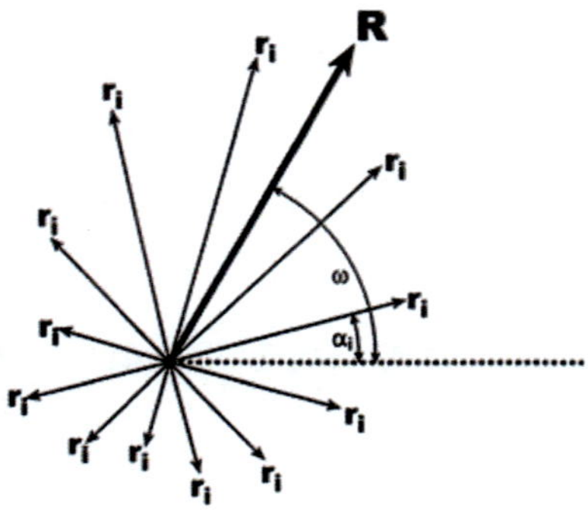

Fig. 12.9 Construction of Markham's seasonality measures. r_i – vector representing drought streamflow deficit per month *i*, R – resultant vector for vectors r_i, α_i – angle representing the middle of the month in relation to the beginning of hydrological year, ω – angle indicating concentration date of hydrological drought in relation to the beginning of hydrological year

$$SI = \frac{|R|}{\sum_{i=1}^{12} |r_i|} \cdot 100\% \tag{12.6}$$

The obtained measure assumes values between 0 and 100% and increases along with the rise of the degree of seasonality for hydrological drought. A result equal 0% may indicate not only a total uniformity of drought streamflow deficit over the year but also a situation when hydrological drought occurs only in two opposite months (e.g., November and May). Both cases are extreme and theoretical but point to the need for careful interpretation of the calculation results. The angle of inclination of the resultant vector R (ω) serves as an estimator of hydrological drought concentration date (Fig. 12.9), and the value of the seasonal concentration date index (CI) is calculated according to the following formula:

$$CI = \mathrm{arctg}\left(\frac{\sum_{i=1}^{12} |r_i| \cos\alpha_i}{\sum_{i=1}^{12} |r_i| \sin\alpha_i}\right) \cdot \frac{365}{360} \tag{12.7}$$

The final value, usually represented by a date, indicates the resultant time of concentration of drought streamflow deficits which does not always coincide with the month of their maximum intensity [7, 29, 30]. Multiannual stability of the concentration date for hydrological drought was evaluated using the seasonal concentration date frequency coefficient C. The measure indicates the percentage contribution of the number of years in which the hydrological drought concentration date occurred in a month of typical seasonal concentration date in relation to the total number of years in the multi-year period.

In most catchments investigated in this study, the hydrological drought concentration date occurred in the summer-autumn period (Fig. 12.10). Differentiation of CI revealed a clear spatial order (Fig. 12.11). In the Carpathian catchments, the

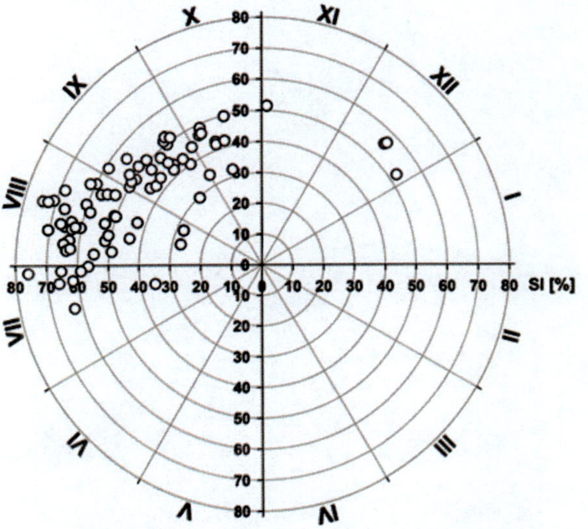

Fig. 12.10 Position of the tips of the vectors denoting seasonality index and the index of seasonal concentration date of hydrological drought in the investigated catchments (1985–2014)

latest CI (second half of December) occurs in the upper part of the Dunajec catchment (Fig. 12.10). This means that the streamflows deficits during winter low flow markedly prevail over those from the warm half-year. This feature of flow seasonality is maintained in the Dunajec until its mouth, but entering of consecutive tributaries moves the seasonal concentration date in its lower section back to the beginning of November. The increased importance of summer drought streamflow deficits means that in the majority of remaining Carpathian tributaries of the Vistula, the hydrological drought concentration date occurs in the first half of October, and for the San even in the second half of September.

In the upper section of the Vistula, above the mouth of the Przemsza, the concentration date falls already in July. This is due to the effects of the Goczałkowice Reservoir, in which water reserves for municipal needs are renewed mainly in the spring and during summer floods. As a result, low flows are observed downstream of the dam in the periods of typical discharge increase, which modifies flow regimes of this section of the river [31]. Catchments of the rivers dominated by one genetic type of low flows (e.g., the Dunajec and the San systems) show relatively stable hydrological drought concentration date within any multi-year period. Frequency coefficients of the seasonal concentration date C in these catchments reach 40–50%. In the other Carpathian rivers, the dynamics of summer and winter low flows remain very high. This results in relatively low values of the C coefficient – in extreme cases, the seasonal concentration date in a typical month of low-flow occurs only once every 7 years (C ≈ 15%). The seasonality index of hydrological drought in the Carpathians is pretty stable and fits a narrow interval of 40–55%. This indicates a high similarity of hydrometeorological and hydrogeological conditions that determine the

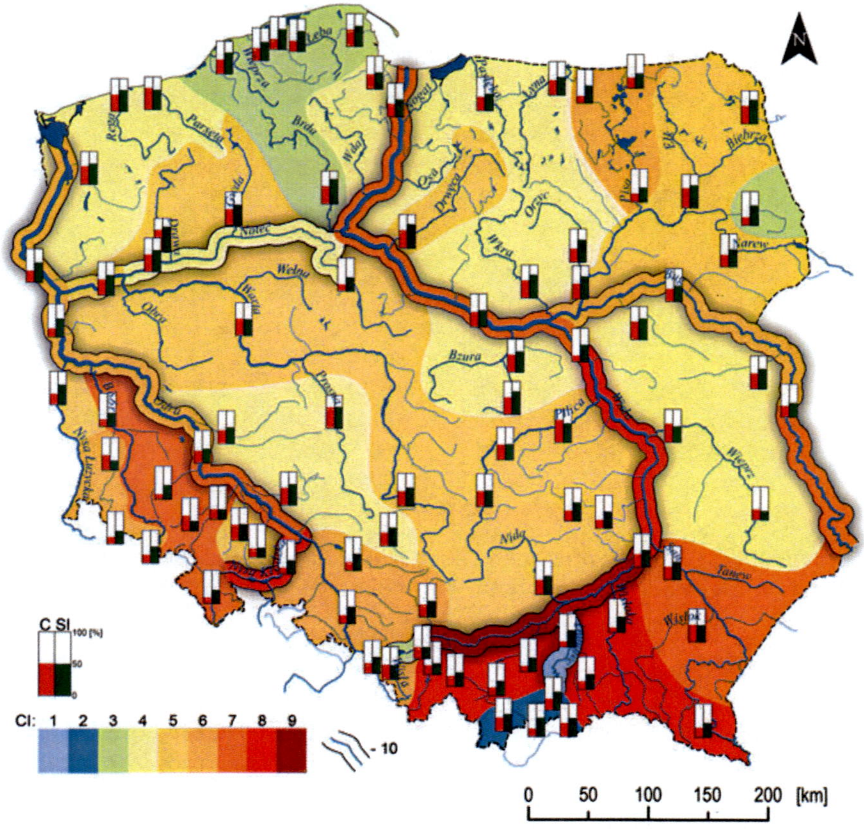

Fig. 12.11 Seasonal concentration date of hydrological drought in Poland (1985–2014). CI – index of concentration date of hydrological drought: 1 – 1–15 November, 2 – 16–31 December, 3 – 16–31 July, 4 – 1–15 August, 5 – 16–31 August, 6 – 1–15 September, 7 – 16–30 September, 8 – 1–15 October, 9 – 16–31 October; 10 – CI changes in larger transit rivers (outside the map scale), the legend colour according to the periods: 1–9; C – frequency coefficient of the seasonal concentration date, IS – seasonality index of hydrological drought

occurrence of river low flows in the region as well as recession and renewal of water resources. The only exceptions are the Soła and a section of the upper Vistula, where water management activities in the Soła cascade and the Goczałkowice Reservoir lower SI down to 33–35%.

Typical concentration date of hydrological drought in the Sudeten rivers precedes that in the Carpathian watercourses by ca. 15 days. However, mean CI in the rivers of the Kłodzko Valley, the Karkonosze Mountains, and the Kaczawskie Mountains falls in the second half of September, thus suggesting a higher contribution of winter low flows than in the other watercourses of the region. A delay in the seasonal concentration date in the Nysa Kłodzka due to the effects of the Otmuchów and Nysa reservoirs is also clearly visible. These effects are strong enough to cause a similar delay in the Oder. Frequency coefficients of CI in the Sudetes are more variable than in the Carpathians and range from 13 to 45%. The Sudeten rivers also show slightly higher spatial diversity of the hydrological drought seasonality index (30–60%) than the Carpathian ones. Low SI is typical for cross-sections located on the rivers with abundant base flow from capacious aquifers (the Oława, the Bystrzyca) or placed downstream of the dam reservoirs (the Nysa Kłodzka).

In most upland and lowland rivers, seasonal concentration date of hydrological drought occurs in August (Fig. 12.11). CI dates from the first half of this month are typical for rivers the water resources of which are slightly smaller, e.g., for right tributaries of the middle Oder, left tributaries of the upper and middle Warta, the Bzura system, the Vistula and the Bug inter-fluve area, and some rivers of the lake districts. Rapid depletion of small water resources results in a relatively early seasonal concentration date of hydrological drought in these rivers. In lake systems with high lake density and a large number of flow-through and outflow lakes, the drought concentration date occurs slightly later (the Pisa, the Drwęca, the Gwda, the Drawa, the upper Noteć) [32]. In the coastal rivers, CI indicates a relatively early seasonal concentration date of hydrological drought, particularly in the eastern part of the region. This is directly linked to an early seasonal concentration date of total flow [33].

Seasonal concentration date of hydrological drought in lowland rivers is highly stable for a multi-year period. The reason for this is the dominance of summer half-year low flows. At the same time, seasonality of hydrological drought is high or very high due to the high similarity of factors determining the formation and severity of the low flows. In many lake and upland river systems characterized by a large contribution of groundwater flow, the seasonal concentration date is also stable for a multi-year period, as permanent base flow effectively buffers the impact of random precipitation events. This is the reason why upland catchments with the large and stable contribution of groundwater flow (e.g., the Wieprz) feature very low seasonality indices of hydrological drought.

The situation may differ in some large transit rivers, in which low flow trends depend on characteristic features of their tributaries (Fig. 12.11). Mean multiannual seasonal concentration date of hydrological drought for the Vistula downstream of the Goczałkowice Reservoir depends on the dynamics of the Carpathian tributaries and typically falls on 16 October. Downstream of the San entrance, CI occurs 10 days earlier, and from the mouth of the Bug and the Narew until the mouth of the Vistula, the date is shifted to 22 September. The difference in mean seasonal concentration date between the lower Vistula and its autochthonous tributaries may even exceed two months. Slightly smaller differences are observed for the Oder and the Bug, where the gradual impact of subsequent tributaries is visible, but the distinctiveness of low-flow regime is maintained until the estuaries. Although the middle and lower Noteć "inherits" low flows from the upper part of the river system and tributaries from catchments of high lake density, its seasonal concentration date of hydrological drought falls relatively early. In the transit rivers, the frequency coefficient C along the river course is lowered when low flows in the tributaries are not synchronized with the low flows of the main river [31]. Such situations happen for the Vistula downstream of the San entrance and for the Oder after it is joined by the Bóbr and the Nysa Łużycka and then the Warta.

12.6 Summary and Conclusions

The study demonstrates that employing low flows as indicators of hydrological drought development seems promising. The adopted identification and separation criteria allowed for unambiguous isolation of mild and severe hydrological droughts in Poland in the years 1985–2014. Transformation of the parameters describing the duration and the relative deficit of streamflow and the share of the catchment in the entire investigated area enabled the assessment of hydrological droughts in terms of their duration, severity, and range. It also allowed for pointing out periods differing in the intensity of the assessed parameters. The evaluation of seasonal and multiannual variability, as well as analysis of genetic relationships between selected estimators of hydrological drought, provided new and valuable cognitive insights and made it possible to identify a group of factors the determine drought development. Practical conclusions, especially those concerning regional barriers for development and regularities of the duration and range of hydrological drought can significantly expand water management activities aimed at mitigation of drought effects at the national and regional scale.

The analysis proves that both the seasonal concentration date and intensity of hydrological drought in Poland show significant and multidirectional variability. In the mountain rivers, low-flow regimes are significantly affected by genetically different summer and winter low flows (CI – September-December), while the rest of the country is dominated by summer half-year deficits (CI – July–September). Apart from hydrometeorological conditions, seasonal distribution of hydrological drought is also determined by local factors associated with water resources and the rate of their exchange in the hydrologically active zone and with some aspects of water management. Large transit rivers gradually change the features of their low-flow regime along with the entrance of successive tributaries but retain their regime specificity up to the estuary. Allochthonous features of low-flow regime inherited from the upper part of the basin, and larger tributaries are present in the regimes of the Vistula, the Oder, the Bug, the Noteć, the Nysa Kłodzka and the Dunajec. The high degree of hydrological drought seasonality occurs when its average seasonal concentration date falls in the summer months (Fig. 12.10). When the CI is shifted towards autumn months, the seasonality index decreases as droughts of the winter half-year become more and more important.

12.7 Recommendations

Presented results of analyses have finished some stage of the study. Knowledge of hydrological droughts spatial pattern and time variability in Poland as well as their determinants allows to use some conclusions in water management planning and water resources assessment. It also might effectively support the tools and strategies of optimal reduction of drought results, its prevention, and prediction.

When the defined scientific problems have been solved, new questions appeared. It is possible to express it in one statement: does present level of knowledge allow to construct the regional model of hydrological drought? This does not concern operational active model but the procedure predicting general level of drought streamflow deficit expected in annual or half-yearly advance. In authors opinion, features of investigated time series and identified regularities are so promising that the trial of such study might be realized soon.

References

1. Maunder WJ (1992) Dictionary of global climate change. Chapman & Hall, New York
2. Nagarajan R (2009) Drought assessment. Springer, Dordrecht
3. Lloyd-Huges B, Saunders MA (2002) A drought climatology for Europe. Int J Climatol 22:1571–1592
4. Łabędzki L (2007) Estimation of local drought frequency in central Poland using the standardized precipitation index SPI. Irrigat Drainage 56:67–77
5. Łabędzki L, Bąk B (2004) Zróżnicowanie wskaźnika suszy atmosferycznej SPI w sezonie wegetacyjnym w Polsce (differentiation of the atmospheric drought index SPI in the vegetation period in Poland). Woda-Środowisko-Obszary Wiejskie 11(2a):111–122
6. Somorowska U (2016) Changes in drought conditions in Poland over the Past 60 years evaluated by the standardized precipitation-evapotranspiration index. Acta Geophys 64(6):2530–2549
7. Tomaszewski E (2012) Wieloletnia i sezonowa dynamika niżówek w rzekach środkowej Polski (multiannual and seasonal dynamics of low-flows in rivers of central Poland). Wyd. Uniwersytetu Łódzkiego, Łódź
8. Wilhelmi OV, Hubbard KG, Wilhite DA (2002) Spatial representation of agroclimatology in a study of agricultural drought. Int J Climatol 22:1399–1414
9. Dingman SL (2002) Physical hydrology. Prentice Hall, New Jersey
10. Hisdal H, Tallaksen LM, Peters E, Stahl K, Zaidman M (2001) Drought event definition. In: Demuth S, Stahl K (eds) Assessment of the regional drought impact of droughts in Europe. Institute of Hydrology, University of Freiburg, pp 17–26
11. Sene K (2010) Hydrometeorology. Forecasting and applications, Springer, Dordrecht-Heidelberg-London-New York
12. Wilhite DA (2005) Drought. In: Olivier JE (ed) Encyclopedia of World Climatology. Springer, Dordrecht, pp 338–341
13. Jokiel P (1994) Zasoby, odnawialność i odpływ wód podziemnych strefy aktywnej wymiany w Polsce (groundwater resources, renewal and flow in the active exchange zone in Poland). Acta Geographica Lodziensia 66–67:1–236
14. Smakhtin VU (2001) Low flow hydrology: a review. J Hydrol 240:147–186
15. Tallaksen LM (1995) A review of baseflow recession analysis. J Hydrol 165:349–370
16. Laaha G (2002) Modelling summer and winter droughts as a basis for estimating river low flows. In: FRIEND 2002 – regional hydrology: bridging the gap between research and practice

(proceedings of the fourth international FRIEND conference held at Cape Town, South Africa, March 2002), IAHS Publ., vol 274, pp 289–295
17. Loon AF, van Lanen HAJ, van Hisdal H, Tallaksen LM, Fendeková M, Oosterwijk J, Horvát O, Machlica A (2010) Understanding hydrological winter drought in Europe. In: FRIEND 2010 – global change: facing risks and threats to water resources (Proceedings of the Sixth World FRIEND Conference, Fez, Morocco, October 2010), IAHS Publ., vol 340, pp 189–197
18. Pfister C, Weingartner R, Luterbacher J (2006) Hydrological winter droughts over the last 450 years in the Upper Rhine basin: a methodological approach. Hydrol Sci J des Sciences Hydrologiques, 51(5):966–985
19. Wilhite DA, Hayes MA, Knutson C, Smith KH (2000) Planning for drought: moving from crisis to risk management. J Am Water Resour Assoc 38(4):697–710
20. Tokarczyk T, Szalińska W (2018) Drought hazard assessment in the process of drought risk management. Acta Sci Pol Formatio Circumiectus 18(3):217–229
21. Kozek M, Tomaszewski E (2018) Selected characteristics of hydrological drought progression in the upper Warta river catchment. Acta Sci Pol Formatio Circumiectus 18(3):77–87
22. Tokarczyk T (2013) Classification of low flow and hydrological drought for a river Basin. Acta Geophys 61(2):404–421
23. Tomaszewski E (2011) Defining the threshold level of hydrological drought in lake catchments. Limnol Rev 11(2):81–89
24. Dębski K (1970) Hydrologia (Hydrology). Arkady, Warszawa
25. Hisdal H, Tallaksen LM, Clausen B, Peters E, Gustard A (2004) Hydrological drought characteristics. In: Tallaksen LM, van Lanen HAJ (eds) Hydrological drought. Processes and estimation methods for streamflow and groundwater. Developments in Water Science, vol 48, pp 139–198. Elsevier, Amsterdam
26. Ozga-Zielińska M (1990) Niżówki i wezbrania – ich definiowanie i modelowanie (droughts and floods – their definition and modelling). Przegl Geof 1–2:33–44
27. Yevjevich V (1967) An objective approach to definitions and investigations of continental hydrologic drought. Hydrology Paper, No. 23, Colorado State Univ., Fort Collins, Colorado
28. Stachý J (2011) Występowanie lat mokrych i posusznych w Polsce (1951–2008) (Wet and dry years occurrence in Poland (1951–2008). Gospodarka. Wodna. 8:13–321
29. Markham CG (1970) Seasonality of precipitation in the United States. Ann Assoc Am Geogr 60(3):593–597
30. Tomalski P, Tomaszewski E (2015) Metody, formuły i wzory obliczeniowe zastosowane w pracy (Methods, formulas and computational equations). In: Jokiel P (ed) Metody statystyczne w analizach hydrologicznych środkowej Polski (statistical methods in hydrological analyses of central Poland). Wyd, Uniwersytetu Łódzkiego, pp 215–272
31. Tomaszewski E (2017) Niżówki i susze (Low-flows and droughts). In: Jokiel P, Marszelewski W, Pociask-Karteczka J (eds) Hydrologia Polski (Hydrology of Poland). Wydawnictwo Naukowe PWN, Warszawa, pp 175–182
32. Tomaszewski E (2016) Impact of Lake Gopło on low-flow regime of the upper Noteć river. Limnol Rev 16(2):95–103
33. Bogdanowicz R (2009) Zasoby rzek Przymorza i ich zmienność (Water resources of coastal rivers and their variability). In: Bogdanowicz R, Fac-Beneda J (eds) Zasoby i ochrona wód. Obieg wody i materii w zlewniach rzecznych (Water resources and water protection. Water and matter cycling in river basins). Fundacja Rozwoju Uniwersytetu Gdańskiego, Gdańsk, pp 47–62

Chapter 13
Flood Marks in Poland and Their Significance in Water Management and Flood Safety Education

Marcin Gorączko

Abstract The tradition of permanent marking of maximum water levels reached during catastrophic floods in Poland dates back at least to the 2nd half of the sixteenth century. Currently known are more than 300 high water marks, in most cases in a form embedded onto various kinds of structures. The existence of these objects makes it possible to attempt forming a comprehensive body of knowledge about high water marks, including details such as their geographical and hydrological positioning, place of attachment, method of assembly, data they display, accessibility, etc. Flood marks have a particularly significant value for the study of historical hydrology, as they allow the recreation of the height and range of floods that occurred prior to the introduction of the systematic measurements of the water levels conducted using special instrumentation. Wherever existing in publicly accessible spaces, flood marks have the potential to serve an educative role, informing the population inhabiting and working in river valleys about the potential flood threat in the given area. Finally, old flood marks represent an element of the cultural heritage, serving as a remembrance of the natural cataclysms which occurred in places which for centuries have been inhabited by mankind. They allow the improved understanding of the way the settlement network developed on the areas threatened by floods. The work presents the current state of research on the high watermarks existing in Poland and its line of advancement. Utilizing specific examples, the work presents arguments in support of the preservation of the old high watermarks and their active protection from devastation. It also points to the necessity for the development of a unified methodology for their cataloging, providing certain recommendations for any new objects of that kind.

Keywords Flood marks · High watermarks · Poland · Flood risk · Historical flood events

M. Gorączko (✉)
Faculty of Civil and Environmental, Engineering and Architecture, University of Science and Technology, Kaliskiego 7, 85-796 Bydgoszcz, Poland
e-mail: gorgon@utp.edu.pl

M. Zeleňáková et al. (eds.), *Management of Water Resources in Poland*, Springer Water,
https://doi.org/10.1007/978-3-030-61965-7_13

13.1 Introduction

Withdrawing flood waters leave on the surface of the terrain and structures affected by its visible signs, allowing an accurate visual determination of the maximum levels reached by the water [1]. These maximum heights are even more recognizable, as during floods, rivers carry significant amounts of both mineral and organic material (including ice during the cold seasons), as well as various objects captured by the river. In most cases, these marks gradually vanish as the terrain or structure continue to be exposed to sunlight and wind, as well as undergo repairs conducted by the population of the affected area. In some cases, however, commemorating the tragic and yet spectacular events of the flood, the maximum height reached by the water is being marked by people – by leaving a scoff mark, painting or drawing a line, etc. These provisory marks have often been transformed into more permanent high watermarks indicating in a clear and obvious way the maximum level reached by the water. High watermark is defined as a permanent sign marking of the height reached by the flood (the maximum water level), including an obvious indicator – a line, an arrow, edge of the plaque, etc. [2, 3].

13.2 The Current Stage of Research of High Watermarks in Poland

Mentions of high watermarks can be found in dozens of articles discussing the flood risks in Poland. They are particularly present in the ones that present the problem of floods in the context of a specific region or locality [4, 5]. This fact should not come as a surprise, but rather as a proof that the problem of floods has been present in the life of communities inhabiting the river valleys in Poland for centuries. It is worth mentioning here the authors who in their research focused on cataloging the high watermarks. In terms of the administrative regions, the most detailed works have been made for Krakow [6, 7], Poznań [2, 8–12], Gdansk [13, 14], Toruń [3] and Bydgoszcz [15], while the ones describing the phenomenon from the perspective of hydrographic regions are most detailed for the Vistula River [3, 16–20].

In the past, gathering information about the existing flood marks came was one of the tasks performed by the national hydro-meteorological service. Records were being created for all the existing and newly erected high watermarks, describing the dimensions and characteristics of these marks, situational map and the origins of the flood. The standard was to relate the high watermarks to the national terrestrial reference system [22]. Archives created according to this methodology can be found in the Institute of Meteorology and Water Management of the National Research Institute in Warsaw, where they are still pending a proper comprehensive elaboration. In later years, a significant effort toward the popularization and increase of knowledge about high water marks was the project 'Znaki Wielkich Wód' ('High Water Marks'), conducted by professor Marek Grześ from Nicolaus Copernicus University in Toruń. The goal of that project was to catalogue high water marks in Poland in order to determine the frequency of historical floods and make viable prognosis for their occurrences in the future. It also included the assessment of the efficiency of the hydro-technical installations and flood-protection systems, as well as the assessment of the flood risks in the conditions of climate change, transformations in the management and utilization of drainage basins, and bringing awareness to the significance of the threat presented by floods. The platform used for the gathering and the exchange of information was a website [21], where any person interested in the subject could contribute to the creation of a comprehensive database, which would be used for further research (see Fig. 13.1). The process of organizing the data included the creation of records for each of the high watermarks. They contained: the name of the contributor, date of registration, name of the river, point of the river where the mark was placed, date of the flood and height of the culmination wave above the sea level and relative water level (in centimeters). The information such as name of the locality, date displayed on the mark, brief description of the sign, its positioning and history, information on other installations of that kind in the area, name of the owner or caretaker of the mark, as well as the photographic documentation and the localization sketch were also put into the records. Although this initiative quickly lost its initial impetus, it set a clear direction for further research of high water marks – making it clear that it ought to be necessarily a subject of a systematic, nation-wide research initiative.

The de facto successor of that initiative is the project 'Flood Marks' initiated by the Tadeusz Kościuszko University of Technology in Krakow, under Robert Szczepanek [23]. It is a highly advanced enterprise, which aims to create an open-access internet

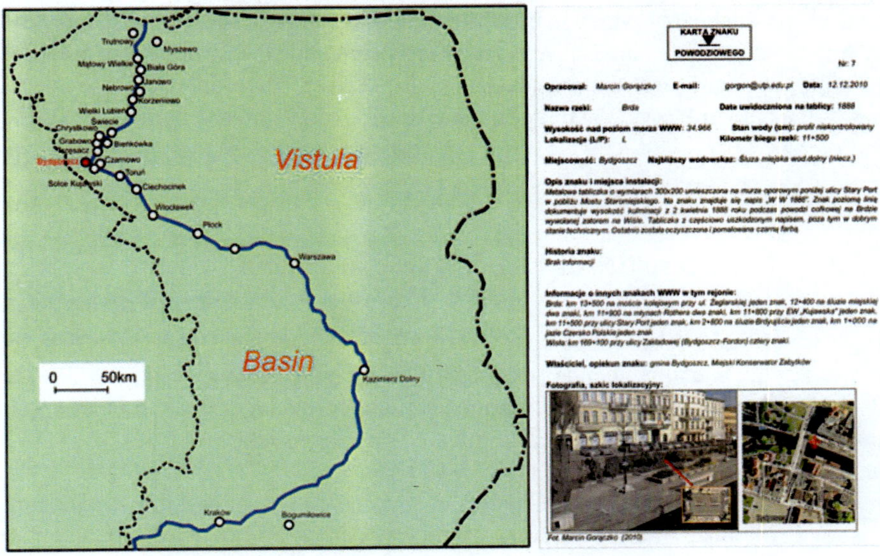

Fig. 13.1 Website of the 'High Water Marks' Project – examples of data visualization [21]

Fig. 13.2 Website of the 'Flood Marks' Project – examples of data visualization [23]

database, based on the OpenStreetMap (Fig. 13.2). Initially, the database was gathering the data on the high watermarks located in Krakow alone [24], but it quickly expanded to include records of 261 objects from the entire country [25] – proving the great accessibility and functionality of the IT tools employed for that project. It is worth mentioning that the majority of the entries came as a result of promoting the initiative through various national internet portals and message boards – not necessarily related to the subject of hydrology or water management. It turned out that many internet users are sufficiently interested in knowing their local surroundings to spot the existing high watermarks and to share their discoveries with others. This was the case of the currently oldest high watermark in Poland, placed on St. Lawrence Church in Głuchołazy, which marks the culmination wave of the Biała River on 2nd of July 1472. Prior to that discovery, it was considered that the earliest flood recorded by high watermarks was the one caused by Vistula River, in Torun, in 1570 [3]. As of today, the "Flood Marks" database contains records of 297 high watermarks existing in Poland.

13.3 Characteristics of Flood Marks in Poland

High watermarks (also called 'flood marks') are defined as a permanent way of indicating the maximum heights reached by the rising water levels [2] as well as the date of that occurrence. Due to a significant number of such objects in Poland, it is possible to attempt their categorization based on various characteristics (Figs. 13.3 and 13.4).

The majority of the marks are made out of metal – usually, cast iron but in some cases steel, bronze or brass. In some cases, we encounter flood marks made out of stone (sandstone, granite, marble, limestone), ceramic elements (brick), concrete, wood, or even glass or plastic. The level of high water is indicated most commonly

Fig. 13.3 Examples of flood marks in Poland (P. Kastner, M. Gorączko, [26])

with a line or in other cases with a point, arrow, edge of the plaque, or through referring it to a specific element of the structure, it is embedded into (like the floor or a mantelpiece). The marks include the date of the flood – in many cases only the year, but sometimes are more detailed, including the month and the day. The high watermarks are made distinct from other similar marks (such as geodetic marks, advertisements, etc.) by adding a text such: "flood", "level of water", "high water", "deluge" or their abbreviations. Due to changes in the national affiliation of the given region, the inscriptions found on the high watermarks in Poland are either in Polish or German. Furthermore, the marks often contain information about their sponsors (administrative institutions, private entities). Depending on the material the mark was made from, the inscriptions and symbols were moulded, engraved (both in metal and in stone), impressed (on brick and concrete), painted (on the mark itself or a smooth surface of the plaster, concrete or asphalt). The dominant form of high watermarks was changing over the years. While the flood marks from

Fig. 13.4 Examples of flood marks in Poland – continuation (M. Gorączko, P. Kastner, [26])

the beginning of the nineteenth century display many characteristics common with sepulchral monuments, in the decades that followed, they gradually moved towards the universal form of a rectangular metal plaque. The attempts to standardize the form, content and placement method are visible particularly in the case of marks erected in the 1960s from the initiative of the national hydrological-meteorological service. It implemented a uniform country-wide pattern for the flood marks, with the date of the flood was minted using a special dater [22].

The high watermarks can be found both outdoors and indoors, although the formers are decisively much more common. They are usually placed on objects whose structure can withstand the passage of time, natural or man-made disasters – such as fires, war, or floods [12]. They are usually located on walls of buildings, particularly churches – which for centuries had been ones of the few stone/brick structures existing in the rural areas. Since the beginning of the eighteenth century, when the water regulation and channeling work intensified, they began to appear on hydro-technical structures, such as locks (Fig. 13.5), embankments (Fig. 13.6), water stations, etc. A particular mention here should be made in regards to bridges, as their relation to rivers is the most obvious. High watermarks can be found on bridgeheads and piers located on flood plains. Only in very seldom cases marks were placed on the piers located within the riverbed, as they are naturally less accessible or not accessible at all, even when the level of the water is low. Some of the high watermarks in Poland are standalone structures, such as monuments, obelisks, posts, boulders (bearing plaques or carved inscriptions) and as small architectural objects (such as wayside shrines).

Fig. 13.5 Flood mark on navigation lock – Noteć River in Pakość (S.Grzegorzewicz)

Fig. 13.6 Flood mark on an embankment – Vistula River in Kiezmark (P. Kastner)

As previously mentioned, a high watermark could be erected right after the flood, based on the temporary impression left by the water. In some cases, however, its placement occurred a long time after and was based on witness' testimonials, analysis of photographic archives, the study of descriptions and mentions in source materials. The indispensable condition for ensuring the validity of a mark erected in such a manner is the preservation of the reference points in the terrain according to which the level of a historical flood can be recreated (Figs. 13.7 and 13.8).

Despite a significant number of catastrophic floods occurring in Poland, they tend to be registered by a single high watermarks, sometimes two or three. Locations, where a larger number of them has been placed, are seldom. It would seem that the most efficient and useful way to place a high watermark, would be to display it on a visible spot, at the level of the eyes of a passer-by. However, a significant number of these objects are placed either close to the ground, or on significant heights. Which

Fig. 13.7 Flood mark based on an old hand-made inscription from 1888 – Vistula River in Chrystkowo (M. Gorączko)

Fig. 13.8 Flood marks based on old photography (Vistula River in Ciechocinek on the left side) and parish chronicle (Vistula River in the Wielkie Walichnowy one the right side) (M. Gorączko)

is why even having the precise geographical coordinates of their location does not mean they can be easily found in the terrain.

The first and only chronological tally of the high watermarks in Poland was conducted by Bożek [25]. Decisively the most numerous marks (59 objects) were erected in the aftermath of the 'flood of the millennium' in July 1997, caused by the long-lasting downpour in the upper and middle part of the Vistula's basin. The second most documented (with high water marks) flood (19 objects) was the one that occurred in the period of March–April of 1888. It was a snow-melt flood which in many locations of the country's lowlands (i.e. on Vistula, Noteć, Warta, and Słupia) resulted in ice jams. In 1903 13 high water marks were erected (mainly in Wroclaw and Krakow) to document the flood caused by the rainfalls in July of that year in the upper parts of the Vistula's and the Odra's basin. Similar in number are the marks from 1829 (related mainly to the snow-melt flood on Żuławy Wiślane), 2010 (the largest so far flood in twenty-first century Poland), 1813 and 1934 (floods caused by rainfall in the upper and middle part of the Vistula's basin). In most cases, however, Bożek's research (2017) has revealed until now that the majority of floods have been documented by one or two high watermarks.

13.4 Hydrological, Historical and Educational Significance of Flood Marks

The use of high watermarks on the terrains that constitute modern Poland developed much earlier than the periodic (and later constant) water measurements conducted through stream gauges [27]. The beginning of measuring using stream gauges, according to the existing records, dates back to 1760 – when they were used to measure the levels of Vistula River in Toruń [22, 28]. In 1799 measuring also begun for the levels of Vistula River in Warsaw. The first measurements of Vistula River in Krakow begun in 1813. In case of the second largest river in Poland – Odra River – the first existing measurements come from 1809 in Krosno and 1810 in Koźle, Opole,

Brzeg, Głogów, Słubice, Kostrzyń, and Szczecin. In the same period, measurements began on Warta River – in Gorzów Wielkopolski (1809) and Poznań (1818) and Noteć River in Drezdenko (1811). These, however, are exceptions, as the overall history of a systematic measuring of water levels of Polish rivers rarely goes back more than 200 years. For example, the first measurements on Podkarpacie rivers such as San, Raba, Dunajec, Wisłoka and Wisłok were made in 1887. Stream gauges on rivers such as Bzura, Drwęca, Wierzyca, Rega, Parsęta, Słupia, Pasłęka and Łyna appeared only in the twentieth century [22]. Effectively, the current knowledge about floods that occurred in the country prior to the late nineteenth century comes from annals, lexicons, administrative documents, newspapers and other archives, i.e. [29–31]. The main flaw of these records is that they lack the precise measurements of the flood wave, consequently lacking the precise indication of the range of the floods. This is exactly where high watermarks provide an exception, as by marking the maximum levels of historical floods, they manage to provide a valuable reference to assess the possibilities of high water rises in the area in regards to the current conditions of the terrain. Placing high watermarks has value even today, despite the existence of over 800 hydrological stations throughout the country [32] and the great progress made in the last 20 years, in fields of methodology and technology of hydrological measurement. It is significant in spite of the developments in interpretation and implementation of the obtained data (use of the modern measurement tools, development of aerial and satellite terrain mapping, advancement of hydrological modeling, creation of flood risk maps etc.). High watermarks placed along the river flows, amongst the controlling bed cross-section, allows a credible verification of flood forecasting.

Where placed in an open public spaces, high watermarks can bring awareness to the potential threats posed by floods. By displaying the level once reached by the water, they serve as a valuable referential for the entire surrounding terrain and the structures existing within it. This is particularly significant on the areas directly next to the river embankment. In modern times, due to the progressive process of suburbanisation, these areas are commonly being transformed from farmlands into residential zones. The new inhabitants of these terrains are rarely aware of the risks they had taken by moving into the vicinity of the river, perceiving it as a mere attraction of the landscape. Some citizens place excessive trust into the flood protection walls, even though the conclusion coming from the larger floods that occurred in modern times in Poland is unequivocal that the efficiency of any technology and methods for flood protection diminishes proportionally to the increase of the levels of the water and duration of the increased water flow. It can even be concluded that the existence of the flood barriers leads many into a false sense of security. Therefore, sharing the knowledge about the existence of high watermarks, especially in schools is very important. Maintaining their good technical and aesthetic condition, and even erecting new objects of that type immediately after an occurrence of a maximum flood ought to be elements of the fundamental flood-protection education in the given localities. Involvement in such actions is in the very interest of the local population.

Catastrophic floods, just as armed conflicts often constituted major events in the history of settlements on the Polish lands, in some cases being the very turning points for their existence. Often they resulted in an economical stagnation and depopulation, while in other cases they led to specific planning and technological actions aiming at lowering the damaging potential of the future occurrences. This refers especially to cities, where a large number of historical high watermarks can be found. For example, the largest (documented by three high watermarks) flood in Gdansk occurred in 1829, when water spread over 75% of the city, leaving 12,000 of its inhabitants without shelter. The so-called "flood of the millennium" which is particularly remembered by the citizens of Wroclaw (over a dozen of high watermarks) had 40% of the city area flooded. The largest known flood of Brda River in Bydgoszcz occurred in 1888 and was marked by four high water marks [15] when due to an ice jam on Vistula River, the backflow of Brda spread on the area nearly 20 km from the estuary. The maximum levels of floods had also been recorded in smaller cities. In 1979, a snow-melt flood on Narew River flooded 50% of Pułtusk (two high water marks) A similar event occurred there in 1958 (four high water marks). In January of 1982, an ice jam near Włoclawek Reservoir led to the largest flood in the history of Płock. The reservoir, created in 1970 has since then been recognized as one of the places most vulnerable to ice jams in Poland. The 1982 flood led to significant changes in the approach to ice-breaking on the whole lower reaches of Vistula.

An interesting story written in high water marks can be found inside the St. Stanislaw Church in Świecie, where the increase in the frequency of floods in nineteenth century led this medieval city located at the Wda-Vistula estuary to be relocated to higher grounds [20].

13.5 Factors Threatening the Preservation of Flood Marks

It can be assumed that a large number of high watermarks was destroyed during the WWII – not only within the premises of large cities such as Warsaw, Wroclaw, Opole, Gdansk or Szczecin but also in dozens of smaller localities where warfare led to the damage or destruction of a large number of buildings. We also know that upon regaining Polish independence, a significant number of high watermarks was removed only because they contained inscriptions made in German.

Despite the more favourable and stable conditions of modern times, the risk to the existence of the high watermarks remains high. Unfortunately, theft and devastation are the main contributors to this situation. Particularly exposed to such threats are the marks placed in secluded areas, distant from public locations. Other potential problems are the repairs and improvements (painting, plastering, thermo-isolation, expansions, etc.) of the structures on which the marks were placed. During such procedures, the marks are often removed and often placed back incorrectly or not at all once the works have concluded. Many contractors are unaware of the fundamental importance of the placement of the mark. It is worth mentioning that according to the current conservation laws, even the 300-year-old flood marks are not subject to

Fig. 13.9 Deterioration of flood marks (M. Gorączko, P. Kastner)

protection in themselves, as the regulations apply only to the structures on which they are mounted. Flood marks, mainly those made out of metal and stone, exposed to environmental factors (fluctuating temperature, moisture, salinity, pollution) are deteriorating when deprived of conservatory works (Fig. 13.9). The inscriptions wear off, leading often to the removal of now unintelligible high watermarks, which are no longer recognizable for their purpose and value. It also happens that the local authorities are not particularly interested in preserving in the communities' awareness the history of catastrophic floods – in order not to discourage potential investors, influx of population or decisions to start a business in the area. Each subsequent, not necessarily catastrophic, larger flood proves that this attitude is nothing else than short-sighted.

13.6 Summary and Conclusions

Due to their great hydrological, educational and historical significance, currently high watermarks incite interest of hydrologists and historians, not only in Poland, but also in other European countries [33–40]. It has to be noted though that these initiatives have been inadequately coordinated and have been rarely comprehensive in nature. A serious approach seems to be a necessity in regards to the research of high watermarks, due to the significant fluctuations in their number and distribution. After each new flood, new marks are being placed. Simultaneously, due to various factors, the existing historical marks vanish. Therefore, the situation calls for the intensification of research aiming at the cataloging of these objects, to the level of detail which would enable their precise recreation whenever the need arises.

Obvious is the lack of a complete database containing the information about the high watermarks already identified in Poland, with the possibility of adding newly erected, or newly discovered ones. The "Flood Marks" (www.openhydrology.org/maps/flood_mark) project seems to be a step in the right direction. Its strength is the utilization of the popular and still growing internet community of OpenStreetMap users. It gives hope to extend the lifespan of that project beyond what was accomplished by any of the previous initiatives of this kind. Since the database is being developed using a free and open-access map of the entire globe, chances are the used methodology will also be utilized in other countries.

Regardless of the direction which the documentation and popularization of high watermarks will take, the leading role in the proper assessment of their credibility should be taken by hydrologists.

13.7 Recommendations

Flood marks are an important source of information about catastrophic phenomena in river valleys. That is why for many years they have been of interest to hydrologists in various countries. In future research, it is worth paying attention to the popularization of this topic among policy planners, officials, engineers, representatives of local communities, etc. Very often, it depends mainly on their decision whether and how long high watermarks will be protected against destruction and devastation.

Acknowledgements In memory of Prof. Marek Grześ (1946–2020).

Many thanks to Paweł Kastner and Szymon Grzegorzewicz for permission to use their photos in this work.

References

1. Gorączko M (2019) Znaki wielkiej wody – obiekty budowlane jako nośnik informacji hydrologicznej, [in] Mosty. In: Sobczak-Piąstka J, Podhorecki A (eds) Tradycja i nowoczesność. WU UTP, Bydgoszcz, pp 37–47
2. Wosiewicz BJ (2002) Zachowane znaki i tablice powodziowe w obrębie Poznania. Gosp Wod 10:416–421
3. Pawłowski B, Gorączko M (2014) Z badań nad znakami powodziowymi w dolinie Wisły. Gosp Wod 2:57–63
4. Kaniecki A (2004) Poznań miasto wodą pisane. Poznańskie Towarzystwo Przyjaciół Nauk, Poznań
5. Cyberski J, Grześ M, Gutry-Korycka M, Nachlik E, Kundzewicz Z (2006) History of floods on the River Vistula. Hydrol Sc J 51(5):799–817
6. Skarżyńska K. (1961). O tablicach powodziowych na obszarze Krakowa. Przyczynek do dziejów wezbrań Wisły pod Krakowem. Przegl Geofiz 6(XIV), 4:271–277
7. Opyrchał L, Opyrchał U, Bąk A (2018) Tablice powodziowe na terenie Krakowa. Gosp Wod 7:211–220
8. Kawka M, Wosiewicz BJ (2003) Jeszcze o znakach i tablicach powodziowych w Poznaniu. Gosp Wod 6:244–245
9. Stasiak K (2009) Historyczne powodzie w świetle znaków wielkich wód na przykładzie Warty w Poznaniu, Master's Thesis, UMK, Toruń
10. Wosiewicz BJ (2017a) Znaki wód wielkich jako źródła informacji o powodziach. Przegl Bud 9:18–24
11. Wosiewicz BJ (2017b) Niekonwencjonalne źródła informacji o powodziach. Znaki powodziowe i pocztówki z treściami powodziowymi w Poznaniu. In: Buczkowski W,

Szymczak-Graczyk A (eds) Rewitalizacja obszarów zurbanizowanych. Powodzie w miastach - przyczyny, skutki, zapobieganie, Prodruk, Poznań, 7–42
12. Wosiewicz BJ (2018) Raz jeszcze o znakach powodziowych w Poznaniu. Gosp Wod 8:247–254
13. Majewski A (1970) Znaki wielkiej wody w Gdańsku i na obszarze Żuław Wiślanych, Przegl Geofiz XV (XXII), 1, Gdynia
14. Cieśliński R (2017) Znaki wielkiej wody na terenie miasta Gdańska. Inż Mor I Geotech 1:7–12
15. Gorączko M, Gadomski J (2011) Znaki wielkiej wody na terenie Bydgoszczy. Gosp Wod 7:295–300
16. Grześ M (2006) Historyczne powodzie na Wiśle w świetle znaków wielkich wód, [in] Funkcjonowanie geoekosystemów zlewni rzecznych 4. Procesy ekstremalne w środowisku geograficznym, UAM, Poznań-Storkowo
17. Prabucka M (2009) Znaki Wielkich Wód na Żuławach Wiślanych, Master's Thesis, UMK, Toruń
18. Dziechciarz M (2010) Historyczne powodzie na Wiśle Lubelskiej w świetle znaków i tablic wielkich wód, Master's Thesis, UMK Toruń
19. Kuźniar P (2017) Znaki największych powodzi na Wiśle - ciekawostka historyczna czy przestroga? In: Kałuża T, Wosiewicz BJ (eds) [in] Wybrane zagadnienia dokumentowania i ochrony zabytków hydrotechniki. Bogucki Wydawnictwo Naukowe, Poznań, pp 39–50
20. Gorączko M (2017) Wezbrania i powodzie na kociewskim odcinku dolnej Wisły. In: Pająkowska-Kensik M, Tobolski K (eds) [in] Kociewie Kraina nad Wisłą. Towarzystwo Przyjaciół Dolnej Wisły, Gruczno, pp 62–75
21. www.wielkawoda.umk.pl, DOA: 22.02.2019
22. Pasławski Z (1973) Metody hydrometrii rzecznej. WKiŁ, Warszawa
23. www.openhydrology.org/maps/flood_mark, DOA: 22.02.2019
24. Mleczek F (2016) "High water marks" - projekt kartowania znaków wielkiej wody w OpenStreetMap, Master's Thesis. Polit. Krak, Kraków
25. Bożek P (2017) Kartowanie znaków wielkich wód na obszarze Polski, w ramach projektu OpenStreetMap, Engineer's Thesis. Polit. Krak, Kraków
26. www.wikimedia.commons.org, DOA: 22.02.2019
27. Kolberg W (1861) Wisła, jej bieg, własności i spławność. Drukarnia Jana Jaworskiego, Warszawa
28. Makowski J, Tomczak A (2002) Stany wody Wisły w Toruniu w świetle pomiarów z ostatnich dwóch stuleci. Stud. Soc. Scient. Torun, TNT, Toruń
29. Słownik Geograficzny Królestwa Polskiego (1880–1902) Kasa im. Józefa Mianowskiego, Warszawa
30. Girguś R, Strupczewski W (1965) Wyjątki ze źródeł historycznych o nadzwyczajnych zjawiskach hydrologiczno-meteorologicznych na ziemiach polskich w wiekach od X do XVI, Instrukcje i podręczniki, 87. PIHM, WKiŁ, Warszawa
31. Majewski A (1993) Kronika powodzi w delcie Wisły. In: Churski Z (ed) [w:] Uwarunkowania przyrodnicze i społeczno-ekonomiczne zagospodarowania dolnej Wisły. IG UMK, Toruń, pp 13–28
32. Malota A, Nadolna A (2015) Vademecum. Pomiary i obserwacje hydrologiczne, IMGW PIB, Warszawa
33. Bürger K, Dostal P, Seidel J, Imbery F, Barriendos M, Mayer H, Glaser R (2006) Hydrometeorological reconstruction of the 1824 flood event in the Neckar River basin (southwest Germany). Hydrol Sc J 51(5):864–877
34. Munzar J, Deutsch M, Elleder L, Ondracek S, Kallabova E, Hradek M (2006) Historical floods in Central Europe and their documentation by means of floodmarks and other epigraphical monuments. Morav Geograph Rep 14:26–44
35. Deutsch M, Pörtge K-H (2009) Hochwassermarken in Thüringen, Erfurt
36. Balasch JC, Ruiz-Bellet JL, Tuset J, Martin de Oliva J (2010) Reconstruction of the 1874 Santa Tecla's rainstorm in Western Catalonia (NE Spain) from flood marks and historical accounts. Nat Hazards Earth Syst Sci 10:2317–2325

37. Pekárová P, Halmová D, Bačová Mitková V, Miklánek P, Pekár J, Škoda P (2013) Historic flood marks and flood frequency analysis of the Danube River at Bratislava, Slovakia, J Hydrol Hydromech 61, 4:326–333
38. Benito G, Brázdil R, Herget J, Machado MJ (2015) Quantitative historical hydrology in Europe. Hydrol Earth Syst Sci 19:3517–3539
39. www.matdeutsch.de, DOA:26.02.2019
40. www.real-planet.eu/hochwasser, DOA: 22.02.2019

Chapter 14
Flood Potential of Polish Rivers

Artur Magnuszewski

Abstract Comparison of the highest precipitation totals shows that they are lower in Poland than in Europe. This can be explained by moderate and transitional to continental climate of Poland and lowland relief of the country which do not enhance precipitation. Higher precipitation can be obtained by the accumulation of the snow and rapid thawing in the spring time. The different reaction of the rivers to precipitation can be measured by the use of flood potential index *k* proposed by J. Françou. Flood potential index is calculated using long term discharge measurements at hydrological posts located at major rivers of Poland. Highest values of *k* index represent Odra river mountain tributaries and upper Vistula tributaries. Lowest values of *k* index are characteristic for the lowland rivers with a high percentage of lakes and bogs in their catchments. It can be observed a change of flood potential index in the longitudinal profile of Odra and Vistula river reflecting their potential for flood hydrograph transformation. Flood potential index can be used in regional plans of flood risk management.

Keywords Poland · Vistula river · Odra river · Flood potential index

14.1 Introduction

Floods on the rivers belong to the most severe natural hazards in Poland. Recently after the two major floods in 1997 on the Odra river and 2010 on the Vistula river, there is a growing interest in the temporal and regional distribution of that phenomena. There is also a discussion in the mass media as well as among scientists about contributions of climate or anthropogenic processes to those floods. To be prepared for floods, it has been started in Poland a program called ISOK (informatics system of risk assistance and evaluation) including also flood risk management plan based on flood risk maps. The plan of flood risk management proposes different actions which should minimize flood risk-taking in to account the differences between rivers

A. Magnuszewski (✉)
Department of Geography and Regional Studies, University of Warsaw, Krakowskie Przedmieście 30, 00-927 Warsaw, Poland
e-mail: asmagnus@uw.edu.pl

M. Zeleňáková et al. (eds.), *Management of Water Resources in Poland*, Springer Water,
https://doi.org/10.1007/978-3-030-61965-7_14

in generating floods. To analyze the flood occurrence and spatial distribution, we assume that the processes of flood generation are statistically stationary and are not a subject of trend. In the study of [1] it has been proved that Elbe and Odra rivers do not show any increase of flood frequency in long time horizon. Based on long-term time series they found decreasing winter flood occurrence, while summer floods showed no trends [2] so the process of flood generation can be considered as random and not showing any trend. Also, in the study of [3] based on the data from period 1951–2006, it has been found no evidence of a trend in frequency and magnitude of floods.

New study results of potential effects of a global warming on meteorological and hydrological extremes at regional scales are given at [4]. In their work an ensemble of EURO-CORDEX RCP8.5 scenarios is used in connection with distributed hydrological model to assess the projected changes in flood hazard in Europe through the current century. A positive trend for the maximum daily precipitation is found in most of the study region, with scale and statistical significance becoming stronger moving toward Eastern and Northern Europe. Hydrological response to increase of precipitation is more complicated due to a interaction of different runoff formation processes, reduced snow cover accumulation, increase of evapotranspiration. In the study it was found reduction of peak discharges in southern Spain, Scandinavia and Baltic countries, while a large portion of central Europe and the British Isles are likely to experience a progressive increase in the magnitude and frequency of maximum discharges.

According to report of European Environmental Agency [5] there is observed an increase of flood losses in Europe since the 1970s. In Poland recent catastrophic floods of 1997 on the Odra River and 2010 on the Vistula River had more negative consequences much more losses than previous events that they led to significant changes in flood risk estimation methods. The trend for increasing losses from river floods is primarily attributable to increase of society wealth, but also growing exposure to risk due to improper spatial planning and growth of settlement at flood prone areas. In Poland after the flood of 2010 there has been initiated a national program of flood reduction risk called ISOK (informatics system of risk assistance and evaluation). As the first step of the program it was made inventory of the major historical floods and their range. In this program a whole country has been covered by the accurate measurements of elevation DEM (Digital Elevation Model). This data are indispensable for calculations of flood range of given probability with the use of hydrodynamic models. In ISOK terrain elevation has been measured by the lidar altimeter in the basic field element of 1×1 m, with the density 4 points/m^2 at the open areas (including forest), and in the cities 12 points/m^2. Accuracy of elevation measurement is on a flat country surfaces in the range of $\leq$0.15 m, and for the cities $\leq$0.10 m. In the forest area accuracy is lower – 0.25–0.30 m.

Accurate DEM and hydrodynamic modeling have been applied to prepare for the major river valleys maps of flood range and flood risk. The ISOK flood risk maps can be obtained in pdf. format from https://mapy.isok.gov.pl/imap/. Example of ISOK map index is shown at Fig. 14.1.

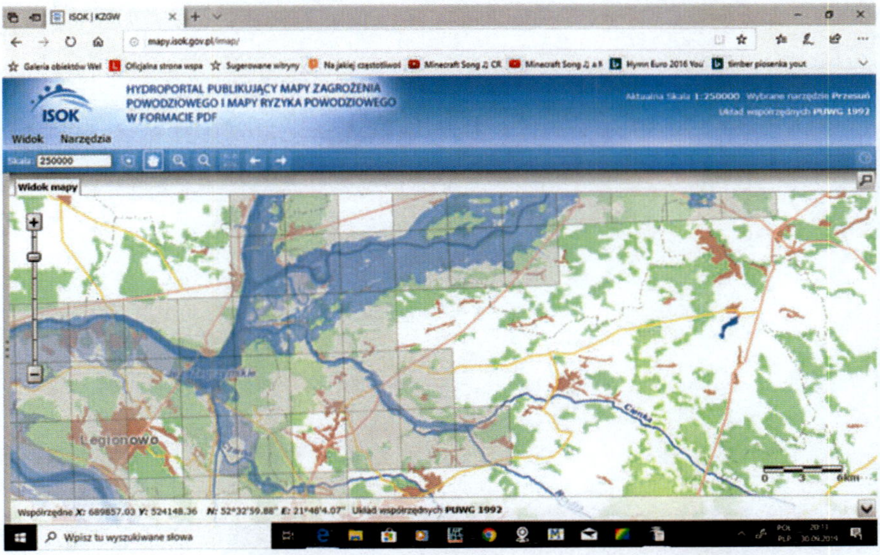

Fig. 14.1 Example of ISOK flood range and flood risk maps in the area of confluence of Bug and Narew rivers

There has been performed also the study of rainfall induced flood frequency in Poland in the period 1946–2001 [6]. Data were obtained from different flood reports, mass-media articles and hydrological observations. The distribution of flood frequency shows that the most frequent floods (40–80 cases) occurs in the south-western part of Poland on the tributaries of the Odra river which drain foot slopes of the Sudety Mountains (Fig. 14.2). Second area most exposed to floods are tributaries of the upper Vistula River draining slopes of the Carpathian Mountains. Isolated area of the higher flood frequency is in the area of Holly Cross Mountains. There is visible difference in flood frequency between mountains and lowland area of Poland where the number of floods observed was in the range 10–20. This area has also lowest precipitation totals, and is drained by lowland rivers with large floodplains. The lowest flood frequency is in the northern part of Poland where there is a belt of post-glacial landscape with many lakes forming natural storage for run-off.

Magnitude of floods and scale of losses have raised the concern that there must be a shift from a purely civil-engineering measures of flood defense toward a more integrated flood risk management system which contains also nonstructural measures such as adaptation and reduction of infrastructure vulnerability. In the flood risk reduction studies we need to know the magnitude of risk expressed by probability of frequency and also by the flood dynamics.

In a spatial studies of floods it is assumed that in a given region there is a similarity in variance of maximum discharges $c_v[WWQ(A)]$, where c_v – coefficient of variance, WWQ – maximum discharge, A – catchment area. To fulfill this similarity in a large catchments it is difficult due to a difference in landscape relief of upper and lower

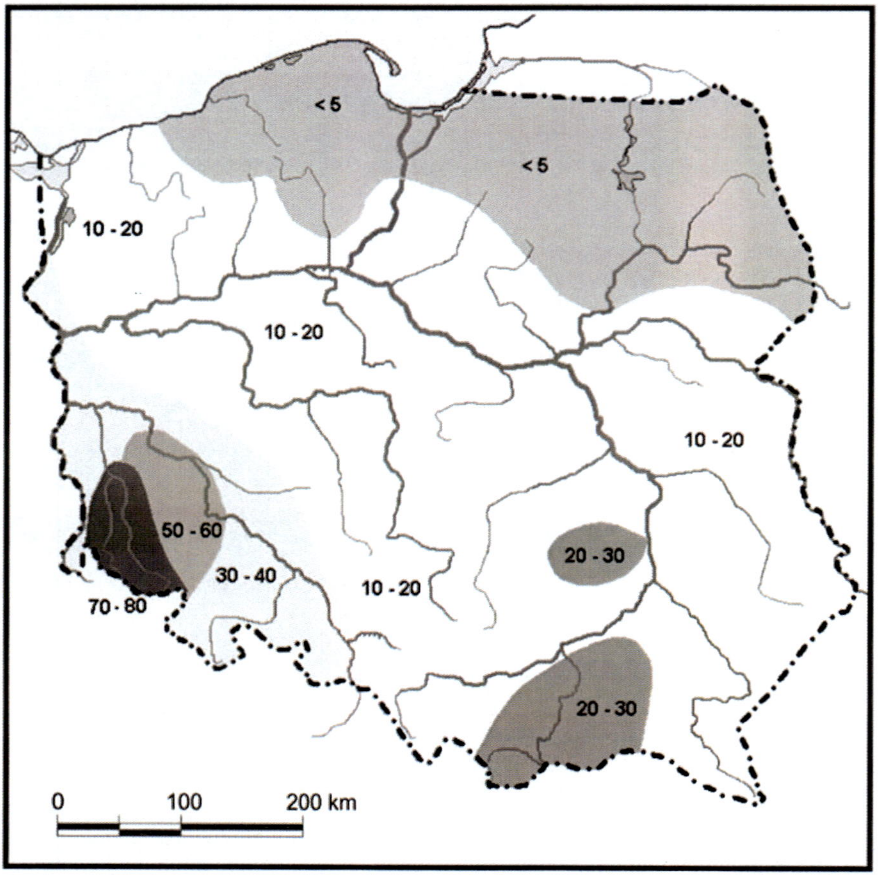

Fig. 14.2 Area and frequency of rainfall induced flood occurrence in Poland in the period 1946–2001 after [3]

part of the catchment [7]. To be able to compare the catchments located at different altitudes we should add one more variable of land elevation above the sea. For a practical reasons it is more feasible to look for a index based only on hydrological data.

One of the methods how to measure the dynamics of catchment response to rainfall is the flashiness which reflects the frequency and rapidity of short term changes in stream flow, especially during high runoff events [8]. Flashiness can be based on mean daily flows. Calculated by dividing the path-length of flow oscillations for a time interval (i.e. the sum of the absolute values of day-to-day changes in mean daily flow) by total discharge during that time interval. This index can be used for comparison of subcatchments having different physiographic properties. Calculation of flashiness index requires a long data series of daily discharges which are difficult to obtain in the study of large number of catchments.

Every flood has a beginning in the processes acting in the atmosphere since it is induced by intensive precipitation of rain or snow. In atmospheric processes, it is visible certain cascade of processes acting in different time scales from cyclonic circulation to convective cloud. Every scale of the atmospheric processes has a certain dynamics and time of duration. Very intensive torrential rainfalls from convective clouds cause flash floods which have limited spatial coverage on the other hand, rain at the atmospheric front has a lower intensity but can affect large catchments. The study of rain effectiveness in creating floods in Poland has shown that according to a ratio of rain maximum to rain duration three types of rainfalls can be defined: convective, frontal and wide convergence zones [9, 10].

The isolated system of convective clouds can produce the volume of rain which can be calculated as a product of cloud area, rain duration, and height of precipitation. The total volume of rain depending on the size of cloud systems can vary from $1.8 \times 10^5 - 2.1 \times 10^7$ m^3 at small convective clouds to 6.1×10^9 m^3 at meso-scale cloud systems [11].

In Poland, rain amount is a function of air masses properties of different origin. Dominating advection of air masses to Poland is from the Atlantic Ocean, a long distance from the ocean creates conditions of air masses transformation on the way over the continent. Additionally, the lowland relief of the country with a mountain ranges of Sudety and Karpaty in the south do not induce high precipitation totals by the orographic effect. This explains why in Poland maximum precipitation totals compared to world data are considerably lower [11].

It has been shown that among catchment's morphometric variables controlling hydrologic response to the intensive rain the most important is the catchment area. The area of the catchment is related to the spatial scale of meteorological processes which are responsible for the creation of rain. According to the study of [12] and [13] based on data from the period 1951–1990 the highest river run-off layer is created in Poland from mountain catchments having an area less than 300 km^2. For example mountain catchment of upper Vistula at Skoczów profile was able to create the layer of run-off of 275 mm. The river run-off layer considerably gets smaller with the increase of the catchment area; this is especially visible in the longitudinal profile of transit rivers. The increase of the river catchment area means also change in the scale of rain inducing processes, from intensive convective cells to longer duration and less heavy frontal rain.

In climatic conditions of Poland, high river run-off layer can be created by the melting of the accumulated snow layer [13, 14]. The highest snow melting floods of 1970 and 1979 were observed in central Poland from the catchments of area around 1000 km^2 (Skrwa – 144 mm, Wkra – 123 mm). At the mountain catchments of Poland snow melting floods were at the same time producing river run-off layer of less than 80 mm. This difference between lowland and mountain catchments can be explained by the fact that in the mountains accumulated snow changes its physical state slowly due to a diurnal pattern of thawing and freezing. The highest thawing floods occur on the large lowland rivers Bug and Narew as well as in the lower and middle reach of the Vistula river. Time of thawing floods occurrence at the lowland catchments is

March and April. If they occur in the middle of winter, they have a limited magnitude due to thermal conditions controlling the rate of snow cover thawing.

The frequency of floods in Poland has been a subject of study of [3] who analyzed hydrological data from the period 1946–2001. They have outlined the following regions of the highest frequency of catastrophic floods occurrence:

- Sudety mountains – upper Nysa Kłodzka, Bystrzyca, Kaczawa.
- Western and Central Carpathian Mountains – Dunajec, Soła.
- Holly-Cross Mountains – Kamienna, upper Pilica, upper Warta.
- Valleys of transit rivers especially at the confluence of major tributaries – Vistula, Odra.

This chapter focuses on method how to calculate flood potential index for Polish rivers and how to interpret its value in the context of flood risk. Flood potential index can be understood as a measure showing the effectiveness of the catchment to produce maximum run-off. The area of the catchment is the most important parameter, but catchment properties especially altitude and storage capacity are also very important.

14.2 Study Area

Poland is located in a East-Central Europe, it has an area of 312 thousand km^2. Relief of the country is dominated by the lowlands, only 3% of Poland is located at the altitude higher than 500 m. amsl Southern border of the country is separated by the two mountain ranges of Sudety Mountains and Western Carpathian Mountains. The highest elevation is Mount Rysy, which rises 2499 m and is located in the Polish Tatra Mountains. The largest territory is occupied by the central lowlands, north of the country has a relief of lake land shaped during last Vistulian glacition. Poland has a moderate climate with transitional properties between maritime and continental. The average annual precipitation sum for Poland is 600 mm. Western Carpathian Mountains and Sudety Mountains enhance precipitation totals which reach maximum of 1300 mm. Lowest precipitation sums of 550 mm are in the central part of the Poland in the area of lowlands. Major Polish rivers are Vistula Rivera and Odra River, both have their mouths in the Baltic Sea. The Vistula Basin is located in the eastern and central part of Poland its major tributaries are San River, Bug River, Narew River. Drainage density is high in the mountains and poorly developed in the lowland area. In the belt of the northern lakelands river run-off is attenuated by the large share of lakes and swamps in the catchment area. Floods at Polish rivers are formed in a spring time March–April due to a snow melt and in summer June–August caused by intensive and long lasting rains of frontal origin.

14.3 Flood Potential

Starting from XIX century many studies were undertaken to quantify river basins characteristics and their influence on rainfall-runoff response. First empirical formulas used in hydrology were using only area of the whole catchment as a independent variable. Area of the catchment affects not only the time of concentration but also the total volume of run-off, since the size of the catchment correspond to the size of convective clouds and frontal systems. Maximum discharge of the flood peak is inversely related to size of the catchment; this is because the intensive storms caused by convective clouds have a dimension of first-order catchments. Area of the catchment is also correlated to almost all topographic characteristics such as stream order, stream network length, main channel length, basin length. Because of this correlation with other characteristics, it is possible to use only catchment area to represent river size. There had been many attempts to find the relations between catchment size and maximum discharge at the flood peak, such characteristic can be used as a measure of flood potential. One of the better known is the index of flood potential proposed by J. Françou and applied by [14] in the World Catalogue of Maximum Observed Floods.

Flood potential k is a dimensionless index showing the efficiency of producing maximum discharge from the given catchment area. Since the k index is used for comparison of different size catchments of the world, it is standardized by maximum discharge 10^6 m^3s^{-1} and maximum catchment area 10^8 km^2 physically possible on the Earth. It has the form of Eq. (14.1).

$$k = 10 \cdot \left(1 - \frac{\log WWQ - 6}{\log A - 8}\right) \tag{14.1}$$

where: A – catchment area km^2, WWQ – highest observed flood discharge m^3s^{-1}.

Application of k index for Polish rivers has been first time done by [15] later improved by the application of new data by [16]. Data for calculation of flood potential index come from [17]. Major floods of 1997 and 2010 on Odra and Vistula river have been updated from monograph [6]. The number of gauges which were analyzed is 624, their location is shown at Fig. 14.3. The hydrologic data were entered to a spreadsheet and combined with the shape file containing geographical coordinates of the hydrological gauges. Shape file is one of thematic layers from digital hydrographic map of Poland [18]. To obtain spatial coverage of flood potential index k it has been used a procedure of inverse distance interpolation, the result is shown at Fig. 14.4.

The distribution of k index shows the difference between the mountain catchments (Sudety and Karpathian) and the lowland part of the country. Lowest values of the k index are observed in the lowland part of Poland also characterized by to lower totals of annual precipitation. In the mountain catchments, highest flood potential is characteristic for Dunajec, Soła, Skawa, Raba. It is interesting that the higher flood potential index is observed down the middle reach of the Vistula river up to Zawichost

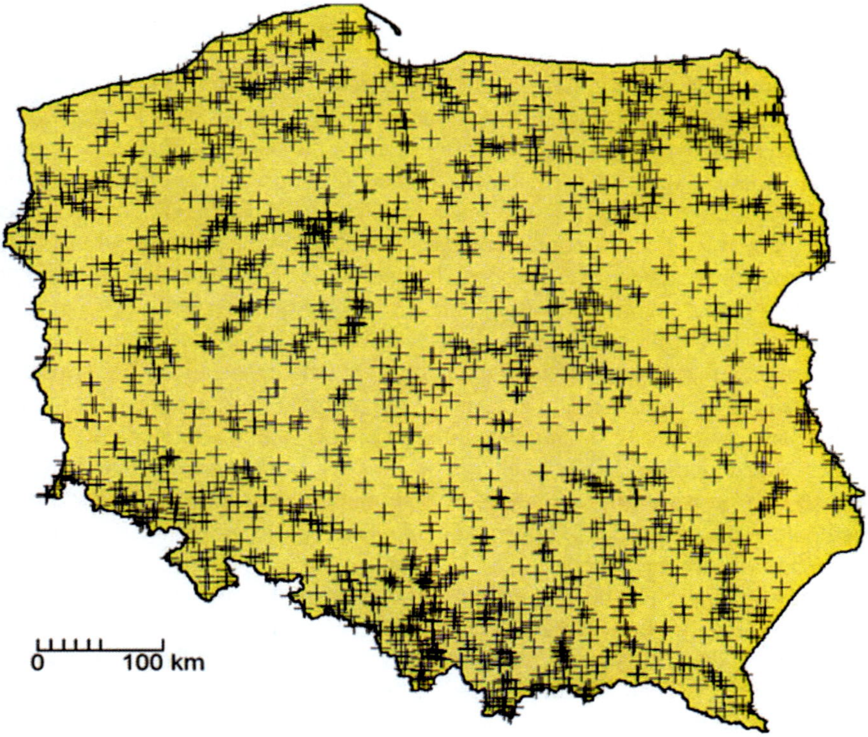

Fig. 14.3 Location of river gauges used in the study to calculate flood potential index *k* in Poland

gauge. This feature can also be observed down the upper reach of Odra river. Such pattern is characteristic for transit rivers where flood peak discharges are lowered in a flood transformation process.

Small catchments in Poland having area 10–100 km^2 produce maximum discharge 10.8–12.2 times less than similar catchments of the World, while catchments of the area 100,000–190,000 km^2 are 8.6–9.5 times less efficient in generating maximum discharge. This phenomenon can be explained by the continental properties of Polish climate and relief of the country with the dominating lowland except for southern border with Sudety and Carpathian mountains.

The highest flood indexes in Poland are characteristic for Carpathian tributaries of the upper Vistula river, such as Soła/Żywiec gauge $k = 4.31$, Dunajec/Nowy Sącz gauge $k = 4.31$ (Table 14.1). Comparison of maximum values of k index of Polish rivers and rivers listed in the World Catalogue of Maximum Observed Floods shows that Polish territory has much lower flood potential. For example, highest flood potential of the catchments having area around 1000 km^2 in Poland is equal $k \approx 4$ while for the similar size catchments of the rivers listed in the World Catalogue of Maximum Observed Floods $k \approx 6$.

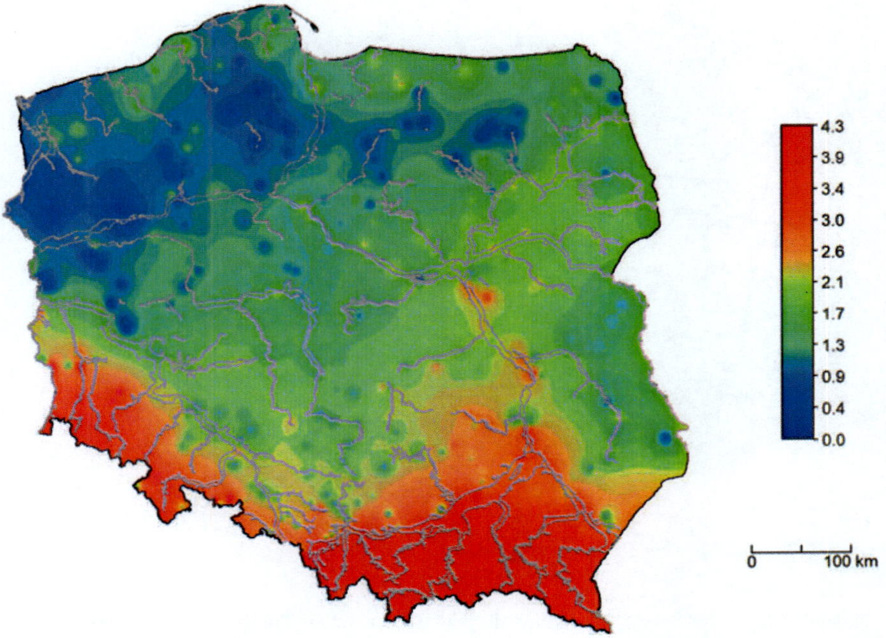

Fig. 14.4 Distribution of flood potential index *k* in Poland on major rivers with gauge control

Table 14.1 Maximum values of flood potential index *k* of Polish rivers

Gauge	River	*WWQ*	*A*	Index *k*
		m^3s^{-1}	km^2	
Żywiec	Soła	1250	784.8	4.31
Nowy Sącz	Dunajec	3300	4341	4.31
Łabowa	Kamienica	281	65	4.26

The lowest value of flood potential index was calculated for Noteć/Krzyż gauge where $k = 0.18$. This lowland river is controlled by the cascade of weirs build in XIX century for inland transport purposes. Low values of index k are characteristic for catchments with large natural storage in the lakes, for example, Pisa/Pisz $k = 0.31$, Brda/Swornegacie $k = 0.43$.

Distribution of the flood potential index can be compared with the map of historical floods range map created by merging two maps of ISOK program and PIG map [19] of the inundation area (Fig. 14.5). ISOK program produced the map of historical floods and floods of 1% probability magnitude. Map [19] shows the range of potential floods taking in to account coverage of Holocene fluvial deposits at Accurate Geological Map of Poland in scale 1:50,000. It is visible that the area of valleys exposed to high flooding potential is high at the upper reach of the Vistula river and Odra river. High

Fig. 14.5 Map of the flood range in the valleys of major rivers in Poland after ISOK program [18, 19]

flood potential index values are characteristic also for middle reach of the Vistula river. Vistula river tributaries draining footslopes of the Carpathian mountains transfer their flood potential down the main river.

14.4 Conclusion

Flood potential index *k* is useful non dimensional parameter which can be used for comparison of different river catchments. Together with the map of flood range, it shows an exposition to flood in different geographical regions of Poland. To analyze the flood risk, it should be added a probability of the flood. Flood potential is highest in the Carpatian mountains and is transferred down the Vistula river. Value of *k* index gets smaller with the increase of catchment area. This can be explained by the fact that small mountain catchments are exposed to highest torrential rains from convective clouds, and by the fast response of the catchment to rain. As catchment gets bigger, the flood potential index is smaller due to the influence of flood wave transformation in the river channel and the storage on the floodplain. Flood potential index k can be used for comparison of catchment reactions to rain in different regions. The pattern of flood potential index *k* in cartographic representation can by used for comparison of historical versus projected floods magnitude. It has been obtained similar pattern of flood potential and flood frequency [6]. The lowland part of the Poland has lowest flood potential and also floods are less frequent than in the mountains. Important for flood risk management is the observation that major rivers like Odra and Vistula transfer the high flood potential from their upper catchement tributaries to the middle reach. This is a feature that is relevant in flood protection designs for instance in case of City of Warsaw. The major floods in Warsaw have been created by the high runoff at the Vistula river mountain tributaries draining foot slopes of the Carpathian mountains. The flood wave propagates down the middle and lower reach of the Vistula river and it can lower its peak flow only by the flood transformation due to natural river valley storage.

Lowland rivers like for example Bug and Narew (major tributaries of the Vistula river) have a run-off regime controlled by the snow melt [20]. This makes the difference between flood index calculated for rainfall induced floods (years 1958 and 1970) where $k < 2$ and snowmelt induced flood (year 1979) with $k > 2$. This is important feature especially in the context of expected climate change. If in the winter time there will be more rains than snow-falls the flood potential will be lower. The catastrophic flood of 1979 was induced by rapid melting of accumulated large masses of snow. This sudden release of water accumulated in the snow-cover set an absolute maximum of discharge at many lowland rivers in Poland. Rainfall induced floods will be dominating at the mountains while lowland catchments will react less abruptly due to their natural storage capacity, higher than in the case of mountain catchments.

14.5 Recommendations

This study has shown spatial difference in flood potential index on the territory of Poland. This measure could be applied also for other geographical areas and could be useful metric to compare for example EU countries. In case of Poland map of flood potential index was used in the preparation of flood risk management plans for the Vistula river. It has been shown that floods generated in the Carpathian mountains propagate down the middle reach of the Vistula river. It means that the Vistula's river a middle course can propagate the floods with the magnitude characteristic for mountain tributaries. To get a consistent results it is recommended to use the common length of maximum discharge records for all analyzed hydrological gauges. Presented method of flood index calculation can be used for making plans of flood risk management. Index is showing in a quantitative way of the catchment's potential to produce maximum discharge at the flood peak.

References

1. Mudelsee M, Börngen M, Tetzlaff G, Grünewald U (2003) No upward trends in the occurence of extreme floods in central Europe. Nature 425:166–169
2. Radziejewski M, Kundzewicz ZW (2004) Detectability of changes in hydrological records. Hydrol Sci J 49(1):39–51
3. Dobrowolski A, Mierkiewicz M, Ostrowski J, Sasim M (2010) Regiony polski najbardziej zagrożone powodziami katastrofalnymi. In: Magnuszewski A (ed) Hydrologia w ochronie i kształtowaniu środowiska. Monografie Komitetu Inżynierii Środowiska PAN, vol 69. pp 55–69
4. Alfieri L, Burek P, Feyen L, Forzieri G (2015) Global warming increases the frequency of river floods in Europe. Hydrol Earth Syst Sci 19:2247–2260
5. EEA (2016) What is the trend in river floods across Europe?
6. Dorzecze Wisły – monografia powodzi maj-czerwiec 2010 (2011) In: Maciejewski M, Ostojski M, Walczykiewicz T (eds) IMGW, Warszawa
7. Gupta VK, Waymire EC (1998) Spatial variability and scale invariance in hydrological regionalization. In: Sposito G (ed) Scale dependence and scale invariance in hydrology. Cambridge University Press, Cambridge
8. Baker DB, Richards RP, Loftus TT, Kramer JW (2004) A new flashiness index: characteristics and applications to midwestern rivers and streams. J Am Water Resour Assoc (JAWRA) 40(2):503–522
9. Kupczyk E, Suligowski R (1997) Statystyczny opis struktury czasowej opadów atmosferycznych jako elementu wejścia do modeli hydrologicznych. In: Soczyńska U (ed) Predykcja opadów i wezbrań o zadanym okresie powtarzalności Wyd. UW, Warszawa, pp 17–82
10. Kupczyk E, Suligowski R, Kasprzyk A (2005) Typowe warunki meteorologiczne pojawiania się wysokich opadów i wezbrań rzek zachodnich Beskidów i środkowych Sudetów. In: Bogdanowicz E, Kossowska-Cezak U, Szkutnicki J, Monografie S (eds) Ekstremalne zjawiska hydrologiczne i meteorologiczne. IMGW, Warszawa, pp 131–152
11. O'Connor JE, Grant GE, Costa JE (2002) The geology and geography of floods ancient floods, modem Hazards: principles and applications of paleoflood hydrology water science and application volume 5, American Geophysical Union. pp 359–385
12. Stachy J, Fal B, Dobrzyńska I, Hołdakowska J (1996a) Wezbrania rzek polskich w latach 1951–1990. Gospodarka Wodna 9:261–268
13. Stachy J, Fal B, Dobrzyńska I, Hołdakowska J (1996b) Wezbrania rzek polskich w latach 1951–1990. Gospodarka Wodna 10:296–301
14. Rodier JA, Roche M (1984) World catalogue of maximum observed floods. IAHS Publ. no. 143
15. Bartnik A, Jokiel P (2012) Indeksy powodziowości (Francou-Rodiera) i indeksy wysokiej wody w Karpatach i na nizinach, w przekroju wieloletnim. Gospodarka Wodna 5:204–208
16. Magnuszewski A, Porczek M (2015) Wskaźnik potencjału powodziowego i względna ekspozycja na niebezpieczeństwo powodziowe gmin w Polsce. Pr I Stud Geograficzne 57:55–65
17. Atlas posterunków wodowskazowych dla potrzeb Państwowego Monitoringu Środowiska (1996) Państwowa Inspekcja Ochrony Środowiska. Warszawa
18. MPHP – Mapa Podziału Hydrograficznego Polski (2010) IMGW, Warszawa
19. Mapa obszarów zagrożonych podtopieniami w Polsce (2007) Państwowy Instytut Geologiczny, Warszawa
20. Bajkiewicz-Grabowska E, Markowski M, Golus W (2017) Polish rivers as hydrographic objects. In: Korzeniewska E, Harnisz M (eds) Polish river basins and Lakes – Part I. The handbook of environmental chemistry book series, vol 86. HEC, Springer

Chapter 15
Flood Risk Management System in Poland

Renata Graf

Abstract Flood, as a hydrological phenomenon with socio-economic significance, is classified among the most dangerous and costly extreme occurrences in the world and Europe. Because of the growing flood risk recorded in the European Union countries, in 2007, there was implemented the Floods Directive on the assessment and management of flood risks. The new flood risk management system in Poland, based on a technical protections system and non-technical flood prevention methods includes preliminary risk assessment, preparation of flood hazard and flood risk maps, and risk management plans. Preliminary assessment involves the identification of places with significant flood risk, which is determined by analysis of the immediate impact of the flood on human life and health, and various areas of economic activity with the infrastructure, as well as the effectiveness of existing flood control structures. Flood risk management plans are a planning document delineating actions which are to realise the main flood risk management objectives. The models dealing with flood risk management implemented in Poland, in connection with the Floods Directive, represent the strategies: "Moving flood away from the people", "Moving people away from flood" and "Learning to live with flood". The first strategy is altering flood to meet the requirements of human activity, the second accommodate human economic activity to flood and the third proposes reducing an area's flood and inundation susceptibility during a flood. Considerable significance is now attributed to the so-called non-technical methods, organisational-legal and planning measures, with on the other hand stressing the efficacy of comprehensive flood protection solutions, which are adapted to the hydrological conditions and the extent of land development of catchments. Comparison of elements of flood risk management strategies and the Polish experience in flood loss reduction has shown that a reasonable avenue is a compilation of technical and non-technical actions, and selection of an optimal one, which will make possible the realisation of the directive's guidelines.

R. Graf (✉)
Department of Hydrology and Water Management, Institute of Physical Geography and Environmental Planning, Adam Mickiewicz University, Bogumiła Krygowskiego 10, 61-680 Poznań, Poland
e-mail: rengraf@amu.edu.pl

M. Zeleňáková et al. (eds.), *Management of Water Resources in Poland*, Springer Water,
https://doi.org/10.1007/978-3-030-61965-7_15

Keywords Flood risk · EU floods directive · Management · Flood hazard map · Flood risk map · Risk management plan · Mitigation · Poland

15.1 Introduction

Extreme natural phenomena are the cause of many disasters, and the greatest on the world scale is water-related hazards [1–3]. At present, there is observed an increase of the incidence and magnitude of floods, to which contributes the reduction of catchment retention, increasing the exposure and susceptibility of an area to the risk of flood occurrence and the rising number of people inhabiting flood-prone areas [4–7]. Susceptibilities arise and increase for many reasons, including population growth, urban development in risk-prone locations, land-use changes, environmental degradation, weak governance, and climate change. Flood, as a hydrological phenomenon with socio-economic significance [8] is classified among the most dangerous and costly extreme occurrences in the world accounting for over 43.4% of total catastrophic events (for 1998–2017, source: United Nations International Strategy for Disaster Reduction UNISDR) – Fig. 15.1.

Observed increases in heavy precipitation and extreme coastal water levels have increased the risk of river and coastal flooding in many European regions. The number of large inland floods in Europe has been increasing since the 1980s. Since 2000 river and coastal flooding have affected many millions of people (source: The European Environment Agency). Because of the growing flood risk recorded in the European Union countries [9, 10], in 2007 there was implemented the Floods Directive on the assessment and management of flood risks [11]. The analysis of the criteria

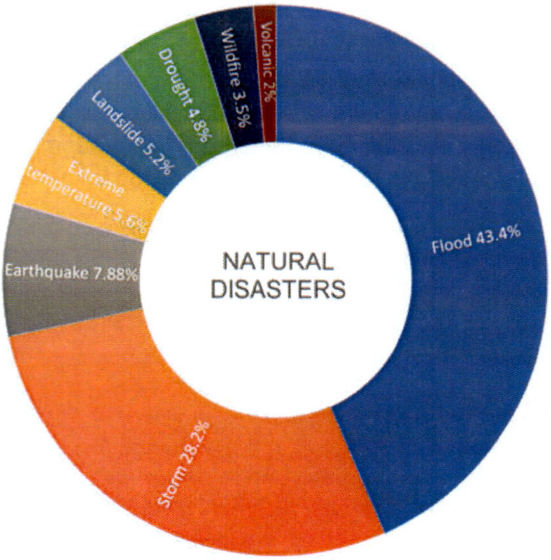

Fig. 15.1 Percentage of occurrence of natural disasters by disaster type (1998–2017) (*source* of the data: UNISDR/CRED https://www.unisdr.org/2016)

and methods of flood risk identification includes a definition of the flood, which is interpreted as a hydrological event with a specific origin, magnitude, duration and active coverage [12]. During the flood water in natural watercourses, water reservoirs, channels or sea after exceeding the bank level floods river valleys, depression areas and causes a hazard to persons and property [13]. According to the Directive [11], a flood is an instance of water temporarily covering an area which normally is not covered by water, caused by rivers and mountain streams, and storm floodings in coastal areas.

Flood risk is interpreted as a measure of the extreme event state that leads to a threat or loss. It is a combination of risk, that is flood occurrence probability, as well as potential losses, which are the sum of exposure and susceptibility, resulting from natural and socio-economic circumstances. Previous flood prevention models were mainly based on a technical protections system whose purpose was primarily to reduce flood wave culmination, delay it or confine the flood waters (flood control reservoirs and levees), and immediate warning and response at the time of flood occurrence. The new risk management system, also based on non-technical flood prevention methods [14, 15], includes preliminary risk assessment, preparation of flood hazard and flood risk maps, and flood risk management plans [11].

A vital stage of risk management in the event of a flood, mainly in terms of reducing the consequences of its occurrence: the hazard, damage and losses, beside flood protection, is also flood prevention and mitigation [7, 11, 16–19]. Prevention of flood risks focuses on the design of flood management schemes taking into account economic development and climate change [20–23]. Flood risk mitigation (reduction) as a management phase includes actions aimed at decreasing of risk-generating threats and protection against adverse effects [11]. It requires another type of measure, taking into account spatial planning: development of new urban areas, flood proofing and compartmentalisation and dedicated protection of vital infrastructure [15, 24–26]. Mitigation, as a risk management phase, meaning efforts to lower the probability of flood hazard occurrence and minimise the consequences of a flood situation, is implemented primarily in the time of 'dormant public vigilance' between floods. It is defined by a number of mechanisms, so-called mitigators, used to lower the probability of a threat occurring and reduce the after-effects of a flood situation [11]. In analysing it, of help are evaluations of area flood susceptibility and the ability to adapt to changes resulting from the flood risk increasing under the influence climatic and anthropogenic factors [3, 4, 10, 27–30]. The models dealing with flood risk management implemented in Poland, in connection with the Floods Directive, represent the three strategies. The strategy "Moving flood away from the people" is altering flood to meet the requirements of human activity and the next strategy "Moving people away from flood" accommodate human economic activity to flood. In turn the strategy "Learning to live with flood" proposes reducing an area's flood and inundation susceptibility during a flood.

15.2 European Climate Change Projections and Flood Risk

The European continent is characterised by a varying susceptibility to climatic effects and the ability to adapt to its changes [1, 7, 10]. This is expressed in the hydrological characteristics, whose primary determinant is the growth of the incidence of extreme events: floods and droughts [31, 32]. The climate change mechanism is related to multiple factors, among which of key significance are: solar radiation fluctuations, Earth orbit movement shifts in a time scale of tens of thousands of years, oscillations in the ocean–atmosphere system (ENSO, NAO, AO, AMO) and changes of the chemical composition of the atmosphere and the Earth surface properties [6, 33]. The last two factors are the side effects of anthropogenic activity, particularly in terms of changes in and use and the increase of impermeable surfaces, which accelerates water outflow from catchments with simultaneous decrease or absence of retention which is a determinant of catchment sensitivity to climatic change [34].

Projections of climate change in Europe with a forecast until the end of the twenty-first century assume changes both of air temperature and precipitation (Fig. 15.2), which may contribute to generating a greater number of extreme events and an increase of their incidence [30, 35]. The forecasts show a positive, by a maximum of 6 °C, the trend of air temperature rise, particularly in Northern and Southern Europe (the Mediterranean region) and a characteristic duality of the European territory with respect to precipitation. The northern part of Europe is a region with a projected precipitation growth of more than 30%, whereas the southern part

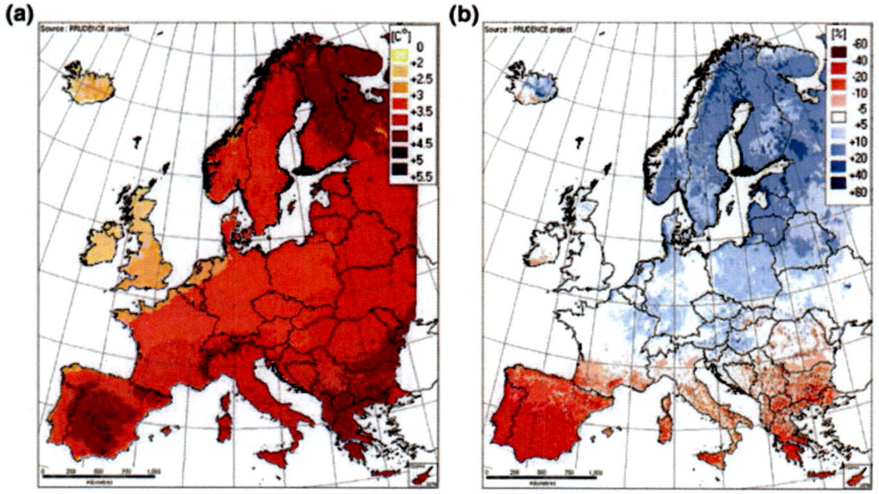

Fig. 15.2 Absolute change in mean annual temperature **a** and mean annual precipitation **b** between control period 1961–1990 and 2071–2100, under the IPCC SRES scenario A2. Data have been provided through the PRUDENCE data archive, funded by the EU through contract EVK2-CT2001-00132 [https://ec.europa.eu/jrc/en/peseta/peseta-i-results/change-mean-annual-temperature-and-precipitation-end-century]

will be characterised by a precipitation shortfall (over 30%) in relation to normal precipitation.

In climate change forecasts for Poland, there is observed an increase of the mean annual air temperature, however, there are not seen any distinct changes in the area of precipitation [36, 37]. According to E. Żmudzka [38], precipitation in lowland Poland did not show any significant direction of change in the second half of the twentieth century. The results of analyses and projections show that climate change in Poland will be moderate, both in the short-term as well as a long-term perspective, and will involve: precipitation volumes, maximum flows, and extreme event incidence. In the short-term (2011–2030), a moderate precipitation growth is expected in Poland (up to 10%), whereas in the perspective of the period 2081–2100, up to about 15% [39]. European projects [40] predict for the period 2070–2100, among others, an increase of more than 10% in maximum flows with a probability of exceeding 1% in upper and medium Vistula and its tributaries, and in the Oder along its entire course we well as in its tributaries up to the Warta estuary.

Models prepared under the European ESPON Climate Project [41], dealing with climate change impact in the area of susceptibility and ability to adjust to the changes in particular regions, have revealed some peculiar features of the European continent. It is potentially assumed that the greatest negative impact of the climate will be recorded in Southern and Western Europe, which corresponds with the equally high susceptibility of that part of the region to climate change and generation of extreme events, also floods (Fig. 15.3). In a hypothetical scenario without climate change,

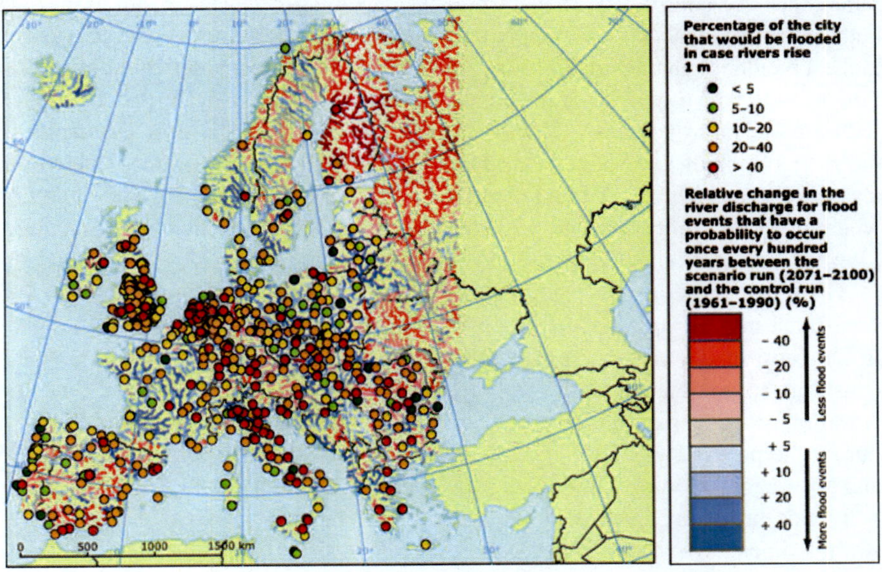

Fig. 15.3 Percentage of the city that would be flooded in case water in rivers rises 1 m (only cities >100 000 hab) and the relative change in 100-year return level of river discharge (*source* https://www.eea.europa.eu/legal/copyright)

total flood losses will continue to increase as a consequence of societal and economic factors such as an increase in exposure and vulnerability [42]. Flood disasters in Europe increased in number and amount of loss from the 1970s to the 2000s.

The model defining the total ability of an area to adapt to climate change indicates a high potential in this regard of the regions of North-Eastern and Central Europe, except the areas of Eastern Europe and the Mediterranean. The gauges taken into account in the models: the aggregate potential impact of climate change, the overall capacity to adapt to climate change and the potential vulnerability to climate change are an expression of compiling natural environment characteristics with socio-economic circumstances.

15.3 Diagnosis of the Flood Situation in Europe and Poland

A flood is an event localised in a specific time and space [43], and its extent is dependent on the 'entry state', that is the cause, the sequence of changes and the 'exit state' in the form of the consequences and after-effects – hazards, casualties, displacements, damage or synergic events. The determinant of the magnitude, coverage, and recurrence of floods is the probability of exceeding culmination flows, which for common floods is 10%, great floods – 1%, and catastrophic – 0.2% or 0.5%.

An analysis of observation series dealing with flood events in Europe and Poland (Dartmouth Flood Observatory), revealed that floods are is caused primarily by heavy rains and, increasingly more often, by brief, heavy rainfalls. The floods duration range between several days up to a month, but 9 days on average, most often 5 days. The highest incidence characterise floods with a probability of exceeding the culmination flow of 5–10%. An analysis of the number of floods occurring in Europe, correlated with long-term projections [44] showed that they occurred the most frequently in the river basins of the Rhône, the Rhine, the Danube and the Thames, which are classified among the most flood-prone areas in Europe (Fig. 15.3). The countries where great floods occurred the most frequently are Romania, the Czech Republic, Slovakia, Great Britain, Germany, and Austria.

The oldest piece of information about a flood in Poland, which occurred in 988 was found in the Długosz Chronicle [45]. Evidence of the flood threat also comes in the form of old cartographic studies, flood boards or high water signs, which constitute approximate hydrological information comprising the location, time and elevation of a water rise culmination. An analysis of flood data recorded in Poland during the period 1946–2001 revealed 590 instances of floods with varying origin and coverage, including 15 regional catastrophic events [46].

Floods in Poland, just as in every European region, can occur every year, or several times a year, in the form of genetically varied extreme events, which is conditioned by multiple adverse factors, e.g.: intense precipitation, sudden warming resulting in rapid melting, or an unfavourable baric situation [47, 48]. Precipitation type floods occur mainly in the southern part of the country, whereas the other areas are susceptible to the occurrence of snowmelt floods, and locally in the Baltic coast area storm, floods are recorded. In the winter season, the risk of ice-jam floods is increased. Data on floods in Poland confirms that the greatest number of floods have been recorded on the upper Vistula and Oder, and their mountain tributaries (see Fig. 15.4).

Studies of the long-term flood occurrence trend in Europe have shown an increasing number of great floods (in the period 1985–2007), which overlap with a high variability of the events (Dartmouth Flood Observatory). This means that the character, intensity, and incidence of floods is dependent mainly on the climatic and geographic circumstances, as well as human activity [49]. These factors determine the level of susceptibility and exposure of an area to flood, which, besides a threat, are also flood risk attributes.

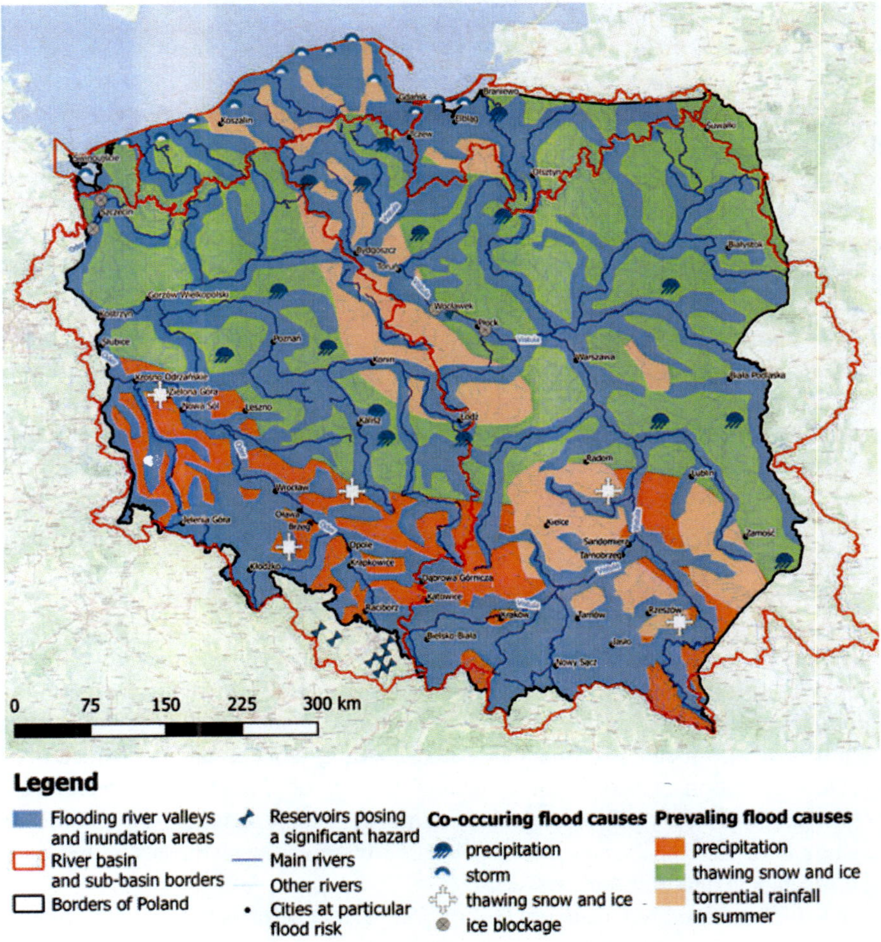

Fig. 15.4 Flood hazard in Poland (on the basis of [50])

On the global scale, an increasing flood risk level is a determinant of the multi-risk which also comprises the threat and consequences of the occurrence of extreme climatic, meteorological and geophysical events, e.g., droughts, storms, fires, slides, and earthquakes. The risk profile for Poland, prepared by the United Nations International Strategy for Disaster Reduction (*source: UNISDR/CRED*), shows that the highest human exposure is with regard to the occurrence and aftermath of drought, to a lesser extent flood, whose consequences apply to the property and persons living in the flood prone-areas, located mostly within river valleys.

15.4 Flood Risk Management in Poland

15.4.1 Criteria for Flood Risk Identification

The Flood Directive underscores that complete flood protection is impossible, flood damage and losses only being able to be reduced in their severity and that instead of reacting during a flood, flood risk should be controlled and managed in advance [11]. According to Kaczmarek [51], the risk is the measure of the condition of an extreme event which leads to risk or loss. The risk means the measure of a threat or hazard, which results from probable events independent on human activity or from possible consequences of making a decision [52, 53]. Flood risk in the meaning of the directive is "a combination of the probability of a flood event and the potential adverse consequences to human health, the environment, cultural heritage, and economic activity", and the goal of risk management is the reduction of the risk occurrence probability and losses due to flood by introducing appropriate mechanisms and instruments. The character of each type of risk is determined by three basic factors: the event, that is risk occurrence circumstances, probability, and consequences. The flood hazard problem becomes particularly important, especially during a flood occurrence when interest in flood protection is increased.

Flood risk is a combination of flood occurrence probability and potentially loses, which are the sum of exposure and susceptibility resulting from natural and socio-economic circumstances. Exposure means the state of land development of a threatened area, defined by the structures, infrastructure, and communities present in a threatened area [54]. Whereas flood susceptibility understood as the resistance of infrastructure located in inundation and flooding-prone areas, is characterised by the preparation of people and structures for a flood. Flood risk increases as a consequence of heavy meteorological conditions. Also, it increases due to reduced catchment retention and improper floodland development, which leads to increased hydrotechnical and economic risk [55].

The highest flood risk category (I – very high risk) pertains to a threat to human life, flooding of residential and utility buildings, and damage to motorways and main bridges. In spatial terms, it is assigned to embanked rivers in cities and areas being urbanised as well as built-up river valleys without protection. Moderate flood risk (category II) is related to losses mainly in agriculture (crop destruction) and on local thoroughfares (roads, bridges). This risk category pertains to river valleys used for agriculture or not provided with flood protection and floodland without agriculture and industrial infrastructure. Low flood risk (category III) is recorded when economic losses are insignificant and involve mainly local damage to roads and river infrastructure. Potentially, flood risk could reach this level in unprotected river valleys and meadows and pastures.

15.4.2 *Stages of Flood Risk Management*

Flood risk management is viewed in the aspects of an event, reaction and response, recovery- assessment, mitigation, and warning. This system encompasses actions aimed at alleviating hazards which generate risk and preventing their adverse outcomes [11, 56] – Fig. 15.5.

A vital management stage is risk mitigation, involving many aspects of minimising the effects of a flood situation, which is conducted both in the area of prevention as well as reduction of the negative impact of events. The former aspect pertains to implementing a systemic approach to management, which is aimed at the identification of potential hazards and development of procedures involving minimisation of the probability of risk manifestation, whereas in the latter area there are formed plans to address the risk.

Fig. 15.5 The cycle of flood risk management (*source* of the data [11, 56])

The flood risk management strategy, implemented in Poland by the National Water Management Authority (Krajowy Zarząd Gospodarki Wodnej KZGW), at present – State Water Holding Polish Waters [57] in agreement with the competent minister for water management affairs, entails: preliminary flood risk assessment (*in polish WORP*) (Fig. 15.6), preparation of flood hazard and flood risk maps (*in polish FHMs and FRMs*). As planning documents related to the Floods Directive were source material for the preparation of flood risk management plans (*in polish PZRP*) at the level of a river basin or management unit area. The collection, processing and making available of information related to flood risk and flood protection is carried out by Operating Centres of Regional Water Management Authorities (organisational units of Polish Waters), aided by the National Information Protection System (*in polish ISOK*) [58] and the Computer Information System Supporting Flood Risk Management.

Preliminary flood risk assessment (Fig. 15.6) involves the identification of places with significant flood risk, which is determined by analysis of the immediate impact of the flood on human life and health, and various areas of economic activity with the infrastructure, as well as the effectiveness of existing flood control structures. On the basis of information on past, probable and current flood events, there are designated

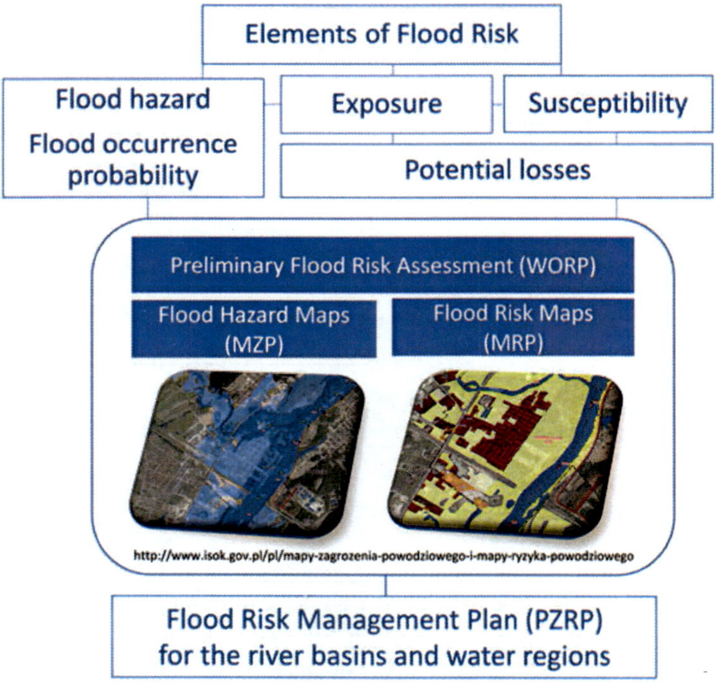

Fig. 15.6 The flood risk assessment level in Poland (on the basis of [58]); (*source* of the data: https://www.isok.gov.pl/en/preliminary-flood-risk-assessment; https://www.isok.gov.pl/en/flood-hazard-maps-and-flood-risk-maps)

so-called "Areas with Potential Significant Flood Risk (APSFR)", also referred to as "Flood Hazard-Prone Areas FHPA" (*in polish: ONNP*) – Fig. 15.7, among which in the two planning cycles there were classified in Poland 839 rivers [57]. A total of 29,435.8 km of rivers have been designated as APSFR by the preliminary flood risk assessment (WORP) review and update process. For 14,940.4 km of river sections, identified in the preliminary flood risk assessment in 2011, flood hazard and flood risk maps were developed in the first planning cycle of Flood Directive and for 13,334.4 km of the river sections (identified in 2011), maps are developed in the second planning cycle. The 1161.0 km are the river sections identified in the 2018 review and update of the preliminary flood risk assessment, for which hazard and flood risk maps will be developed in the second planning cycle of Flood Directive [59].

Areas with Potential Significant Flood Risk were designated based on detailed spatial data: a numerical terrain model, river bed cross-sections and hydrological

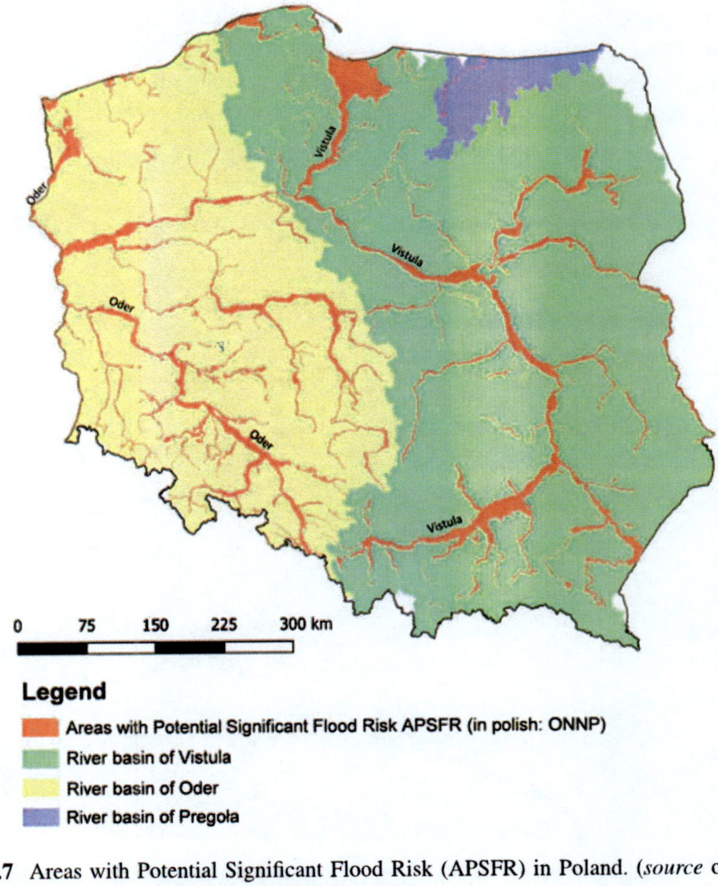

Fig. 15.7 Areas with Potential Significant Flood Risk (APSFR) in Poland. (*source* of the data: https://powodz.gov.pl/pl/worp_I_cykl_planistyczny)

data, which are indispensable for deriving a hydrological model. Maps presenting the results of the preliminary flood risk assessment were made for Poland in the sale of 1:800,000 and in province division in the scales of 1:250,000, 1:300,000 and 1:350,000 [57]. For the areas designated there were prepared flood hazard (Fig. 15.8) and flood risk maps (Fig. 15.9) show the levels of hazard and losses for every scenario of occurrence of probable water conditions. Flood hazard and risk maps are available at the Hydroportal of the National Water Management Authority [60].

The flood hazard maps cover areas with a particularly high flood hazard, where flood may occur based on one of the probability scenarios: low (0.2%), medium (1%) and high (10%). For every scenario, there is presented flood coverage, water depth, or water surface level and water flow speed (Fig. 15.8). The flood risk maps, prepared at the next stage, show potential negative consequences related to floods which could occur based on one of the probability scenarios (Fig. 15.9). By integrating the relation of risk and its consequences to the natural environment elements, land use type and development structure as well economic activity and cultural heritage, these maps make it possible to pinpoint zones of the highest flood risk. In the maps, there are included objects of high historical or natural value, and strategic objects with regard to flood-induced effects, e.g. water intakes, sewage treatment plants, and ones which could generate synergic occurrences. By analysing a map's contents, there can be indicated locations of buildings with a specific purpose and costs of potential flood

Fig. 15.8 Flood hazard map. Map fragment Poznań-Naramowice N-33-130-D-b-4; *source* [60] *Flood hazard areas – blue marking (water depth)

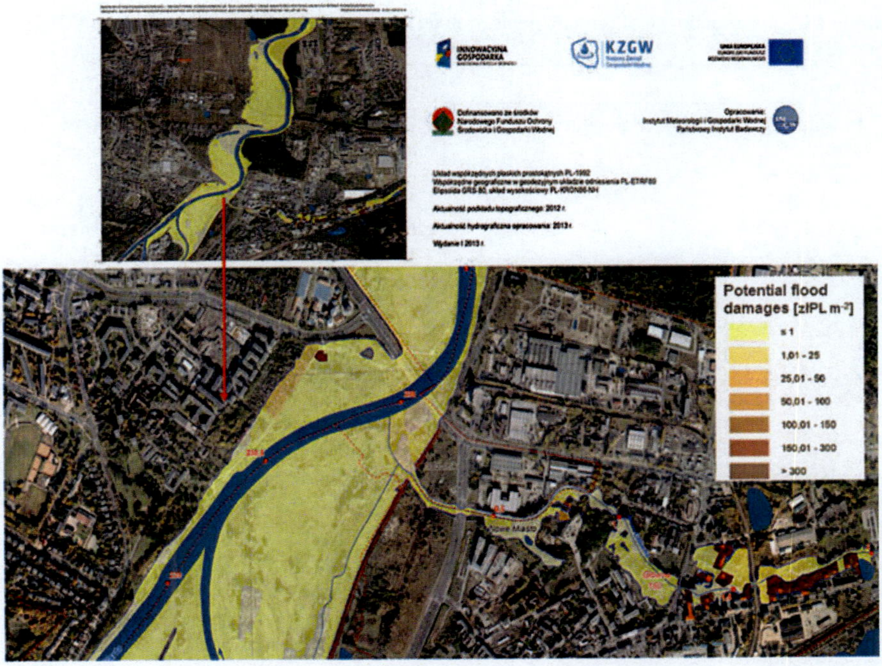

Fig. 15.9 Flood risk map. Map fragment: Poznań-Naramowice N-33-130-D-b-4; *source* [60] *Flood risk areas – colour marking (potential flood damages)

damages (Fig. 15.9), which constitutes a basis for damages assessment and receiving compensation.

By 2011 there was prepared, in the form of a cartographic description and studies, a preliminary assessment of flood risk throughout the country, divided by provinces, and by the end of 2013 – flood hazard and risk maps [60]. The spatial databases compiled were a basis for the preparation of flood risk management plans (PZRP) in river basin and water region planning areas, whose completion deadline was slated for the end of 2015 [11]. The plans for the river basins of Odra, Vistula as well as Pregoła, were adopted in 2016. Flood risk management plans are a planning document delineating actions which are to realise the main flood risk management objectives, including, e.g. threat containment (flood coverage), reduction of flood sensitivity of threatened areas and improvement of flood threat mitigation capabilities.

15.4.3 Primary Goals of Flood Risk Management

The catalogue of primary goals of flood risk management for a river basin area includes groups related to inhibiting the growth of flood risk, minimisation of existing flood risk and improvement of the Polish flood risk management system (Table 15.1).

Table 15.1 The catalogue of primary goals of flood risk management for a river basin area (on the basis of [54])

Primary goals	Groups related
1. Inhibiting the growth of flood risk	1.1. Maintaining and increasing the existing retention capacity of catchments in a water region 1.2. Elimination of, avoiding land development growth in areas with particular flood hazard 1.3. Defining the conditions for possible development of land protected with levees 1.4. Avoiding the growth of and defining the conditions for land development in areas with a low ($p = 0.2\%$) probability of flood occurrence
2. Minimisation of existing flood risk	2.1. Containment of existing flood risk 2.2. Reduction of existing land development 2.3. Decreasing the susceptibility of structures and communities
3. Improvement of the flood management system	3.1. Improvement of forecasting and warning of meteorological and hydrological hazards 3.2. Increasing the effectiveness of response of persons, companies, and public institutions 3.3. Increasing the effectiveness of rebuilding and restoration of the conditions from before a flood 3.4. Implementation and improvement of the efficacy of post-flood analyses 3.5. Building legal and financial instruments discouraging or encouraging specific behaviours improving flood safety

Source of the data: https://www.kzgw.gov.pl/index.php/pl/materialy-informacyjne/plany-zarzadzania-ryzykiem-powodziowym

Achievement of flood risk management objectives leads to determining flood risk reduction indices, which is a result of compiling two effects: identifying measures in flood risk management plans and projected measures in future versions of the plans (update of plans). Identified non-technical measures (natural flood protection) in a catchment/river basin indicate the need to employ risk quality and quantity reduction estimation methods, whereas with identified technical measures determining risk quantity reduction can be done by hydraulic modelling. In the area of future interventions, there is stressed the need to utilise the appropriate methodology of evaluating future flood control projects, projected in the course of updating the flood risk management plans.

The documents of strategic and overriding significance for all national and regional sector plans and programmes which existed in Poland until the implementation of flood risk management plans, under which there were projected measures or investments influencing the condition of water resources and water conservation objectives resulting from the Framework Water Directive [61], were MasterPlans. They were prepared for river basin areas, and they included measures in the sectors of water management, flood protection, hydropower engineering as well as inland and sea navigation. MasterPlans detail the attributes and level of multi-risk with regard to: the flood hazard and risk level, the risk of not meeting strategic (environmental) objectives, the extent of pressure (human impact) on the condition of surface and ground waters, and the risk of disturbing the water balance, particularly with respect to the utilisation of groundwater resources. In Poland and other EU countries there are expected the best results in terms of simultaneous achievement of flood risk management objectives, with taking into account their priority (Floods Directive) and environmental goals (Framework Water Directive).

15.5 Flood Risk Management Strategies

Flood risk is a function of the threat, exposure and sensitivity to flooding, which determine the character of contemporary strategies of its mitigation – "Moving flood away from the people," "Moving people away from flood" and "Learning to live with flood" (Fig. 15.10). Flood risk minimisation means the reduction of the probability of maximum flow as well as losses and social sensitivity to a crisis situation. Before the implementation of the Floods Directive [11] in Poland, measures were taken in this

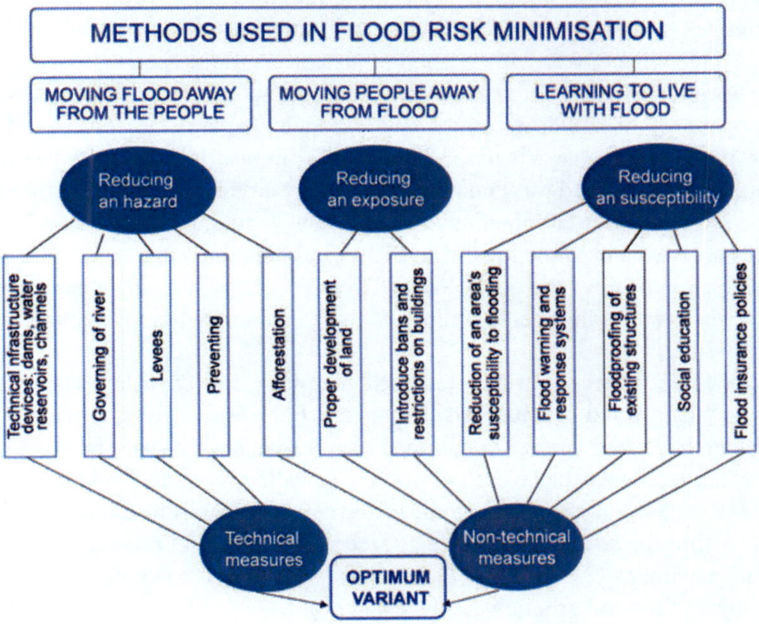

Fig. 15.10 Technical and non-technical methods used in flood risk minimisation (on the basis of [54, 62, 63], partially modified)

regard through flood protection strategies, which, assigned until a flood occurrence, were based on technical precautions through a system of flood control structures and devices as well as immediate warning and response.

Most floods are caused by lowering valley retention with levees and increasing maximum flows, or making the cross-section smaller and decreasing the river bed flow carrying capacity, which further raises the maximum water levels. Levees which protect flood-prone areas used by people during flood are often broken and destroyed on stretches of many kilometres, which limits the possibility of controlling the course of a flood and minimising the extent of the disaster with a technical flood protection system (Fig. 15.10). Potentially, there is a risk of a higher rise, for which there are designed facilities protecting against flood, and as a consequence, water overflowing the levee or breaking it altogether. Too low flow carrying capacity of inter-levee zones in urban areas, caused, e.g., by a low levee, without any possibility of making it higher, requires the introduction of additional protections in the form of a flood relief channel or polder upstream of the city. A flood protection system with a similar scope and components whose purpose is to lower the flood wave is planned for the city of Cracow [64].

The strategy "Moving flood away from the people" (Fig. 15.10) is a continuation of the traditional, technical mode of threat containment and flood protection. It is implemented chiefly through: water storage reservoirs, levees, dry flood control reservoirs, polders, flood relief channels, flood and storm gates, as well as measures aimed at increasing natural retention (e.g. afforestation). Flood protection in a river valley gives rise to the possibility of settling multiple areas, which, however, are characterised by a high risk of flood occurrence, thus jeopardising society with great losses.

The technical direction of protection entails eliminating or reducing flood damage by the operation of a hydrotechnical system. An example is the complex of water storage reservoirs in the Vistula Basin, whose purpose is outflow reduction and maximal spacing of flood wave culminations. The reservoir functioning in the Goczałkowice system delays culminations on the Small Vistula, the Soła Cascades reservoirs distance Soła culminations from Skawa culminations, in a similar manner as the Świnna Poręba reservoir on the Skawa River, which, by stopping about 60 million m^3 of water even during the 2010 flood, delayed a culmination of the wave on the Vistula [64].

To flood rise reduction, besides technical measures, also contributes to soil retention, which is lowered in urbanised catchments. For a leveed river, maximum flows are regulated by active surface retention, which determines the space between levees, whereas the land sides of levees provide for so-called 'confined' retention. Proper management of urban space, biotechnical land development by forestation or limiting surface sealing in urbanised areas (e.g. green car parks) and rational precipitation water management can redound to increasing the active catchment surface area regulating surface and groundwater levels [22, 65].

The unreliability of many technical infrastructure devices recorded during the recent years' floods [46], and the manifestation of the so-called apparent safety syndrome in a leveed valley, have indicated to the need to introduce another flood risk minimisation strategy, "Moving people away from flood" (see Fig. 15.10). This conception involves adapting human economic activity to flood mainly by new legal regulations provided for in spatial planning, which are related to forbidding construction and engaging in any activities in flood-prone areas. Among non-technical methods of flood risk mitigation are included legislative, administrative and social actions as well as the insurance policy, which are aimed at reducing potential threat to leveed and non-leveed areas in a river valley, mainly by altering its development pattern and reducing the so-called flooding-sensitive development.

Preparation of flood risk and hazard maps imposes on local governments the obligation to include areas indicated on maps in local zoning plans and introduce bans and restrictions on construction in those areas. Comparison of elements of flood risk management strategies and the Polish experience in flood loss reduction has shown that a reasonable avenue is compilation of technical and non-technical actions, and selection of an optimal one, which will make possible the realisation of the directive's guidelines over a water region, and not only within a single catchment.

Reducing an area's flood exposure by indicating floodland development possibilities in a manner not sensitive to consequences of flooding (e.g. parks, recreation areas) is now considered among the most effective ways of mitigating potential flood damage. It is also an alternative (ecological) measure for a technical flood protection system which has a negative impact on the natural environment condition. Within engineered and leveed river valleys can be observed multidirectional transformations of flow conditions (increased water speed and depth in the inter-levee zone) and water outflow from the catchment. Changes also include hydromorphological characteristics (straightening of a river's course, clearing tall vegetation in river bed), ecological conditions (isolation of valley ecosystems, disrupting the functioning of habitat corridors) and landscape features (removal of wetlands and oxbow lakes). According to the guidelines of the Framework Water Directive [61], these elements are used to classify the ecological condition, which describes the quality of the structure and the functioning of ecosystems connected with surface waters, and to determine some surface waters which are artificial or highly altered as a result of human activity, whose ecological potential is being estimated. It is emphasised that negative effects of flood control investments, constituting a factor hindering the achievement of the fundamental environmental objective which is a good ecological condition or potential of part of surface waters, should be minimised, and environment-friendly measures of flood prevention, for full protection, supported with technical methods [66].

A lot of damage and losses are recorded during and after a flood, which is why an immediate warning and response system is activated throughout. The strategy "Learning to live with flood" (Fig. 15.10) involves reduction of an area's susceptibility to flooding and inundation during a flood through direct protection as well as emergency damage removal and loss minimisation. Methods of reducing flood susceptibility of land and society include, among others: flood warning and response systems, floodproofing of existing structures, and flood insurance policies.

15.6 Conclusions

Among the legal and environmental regulations currently in force in Poland, there is sought an answer to the questions of effective flood risk minimisation methods, with the growth in recorded flood occurrences, especially since the 20th c.[43], and of what actions are presently indispensable or, 'optimal,' to make society safe from the flood. The adage *"Those who live close to water, let them prepare for loss"* [67] can be related to the past and the present. This olden form of warning against flood disaster is also reflected in the contemporary system of flood risk assessment and management, whose guidelines are defined for the EU countries under the so-called Floods Directive [11]. The efficacy of these measures should be based on proper 'best practices' and 'best technologies available' which do not generate excessive costs. Technical methods, utilised so far in many countries in order to alleviate flood consequences, have proved inadequate or negatively affecting the natural environment. As per the execution of environmental objectives under the Framework Water Directive [61], flood protection activities in river valleys should be limited due to their deleterious impact on water ecosystems [68].

Considerable significance is now attributed to the so-called non-technical methods, organisational-legal and planning measures, with on the other hand stressing the efficacy of comprehensive flood protection solutions, which are adapted to the hydrological conditions and the extent of land development of catchments and include technical measures (where reasonable) as well as non-technical ones (e.g. increasing the catchment retention). In order to effectively realise the above objectives, flood risk minimisation strategies used in practice should take into account all the risk categories: hazard, exposure, and sensitivity [7]. Proper development of land exposed to flood, conforming to planning documents, constitutes the imperative goal of flood risk management, and minimisation strategies in the European Union countries (FD, 2007/60/EC). With the functioning of now advanced flood forecasting systems, utilising projection results, hydrological modelling, and monitoring, it is possible to minimise the risk and consequences of floods and properly administer flood protection measures.

The development of the flood risk management plans under the Floods Directive [11] and the "Water management plans for river basins" under the Framework Water Directive [61] is part of an integrated system of water management in river basins in Poland. Both these processes should utilise the potential of interaction and mutual benefits with taking into consideration the environmental goals, ensuring effectiveness and prudent use of water resources.

15.7 Recommendation

(a) As regards flood risk management, it is necessary to work out measures for coordinating both flood protection and flood prevention, and also for mitigating the effects of floods.
(b) In order to reduce the probability of occurrence of the risk and limit the effects of a flood situation, it is indispensable to assess the susceptibility of an area to flooding, and also its capacity to adapt to changes resulting from an increase in flood risk due to climatic and anthropogenic factors. Models determining an area's capacity to adapt to climate change should take into consideration indices based on a compilation of features of the natural environment and of socio-economic conditions, for example the aggregate potential impact of climate change, the overall capacity to adapt to climate change, and the potential vulnerability to climate change.
(c) Due to the increasing frequency of occurrence of floods and the spatial expansion of urbanised areas, it is necessary (particularly for municipal catchment areas) to update flood hazard and risk maps and flood risk management plans.
(d) An important requirement concerns the implementation of a flood risk management system that would integrate strategies aimed at distancing floods from people and people from floods, and also teaching people how to live with floods. The objective of such an approach would be to support non-technical methods of protection against floods (legislative, administrative and social measures, and a broad insurance policy) with technical methods.
(e) The elaboration of a flood risk management plan obligates local self-government authorities to take flood prone areas and areas exposed to risk into consideration in local zoning plans, and further to impose prohibitions and limitations on development conducted in these areas.

Acknowldgement The author thanks the State Water Holding Polish Waters – National Water Management Authority for sharing materials and consent to their publication.

References

1. Evans EP, Ashley R, Hall J, Penning-Rowsell E, Sayers P, Thorne C, Watkinson A (2004) Foresight. Future flooding. Scientific summary, vol 1. Future risks and their drivers, Office of Science and Technology, London
2. Kang J-L, Su M-D, Chang L-F (2005) Loss functions and framework for regional flood damage estimation in residential area. J Mar Sci Technol 13:193–199
3. Messner F, Meyer V (2005) Flood damage, vulnerability and risk perception – challenges for flood damage research. UFZ Discussion Papers, Department of Economics, p 13
4. Berz G, Kron W, Loste T, Rauch E, Schimtschek J, Schmieder J, Siebert A, Smolka A, Wirtz A (2011) World map of natural hazards - a global view of the distribution and intensity of significant exposures. Nat Hazards 23:443–465

5. Kaźmierczak A, Cavan G (2011) Surface water flooding risk to urban communities: analysis of vulnerability, hazard an exposure. Landscape Urban Plan 103(2):185
6. Kundzewicz ZW ed (2012) Changes in flood risk in Europe, IAHS special publication 10, CRC Press, Taylor and Francis Group
7. Schanze J (2012) Dealing with future change in flood risk management. J Flood Risk Manag 5(1):1–2
8. Toya H, Skidmore M (2007) Economic development and the impacts of natural disasters. Econ Lett 94(1):20–25
9. Barredo JI (2007) Major flood disasters in Europe: 1950–2005. Nat Hazards 42:125–148
10. Lerner LA (2007) Assessing global exposure to natural hazards: Progress and future trends. Environ Hazards 7:10–19
11. Floods Directive (2007) FD 2007/60/EC, Directive of the European parliament and council of 23 October, 2007 on the assessment and management of flood risks, Official Journal L 288, 06/11/2007. pp 0027–0034
12. Ciepielowski A (1999) Basics of water management [Podstawy gospodarowania wodą]. SGGW Publishing House, Warsaw
13. Water Law Act (2001) Art. 9 item 1 pt. 10 of the Act of 18 July, 2001, An act on the amendment of the water law act and certain other acts, Official Journal 2005 No. 130 item 1087
14. Burton AJR, Shepard MA, Riddell KJ (2003) Land use and flood risk through catchment flood management plans. Water Environ J 17(4):220
15. Van der Hoeven E, Aerts J, Van der Klis H, Koomen E (2009) An Integrated Discussion Support System for new Dutch flood risk management strategies. In: Geertman S, Stillwell JCH (eds) Planning support systems: best practices and new methods. Springer, Berlin
16. DEFRA EA (2006) R&D outputs: flood risks to people. Phase 2. FD2321/TR1 The flood risks to people methodology, department for environment food and rural affairs and the environment agency, London
17. Sutanta H, Rajabifard A, Bishop ID (2013) Disaster risk reduction using acceptable risk measures for spatial planning. J Environ Planning Manage 56(6):761–785
18. Stevens MR, Shoubridge J (2015) Municipal hazard mitigation planning: a comparison of plans in British Columbia and the United States. J Environ Plann Manage 58(11):1988–2014
19. Lindsey WL, Berke P, Smith G (2014) Do planners matter? Examining factors driving incorporation of land use approaches into hazard mitigation plans. J Environ Plann Manage 57(5):792–811
20. Evans EP, Ramsbottom DM, Wicks JM, Packman JC, Penning-Rowsell EC (2002) Catchment flood management plans and the modelling and decision-support framework. Proc ICE – Civ Eng 150(5):43–48
21. Walczykiewicz T (2002) The decision priorities in the field of realization of flood protection systems [Priorytety decyzyjne w zakresie realizacji systemów ochrony przeciwpowodziowej]. Gospodarka Wodna 2:61–64
22. Plummer R, Fitzgibbon J (2004) Co-management of natural resources: a proposed framework. Environ Manage 33:876–885
23. Richert E, Bianchin S, Heilmeier H, Merta M, Seidler C (2011) A method for linking results from an evaluation of land use scenarios from the viewpoint of flood prevention and nature conservation. Landscape Urban Plann 103(2):118–128
24. Wheater H, Evans E (2009) Land use, water management and future flood risk. Land Use Policy 26(1):251–264
25. Koomen E, Koekoek A, Dijk E (2011) Simulating land-use change in a regional planning contex. Appl Spat Anal Policy 4(4):223
26. Turcu C (2012) Re-thinking sustainability indicators: local perspectives of Urban sustainability. J Environ Plann Manage 56:695–719
27. Downing TE, Butterfield R, Cohen S, Huq S, Moss R, Rahman A, Sokona Y, Stephen L (2001) Vulnerability indices: climate change impacts and adaptation. UNEP Policy Series, UNEP, Nairobi

28. Graf R, Olszewski R (2006) Use of the HYDRO database for the purposes of water resource management in critical situations [Wykorzystanie bazy danych HYDRO dla potrzeb zarządzania zasobami wodnymi w sytuacjach kryzysowych]. Ann Geomatics 4(1):85–94
29. Parry ML, Canziani OF, Palutikof JP, Van der Linden PJ, Hanson CE, red (2007) Climate change 2007: impacts, adaptation and vulnerability. Working group II contribution to the fourth assessment report of the intergovernmental panel on climate change, Cambridge University Press, Cambridge
30. EEA Report No 12/2012 "Climate change, impacts and vulnerability in Europe 2012en". https://www.eea.europa.eu/pl/themes. Accessed on 23.09.2018
31. Dankers R, Feyen L (2008) Climate change impact on flood hazard in Europe: an assessment based on high resolution climate simulations. J Geophys Res- Atmos 113:D19105
32. Feyen L, Barredo JI, Dankers R (2009) Implications of global warming and urban land use change on flooding in Europe. Water and urban development paradigms-towards an integration of engineering. In: Feyen J, Shannon K, Neville M (eds) Design and management approaches. CRC Press, Boca Raton, FL, pp 217–225
33. Lang M, Renard B, Sauquet E, Bois P, Dupeyrat A, Laurent C, Mestre O, Niel H, Neppel L, Gailhard J (2006) Climate variability and change hydrological impacts. IAHS Publ. 308
34. Arnell NW (2004) Climate change and global water resources: SRES scenarios and socio-economic scenarios. Global Environ Change 14:31–52
35. Kundzewicz ZW, Kanae S, Seneviratne SI, Handmer J, Nicholls N, Peduzzi P, Mechler R, Bouwer LM, Arnell N et al (2014) Flood risk and climate change: global and regional perspectives. Hydrol Sci J 59(1):1–28
36. Kożuchowski K, Żmudzka E (2001) Warming in Poland: scale and seasonal pattern of changes in the air temperature during the second half of the 20th century [Ocieplenie w Polsce: skala i rozkład sezonowy zmian temperatury powietrza w drugiej połowie XX wieku]. Przegląd Geofizyczny 46:81–90
37. Kundzewicz ZW, Przymusińska I (2002) The changes in the air temperature in Poland 1961–2000, Facts and interpretation [Zmiany temeperatury powietrza w Polsce w okresie 1961–2000, Fakty i interpretacja]. Wiadomości IMGW, 4: s. pp 63–72 [in Polish]
38. Żmudzka E (2002) On a variability of atmospheric precipitation over Polish lowlands in the second half of the 20th century [O zmienności opadów atmosferycznych na obszarze Polski nizinnej w drugiej połowie XX wieku]. Wiadomości IMGW, 25(46): 4, 23–37
39. Project CLIMATE "The influence of climate change on the environment, economy and society" IMGW-PIB. https://klimat.imgw.pl, Accessed on 25 Sep 2018
40. PESETA project "Projection of economic impacts of climate change in sectors of the European Union based on bottom-up analysis", European environment agency (EEA). https://peseta.jrc.ec.europa.eu, Accessed on 25 Sep 2018
41. ESPON Climate Project (2011) "Climate change and territorial effects on regions and local economies in Europe". https://www.espon.eu/Menu_Projects, https://www.espon-climate.eu, Accessed on 30 Sep 2018
42. Pielke Jr RA (2005). Making sense of trends in disaster losses. Bridges – The OST's publication on science and technology policy, vol 7. https://sciencepolicy.colorado.edu/admin/publication_files/resource-1771-2005.38.pdf. Accessed on 25 Sep 2018
43. Hirschboeck KK (1988) Flood hydroclimatology. In: Baker VR, Kochel RC, Patton PC (eds) Flood geomorphology. Wiley, New York, pp 27–49
44. Eionet-European topic centre on spatial information and analysis ETC/TE. https://sia.eionet.europa.eu/activities/announcements/2ndEIONETws. Accessed on 30 Aug 2018
45. Girguś R, Strupczewski W (1965) Exceptions from historical sources on extraordinary hydrological-meteorological phenomena in the Polish territory in the period from the 10th to 16th century [Wyjątki ze źródeł historycznych o nadzwyczajnych zjawiskach hydrologiczno-meteorologicznych na ziemiach Polskich w wiekach od X do XVI]. PIHM, Warsaw
46. Żelaziński J (2005) General news on rises and floods. Training: environment-friendly flood protection strategies [Ogólne wiadomości o wezbraniach i powodziach. Szkolenie – przyjazna środowisku strategia ochrony przed powodzią]. IMGW-PIB, Warsaw. https://www.tnz.most.org./pow/imgw_w2.htm. Accessed on 25 Sep 2018

47. Kowalewski Z (2006) Floods in Poland. Types, occurrence and the system of protection against their effects [Powodzie w Polsce - rodzaje, występowanie oraz system ochrony przed ich skutkami]. Woda-Środowisko-Obszary Wiejskie 1(16): 207–220
48. Matczak P, Lewandowski J, Choryński A, Szwed M, Kundzewicz ZW (2016) Analysing and evaluating flood risk governance in Poland. Looking for strategic planning in a country in transition, STAR-FLOOD receives funding from the EU 7th Framework programme (FP7/2007–2013) under grant agreement 308364 Document Dissemination Level
49. Kundzewicz ZW, Pińskwar I, Brakenridge GR (2013) Large floods in Europe, 1985–2009. Hydrol Sci J 58(1):1–7
50. National Spatial Development Concept 2030 (2012) Ministry of regional development, Warsaw. https://www.espon-usespon.eu/dane/web_usespon_library_files/682/national_spatial_development_concept_2030.pdf. Accessed on 30 Aug 2018
51. Kaczmarek TT (2005) Risk and risk management. An interdisciplinary work [Ryzyko i zarządzanie ryzykiem]. Difin, Warsaw
52. Chong YY, Brown EM (2001) Project risk management [Zarządzanie ryzykiem projektu]. Economic Publishing House, Cracow
53. Pritchard CL (2001) Risk management: concepts and guidance, 2nd edn. ESI International, Arlington, VA, USA
54. The methodology of development of flood risk management plans for river basins and water regions [Metodyka opracowania planów zarządzania ryzykiem poowdziowym dla obszarów dorzeczy i regionów wodnych] (2013) National water management authority, https://rzgw.gda.pl/cms/fck/uploaded/BKOP/PZRP/METODYKA-PZRP.pdf, Accessed on 23 Aug 2018
55. Ozga-Zieliński B (2010) The risk of extreme hydrological events [Ryzyko hydrologicznych zdarzeń ekstremalnych], In: Więzik B (ed) Hydrology in water engineering and management [Hydrologia w Inżynierii Gospodarce Wodnej]. Monografie Komitetu Inżynierii Środowiska. PAN, Warsaw, 68(1): 319–325
56. Disaster Risk Management DRM. https://www.drmonline.net, Accessed on 25 Sep 2018
57. State Water Holding Polish Waters – National Water Management Authority (Państwowe Gospodarstwo Wodne Wody Polskie – Krajowy Zarząd Gospodarki Wodnej KZGW). www.kzgw.gov.pl, Accessed on 30 Sep 2018
58. National Information Protection System for extraordinary hazards (ISOK). https://www.isok.gov.pl. Accessed on 25 Sep 2018
59. Report on the Review and Update of the Preliminary Flood Risk Assessment (2018) Review and update of the preliminary flood risk assessment. Task 1.3.5, Preparation of the final report on the review and update of the preliminary flood risk assessment, State Water Holding Polish Waters, KZGW, Warsaw, Poland
60. Hydroportal of the National Water Management Authority (KZGW) https://mapy.isok.gov.pl. Accessed on 26 Aug 2018
61. Framework Water Directive (2000) FWD 2000/60/EC, Directive of the European parliament and council of 23 October, 2000 establishing a framework for community action in the field of water policy, Official Journal L 327, 22 Dec 2000
62. Penning-Rowsell EC, Parker D, Harries T (2008) Systematisation, evaluation and context conditions of structural and non-structural measures for flood risk reduction. RUE Research Report No I-1, FLOOD-ERA Report for England and Scotland. Middlesex University. https://evidence.environment-agency.gov.uk/FCERM. Accessed on 30 Sep 2018
63. Graf R (2014) Mitigation of flood risk in Poland [Mitygacja ryzyka powodziowego w Polsce]. In: Magnuszewski A (ed) Monografie Komitetu Gospodarki Wodnej PAN. 20(2): 247–259
64. An evaluation of the condition of flood protection of Cracow City (2012) [Ocena stanu zabezpieczenia przeciwpoowdziowego miasta Krakowa]. District crisis management group in Cracow, Cracow City Hall, Crisis safety and management department, Cracow. https://www.bip.krakow.pl. Accessed on 27 Sep 2018
65. Ellis JB (2013) Sustainable surface water management and green infrastructure in UK urban catchment planning. J Environ Planning Manage 56(1):24–41

66. Pierzgalski E, Żelazo J (2008) The conditions and directions of flood protection [Uwarunkowania i kierunki ochrony przed powodzią]. Wiadomości Melioracyjne I Łąkarskie 1:15–20
67. Adalberg S (1894) The book of Polish proverbs, parables and proverbial phrases [Księga przysłów polskich, przypowieści i wyrażeń przysłowiowych polskich]. Warsaw, Printed by Emil Skiwski, pp 1889–1894
68. Bourblanc M, Crabbé A, Liefferink D, Wiering M (2013) The marathon of the hare and the tortoise: implementing the EU Water Framework Directive. J Environ Planning Manage 56(10):1449–1467

Chapter 16
Water Management in the Pomeranian Rivers Estuary Zone on the Background of Hydro-meteorological Conditions

Joanna Fac-Beneda, Izabela Chlost, and Alicja Olszewska

Abstract The modern hydrographic system of Pomerania has been inherited from the postglacial water system. The outflow system, transformed today as a result of erosion and accumulation of rainwater and snowmelt, overlaps with both this inherited drainage system and human activities. A key role in forming the water conditions of the Pomeranian rivers, and especially their estuary sections, is played by the main (Baltic Sea) and regional (coastal lakes) drainage bases. The temporal variability of water resources manifested in their seasonal abundance and resulting from the excessive inflow of allochthonous waters or inhibition in free flow, characteristic for the conditions of the geographical environment of Pomerania, bring flood risks. In the preliminary flood risk assessment, it was necessary to designate risk areas, i.e. those where there is a significant risk of flooding or where high risk is likely to occur. From the point of view of flood prevention, buffer and strategic investments are of great importance which include the reconstruction of old and construction of new flood embankments, engineering channels and construction of flow regulating devices, strengthening of river banks, as well as the construction of retention reservoirs and development of valley retention. No less important are the increase of retention through afforestation, counteracting water erosion in lowland areas, restoration of riverbeds and riverbanks, increasing retention in urban areas.

Keywords Water management · Poland · Pomeranian catchments · Estuary zone · Flood risk · Water pollution

J. Fac-Beneda (✉) · I. Chlost · A. Olszewska
Department of Hydrology, Institute of Geography, Faculty of Oceanography and Geography, University of Gdańsk, Bażyńskiego str.4, 80-309 Gdańsk, Poland
e-mail: joanna.fac-beneda@ug.edu.pl

I. Chlost
e-mail: izabela.chlost@ug.edu.pl

A. Olszewska
e-mail: alicja.olszewska@ug.edu.pl

M. Zeleňáková et al. (eds.), *Management of Water Resources in Poland*, Springer Water,
https://doi.org/10.1007/978-3-030-61965-7_16

16.1 Characteristic of the Pomeranian Catchments

The catchments located between the basins of the Odra and Vistula are referred to as Pomeranian [1] and cover part of the Coastal region and the South-Baltic Lake Districts [2] (Fig. 16.1, Table 16.1).

The area includes northern slopes of the Pleistocene moraine plateaus, often in the form of patches, but above all coastal lowlands, ice-marginal valleys and spit forms. This results in an extensive geological and hypsometric diversity conditioning the shaping of water supply areas, and transit and drainage of surface and underground waters. The end moraine hills are built of glacial tills and sands, and their surfaces are often cut with subglacial channels and river valleys [4]. Other depressions are melt-out basins filled with clays and silts or numerous peat-based drains with no outflow. The landscape has pronounced ice-marginal valley forms, among which the Reda-Łeba Ice-marginal Valley dominates. The genetic structure of the ice-marginal valleys is associated with the accumulation of sand-gravel materials, entirely or partially covered with a complex of organic deposits. The coastal lowlands are flat areas built of sands, elevated 2–10 m above sea level. From the north, they are bordered by a range of dunes which separate them from the sea and culminate at 56 m above sea level (Łeba Spit). There are large but shallow coastal lakes in the lowlands. In places where the plateaus reach the sea, the shore shows the cliff nature.

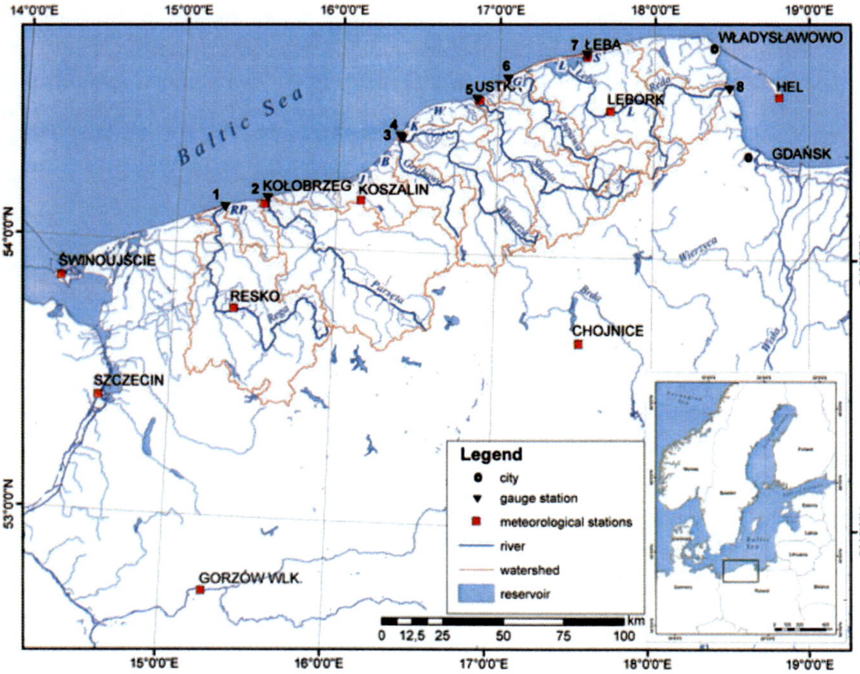

Fig. 16.1 Location of the research area

Table 16.1 Selected characteristics of the analysed rivers [3] (changed)

	River	Gauge station	Distance of the sampling point from the outlet [km]	Catchment area A [km^2]	Drainage Basin A_b [km^2]	$(A/A_b) \cdot 100$ [%]
1	Rega	Trzebiatów	12.9	2628	2724	96.5
2	Parsęta	Bardy	25.0	2955	3151	93.8
3	Grabowa	Grabowo	18.0	439	535	82.1
4	Wieprza	Stary Kraków	20.6	1519	1634	93.0
5	Słupia	Charnowo	11.3	1599	1623	98.5
6	Łupawa	Smołdzino	13.3	805	924	87.1
7	Łeba	Cecenowo	25.2	1120	1801	62.2
8	Reda	Wejherowo	20.9	395	485	81.4

The estuaries of coastal rivers form various hydrographic systems. Some of them go directly to the sea or bay (the Rega, Parsęta, Wieprza with the Grabowa, Słupia, Reda), while others to large coastal lakes (the Łupawa, Łeba).

The modern hydrographic system of Pomerania has been inherited from the postglacial water system [5]. The outflow system, transformed today as a result of erosion and accumulation of rainwater and snowmelt, overlaps with both this inherited drainage system and human activities. The river network can be described as polygenetic. Its diversity is a consequence of a complex deglaciation process, local morphological, lithological and hydrogeological conditions, climate change, contemporary vertical movements, and anthropogenic stress. One of the characteristics of the area is a large share of periodic river network [6]. It mainly concerns spring watercourses. For example, 25% of watercourses in the Łeba catchment are of this type [7]. In the characteristics of the Pomeranian catchment, lakes occupy an essential place (according to Lange 1986 the average lake coverage is 2%). Most often them are flow lakes, which is the evidence of the young age of the river network [8]. The share of lakes (lentical) along river courses, i.e., the limnic index for the Słupia and Łupawa, is about 7% [7]. On the Rega, Parsęta, and Słupia there are large artificial reservoirs that were created at the beginning of the twentieth century. Their characteristic feature is a very high water exchange rate, and the water quality in them directly depends on the quality of river waters. In addition to glacial lakes, in the mouth sections of the Łupawa and Łeba, there are large shallow coastal lakes (Gardno and Łebsko, respectively). There are small watercourses feeding the lakes, most often as canals or drainage ditches. The area surrounding the lakes has been turned into polders. The particular hydrographic attributes of the Pomeranian catchment are natural outflows of groundwater [9–11]. In ice-marginal valleys, which drain deep groundwater horizons, the surplus water that has already been part of the underground water cycle is here again included in the surface cycle. A perfect example can be a fragment of the Reda-Łeba ice-marginal valley near Lębork. In its bottom at the foot of the plateau, edge is Lubowidzkie Lake, which owes its existence to the occurrence of groundwater outflows. Their yield oscillates from just under 0.1 to about 30 $dm^3\ s^{-1}$ [12]. The outflow waters feed the lake in a significant way and quadruple the volume of the flow of the Węgorza (the tributary of the Łeba) flowing through Lake Lubowidzkie.

The catchments of the Pomeranian rivers belong to the richest in flowing water resources in Poland with the relatively stable outflow and high unit outflows [3]. The underground structure has a significant share (over 70%) of underground outflow [7, 13]. This value is much above the average for Poland. For the catchment of the Słupia, the share of underground water supply significantly exceeds the average for Poland and amounts to about 80%, while the average annual value of unit runoff exceeds $q = 10\ dm^3\ s^{-1}\ km^{-2}$ [3]. A high share of underground outflow should be attributed to good conditions of infiltration in numerous areas lacking outflow, and a large thickness of well-permeable surface rock layers. High importance is also attributed to subglacial channels, ice-marginal valleys, and deep-cut river valleys draining deep groundwater [14]. For the South Baltic Lake Districts, including the Pomeranian catchment, the hierarchical organisation of outflow is characteristic. The water cycle here depends more on the zonal factors, and less on local factors. The main river is the axis of the system; in its lower course, it has allochthonous character, while the main river's function consists in draining surplus water coming from the source catchments. Often also several catchments participate in the process of discharging water surpluses. It is the case in the Kashubian hydrographic system (ksh), the Bytów hydrographic system (bsh), and the Drawsko hydrographic system (dsh) [15].

16.2 Atmospheric Inflow

The climate of Pomerania exhibits features attributed to mid-latitudes showing the transitional nature of oceanic and continental influences. What is strongly pronounced here is the predominance of advection of moist air masses from the west, as well as instability, expressed in the variability of weather conditions all year round [16]. However, the range of the Baltic Sea influence and the morphological contrast between the raised and diversified relief of the lake district and the coastal plains are of vital importance in shaping the climate conditions of Pomerania. It is reflected in the distinct climatic regions of this area. It is assumed that the direct impact of the sea is limited to a 20–30 km wide strip of land [17]. The influence of the sea determines radiation and aerodynamic, thermal and hygrometric features of the climate. On the coast, one of the most significant amounts of solar energy in Poland is recorded, both on a yearly and summer basis (from May to August). It is favoured by the summer length of the day and relatively small convection-type overcast. The development of overcast, especially in May and June, is inhibited by the occurrence of the sea breeze circulation. As a result, in the coastal zone, the average annual real sunshine duration is 100 h longer (1700 h) than in the watershed zone separating the northern and southern slopes of Pomerania (1600 h).

Varied wind conditions are the specificity of the studied area. The wind directions and their seasonal variability are typical of the Polish Lowlands, which means that the winds with the component of the western sector (SW, W, and NW) dominate. However, there are relatively strong winds and a small number of windstill days. The dichotomy of the region regarding wind speed is marked. The coastal zone (the Koszalińskie, Słowińskie and Kaszubskie Coastlands) has a high number of days with strong and very strong wind, i.e., ≥ 10 and ≥ 15 m s^{-1}. Winds of this type appear mainly in the autumn and winter seasons and constitute over 19% of days a year. Their share decreases towards the west coast (Kołobrzeg, Świnoujście) to 12–15%. The strongest winds are recorded in Hel, Ustka, and Łeba. Inland, in the Pomeranian Lake District, especially in ice-marginal valleys, there are generally winds of lower velocity, thus the share of weak winds and wind-stagnant days increases. The incidence of days with winds exceeding the threshold of 10 m s^{-1} decreases to about 3%.

Diversification of the relief and the influence of the Baltic Sea determine the spatial distribution and fluctuations in the air temperature. The average annual temperature from the years 1951–2015 is 8.0 °C (Table 16.2) and decreases from SW to NW, ranging from 8.7 °C in the Szczecin area to 7.7 °C in Łeba and Lębork [18]. A clear drop in temperature is recorded in the Kaszubskie Lake District (7.2 °C). The coldest month is January (−0.8 °C), while the warmest month is July (17.3 °C). In the coastal belt, the average annual air temperature ranges are lower than in other parts of the country [19, 20]. It is because the chilled water masses of the Baltic Sea delay the arrival of spring and make the summer cooler. In turn, after warming up, they cause an extension of autumn and a milder course of winter [21]. Thus, the average spring temperature (March–May) is lower than the average autumn temperature (Sept–Nov) even by 3–4 °C [22, 23].

Table 16.2 Annual and seasonal mean of air temperature, the coolest and the warmest month temperature in Pomerania in 1951–2015 [18] (changed)

Station	In Year	January	July	Amplitude	Spring	Summer	Autumn	Winter
Świnoujście	8.4	−0.1	17.5	17.6	7.0	16.7	9.4	0.5
Kołobrzeg	8.1	−0.3	17.1	17.4	6.6	16.2	9.1	0.3
Ustka	7.9	−0.3	17.0	17.3	6.2	16.1	9.2	0.3
Łeba	7.7	−0.7	16.8	17.5	6.0	15.9	8.9	−0.1
Hel	8.0	−0.3	17.3	17.6	5.9	16.5	9.4	0.3
Szczecin	8.7	−0.5	18.1	18.6	8.1	17.4	9.1	0.2
Resko	8.0	−1.1	17.4	18.5	7.3	16.6	8.4	−0.4
Koszalin	7.9	−0.8	17.0	17.8	6.8	16.2	8.7	−0.2
Lębork	7.7	−1.0	17.2	18.2	6.6	16.3	8.5	−0.5
Gorzów Wlk	8.5	−1.2	18.3	19.5	8.2	17.6	8.8	−0.4
Chojnice	7.2	−2.5	17.1	19.6	6.6	16.9	7.6	−1.8
Mean	8.0	−0.8	17.3	18.1	6.8	16.6	8.8	−0.2

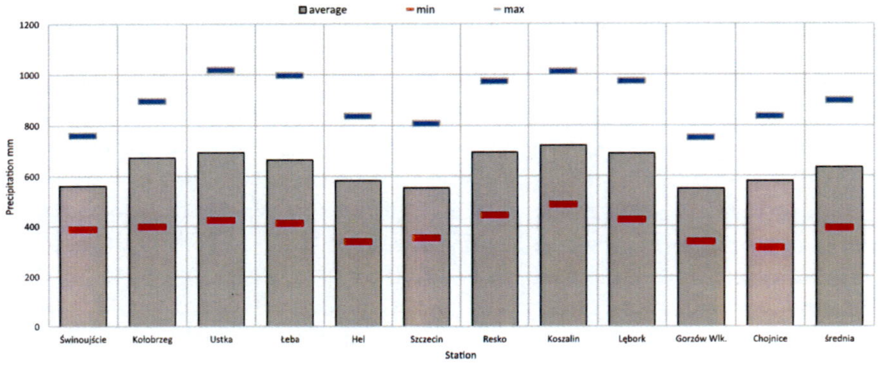

Fig. 16.2 Average, minimal i maximum precipitation Annual and seasonal mean of air temperature, the coolest and the warmest month temperature in Pomerania in 1951–2015 [18] (changed)

The water cycle in Pomerania is determined by atmospheric precipitation. Highly elevated plateaus constitute the main alimentation zone of surface and underground waters in the Pomeranian Lake District and the coastal region. A feature of the distribution of atmospheric precipitation is their temporal and spatial differentiation. The average annual precipitation (1951–2010) reaches 633 mm and ranges from 549 mm in Gorzów Wielkopolski to 721 mm in Koszalin. The variability of annual precipitation totals is on average 62–142% of the multiannual totals, and in individual regions represented by specific weather stations, it is 49–160%. The lowest precipitation is recorded in the eastern and western parts of the coastal area (<640 mm), while the most abundant precipitation is concentrated in the catchments of the Parsęta and Wieprza (>700 mm). Concerning the seasonal distribution, precipitation is mainly represented by the Pomeranian type [18, 24] characterised by descending order: summer, autumn, winter, spring. Thus, the maximum precipitation values are recorded in July or August, while the minima dominate in February, but they also happen in March and April. Another feature of precipitation is the frequency of their occurrence, expressed in the number of days with precipitation $\geq$0.1 mm. In the Pomeranian Lake District, as well as in the area of the South Baltic Coastlands, the incidence of precipitation is significant and exceeds 170 days per year (Fig. 16.2). Most often, precipitation occurs in the Łupawa and Łeba catchments, reaching the maximum near Lębork (52%). Precipitation is the least frequent on the SW edge of the Parsęta basin (<176 days).

16.3 Surface Inflow

Against the rivers of Poland, the Pomeranian rivers are rich in water resources. They present high total outflows and, at the same time, high outflow stability. The dominant share in the supply of rivers of lake districts constitutes the underground supply, making over 70% of the total [13]. From this value, 60–70% comes from the base

Table 16.3 Mean flow, mean specific runoff and outflow index of Pomeranian rivers in 1951–2015 (data base hydrometric monitoring IMGW-PIB)

River	Gauge station	Period observation	Mean Avg. Q [m^3 s^{-1}]	Mean Avg. q [dm^3 s^{-1} km^2]	H [mm]
Rega	Trzebiatów	1951–2015	20.20	7.64	241
Parsęta	Bardy	1956–2015	27.39	9.49	299
Wieprza	Stary Kraków	1951–2015	15.72	10.20	207
Grabowa	Grabowo	1976–2015	6.46	14.14	446
Słupia	Charnowo	1967–2015	17.81	11.15	352
Łupawa	Smołdzino	1951–2015	8.32	10.31	325
Łeba	Cecenowo	1956–2015	11.72	10.66	336
Reda	Wejherowo	1961–2015	4.32	10.54	332

outflow [25] creating the most stable form of underground supply from deep aquifers. The advantage of the underground supply component results from the presence of favourable infiltration conditions, numerous lakes, endorheic areas, wetlands, as well as significant areas with high ground retention capabilities. These include subglacial channels, ice-marginal valleys (the Grabowa Ice-marginal Valley, Łeba-Reda Ice-marginal Valley, Kashubian Ice-marginal Valley) and lowlands (the Gardno-Łeba Lowland, Karwia Lowland) filled with permanent wetlands. The discharge of the Pomeranian rivers varies depending on the size of rivers, from 4.32 m^3 s^{-1} in the Reda to 27.39 m^3 s^{-1} in the Parsęta (Table 16.3).

The seasonal nature of the outflow is modified by the annual cycle of changes in the volume and forms of supply (Fig. 16.3). Higher outflow rates occur in the winter half-year, culminating in March, while the lowest in the summer months. The minimum values of the coefficient are recorded in August (the Rega, Parsęta, Wieprz, Grabowa Łupawa), in June (the Łeba, Reda) or in both mentioned months they balance (the Słupia). The rivers of the western part of Pomeranian, like the Rega and Parsęta, show much higher variability of the outflow than the rivers of the central and eastern parts. The scale of volatility in the first case ranges from 70–150%, while in the second from 80 to 120% [3]. The distribution of the outflow in the annual cycle qualifies the examined rivers to the hydrological regime of the nival type [25] poorly (the Wieprza, Grabowa, Słupia, Łupawa, Łeba, Reda) or medium-developed (the Rega, Parsęta).

The unit outflow and the runoff layer are measures of the water resources of the examined rivers. All rivers have a unit outflow higher than the average for Poland estimated at 5.6 dm^3 s^{-1} km^2. The highest unit outflow is recorded for the Grabowa and Słupia catchments, respectively 14.14 and 11.15 dm^3 s^{-1} km^2 (Table 16.3), while the lowest in the Rega and Parsęta basins (<10 dm^3 s^{-1} km^2).

The remaining rivers have a unit outflow similar to the average outflow value calculated from the entire analysed area of Pomeranian (10.52 dm^3 s^{-1} km^2). In the case of the runoff layer, the boundary of the water resources of the examined rivers' catchment is determined by the Grabowa River with the highest values of

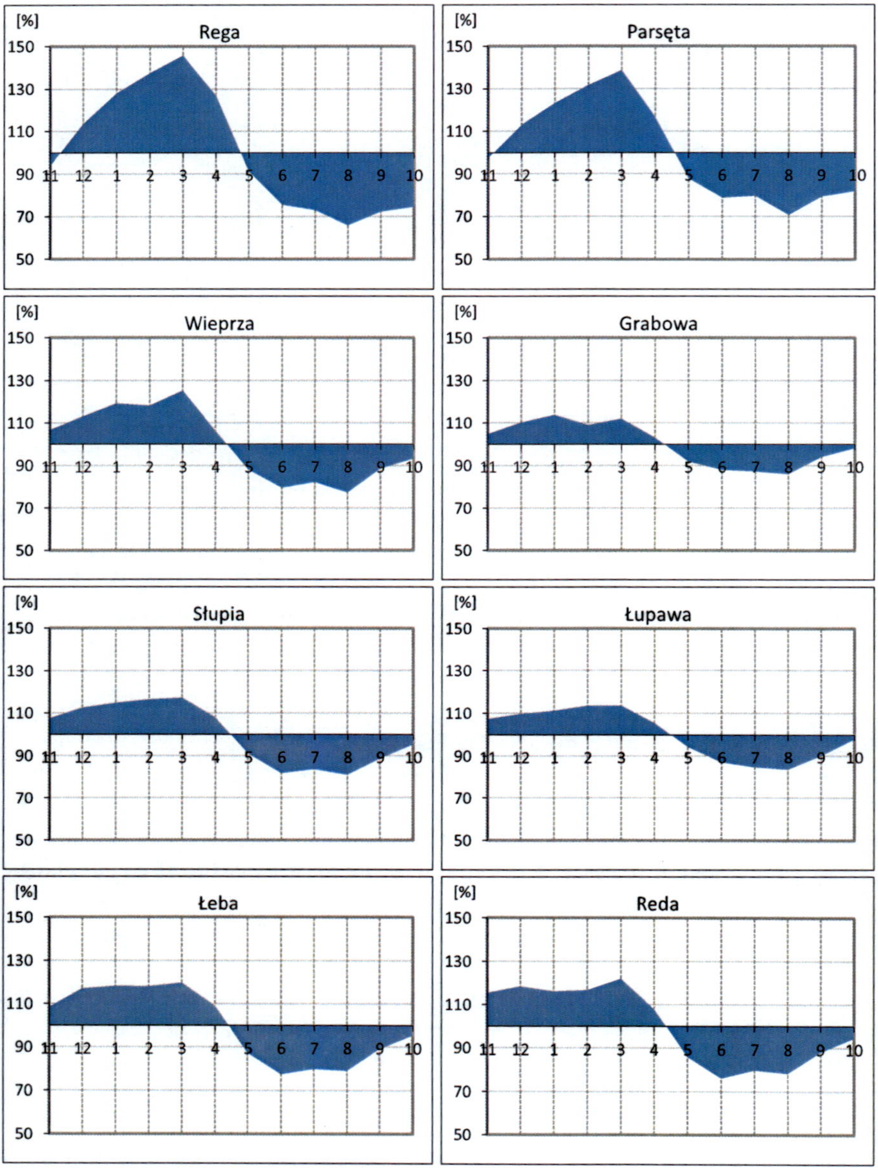

Fig. 16.3 Seasonal variability of outflow in 1951–2015

over 400 mm. To the west of it, the values of the measured indicator are lower and are below 300 mm, with the minimum in the Wieprza catchment (207 mm). In the basins on the eastern side of the Grabowa, the runoff layer is more significant, levelled and oscillates in the range of 325–352 mm.

16.4 Factors of Water Level Variation

A key role in forming the water conditions of the Pomeranian rivers, and especially their estuary sections, is played by the main (Baltic Sea) and regional (coastal lakes) drainage bases. The most critical causes of changes in the level of the southern Baltic waters (Fig. 16.4) include water accumulation and depressions caused by air pressure changes and wind waves as well as hydrological factors associated with the seasonal outflow of land waters.

The average long-term level of sea waters changes from 1951–2000 from west to east in the range from 499 to 502 cm. In the annual cycle, there are elevated water levels in the autumn and winter as well as in summer. In the first case, they are generated by frequent migration of low-pressure centres from the North Atlantic to Europe over the Baltic Sea. They are very dynamic and induce extreme values of sea levels, both maximum and minimum. According to [26] storm surges can reach up to 2 m, even 3 m, which determines the range of absolute amplitudes (Fig. 16.5) [27]. In

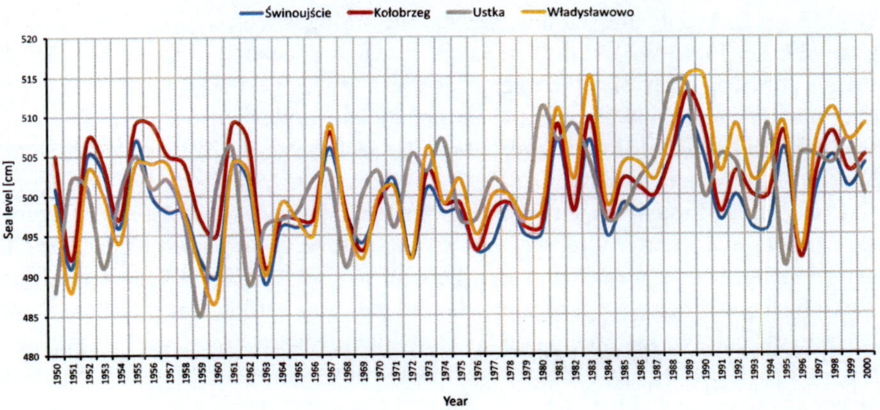

Fig. 16.4 Mean annual sea level on the coast southern Baltic in 1951–2000

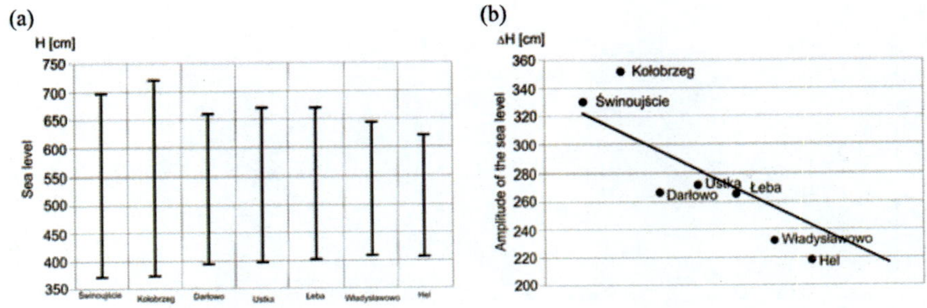

Fig. 16.5 Fluctuation range of the southern Baltic: a – maximum and minimum levels; b – fluctuation amplitudes [29]

the case of the summer, elevated sea levels are related to the increase in precipitation and increase (from the end of runoff of meltwater) inflow to river waters [28]. With spring (April, May), the dominance of the eastern atmospheric circulation causing coastal winds is associated with low average sea levels.

Climate change is accompanied by rising sea levels, and its different rate depends on the location of coastal stations along the Polish coast of the open Baltic Sea [30]. Research indicates that the average annual sea level rise is lower in the western part of the coast (0.07 cm $year^{-1}$ in Świnoujście) and increases towards the east (0.17 cm $year^{-1}$ in Ustka up to 0.23 cm $year^{-1}$ in Władysławowo). On a seasonal basis, the highest increase occurs in winter and spring (especially in February and March), while in autumn and winter months (September, October, and December) there are no upward trends.

The coastal lakes (Resko Przymorskie, Jamno, Bukowo, Kopań, Gardno, Łebsko, Sarbsko), often connected with the sea by engineered sections of rivers (canals), play a unique role in the potamic outflow of the Pomeranian rivers. In practice, this means that the lakes form an extension of the main drainage base on the mainland, constituting a regional base. It is determined by the convergent rhythm of variations in the level of lakes with sea fluctuations, the strongest correlating in the case of Lake Łebsko (Fig. 16.6). At the same time, due to the location of lakes in the lower sections of rivers, they play the role of receivers of water flowing down from entire river basins, but also – which is not very beneficial – pollution.

Additional pressure on these hydrographic objects is exerted by the polder systems with enforced water circulation.

Connection of the rivers with the sea both direct and through lakes has its hydrological and hydrochemical repercussions. The observed, long-term changes in the main drainage base cause an increase in the level of coastal lakes. As a result, they may also influence the decrease of potamic and groundwater outflow towards the sea, and increase the entry of sea waters into the land. Such scenarios are shown by previous research [21]. During short storm surges, sea waters block the free river runoff or invade their course and the lakes located in the coastal zone. So far, the maximum range of intrusions has been noted on the Łeba River for the profile in Cecenów (20 km from the estuary). In storm situations, there may be a direct flood hazard,

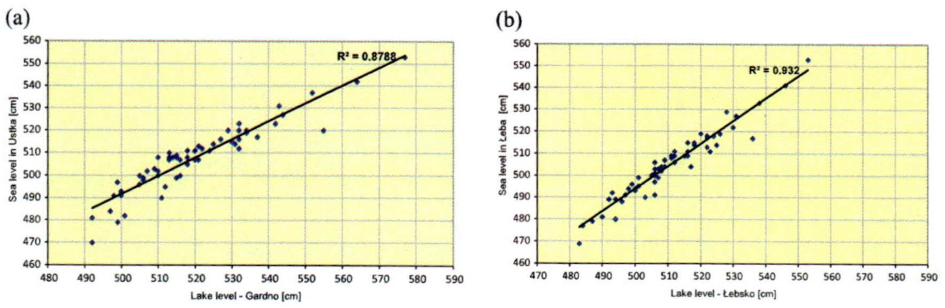

Fig. 16.6 Dependence of water fluctuations of Gardno and Łebsko lakes from the level of the Baltic Sea [31]

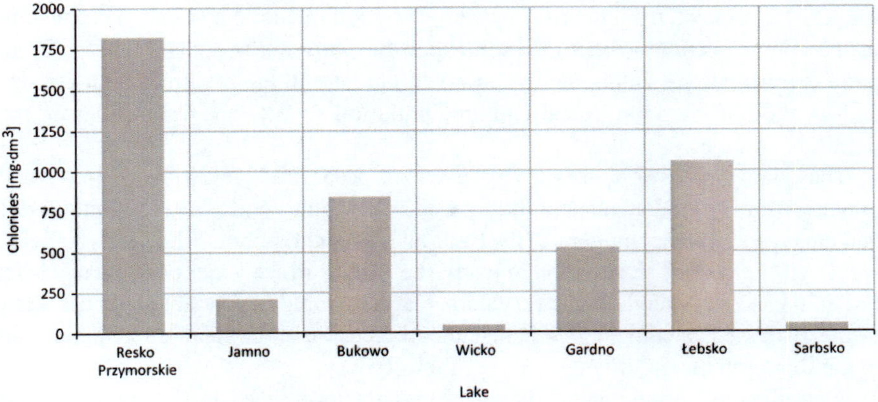

Fig. 16.7 Average values of chloride concentration in coastal lakes in the period 2002–2003 [32]

especially in the areas of coastal lowlands and coastal lakes. It is not uncommon that, as a result of a storm, the estuary sections of the rivers are wholly or partially blocked by sea sands, limiting the potamic outflow to the sea. Such situations are observed at the mouth of the Łupawa or smaller streams of the Karwianka or the channel connecting Lake Kopań with the sea. In unregulated rivers (the Piaśnica), the estuary often changes its location.

Intrusions of sea waters result in changes in physical and chemical properties of the surface and underground waters. The most susceptible to these changes are the waters of coastal lakes, and the primary indicator of the impact of sea waters are the values of salinity (chloride). The highest average salinity is observed in lakes Resko Przymorskie and Łebsko (Fig. 16.7), reaching over 1000 mg·dm^3 Cl$^-$, and the smallest in Lake Sarbsko [32]. The spatial distribution of chlorides shows a decrease in concentrations as we move away from the channels that provide communication between the lakes and the sea. There is a possibility of recording almost entirely fresh water in the reservoir. It applies, among others, to lakes Jamno, Wicko, Gardno, and Sarbsko.

16.5 Pollution Supply

The high content of nutrients and some other substances in the coastal waters of the Southern Baltic is mainly due to the rivers carrying pollution load. Delivery of pollutants (primarily nutrients – forms of nitrogen and phosphorus) to the receiver, which is the Baltic Sea, takes place via rivers, but only a particular part of nutrients from various sources in the basin get to the rivers and then to the receiver. These differences are the result of retention and loss of nutrients in waterways [33]. Retention is understood as temporary retention of biogens within the river valley, and losses as irreversible removal from the river system. Nitrogen retention and loss occur mainly in the source sections (denitrification), and the phosphorus – in the downstream part (sedimentation). Undoubtedly, however, land use in the catchment and the size and location of the point, linear and area pollution sources are responsible for the pollution volume.

The quality of river waters leaves much to be desired. In 2016 the ecological potential of only the Parsęta and Słupia was assessed as good. The remaining rivers had the potential rather moderate (the Reda and Wieprza) or poor (the Łupawa, Łeba, and Rega) [29, 34]. The situation in jcwp (the WFD surface water body) was worse. For all the rivers studied, it was very poor. Causes should be linked to land use. Half of the discussed catchment area is occupied by built-up areas and arable lands, and in the Rega catchment, it is up to 65% (Table 16.4).

According to Corine Land Cover 2012 classification (level 1), the dominant type of catchment land use is the cropland (52.7%) and forest (42.2%) (Table 16.4). In Rega catchment, the total area of cropland is 1705.6 km^2, which is 23.5% of the total area of all discussed catchments. The total area of forest in Parsęta catchmest is 1340.1 km^2. Forests are the dominant type of land use in the Reda and Wieprza catchments (Fig. 16.8). Wetland occupies only 0.2% of the area of all discussed catchments. In the Łeba catchment wetlands covers 17.3 km^2 (which is 1% of the Łeba catchment).

Table 16.4 Landuse in the Pomeranian catchments (according Corine Land Cover 2012)

Landuse		Urban	Cropland	Forest	Wetland	Water	Total
Rega	km^2	90.3	1705.6	931.2	2.5	35.6	2765.2
	%	3.3	61.7	33.7	0.1	1.3	100
Parsęta	km^2	83.9	1634.5	1340.1	2.7	14.8	3075.8
	%	2.7	53.1	43.6	0.1	0.5	100
Grabowa	km^2	8.7	286.6	276.2	1.3	1.4	574.2
	%	1.6	50.0	48.1	0.1	0.2	100
Wieprza	km^2	31.4	768.6	822.5	0.0	16.0	1638.4
	%	1.9	46.9	50.2	0.0	1.0	100
Słupia	km^2	72.1	805.3	719.1	0.4	23.8	1620.8
	%	4.4	49.7	44.4	0.0	1.5	100
Łupawa	km^2	24.0	463.8	388.4	2.2	35.2	913.65
	%	2.6	50.8	42.5	0.2	3.9	100
Łeba	km^2	48.6	914.3	713.2	17.3	86.5	1779.9
	%	2.7	51.4	40.1	1.0	4.9	100
Reda	km^2	47.1	278.8	285.6	2.0	5.7	619.2
	%	7.6	45.0	46.1	0.3	0.9	100
total	km^2	432.9	7266.2	5824.8	30.2	233.0	13,787.1
	%	3.1	52.7	42.2	0.2	1.7	100

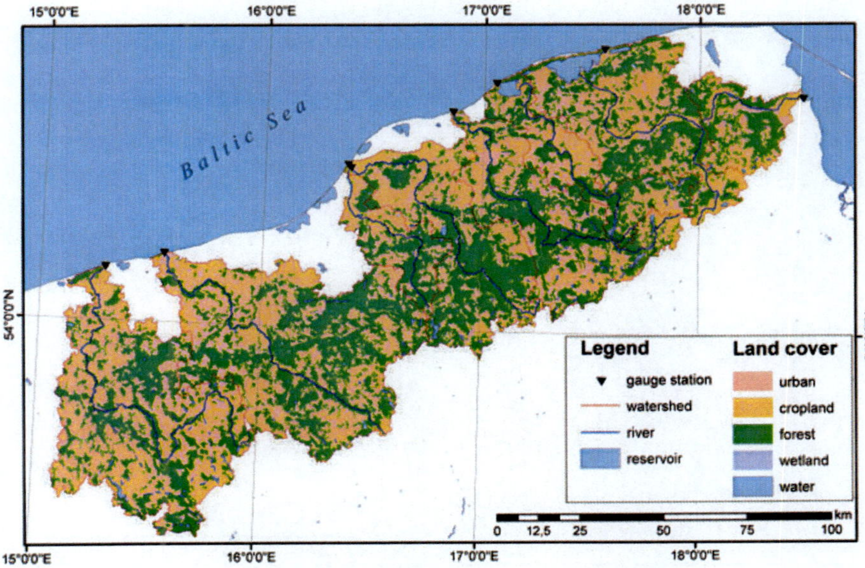

Fig. 16.8 Landuse in the Pomeranian catchments (according Corine Land Cover 2012)

There are three types of pollution sources: point (hotspot), linear and area. The area sources of pollution include: arable land, urban fabric and industrial. The area sources of pollution cover 5387.0 km^2 of the Pomeranian catchments. Only in Rega catchment, the area sources of pollution covers 51.4% of the catchment (1423.1 km^2). The line sources of pollution include: commercial and transport units (road and rail networks). In total the length of primary and secondary roads is 2118.0 km, but the length of rail is 850.3 km. The last group is point sources of pollution from wastewater plant, campsite, gas station or carwash. The biggest threat to water quality is sewage discharges from wastewater plants. There are two wastewater plants in Rega catchment and two in Parsęta catchment (Fig. 16.9). In total, there are at least 185 hotspots in Pomeranian catchments.

16.6 Flood Risk and Prevention

The temporal variability of water resources manifested in their seasonal abundance and resulting from the excessive inflow of allochthonous waters or inhibition in free flow, characteristic for the conditions of the geographical environment of Pomerania, bring flood risks. Due to the formation process, there are different threats from land and sea. Land-induced floods include dangers brought by volatile or widespread atmospheric precipitation and thaw, or by the formation of ice jams on rivers. Sea-induced floods are hazards associated with storm phenomena. Among the mentioned, the primary place is taken by rainfall, usually occurring in the summer (dominance on

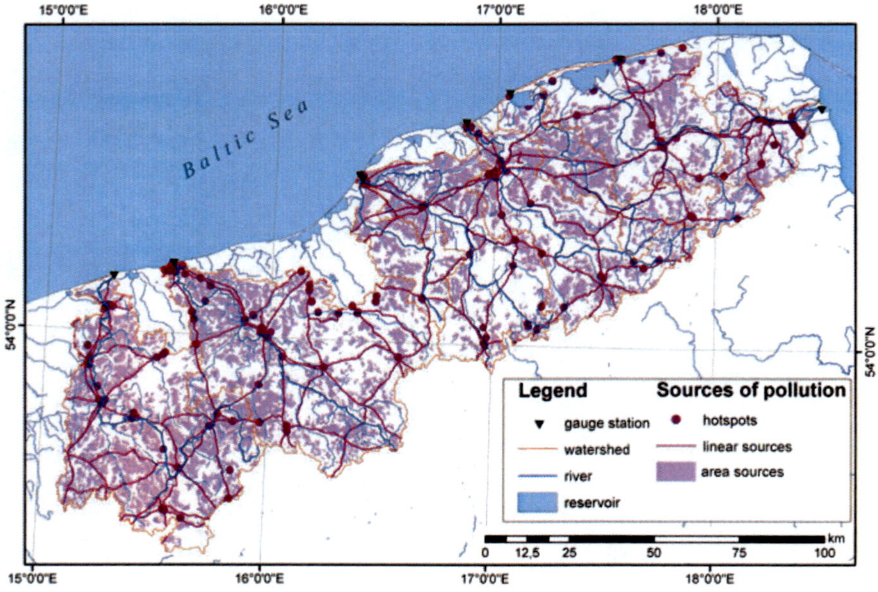

Fig. 16.9 Sources of pollution in the Pomeranian catchments

smaller watercourses), followed by snowmelt and ice jams. In all cases, the situation is highly unfavourable when excess water stagnates in the inter-embankment area, causing the flood banks to wash out. Sea-induced floods are most dangerous during the autumn and winter rapid surges of the Baltic Sea, or the Gulf of Gdansk. Hence, the areas most vulnerable to flooding are the Gardno-Łeba Lowland, Karwieńskie Błota together with the Piaśnica valley, Płutnica Ice-marginal Valley and Kaszubska Ice-marginal Valley. Often, storm surges are accompanied by other phenomena, in combination with which hazards of a mixed, land-sea type with a storm-thaw or storm-ice jam genesis are created. This is when the flood risk is highest. It applies above all to coastal lowlands, the more so because the existing drainage system turns out to be less efficient and useful, due to the opposite directions of water runoff. A separate type of emergency are events related to a failure of hydraulic structures or construction disaster of such structures. It is connected with water and drainage systems based on a gravitational and forced way of draining surplus water using a pump station (the Gardno-Łeba Lowland) or river valleys used for energy purposes (the Słupia, Łupawa). Defects of this kind can be caused by a power shortage or by the recklessness of a human being. The flow of water through the main valleys, especially during surges, is a potential threat to neighbouring areas, both to the risk of flooding and the risk of contamination in case of spilling waters.

In the preliminary flood risk assessment, it was necessary to designate risk areas, i.e. those where there is a significant risk of flooding or where high risk is likely to occur. This assessment was based on hydrography, topography and land use of the area. Information on historical floods and the evaluation of the potential adverse effects of floods that may occur in the future were taken into account. The effect of these activities is the forecast of the long-term development of events, in particular, the impact of climate change on the occurrence of floods. In all catchments discussed, after the year 2000, an increase in the number and location of floods was registered. These were floods on the Reda (2008, 2009), Parsęta (2002), Grabowa (2005), Wieprza (2004), Słupia (2001), Łupawa (2007), Łeba (2002) and Reda (2001). Besides, in 2007 flood occurred on Lake Jamno. At the end of the nineteenth century and the beginning of the twentieth-century floods took place on three rivers: the Parsęta (1874), Łeba (1898) and Wieprza (1914). The Słupia and Łeba experienced high water in 1982 and 1998 respectively.

High and medium (10 and 1%) probability of flooding concerns the Reda. The Słupia, Łupawa, Łeba, Grabowa, Rega, and Wieprza are at risk of flooding with a medium probability (1%) of occurrence (Table 16.5). Low probability applies to the Parsęta and the lakes Jamno, Resko Przymorskie, and Wicko.

All discussed Pomeranian rivers, sections with a total length of 947 km (Fig. 16.10), were classified under the Preliminary Flood Risk Assessments (PFRAs) [35] for the development of flood hazard maps and flood risk maps. The longest section was designated for Rega 0–187 km, then for Słupia (0–148 km), Parsęta (0–142 km), Wieprza (0–133 km) and Łeba (0–126 km). Sections shorter than 100 km were assigned to Łupawa (0–90 km), Grabowa (0–72 km) and Reda (0–49 km).

From the point of view of flood prevention, buffer and strategic investments are of great importance [36]. Buffer investments mainly relate to the reconstruction of old and construction of new flood embankments, engineering channels and construction of flow regulating devices, strengthening of river banks, as well as the construction of retention reservoirs and development of valley retention (Table 16.6). Strategic investments in the catchment areas concerned the increase of their retention through afforestation (comprehensive forest and forestry adaptation projects to climate change), counteracting water erosion in lowland areas, restoration of

Table 16.5 High and average probability of exceedance flow and corresponding water level

River	Gauge station	Avg Q [m^3 s^{-1}] (1951–2010)	High Q [m^3 s^{-1}] (1951–2010)	Maximum flow of exceedance probability Qmaxp% [m^3 s^{-1}] and corresponding water level H [cm]			
				$Q_{10\%}$	$HQ10_{\%}$	$Q_{1\%}$	$HQ1_{\%}$
Rega	Trzebiatów	54.2	87.7	72.9	421	95.2	462
Parsęta	Bardy	80.8	143.0	110.0	435	140.0	460
Grabowa	Grabowo	14.8	21.7	19.1	165	24.5	209
Wieprza	Stary Kraków	41.6	76.7	59.1	442	79.5	504
Słupia	Charnowo	35.9	53.6	44.0	349	58.0	391
Łupawa	Smołdzino	17.1	44.9	24.0	106	36.0	139
Łeba	Cecenowo	28.2	45.9	39.3	284	50.4	327
Reda	Wejherowo	14.0	22.3	19.1	180	24.9	200

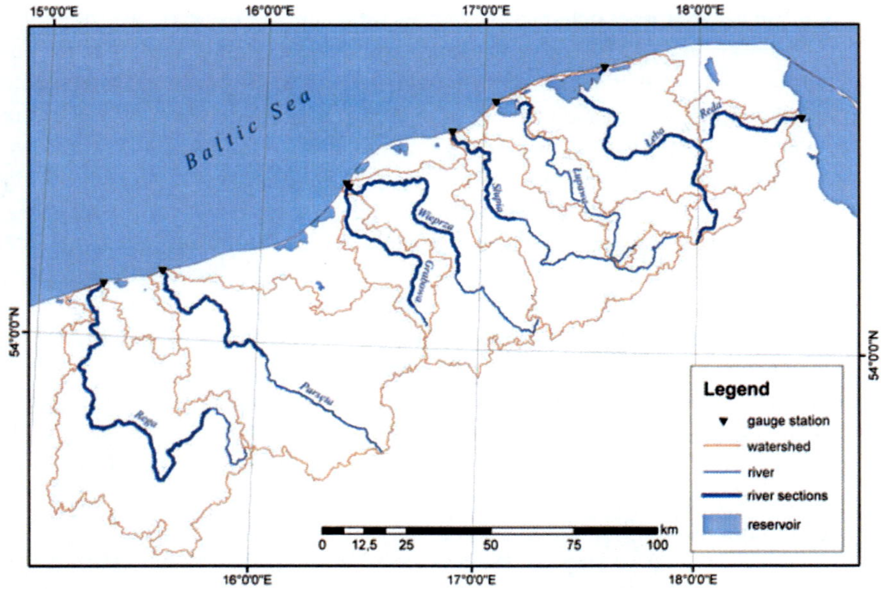

Fig. 16.10 Rivers classified as part of the Preliminary Flood Risk Assessment (PRFAs) [35] for the development of flood hazard maps and flood risk maps

Table 16.6 Buffer investments [34]

Activity	Rega	Parsęta	Wieprza	Reda
Construction of new flood embankments	x			x
Reconstruction of old flood embankments			x	x
Construction of retention reservoirs	x			
Development of valley retention	x			
Engineering channels	x	x		
Construction of flow regulating devices	x			
Strengthening of river banks	x	x	x	

riverbeds and riverbanks, increasing retention in urban areas of over 20 000 residents. For some towns (Resko, Białogard, Sławno, Ustronie Morskie and Trzebiatów), a multi-variant concept of flood protection was implemented. Strategic investments also concern the seashore as a limitation of the identified flood risk, especially in the area of Lake Bukowo and the spit of Lake Jamno.

16.7 Conclusion

The Pomeranian catchments water cycle depends less on local factors is determined by atmospheric precipitation. The favourable infiltration conditions, numerous lakes, endorheic areas, wetlands and significant areas with high ground retention capabilities are providing the underground supply. The annual cycle of changes in the volume and forms of supply modifies the seasonality of outflow and all studied rivers have a unit outflow higher than the average for Poland (estimated at 5.6 $dm^2\ s^{-1}\ km^2$). The highest values (over 400 mm) of the runoff layer of the studied rivers' catchment water resources is determined by the Grabowa River.

The unique role in the potamic outflow of the Pomeranian rivers play coastal lakes which are often connected with the sea by engineered sections of rivers. The lakes form an extension of the main drainage base on the mainland, constituting a regional base. The convergent rhythm of variations in the level of lakes with sea fluctuations also is exerted by the polder systems with enforced water circulation. The long-term changes in the main drainage base cause an increase in the level of coastal lakes, that in result, they may influence the decrease of potamic and groundwater outflow towards the sea, and increase the entry of sea waters into the land.

The high concentration of nutrients and some other substances in the coastal waters is mainly due to the rivers carrying pollution load, but only a particular part of nutrients from various sources in the catchment get to the water.

Retention and loss of the nitrogen occur mainly in the source sections and the phosphorus - in the downstream part. The pollution volume is determined by land use in the catchment and the size and location of the point, linear and area pollution sources.

The flood risks results from temporal variability of water resources manifested in their seasonal abundance and resulting from the excessive inflow of allochthonous waters or inhibition in free flow and because of the existing drainage system turns out to be less efficient and useful, due to the opposite directions of water runoff.

16.8 Recommendation

According to the definition of water management, the work presented includes the determination of water resources in the area of Pomerania, which can be rationally used for the needs of the natural and social environment. Such studies should be repeated periodically, due to unfavorably changing climatic conditions that visibly change also water resources, and thus affect periodic shortages or excess water. The work is addressed to authorities at various levels (local, regional and national) responsible for proper water management, waste management, spatial management and agriculture, which are directly or indirectly dependent on water or affect its quality. The study should be of particular importance for institutions dealing with flood risks and protection, such as Maritime Offices, for water construction enterprises and establishments of municipal economy. Pomeranian rivers are contaminated with nutrients and are the main cause of pollution in the Baltic Sea. Therefore, continuous monitoring of the quality of river waters is important, as well as works aimed at reducing the use of artificial fertilizers in agriculture in Pomerania.

References

1. Stachy J (1980) Odpływ rzek Przymorza na tle odpływu z terenu całej Polski. Stosunki wodne w zlewniach Przymorza i dorzecza Dolnej Wisły ze szczególnym uwzględnieniem gospodarki wodnej jezior. IMGW, Słupsk, pp 13–27
2. Kondracki J (2000) Geografia regionalna Polski. PWN, Warszawa
3. Bogdanowicz R (2004) Hydrologiczne uwarunkowania transportu wybranych związków azotu i fosforu Odrą i Wisłą oraz rzekami Przymorza do Bałtyku. Wyd, UG, Gdańsk
4. Bajkiewicz-Grabowska E, Markowski M, Golus W (2020) Polish rivers as hydrographic objects. In: Korzeniewska E, Harnisz M (eds) Polish river basins and lakes – part I. The handbook of environmental chemistry, vol 86. Springer, Cham. https://doi.org/10.1007/978-3-030-12123-5_2
5. Bajkiewicz-Grabowska E, Golus W, Markowski M, Kwidzińska M (2020) Characteristics of the water network in postglacial areas of Northern Poland. In: Korzeniewska E, Harnisz M (eds) Polish river basins and lakes – part I. The handbook of environmental chemistry, vol 86. Springer, Cham. https://doi.org/10.1007/978-3-030-12123-5_8
6. Bajkiewicz-Grabowska E (2002) Obieg materii w systemach rzeczno-jeziornych. Geografii i Studiów Regionalnych, Warszawa, UW, Wydz
7. Fac-Beneda J (2011) Młodoglacjalny system hydrograficzny. Wyd, UG, Gdańsk
8. Bajkiewicz-Grabowska E, Golus W, Markowski M, Kwidzińska M (2020) The role of lakes in shaping the runoff of lakeland rivers. In: Korzeniewska E, Harnisz M (eds) Polish river basins and lakes – part I. The handbook of environmental chemistry, vol 86. Springer, Cham. https://doi.org/10.1007/978-3-030-12123-5_9
9. Fac-Beneda J (2008) Naturalne wypływy wód podziemnych na obszarach chronionych w północnej Polsce. In: Pociask-Karteczka J (ed) Wody na obszarach chronionych. IGiGP UJ, Kraków, pp 133–141
10. Mazurek M (2010) Hydrogeomorfologia obszarów źródliskowych (dorzecze Parsęty, Polska NW). UAM w Poznaniu, seria Geografia, 92, Poznań
11. Osadowski Z (2010) Wpływ uwarunkowań hydrologicznych i hydrochemicznych na zróżnicowanie szaty roślinnej źródlisk w krajobrazie młodoglacjalnym Pomorza. AP w Słupsku, Bogucki WN, Poznań-Słupsk
12. Fac-Beneda J, Hryniszak E (2006) Wypływy wód podziemnych u podnóży krawędzi pradolinnych. In: Źródła Polski. Wybrane problemy frenologiczne, (ed.) P. Jokiel, P. Moniewski, M. Ziułkiewicz, Wydz. Nauk Geogr. UŁ, Łódź
13. Dynowska I (1972) Typy reżimów rzecznych w Polsce. Zeszyty Naukowe UJ, Prace Geograficzne, 28, Kraków
14. Drwal J (1985) Jeziora w egzoreicznych systemach pojezierzy młodoglacjalnych. Zeszyty Nauk. Wydz. BiNoZ UG, Geogr 14:7–15
15. Fac-Beneda J (2017) Struktura sieci hydrograficznej. In: Jokiel P, Marszelewski W, Pociask-Karteczka J (eds), Hydrologia Polski, Wyd. PWN, Warszawa, pp 116–121
16. Miętus M, Wielbińska D (1996) Średni rozkład ciśnienia atmosferycznego nad Europą i jego modyfikacja w rejonie Morza Bałtyckiego. Wiadomości IMGW 3:85–100
17. Kwiecień K (1959) Próba określania zasięgu wpływu Bałtyku na klimat przyległego Pomorza. Biuletyn PIHM, 9
18. Kirchenstein M (2013) Zmienność temperatury powietrza i opadów atmosferycznych w północno-zachodniej Polsce. Wydawnictwo Naukowe Akademii Pomorskiej w Słupsku, Słupsk

19. Paszyński J (1984) Główne cechy klimatu. In: Augustowski B (ed) Pobrzeże pomorskie. GTN, Gdańsk, pp 169–187
20. Trapp JA (2007) Warunki klimatyczne. In: Czochański J, Lemańczyk J (eds) Aktualizacja opracowania ekofizjograficznego do planu zagospodarowania przestrzennego województwa pomorskiego. Wojewódzkie Biuro Planowania Przestrzennego, Urząd marszałkowski Województwa Pomorskiego, Słupsk-Gdańsk, pp 48–62
21. Miętus M, Filipiak J, Owczarek M (2004) Klimat wybrzeża południowego Bałtyku. Stan obecny i perspektywy zmian In: Cyberski J (ed) Środowisko polskiej strefy brzegowej południowego Bałtyku – stan obecny i przewidywane zmiany w przededniu integracji europejskiej, GTN, Wydział V Nauk o Ziemi, Gdańsk, pp 11–44
22. Friedrich M, Sredzicka B, Jakoniuk J (1980) Klimat województwa słupskiego w świetle potrzeb rolnictwa. Wojewódzki Ośrodek Postepu Rolniczego w Strzelnie, IMGW O/Słupsk, Strzelino-Słupsk
23. Paszyński J, Niedźwiedź T (1999) Klimat In: Starkel L (ed) Geografia Polski, środowisko przyrodnicze. Wyd. Naukowe PWN, pp 288–342, Warszawa
24. Kożuchowski K, Wibig J (1988) Kontynentalizm pluwialny w Polsce: zróżnicowanie geograficzne i zmiany wieloletnie. Wydanie 55 z Acta geographica Lodziensia, Łódź
25. Wrzesiński D, Brychczyński A (2014) Zróżnicowanie reżimu odpływu rzek w północno-zachodniej Polsce. Badania Fizjograficzne, R. V, Seria A, Geografia Fizyczna (A65), 261–274
26. Majewski A (1987) Charakterystyka wód: In: Augustowski B (ed) Bałtyk Południowy. GTN, Wyd. PAN, Wrocław
27. Wiśniewski B, Wolski T (2009) Occurrence probability of maximum sea levels in Polish ports of Baltic Sea coast. Polish Marit Res 16(3):62–69
28. Mikulski Z (1983) Dopływ wód rzecznych do Morza Bałtyckiego. Przegląd Geofizyczny XXVIII 2:65–183
29. Stan środowiska województwa zachodniopomorskiego w roku 2016. Raport 2017 (2017) Wojewódzki Inspektorat Ochrony Środowiska w Szczecinie, Szczecin
30. Girjatowicz JP (2008) Katalog zasolenia i stanów wody polskiego wybrzeża Bałtyku. Wydawnictwo Naukowe Uniwersytetu Szczecińskiego, Szczecin
31. Chlost I (2012) Geograficzne uwarunkowania stosunków wodnych Niziny Gardneńsko-Łebskiej. Rozprawa doktorska, typescript. Katedra Hydrologii UG, Gdańsk
32. Cieśliński R (2011) Geograficzne uwarunkowania zmienności hydrochemicznej jezior wybrzeża południowego Bałtyku. Wydawnictwo Uniwersytetu Gdańskiego, Gdańsk
33. Behrendt H, Opitz D (1999) Retention of nutrients in river systems: dependence on specific runoff and hydraulic load. Hydrobiology 410:111–122
34. Raport o stanie środowiska województwa pomorskiego w roku 2016 (2017) Inspekcja Ochrony Środowiska, WIOŚ w Gdańsku. Biblioteka Monitoringu Środowiska, Gdańsk
35. Raport z wykonania wstępnej oceny ryzyka powodziowego (2011) Projekt ISOK. Informatyczny system osłony kraju przed nadzwyczajnymi zagrożeniami, Tytuł zadania 1.3.1: Wstępna ocena ryzyka powodziowego (WORP), KZGW, Warszawa
36. www.kzgw.gov.pl (2018.11.30)

Part IV
Economic Importance of Water Resources

Chapter 17
Exploitation of Rivers in Poland for Electricity Production – Current Condition and Perspectives for Development

Katarzyna Kubiak-Wójcicka and Leszek Szczęch

Abstract The energy of flowing water is one of the most popular source of renewable energy all over the World. In Poland, there are difficult landscape conditions to build a large hydropower plant, so there are only 13 power plants with nominal capacity over 10 MW. 75% of potential power is produced in pumped-storage hydropower plants. These plants do not produce energy but only allow to store it as potential energy of water pumped back to the upper (head) reservoir. Since the beginning of the twenty-first century, there was no development in that area. However, in Poland, there still is a reserve to build a few large hydropower plants, especially in the lower Vistula and it is possible to build a few hundred of small hydropower plants on smaller watercourses.

Keywords Hydropower plant (HP) · Hydropower capacity · Electricity production · Poland

17.1 Introduction

The energy of wind and water is, apart from thermal energy tamed along with the art of maintaining and igniting the fire, the form of acquiring energy that is the first known and used by humans. Over the centuries, they were used more and more willingly because they improved the quality of life and assisted during hard work in households or facilitated travel (wind energy for sailing). The energy of water began to be tamed with the emergence of agriculture and the settled lifestyle of the recent nomads. Initially, simple water lifting devices were used, whose task was to irrigate fields or supply water to cities in prehistoric times [1, 2]. Along with the

K. Kubiak-Wójcicka
Department of Hydrology and Water Management, Faculty of Earth Sciences and Spatial Management, Nicolaus Copernicus University, Lwowska 1, 87-100 Toruń, Poland
e-mail: kubiak@umk.pl

L. Szczęch (✉)
Faculty of Mechanical Engineering, Military University of Technology, Gen. Sylwestra Kaliskiego 2, 00-908 Warszawa, Poland
e-mail: leszek.szczech@wat.edu.pl

M. Zeleňáková et al. (eds.), *Management of Water Resources in Poland*, Springer Water,
https://doi.org/10.1007/978-3-030-61965-7_17

development of civilization, the technical solutions used had advanced, and water wheels were used on a large scale. Paddle wheels use the power of flowing water to create torque, which was used to drive machines in industrial activities [3]. The wide use of water wheels is: in grain mills, fulling mills, paper mills, oil mills, tanneries and sawmills [4–6]. In medieval Europe, the development process of various water wheel systems enabling the operation of mills and other technical devices began to develop around the eleventh century [7]. The beginnings of water milling on the Polish lands took place in the twelfth century, and in the thirteenth century, they became quite common [8]. The range and pace of use of water mills are evidenced by their number. The water mills were well received, because already in the twelfth century they were present in almost every village in England [9]. The greatest development of milling in Poland occurred in the sixteenth century [4]. It was only from the nineteenth century that water wheels began to be replaced by newly invented water turbines. These, in turn, began to be used to drive generators in power plants, and propulsion of individual power devices was taken over by thermal (steam and combustion) engines and electric motors.

Due to water energy potential, it is the main source of "green" energy on a global scale. In Poland, however, there are no local conditions to build large installations compared to other regions of Europe or the world. However, many watercourses have sufficient potential to be used to generate energy in small hydropower plants (SHP). They are an ideal way to stimulate development on a small scale and have little impact on the natural environment. With the cascade system of watercourse construction, small hydropower plants use the energy potential of the entire watercourse, which can be a source of electricity for several dozen to several hundred households.

This chapter focuses on the role played by hydropower in Poland in domestic electricity production in recent years. It covers the characteristics of the largest hydropower plants (HP) in terms of parametric and semantic. The chapter contains a brief historical overview of the development of hydropower in Poland, its energy potential and the level of its utilization. The statistics used in the yearbooks of the Central Statistical Office [10] and data on selected hydroelectric plants, which originate from websites or were made available by energy utilities, were used for the characterization. On the basis of strategic government documents, plans related to the development of hydropower for the coming years have been presented.

17.2 Energy Potential of Rivers

The distribution of hydropower in the world is uneven and related to the occurrence of water resources. In principle, the power P that can be obtained in a hydropower plant depends on the flow rate of water Q and the height of the water head H. It can be expressed by the formula:

$$P = 9.81 \cdot Q \cdot H \cdot \eta_e$$

where η_e – efficiency of the hydropower plant.

Poland's water resources determined on the basis of larger river flows are at an average level. The Vistula and the Odra have the largest river flows in Poland in their estuary sections. Watercourses with smaller flows can get higher power using a higher head. From the point of view of energy use of rivers, the differences in the height of the terrain in Poland compared to other countries are small, because they result from the prevailing lowland nature of the area. Knowing these two parameters, i.e. flow and head, it is possible to evaluate the energy potential of individual rivers.

Calculations of potential energy resources of the Polish rivers were made at the beginning of the twentieth century and presented in the form of water energy cadastre. The development of the cadastre of the mountain and foothills rivers of the Odra tributaries began in 1909 and was completed in 1914. In a later period, the calculations were made, and the updated cadastral changes were modified [11].

The theoretical water resources of Polish rivers were identified in the sixties of the twentieth century, based on the methodology of the World Energy Council. The calculations made by Hoffmann show that the theoretical energy resources of Poland's rivers are on average 2625 MW of power and about 23 TWh/year [12, 13]. The technical resources of all flowing waters in Poland are 14 TWh/year, which accounts for 61% of all theoretical resources. The Vistula River basin has the greatest potential (9.3 TWh/year) and the Odra basin (2.5 TWh/year). Other technical resources refer to coastal rivers and small hydropower plants. In comparison with other European countries, this potential is small.

In 2016, 2139.4 GWh of electricity was produced in Poland from water energy, which is only 15% of the technical potential of rivers [10]. This shows that the energy resources of the Polish rivers are being used on a low level, and their use is among the lowest in Europe.

17.3 Historical Aspects of the Construction of Hydropower Plants in Poland

The beginnings of the development of hydropower in Poland date back to the end of the nineteenth century. Initially, these were low-power hydropower plants (less than 1 MW).

In total, in the years 1898–1906, in the areas covered by the current Polish borders, the production of 5 hydroelectric plants with a total capacity of 2.1 MW began. Starting from 1907 until the outbreak of World War I, in 1914, a total of 22 hydropower plants with a total capacity of 26.9 MW were created, of which 7 on the left-bank tributaries of the central Odra, including the largest Pilchowice on the Bóbr River at that time with power 7.5 MW. Almost in the time (1908), full development (cascade) of the Radunia River in Kashubia was begun, but it was finished only in 1937. This period is characterized by "rapid development of hydropower for local needs in the western and northern territories". World War I, only 5 objects were created at that time, including mainly those, started previously, with a capacity of only 3.2 MW, among them the largest (2.64 MW) Pierzchała on Pasłęka River [13].

In the first 10 years after the war, there was an increase in the number of hydroelectric power plants in Poland using, in particular, the flood storage reservoirs that were being built at that time. 38 professional hydropower plants were built at that time, with a capacity of 68 MW, including the first two power plants built on the Wda River (Czarna Woda) in Pomerania: Gródek (3.9 MW) and Żur (9.0 MW). The construction of the power plants in Gródek and Żur was closely related to the intensive electrification of the country but also contributed to the development of water tourism. Among others, on the initiative of Hoffmann in 1931, the "Wda" Kayak Club was created at the Żur hydroelectric power plant [14]. The 1930s were characterized by a large increase in power (99.5 MW), thanks to the construction of a larger water power plant in Lower Silesia and Pomerania, and especially a large plant in Dychów on the lower Bóbr in 1936 (79.5 MW). Large shipping reservoirs were built on the tributaries of the Odra River: Otmuchów on Nysa Kłodzka in 1933 (4.8 MW) and Turawa on Mała Panew in 1937 (1.8 MW); the construction of these reservoirs was completed only in 1948.

During the Second World War and in the first post-war years, no new hydropower facilities were created in the Regained Territories. In 1942, the construction of the dam and hydropower plant (HP) in Rożnów (50.0 MW) on the Dunajec River was completed, and in 1949 a similar equalizing reservoir and a hydropower plant (second stage of the cascade) were completed below in Czchów (8.0 MW). The construction of the storage reservoir in Turawa and others, damaged during military operations, was completed. It was not until the 1950s that Dychów was rebuilt and re-opened – the largest power plant on the lower Bóbr – the first mixed, pumped storage HP with 4 independent pump units.

At the turn of the 1950s and 1960s, there was a revival of the energy use of Polish rivers – new hydropower plants were being built. Taking advantage of the necessity to put another Odra channelized stage in Brzeg Dolny, in 1958 the Wały Śląskie power station was built there (9.8 MW). Already in 1960, it was the turn of one of the larger lowland reservoirs: Koronowo on the Brda, where a large power plant was constructed (26.0 MW), and two years later (1962) the Tryszczyn equalizing reservoir with a 3.3 MW power plant [13].

At the end of the 1960s, the construction of the Soła cascade was started, building a dam in Tresna in 1967 above Porąbka, with a 21 MW power station next to it. In 1968, the dam in Solina on the San River was put into use, forming the largest storage reservoir in Poland with a capacity of 472 million m^3, and the mixed, pumped storage power plant installed at it reached the power of 136 MW. Finally, in 1970, the only channelized stage of the lower Vistula in Włocławek was commissioned with a 160.2 MW post-peak power plant. In the same year, the first Polish pumped storage power station of 156 MW was created in the north of the country, using two lakes (Kamienno and Kwiecko) with a head of 80 m [15], for which post-war Dychów was a somewhat experimental facility.

The turn of the 1970s and 1980s is the period of the great Polish pumped storage power plants. Two hydroelectric giants were created at that time: Porąbka-Żar (500 MW) in 1979 and Żarnowiec (680 MW) in 1982. In the first of them, the water of the storage reservoir in Porąbka and the artificial water reservoir created on the top of the Żar mountain were used, using a head of up to 440 m. A facility of European and even world class was created in this way. In the second case, the largest hydroelectric power plant in Poland was built, with an artificial upper storage reservoir, Czymanowo located 125 m above Lake Żarnowieckie, on which the power plant was built.

It was not until 1997 that the last large mixed, pumped storage hydropower plant in Niedzica on the Dunajec was put into operation. Together with the equalizing stage using a storage reservoir of a power plant in the Sromowce Wyżne, launched in 1994, they reached 94.8 MW altogether.

After 2000, only modernization works were carried out in hydropower plants in Poland. In the autumn of 2003, the modernization of the Solina-San power plant was completed, increasing its capacity to 200 MW. In 2005 HP Dychów was modernized, eventually reaching 88 MW. In 2012, the modernization of HP Żarnowiec with a capacity of 716 MW was completed, and in 2013 the modernization of HP Żydowo was completed. After modernization, the power plant's power increased to 167 MW. In 2014, the Pilchowice hydroelectric power plant was modernized, raising its capacity to 13.4 MW and the Otmuchów power plant, increasing its capacity to 6.975 MW.

In addition to hydropower facilities that have been implemented, it is necessary to mention projects that were not completed. Plans for hydro-technical engineering of the Vistula River, which has the highest energy potential (52% of the energy potential of the Polish rivers), were prepared at the beginning of the twentieth century. The first concept of hydro-technical development of the Vistula River was established in the 1950s. According to [16], the entire Vistula River was to be developed with 42 stages, of which several facilities were constructed in the upper reaches of the Vistula and only one in the lower reaches (HP Włocławek). As the analysis of energy management mainly of the lower Vistula was conducted, the best solutions were sought for, and the development concepts were subject to various modifications. As a result, at the end of the 1980s, with a connection to the economic crisis, plans for hydro-technical construction of the lower Vistula were abandoned. The return to the hydro-energetic development of the lower Vistula occurred only at the beginning of the second decade of the twenty-first century, when the location of the next channelized stage below Włocławek was considered. From among 4 possible variants of the location of the new stage, Siarzewo was selected due to the possibility of permanent protection of the Włocławek dam [17]. In the following years, long-lasting administrative procedures for obtaining a water-law permit for the implementation of investment in Siarzewo were started, which eventually ended in 2017.

17.4 Types of Hydroelectric Plants

Hydroelectric power plants in Poland can be divided according to several criteria:

- method of storage and use of water (characteristic of cooperation with the energy system),
- water head (height between the upper and lower level of water),
- the way of water accumulation,
- the location of the power plant,
- level of power production,
- the type of turbines used.

The criterion for the use of water is divided between power plants:

- run-of-river (diversion), these are flow plants that do not have a storage reservoir, working in continuous motion, electricity is generated depending on the amount of water flowing (working time – over 4000 h/year) (HP Włocławek, HP Wały Śląskie),
- storage hydropower (impoundment) equipped with a storage reservoir where water is accumulated (working time 2000–4000 h/year), also usually flows power plants operating in the daily alignment mode,
- mixed pumped storage power plants (working time up to 2000 h/year) with periodic flow regulation; the power plant is equipped with storage reservoir, but a relatively small inflow forces periodic shutdown of the power plant with the purpose of filling the reservoir; depending on reservoir capacity, breaks can range from several hours to several days; the power plant is equipped with pumps which can water back to upper reservoir (HP Niedzica, HP Dychów),
- pumped storage power plants, equipped with an upper and a bottom reservoirs between which water flows; down by turbine operation and power generation; up during pump operation; generators in this phase work in the motor mode and consume electrical energy (HP Żarnowiec, HP Porąbka-Żar); the power plants in this system do not produce energy, but allow it to be stored in the form of potential energy; the energy balance of the pumped storage power plant without its supply is always negative, the storage efficiency is quite high and reach to approx. 90%,
- cascade power plants, in which two or more power plants use the same water, it mince that runoff of the one stage may be used as inflow of the next stage; this is so-called shift work; if the level of the upper reservoir of the lower stage starts at the same level as the lower reservoir of the higher level, it can be named as a compact cascade; if the compact cascade covers the whole river, then it is possible to use all its energy potential (e.g. Radunia cascade, which uses about 60% of its energy potential, the San cascade).

Due to the height of the fall, hydroelectric power plants in Poland can be divided into:

- high head with a difference of water levels reaching 100 m and more (HP Porąbka Żar – 440 m, HP Żarnowiec – 125 m),

- medium head with a difference of water levels in the range of 30–100 m (HP Żydowo – 82 m),
- low head power plants with a water level difference in the 1–30 m range (HP Włocławek – 8.8 m, HP Wały Śląskie – 4.7 m).

Due to the method of water accumulation hydropower plants in Poland can be divided into:

- storage reservoir with an embankment dam (HP Porąbka, HP Koronowo),
- storage reservoir with a heavy (gravity), concrete or stone dam (HP Solina, HP Pilchowice).

Due to the location of the power plant, they can be divided into:

- dam-toe powerhouse (HP Solina) it is located just adjacent to the dam,
- with a pressure derivative pipeline (penstock) (HP Żarnowiec, HP Porąbka-Żar) and a power plant located at the level of the lower (tail) reservoir,
- with an atmospheric derivative channel (HP Dychów),
- run-of-river with turbines located inside weir (HP Włocławek, HP Dębe, HP Wały Śląskie).

Some hydroelectric plants may show the combined features of several technical solutions, e.g. HP Koronowo is a power plant with an embankment dam made on the Brda River, water is directed in a new way, by several lakes. To create a water channel, the ditches are made between lakes. Ultimately water flows out of the Białe Lake thru an atmospheric derivation channel to the water gate from where it just flows down by the penstock to the power house, several kilometers from the dam.

Due to the turbines used to power generating sets, hydropower plants can be divided into:

- hydroelectric power plants with Kaplan turbines (HP Włocławek),
- hydroelectric plants with Francis turbines (HP Solina),
- hydroelectric power plants with Deriaz turbines (HP Niedzica),
- hydroelectric plants with Pelton turbines (HP Bielkowo),
- hydropower plants with Archimedes turbines (SHP).

It is also possible to use other turbines, e.g. Banki-Michell (crossflow) or VLH (Very Low Head turbine), but their use in Poland is quite rare.

The Kaplan turbine is by far the most commonly used turbine in Polish hydropower plants. Decisive in this respect is their optimal range of use for medium and low head in the 1–80 m range, as well as medium and large flows. The Kaplan turbine is a propeller-type turbine with variable blades geometry. The turbine blades can be adjusted during operation. The turbine is placed inside a scroll-shaped case whose task is to swirl the water entering the turbine. At the turbine inlet, there is a wicked gate, the blades of which are adjustable as well.

In the largest hydropower plants in Poland – pumped storage power plants, Francis turbines are usually used. It is due to a large difference in the levels of head and tail water as well as the possibility of their transfer from turbine work to pump operation

(HP Żarnowiec, HP Porąbka-Żar, HP Solina). The Francis turbine is intended for a power plant with a large head and a large flow of water. The Francis turbine runner has fixed, spatially contoured water-driven blades. Water inflow into the turbine is led through a spiral casing and wicket gates controlling the intensity and water flow angle.

The Deriaz turbines can also be used in pumped storage power plants. The turbines are an intermediate solution between a Kaplan and a Francis turbine. The turbine has strongly curved blades that are adjustably mounted in the runner, but at a different angle than in the Kaplan turbine. Thanks to this design, the Deriaz turbine can work both as a turbine and as a pump. In Poland, the Deriaz turbines were used at the mixed, pumped storage power plant in Niedzica.

The Pelton turbine is an impulse turbine. In Poland, it is very rarely used due to its usefulness in power plants with a very large head, which may exceed 1000 m and small flows of water. Among large hydroelectric plants in Poland, it is only used in the hydroelectric power plant in Bielków (head is 44 m) as an auxiliary turbine.

The recently used turbines in small hydropower plants in Poland are the Archimedes turbines, also called screw turbines. They can be used even on small watercourses of low flow and head. The turbine rotor is an Archimedes screw with a length of 4–6 m and a diameter of 1.5–2 m placed in the semicircular trough. Because such a power plant does not require complicated steel or concrete constructions it is also relatively cheap. The power plant with the largest number of Archimedes turbines is placed in HP Rosko on Noteć River, in which 6 turbines with a total power of 0.36 MW have been installed.

17.5 Location of the Largest Hydropower Plants in Poland (Above 10 MW)

The power of hydroelectric plants located in Poland is not large. This is mainly due to the lowland nature of our country and the lack of convenient locations for placing a large hydropower plant. Most of the flowing water belongs to the basins of the Vistula and Odra, which in mountainous terrain do not have too much water (flows are relatively small), but there are more water flow in the lowland areas, but the water head does not exceed a few meters. The result of this situation is the presence in Poland of only a dozen or so hydroelectric plants with an output exceeding 5 MW. The map of the deployment of these power plants is presented in Fig. 17.1. The list of the largest power plants is presented in Table 17.1. Most of them are run-of-river power plants, but the highest power has pumped storage power plants, which in Poland is 6. The largest of them (Żarnowiec, Porąbka-Żar and Żydowo) do not have their inflow, and in the others (Solina, Niedzica and Dychów), the own inflow let to produce of 5–10% of the output power.

The largest hydroelectric power plant in Poland is the Żarnowiec power plant located in Czymanowo in Kashubia. It is a pumped storage power plant built in

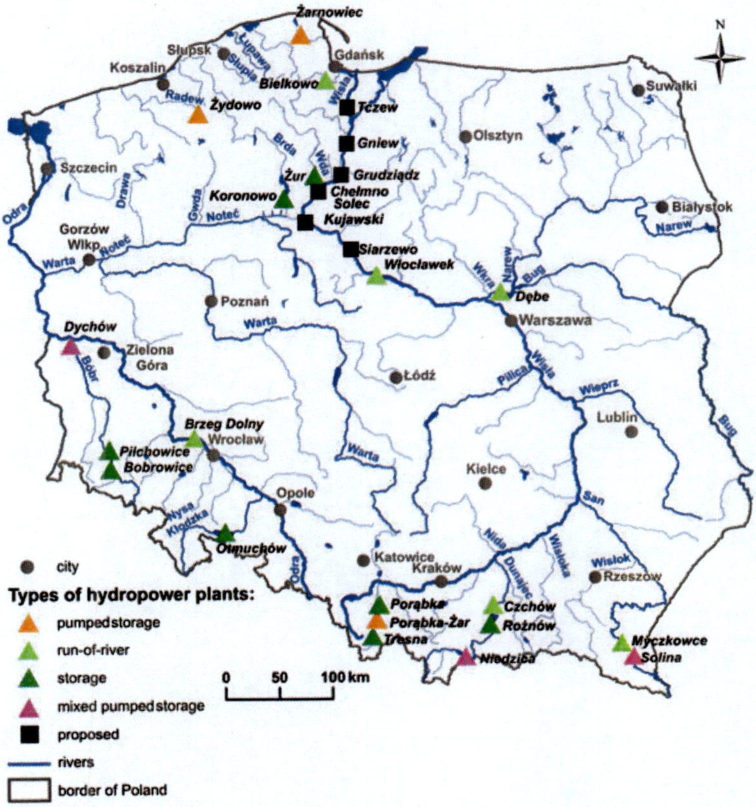

Fig. 17.1 Distribution of hydroelectric power plants with a capacity exceeding 10 MW in Poland

the 1970s as energy storage for a nearby nuclear power plant. The commissioning was carried out in 1983. Since the construction of the nuclear power plant did not materialize after the Chernobyl disaster, the power plant in Żarnowiec is used to equalize the load on the national grid in daily mode. The power plant produces 716 MW [18]. The capacity of the head reservoir is about 13.8 million m^3, which allows working for only 5.5 h. After this time, the Czymanowo reservoir is empty and in order to fill it, the turbines of the power plant must pump water back from the Żarnowieckie Lake, which is the tail reservoir of the Żarnowiec power plant. Four reversible Francis turbines with a diameter of 6 m were used for power generation and pump operation. The difference in levels between reservoirs is 125 m. The water is guided by four penstock pipes with a diameter of 5.4–7.1 m.

The Żarnowiec power plant starts to work only when there are energy shortages in the grid network, and only a fast-running pumped storage power plant can cover them. The starting takes a couple of minutes and is conducted from the National Power Dispatch in Warsaw. The power plant can also be used to support the start-ups of coal-fired power plants and is a network load for failures that can disconnect

Table 17.1 Hydroelectric power plants in Poland with a capacity above 10 MW

Nr	Hydropower plant	River	Year of built/modernisation	Turbine output power (MW)	Pump power (MW)	Head H (m)	Flow $Q_{max}/Q_{śr}$ (m^3/s)	Annual energy production (GWh)	Capacity of head reservoir (mln m^3)	Area of head reservoir (km^2)	Area of catchment (km^2)	Type of turbine/pump
1	Żarnowiec	–	1983/2012	716.0	800.0	125.0	700	–	13.8	1.35	1.35	4x Fpt[a]
2	Porąbka-Żar	–	1979	500.0	542.0	440.0	140	–	2.0	0.14	0.14	4x Fpt
3	Solina	San	1968/2003	200.0	41.4	60.0	400/19.4	85.0	472.0	22.00	1174.5	2x Fpt 2x Ft
4	Żydowo	–	1971/2013	167.0	136.0	82.7	240	–	3.3	1.00	1.00	2x Fpt 1x Ft
5	Włocławek	Wisła	1970	160.2	–	8.8	2190	739.0	408.4	59.20	171,250.0	6x K
6	Niedzica	Dunajec	1997	92.8	89.0	45.6	250/24	70.1	132.8	12.26	1147.0	2x Dpt
7	Dychów	Bóbr	1936/1951/ 2014	88.0	23.3	28.0	357/44	82.5	4.0	1.00	5592.0	3x K 4x ZP
8	Rożnów	Dunajec	1942	56.0	–	29.0	240/61	125.0	165.0	16.00	4865.0	4x K
9	Koronowo	Brda	1961	27.5	–	26.0	120/22	40.8	80.6	15.60	4092.0	2x K
10	Dębe	Narew	1963	21.2	–	5.7	428/270	91.0	90.0	30.30	69,500.0	4x K
11	Tresna	Soła	1967	21.0	–	20.4	122/18.8	2.8	94.6	9.64	1036.6	2x K
12	Pilchowice	Bóbr	1912/2014	13.3	–	38.0	40.7/15.6	30.5	50.0	2.40	1206.0	6x Ft
13	Porąbka	Soła	1953	12.5	–	21.0	64.6/18.8	2.5	26.6	3.80	1082.0	2x K 1x Ft

[a]Fpt – Francis pump-turbine, Ft – Francis turbine, K – Kaplan turbine, Dpt – Deriaz pump-turbine, ZP – independent pump unit

coal-fired power plants from customers, which is associated with a sharp power drop in steam turbines and the possibility of damaging them.

The second largest hydropower plant in Poland is the Porąbka-Żar power plant built in 1979. It is also a pumped storage power plant, which has no natural inflow. The power plant's output power is 500 MW. The head reservoir of the power plant with a capacity of approx. 2 million m^3 was built on the top of the Żar mountain near Bielsko-Biała. Because the difference in levels between the head and tail reservoir is 440 m with less water flow than in Żarnowiec, the Porąbka-Żar power plant achieves little less power. The energy is generated by four reversible Francis units, which are also used for pumping work. All power plant installations, including the powerhouse, are located inside the Żar mountain, and water flows by rock-cut tunnels. Turbine operation can last up to 4 h and pump work 1.5 h longer [19].

The Solina power plant is much smaller than the previous two. Its maximum capacity is about 200 MW, finally obtained after modernization carried out in the years 2000–2003. It is also a pumped storage power plant, but its construction and functioning are slightly different from the previous two. The reservoir of the power plant was built on the San River near the village of Myczkowce. It is, therefore, a pumped storage power plant with its inflow, which on average is about 20 m^3/s (so it should be called mixed pumped storage HP). The power plant's reservoir is maintained by a heavy concrete dam. It allowed the creation of the Solina reservoir in the San River Valley and its tributaries. The reservoir collects water from the Bieszczady region. The water head can reach a maximum of 60 m. The Solina power plant is equipped with four reversible Francis pump-turbine units of 50 MW each. The annual energy production is about 85 GWh from inflow water and about 25 GWh from pumped back water. Below the Solina Power Plant, there is a 8.3 MW Myczkowce hydropower plant that also indirectly uses the water collected behind the Solina dam [20].

The biggest run-of-river plant in Poland is the Włocławek power plant on the Vistula, with a maximum output of 160.2 MW. The channelized stage, which includes the hydroelectric power plant, was officially opened on October 17, 1970. It was the first object of the planned serial construction of the Vistula with 8–9 channelized stages, which were to constitute the so-called cascade of the lower Vistula. The regulation of the river in its lower course was supposed to equalize the flow and secure further reservoirs against the erosion of the Vistula river bed below the dam. Currently, the plans for the construction of a hydroelectric power plant in Siarzewo are planned, the date of commissioning is estimated for 2025.

As a result of the construction of the dam in Włocławek, the head of the power plant up to 8.8 m was obtained. The original dam height calculated in the design phase (11.8 m) has systematically increased as a result of erosion changes of the tail pool, and currently, it is 12.83 m after reconstruction of the dam. Due to the environmental issues and the technical condition of the Włocławek dam, since 2002 the power plant has been operating in a run-of-river system, using the current flow to the size of the maximum flow of the turbines, i.e. 2190 m^3/s [21]. This allows for the production of approximately 740 GWh of electricity per year.

The power plant is equipped with six generator sets driven by vertical Kaplan turbines. Each turbine is equipped with 4 adjustable blades, and the diameter of its rotor is 9 m.

In order to protect the dam in Włocławek against the erosion of the river bed in 1997–2001 500 m below the dam, a stabilizing stone ramp with a height of 2 m was built.

The Żydowo hydropower plant is the third of the classic pumped storage power plants. It was built in 1971 in northern Poland near Koszalin. It uses the particularly favorable location of two natural lakes located less than two kilometers apart, differing by an elevation of the water surface by about 83 m. The upper reservoir of the power plant is used by Kamienne Lake, which does not have any significant natural inflow. Water with a maximum flow exceeding 240 m^3/s flows down with three penstocks with a length of 461 m and a diameter of approx. 5 m to Lake Kwiecko, on whose bank the power house was built. Over 3 million m^3 of water is used to generate electricity. The Żydowo power plant uses three generating sets, two of which can be operated in the pump mode, forcing water back into the Kamienne Lake. All generator sets use Francis turbines. The two reversible units are equipped with FR-25 turbines and the unit in the classical configuration is the FR-30 turbine. The operation of the power plant causes fluctuations in the water level by almost four meters in the upper reservoir and just over three meters in the lower one. The original capacity of the power plant was 150 MW, but as a result of modernization works completed in 2013, its output power increased to 167 MW.

The Niedzica mixed pumped storage power plant is the largest power plant of the Niedzica Hydroelectric Power Plant Company. It is a power plant with its inflow, using the water resources of the upper Dunajec. It was put into use in 1997. The construction of a large hydroelectric power plant on the Dunajec River was planned almost from the beginning of the twentieth century; however, due to the high risk for landscape values and the resistance of the inhabitants of the surrounding areas who had to be displaced, its construction was very long. Eventually, the decision to build a dam in Niedzica was made in the 1950s, but construction work was not started until 1975. The construction lasted for over twenty years in the atmosphere of protests of environmentalists. After completion of the construction, the reservoir was filled in May 1997, and already in July, almost on the day of official commissioning, as a result of the flood of the "century" it underwent a functional test.

The power plant's power is almost 93 MW, obtained from two generating sets using reversible Deriaz turbines. The maximum water flow through the turbines was set at 250 m^3/s. Because the average flow of the Dunajec at Niedzica is less than 24 m^3/s (10%), the remainder of the flow must come from pumping work. For turbine operation, only approx. 20 m of damming is used, from maximum 60 to approx. 40 m (132.8 mln m^3 of water) above the level of the lower reservoir used by the hydroelectric power plant in Sromowce Wyżne. The Niedzica Power Plant, due to small differences in the price of energy between day and night, is currently operating only in turbine mode [22].

The Dychów hydropower plant is the largest power plant of the Dychów Hydroelectric Power Plant Company comprising 17 hydropower plants located on the

Bóbr, Kwisa and Nysa Łużycka rivers in southwestern Poland. Its total output power is 88 MW. The remaining power plants of the Company are small hydropower plants (SHP) with a capacity of 0.8–2.64 MW, reaching a total of approx. 15 MW. The Dychów power plant is the smallest of the Polish mixed pumped storage power plants, in which the power of 4 pumping units is slightly less than 1/3 of the output power of turbine sets. For generating energy, it uses three generating sets equipped with 30 MW Kaplan turbines. It was built in 1933–1936 on the final part of the Bóbr River just before its mouth to the Odra River near Krosno Odrzańskie. The upper reservoir of the power plant is Dychowskie Lake located 28 m above the water level in the lower reservoir located in the Bóbr tributary. The Dychowskie Lake is fed from the Bóbr River with a derivation channel with a length of over 20 km. The beginning of the channel is located in the town of Krzywaniec near Nowogród Bobrzański. Because the Bóbr, on the section between the beginning of the channel and the mouth of the Odra River, strongly meanders, the channel shortens the distance to the Dychowskie reservoir by more than half. Dychowskie Lake is a daily equalizing reservoir, which means it fills almost 24 h. Emptying the upper reservoir with the full power of the turbines takes only three hours. In low water periods and low natural inflow, it is possible to start pump stages with a power of approx. 23 MW [23].

The hydroelectric power plant in Rożnów was built during World War II. Its construction was completed in 1942. It is a classical storage-type power plant with a reservoir creating the Rożnowskie Lake. This reservoir fulfils energy, retention, flood and recreational functions.

The power plant was located in the foothills gorge of the Dunajec River, near the village of Rożnów. Taking advantage of the favorable terrain, the Dunajec trough was partitioned with a heavy dam of a concrete type, based on two nearby mountains, between which the river flowed. Eventually, the dam with a length of 550 m and a height of 32.5 m was built. The width of the dam in the upper side is 9 m and at the base 40. The foundations of the dam are recessed into the ground to a depth of 17 m.

The maximum power of the power plant of 56 MW is obtained from four generating sets of 14 MW each, equipped with Kaplan turbines. The water head is 29 m with an average flow of 61 m^3/s, and the maximum flow is 240 m^3/s. The annual production of electricity is on average 125 GWh [24].

The Koronowo power plant was established in 1961 about 20 km north of Bydgoszcz. The preliminary design, including a dam partitioning the Brda riverbed, was developed already in the 1930s by prof. Karol Pomianowski and engineer Alfons Hoffman. The project envisaged the use of several post-glacial lakes lying along the former Brda bed, but about 20 m above its level. The embankment dam was built in the village of Pieczyska above Koronowo, and the dammed water of the river was led into the connecting lakes Lipkusz, Czarne and Białe delivering water to the area of Samociążek village. The derivation channel created by the lakes allowed to increase the water head from 20 m (on the dam in Pieczyska) to 26 m on the end of 1350 m long channel, led out from the Białe Lake near of Samociążek. The water channel is terminated with a water gate, from which two penstocks with a diameter of 5 m and a length of 96 m are derived. The water flows down 26 m into two generating sets with a total power of 27.5 MW driven by two Kaplan turbines. The maximum flow of the power plant is 120 m^3/s, and the average flow is about 22 m^3/s. The power plant operates in peak mode, generating approx. 41 GWh of electricity annually. Its utilization rate does not exceed 20%. Water outflow from the power plant is led again to Brda River just behind Koronowo.

The Dębe power plant was built on the Narew River about 20 km from its mouth to the Vistula near Modlin, about 35 km north from the center of Warsaw. The construction was completed in 1963. It is a run-of-river hydropower plant with a retention reservoir creating the Zegrzyńskie Lake. The reservoir is created with a 117 m long partially earthfilled (embankment) and partially concrete buttress weir with a five buttress. A bridge was built along the weir, which is followed by provincial road No. 632. The reservoir is fed mainly by the Bug and Narew rivers and several smaller tributaries. Through the Żerański Channel, it is connected with the Vistula at the north part of Warsaw. The dam increase of the water level up to a height of 5.5 m. The total capacity of the Lake is about 81 mln m^3 and the energy-useful capacity is about 15 mln m^3. With an average annual flow of 270 m^3/s and 21.2 MW of power produced by four Kaplan turbines, the power plant produces 91 GWh of electricity per year [25].

The Tresna Power Station was built in 1966 on the Soła River, a tributary of the upper Vistula. It is a reservoir power plant with an embankment dam filled with clay. As a result of the construction of the dam, the Żywieckie Lake was established, gathering approx. 94 mln m^3 of water. The length of the dam is 310 m, and its height is 38 m. The reservoir of the Żywieckie Lake fulfills flood, energy and water regulation functions for the Soła and its tributaries, but also is of great importance for recreation in the Upper Silesia region. The power plant uses two generating sets using Kaplan turbines with a total capacity of 21 MW. The maximum flow is 122 m^3/s, while the average water flow is 18.8 m^3/s, which, with a nominal head of 20.4 m, allows production of 2.8 GWh per year.

The Pilchowice hydropower plant (Fig. 17.2) was built in 1912, and after the modernization completed in 2014, it reached the power of 13.25 MW. The power plant is located on the Bóbr River near Jelenia Góra. The river was partitioned by a heavy arc-masonry dam with a stone structure that accumulates water to a height of 38 m. The height of the dam is 62 m, and its span is 270 m. Behind the dam was created the Pilchowickie Lake with a total capacity of 50 mln m^3 and an active zone capacity of 16 mln m^3. Below the dam, a power house was built, in which 4 generating sets with output power of 2.26 MW each, one set with a capacity of 3 MW and one with a power of 1.21 MW were placed. All generator sets are powered by Francis turbines [27].

The Porąbka Power Station on the Soła River was created in two stages distant by almost 30 years. At the end of the 1930s, as a result of the construction of a water dam, the Międzybrodzkie Lake was created, which was only for flood control functions. The dam project was developed, among others by the later president of Poland, prof. Gabriel Narutowicz. The height of the dam is 38 m and its length 260 m. In the 50s, a hydropower plant was built on the tail of the dam with a capacity of 12.5 MW. About 30 mln m^3 of water is stored in the Międzybrodzkie Lake, of which 19.5 million m^3 can be used for energy production purposes. The average Soła flow

Fig. 17.2 Hydropower plant Pilchowice I [26]

at the level of the dam in Porąbka is 18.8 m^3/s, which gives about 1.35 mln m^3 per day. The maximum flow is 64.6 m^3/s, and the nominal head is 21 m. The power plant in Porąbka is equipped with two generating sets using Kaplan turbines with a total capacity of 12.1 MW and small generating set with a 0.4 MW Francis turbine to supply own purposes. The annual energy production is 2.5 GWh.

17.6 Small Hydropower Plants in Poland

In addition to the largest hydropower plants mentioned, small hydropower plants play an important role in the Polish landscape. The division into categories of small and large hydropower plants is conventional. Large hydropower plants (HP) are those with nominal capacity above the limit value, while small (SHP) are those with a nominal capacity below this value. Up to now, there is still no international agreement on the limit value in the definition of "small" hydropower plants. The upper limits between small and large hydropower plants ranges between 2.5 and 25 MW. Still, in the 1970s, in Poland, a small hydropower plant was considered to be a facility with a nominal capacity below 1 MW [15]. In later years, the maximum limit of 5 MW was assumed to be the upper limit of SHP in Poland, and 10 MW has recently been proposed [28].

In countries such as Brazil, USA [29, 30] and Russia, the SHP limit is set at 30 MW [31]. For example, in China, the upper limit of output power produced in a

Table 17.2 Classification of hydropower generation

Hydrogenerator	Haidar et al. [44]	ESHA [28]	Paish [32]	Feeding
Small (MW)	<25	<10	<25	National power grid
Mini (MW)	<1	<2	<2	National power grid
Micro (kW)	6–100	<100	<500	Small community or remote industrial areas
Pico (kW)	<5	<20	<10	Domestic and small commercial loads

small hydropower plant is 25 MW [32], and recently even 50 MW [33–35]. Also, in Turkey, the limit of 50 MW is being talked about more often [36]. In Canada, the limit value is 50 MW [37].

As a rule, the maximum value of 10 MW is a widely accepted value worldwide [28, 32, 37], among others in Turkey [38, 39], Greece [40], Serbia [41]. In addition, the 10 MW limit is used in the preparation of statistical data and other documents common to all EU member states [42, 43].

Among the small hydropower plants, the division into mini-, micro- and picoelectric power plants is also distinguished. Also, in this case, the division into particular categories is not explicit. Such a division is mainly related to various legal and administrative procedures as well as public aid provided to the RES (renewable energy sources) sector in individual countries.

As going about of small hydropower plants, some authors use the classification limits in Table 17.2. This classification includes only the nominal capacity of hydropower plants. In some countries, financial support depends not on power, but on the amount of electricity produced.

For the purposes of this article, any hydropower plant with a capacity not exceeding 10 MW will be considered as a small hydropower plant.

According to Hoffmann [45], an inventory of all objects powered by water power in Poland carried out in 1954, showed 6630 active and 800 inactive objects. Later, the vast majority of those objects were devastated or immobilized due to insufficient power. Changes in this area occurred in the 1980s when regulations were introduced stating that a small hydropower plant (less than 5 MW) can be owned by a private operator. Most hydro-technical facilities were owned by the state, while very small facilities, usually below 0.5 MW or requiring major renovations, went into private hands. In total, approximately 650 facilities with a total capacity of approximately 80 MW have been reconstructed [45]. As of 2016, a total of 745 hydropower plants with a capacity below 10 MW are registered in Poland, including 579 hydropower plants with a capacity less then 0.3 MW, 98 with a capacity of up to 1 MW, 63 power plants with a capacity below 5 MW and 6 power plants with a capacity of up to 10 MW (Wały Śląskie, Żur, Myczkowce, Czchów, Bielkowo, Otmuchów).

Optimal conditions for the construction of small hydropower plants are primarily in mountain and foothill areas, as well as in the Pomeranian and Masurian areas and Greater Poland area. The rivers Gwda [46], Wda, Mątawa, Radunia and many others are examples of proper energy usage at almost whole river length. These are mostly run-of-river power plants with small reservoirs created as a result of river damming. They accumulate water and use it in a short time. At this point, special attention should be paid to the applied technical solutions in the construction of hydropower plants, which are adapted to the terrain conditions and time of their operation. In many cases turbines installed in hydropower plants have been working for almost 100 years. An example is the Podgaje power plant with derivation pipe built on the Gwda River [47]. Old buildings that have remained after closed water facilities perform various functions, including serving as hotel base, agritourism facilities, museums and others.

Due to the small head and limited flow of water in SHP, Kaplan turbines are usually used. In recent years, Archimedes turbines have become more and more popular due to their simple construction. In power plants with a very small head not exceeding 1 m, VLH (Very Low Head) turbines can be used, which are similar to Kaplan's turbines, but are set at a slight angle and can be built directly into the wall of the weir damming the water.

17.7 Power Capacity and Electricity Production in Hydropower Plants in Poland Against the Background of Renewable Energy Sources in 2002–2016

The Polish hydropower situation compared to other renewable energy sources in recent years is best reflected by the total amount of nominal capacity in hydropower plants, and the amount of electricity generated by them. In the years 2002–2016, there was a steady increase in the capacity of power plants that use renewable energy sources. The largest increase was recorded in the power installation in wind farms (intensive growth since 2010) and power plants based on solid biofuels (since 2012). Since 2011, there has also been an increase in nominal power in solar energy (Fig. 17.3). The exception is the installation of power in hydropower, the increase of which in recent years is less than 1%.

Due to numerous investments in renewable energy sources, especially in wind energy and solar energy, the structure of nominal power in particular types of power plants has changed over the last 15 years. The dominance of the share of nominal capacity in hydropower plants was noted in 2002 and constituted 94.8% share in total power from renewable energy sources (Fig. 17.4), while in 2008 this share amounted to 60.1% (Fig. 17.5). Since 2009, the share of power in hydropower plants exceeded 50% of the share and was systematically falling. As a result, in 2016 it amounted to only 12.3% (Fig. 17.6).

The increase in nominal capacity in renewable energy sources translates into the amount of electricity produced. In total, 166.6 TWh of electricity was produced in Poland in 2016, of which 22.8 TWh (13.7%) came from renewable energy sources

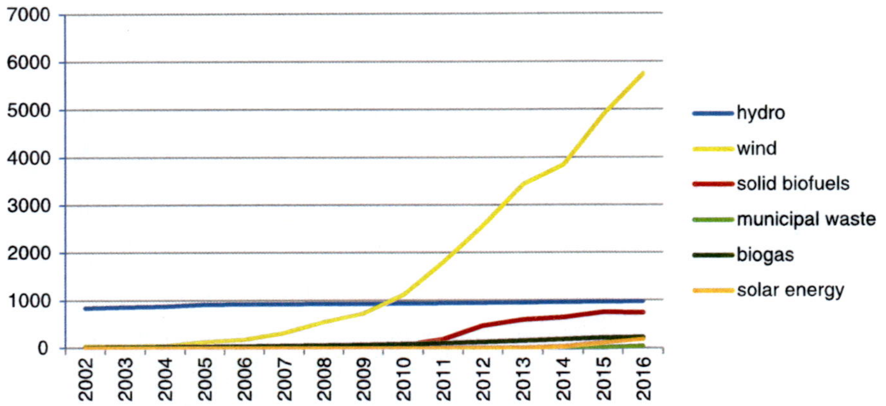

Fig. 17.3 Capacity of power plants using renewable energy in MW in the years 2002–2016 (own study based on the data from [10])

Fig. 17.4 Structure of capacity of power plants using renewable energy in MW in 2002 (own study based on the data from [10])

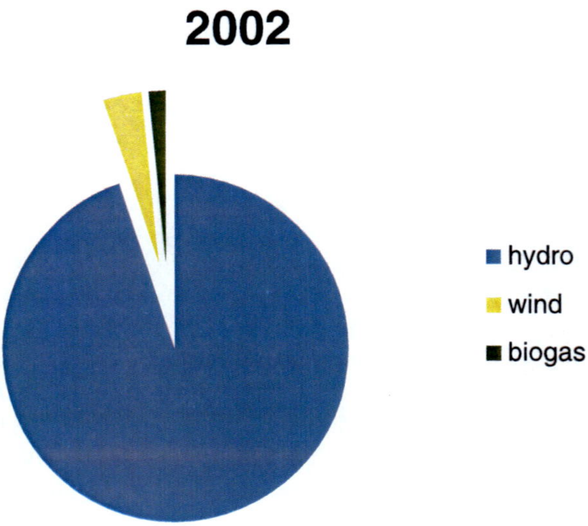

(Fig. 17.7). The energy produced by hydropower represented only 1.6% of total energy produced in 2016 when in 2002 it was 2.7% of total energy produced in Poland. Figures 17.8, 17.9 and 17.10 show the change in the structure of shares of individual carriers in energy production. In 2016, the energy produced in hydropower plants accounted for only 9.3% of energy generated in renewable energy sources, and in 2002 it was 82.4%.

Electricity produced in small hydropower plants (<1 MW) and medium (1–10 MW) hydropower plants shows a relatively stable trend without sudden increases and decreases. The exception is the year 2010, which belonged to extremely rainy years, where the river flow was higher than the average for many years. In high-power

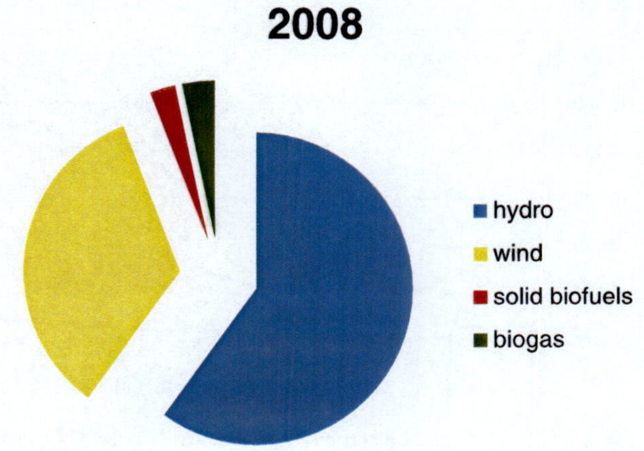

Fig. 17.5 Structure of capacity of power plants using renewable energy in MW in 2008 (own study based on the data from [10])

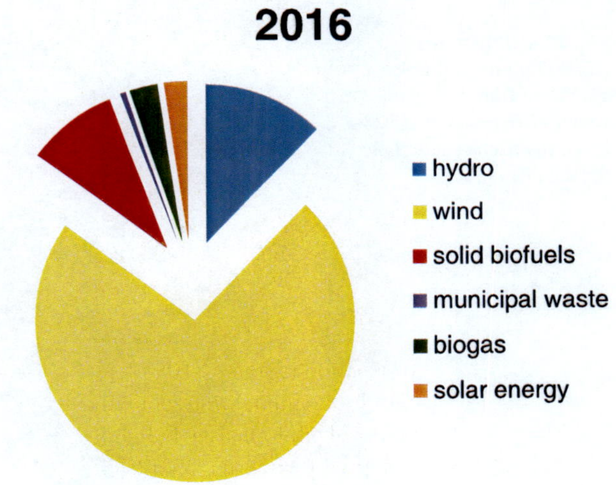

Fig. 17.6 Structure of capacity of power plants using renewable energy in MW in 2016 (own study based on the data from [10])

hydropower plants (>10 MW), significantly higher increases and decreases can be observed compared to the previous year, which are ±30%. The main reason for this situation is meteorological conditions that have an impact on the volume of flow in rivers [48], which was reflected in the amount of electricity produced. In addition to meteorological conditions, the amount of energy produced is affected by renovation work carried out on water installations in large hydropower plants.

The volume of electricity production in hydropower plants, divided into three capacity ranges (<1, 1–10, >10 MW), is shown on Fig. 17.11.

The share of electricity produced in hydropower plants in Poland and other countries in relation to total energy produced from renewable energy sources in 2012–2015 is shown on Fig. 17.12. For comparison, individual European countries were selected, of which particular attention was paid to countries neighboring Poland that

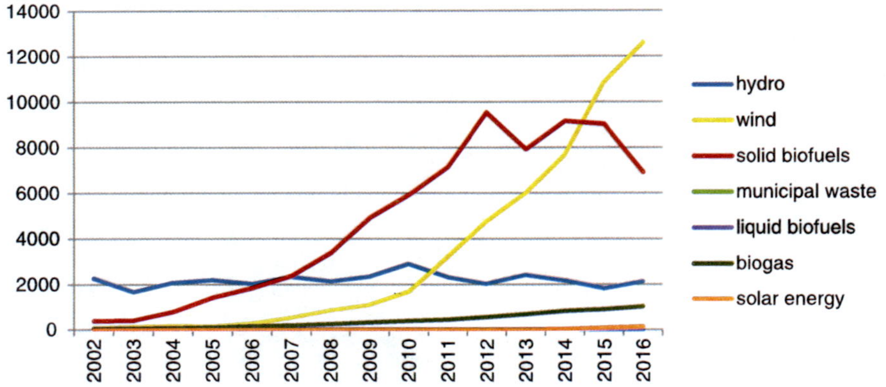

Fig. 17.7 Production of electricity generated from renewable sources in Poland in 2002–2016 in GWh (own study based on the data from [10])

Fig. 17.8 Structure of electricity production generated from renewable sources in Poland in 2002 in GWh (own study based on the data from [10])

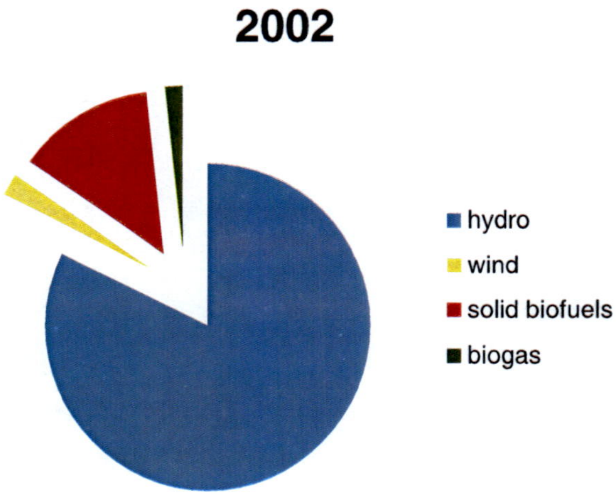

have similar physical-geographical conditions. In the European Union of 28 Member States, the share of electricity produced in hydropower plants is between 36.5 and 43.8%. The largest producers of energy in hydropower plants include Austria and Slovakia (over 65%), the main factor here being landscape conditions. The opposite is the Netherlands, whose share reaches only 1%, and energy derived from wind dominates. Countries neighboring Poland, such as Germany, have a slightly larger share in the production of energy from hydropower (currently dominated by wind and solar energy). Lithuania, in turn, has a share of hydropower in renewable energy sources more than twice as large as Poland. It is because of a small production of electricity from other sources classified as renewable energy sources.

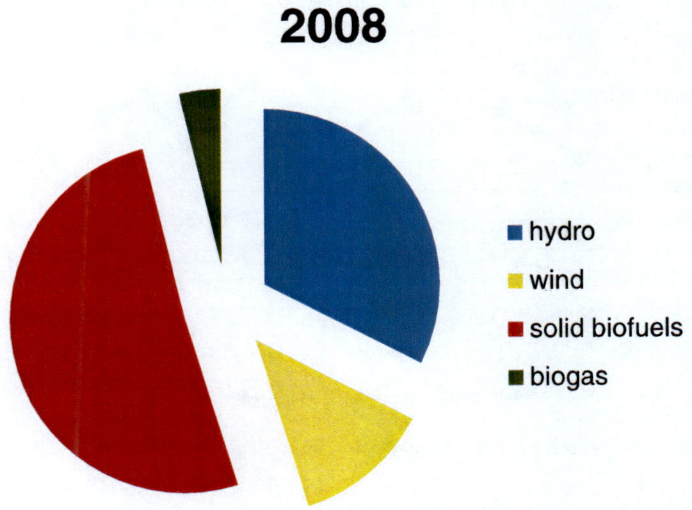

Fig. 17.9 Structure of electricity production generated from renewable sources in Poland in 2008 in GWh (own study based on the data from [10])

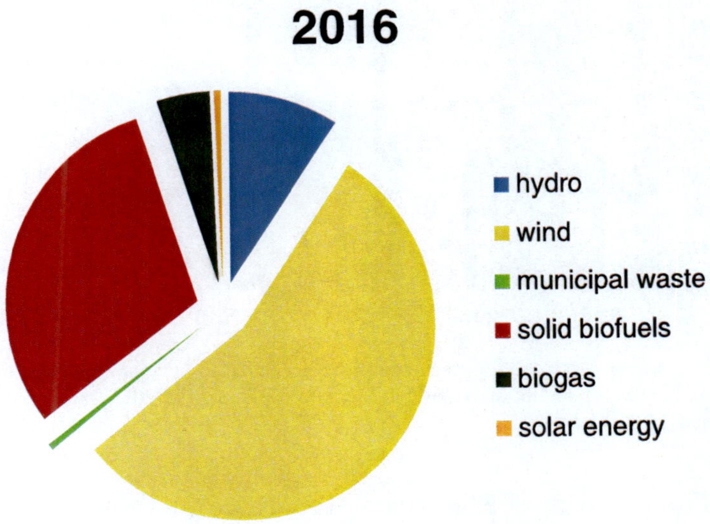

Fig. 17.10 The structure of electricity production from renewable sources in Poland in 2016 in GWh (own study based on the data from [10])

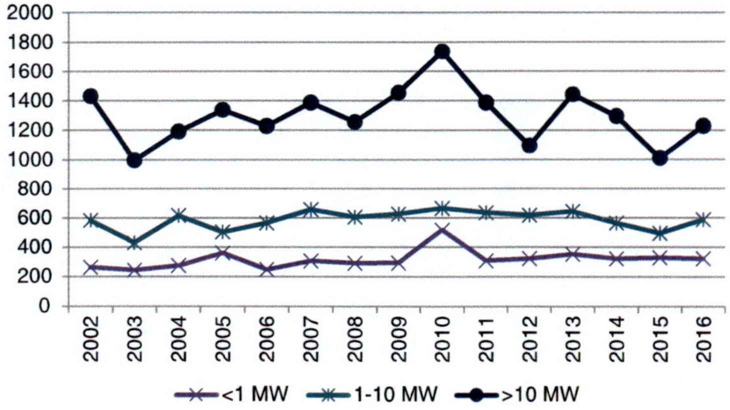

Fig. 17.11 Electricity production in GWh in hydropower plants in the years 2002–2016 (own study based on the data from [10])

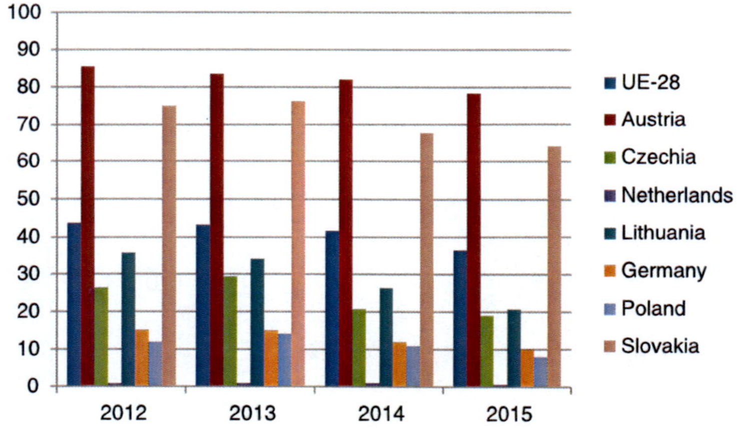

Fig. 17.12 Structure of electricity production in % generated in hydropower plants of selected countries on the background of all renewable energy carriers in 2012–2015 (own study based on the data from [10])

17.8 The Amount of Electricity Generated in Total on the Background of Renewable Energy Sources

Electricity is currently the most universal and easy to use a form of energy, without which it is difficult to imagine everyday existence. We are completely dependent on it, even though we have been using it for over 150 years. The majority of energy in Poland is produced in coal-fired power plants burning hard coal and lignite; small investments are carried out in the use of gas turbines fed with natural gas. Unfortunately, these are objects that use fossil fuels that emit huge amounts of CO_2 and other toxic gases and solid particles during combustion. Due to the threat of climate, renewable energy sources are becoming more and more popular.

The global energy potential is 24,739.3 TW in total, of which the maximum non-renewable energy is 1655 TW, i.e. 6.7% (coal 900 TW, uranium 90–300 TW, oil 240 TW and natural gas 215 TW). The global renewable energy potential is in total 23,084.3 TW, i.e. 93.3% of which is attributable to tides (0.3 TW), waves (0.2–2 TW), geothermal (0.3–2 TW), water (3–4 TW), biomass (2–6 TW), wind (25–70 TW), solar (23,000 TW). 93% of the world's energy potential is renewable energy, and only less than 7% is non-renewable energy. If we were able to use all the solar energy, we would be self-sufficient for many years. The problem with solar energy is that it is quite dispersed (it reaches about 180 W/m^2 to the earth's surface) and it is difficult to use it. Current methods of direct use in photovoltaic cells or solar plants are ineffective. Also, the potential of hydropower, which accounts for 25% of the world's energy consumption compared to other types of renewable energy, is relatively small and amounts to only 0.017%.

The total nominal capacity in hydropower (including pumped storage power plants) in 2016 was 1246 GW (1.246 TW), i.e. about 31% of the world's hydroelectric potential. With this power, it is estimated that in 2016, 4102 TWh of electricity was produced [49]. The growing demand for electricity results from the development of civilization. Hydropower belongs to renewable energy sources that do not pollute the natural environment. Systematic growth of power in newly built hydropower plants in the world is still observed. The largest increase in this area in recent years has been recorded by China and Brazil. It is forecasted that the rapid development of hydropower will take place in African countries due to the lack of sufficient electricity there.

In 2016, the total nominal capacity of hydroelectric power stations, including pumped storage power plants, amounted to 223.008 GW in Europe, of which the six most powerful countries are: Norway (31.626 GW), France (25.405 GW), Italy (21.884 GW), Spain (20.354 GW), Switzerland (16.657 GW), Sweden (16.419 GW). Poland is in the 20th position with the nominal capacity of 2.351 GW (of which almost 75% is pumped storage) [49]. Only approx. 600 MW is generated from flow work. Because the production rate of hydropower plants during a year is around 40% in our country, even if the full energy potential of all rivers in Poland is fully exploited, we will not be able to compete with these countries. The technical potential indicates that it is still possible to increase the power of the Polish run-of river and storage power plants to about 1.6 GW (14 TWh/year). So in the best case, we could increase the power obtained from hydropower plants (without pump plants) by about 1 GW. This means that we can build 5–10 large hydropower plants and several hundred to several thousand SHPs. It can be concluded that despite the lack of potential, there is still much to be done in Poland in the field of hydropower.

17.9 Plans for the Development of Hydropower

Despite many concepts that were set for the purposes of the energy use of Polish rivers, especially the Vistula, many of them have not been implemented or have been implemented to a minimum extent. The stagnation in the construction of large hydropower plants in Poland was associated with the lack of financial resources for the implementation of the investment, and in recent years the environmental issues and objections of many environmental organizations decided on it [50].

An example of a long process associated with the implementation of the investment is the planned construction of the new dam in Siarzewo, below Włocławek, where the hydropower plant will be 80 MW. Public consultations regarding the choice of location (one of the 4 proposed by the investor) began in 2012, within the next 2 years, studies on the impact of investments on the environment were carried out, and at the beginning of 2015, a report on the environmental impact was submitted. After one year from submitting the report, the institution issuing the environmental decision (Regional Directorate of Environmental Protection) refused to give consent for the implementation of the investment. The investor appealed against the decision, and finally a positive environmental decision was issued on December 27, 2017. The construction of the HP stage is planned to start in 2020, and finish in 2025. According to various sources, the cost of building a dam in Siarzewo is estimated at approximately PLN 2.2 billion, i.e. EUR 512 mln. On 12.10.2018, the information about the construction of the stage in Siarzewo was announced by the Prime Minister, Mateusz Morawiecki, himself [51]. This investment will be carried out by the Polish National Water Company and Energy Company. A positive opinion on the construction of the dam in Siarzewo was an impulse to resume work on the energy use of the Vistula River.

In 2018, another concept for the development of the Vistula was presented, which includes new technical solutions for the development of the lower Vistula. The recommended variant of the lower Vistula development covers, beyond the Włocławek and the Siarzewo dams planned for implementation, other 4 dams in Solec Kujawski, Chełmno, Grudziądz and Gniew. Planned water heads will allow for the use of dams for energy production, but also secure the depths guaranteed in the river for the needs of inland waterway transport.

The total power of the lower Vistula cascade, depending on the accepted variant, can range from 507.1 to 857.4 MW, which enables an average annual energy production from 3428.4 to 4221.1 GWh [52]. The obtained results indicate that the possible implementation of a cascade of damming stages together with hydroelectric plants on the lower Vistula will make a significant contribution to energy production in the National Energy System (Table 17.3).

A slight increase in interest in small hydropower and construction of hydroelectric power plants occurred in 2005–2010 and was associated with numerous subsidies for the construction of new facilities and modernization of existing ones. It mainly concerned small objects with a capacity of less than 10 MW. As part of these activities, priority was given to investments that used former damming or water mills. Because

Table 17.3 Variant 2 large cascades [53]

L.p.	Location	Km of Vistula River	NPP (normal level of damming up) in m amsl	Head (m)
1	Siarzewo (environmental permission issued)	706.38	46.00	8.5
2	Solec Kujawski	758.70	37.50	8.5
3	Chełmno	801.55	29.00	7.0
4	Grudziądz	825.00	22.00	7.0
5	Gniew	876.03	15.00	7.2

the location of hydroelectric power plants is related to the location in river valleys, which are mostly covered by the Natura 2000 program, in many cases, the lack of the possibility to obtain an environmental decision was the main barrier to development. The decline in interest in the construction of hydroelectric plants was also due to the lack of financing and the drop in the prices of energy sold. Also, power utilities are not obliged to purchase electricity produced in renewable energy sources. As a result, not all mini hydropower plants can sign a contract with an energy company for the purchase of energy, which is explained by a small amount of energy produced and a high cost of connection to the power grid. The stagnation in the construction of small hydropower plants also results from changes in the law on renewable energy sources, which was introduced in 2015 and concerned the issue of prices of concentrated electricity from their generators. The new law of June 7, 2018, amending the act on renewable energy sources and some other acts [54] gives a chance to change the current situation on the market for the construction of new facilities and functioning of existing small plants.

The main factors stopping the development of hydropower in Poland include long-term administrative procedures, mainly related to obtaining water and construction permits, construction costs of the power plant, including lack of financial support from the state and lack of legal regulations [55, 56]. Similar problems have been noticed in other countries, e.g. USA [30], Great Britain [57].

17.10 Conclusions

The energy potential of the Polish rivers is used to a small extent. In the last 20 years, the share of hydropower plants (over 10 MW) has not been implemented in recent years due to the lack of favorable locations, significant population, environmental objections, as well as investment costs.

Bearing in mind Poland's obligations within the structures of the European Union, it seems reasonable to actively participate in the state supporting the construction of such facilities. The EU climate policy assumes obtaining a 20% share of energy from renewable sources in final energy consumption by 2020 and at least 27% by

2030. Poland in 2017 received only 14% of the share. The construction of hydroelectric power plants should be promoted as directly affecting the increase in the share of renewable energy sources in energy production, and also allowing to increase the diversification of supply sources and create conditions for the development of distributed energy. Currently, there is no state strategy in Poland for the development of small hydropower.

It is important to realize that every method of obtaining energy has its advantages and disadvantages, which entails environmental, economic and social consequences. The most important thing is, therefore, to take an optimal decision, strictly adapted to the prevailing conditions and situations, which is not an easy task.

The main advantage of hydropower engineering is a mature technology, which can be implemented relatively quickly, which allows us to start solving problems related to climate change.

Poland still has the potential to build new power plants and use new, more efficient turbines, and as a result it should translate into an increase in the share of ecological electricity generation that can meet future energy needs.

17.11 Recommendations

The content presented in this chapter is important for the proper management of water resources for the needs of hydropower in Poland, both on a national and regional scale.

The information presented can be used by policy makers to develop an appropriate energy policy for the country. Support from the state administration and establishing a multi-annual program will allow for greater interest in the modernization of existing hydropower plants and the construction of new facilities.

References

1. De Miranda A (2004) Aesthetic tradition and ancient technology: a case study of the waterwheel. In: Collins MW, Brebbia CA (eds) Design and nature II. WIT Press. www.witpress.com. ISBN 1-85312-721-3
2. Yannopoulos SI, Lyberatos G, Theodossiou N, Li W, Valipour M, Tamburrino A, Angelakis AN (2015) Evolution of water lifting devices (pumps) over the centuries worldwide. Water 7:5031–5060. https://doi.org/10.3390/w7095031
3. Bishop P, Muñoz-Salinas E (2013) Tectonics, geomorphology and water mill location in Scotland and the potential impacts of mill dam failure. Appl Geogr 42:195–205. https://doi.org/10.1016/j.apgeog.2013.04.010
4. Baranowski B (1977) Polskie młynarstwo. Zakład Narodowy imienia Ossolińskich, Wrocław-Warszawa-Kraków-Gdańsk
5. Podgórski Z (2009) Młyny wodne w krajobrazie Pojezierza Chełmińskiego. Prace Stud Geogr 41:151–161

6. Kaniecki A, Brychy D (2013) Średniowieczne młyny wodne i ich wpływ na przemiany stosunków wodnych na przykładzie zlewni Obry Skwierzyńskiej. Badania Fizjogr Ser A Geogr Fizyczna A61:145–156. https://doi.org/10.2478/v10116-010-0008-y
7. Gładkowski A (2008) Historia techniki młynarstwa polskiego. In: Mrugalski Z (ed) Inżynierowie Polscy XIX i XX wieku. Polskie Towarzystwo Historii Techniki, Warszawa
8. Dembińska M (1973) Przetwórstwo zbożowe w Polsce średniowiecznej (X–XIV w.). Instytut Historii Kultury Materialnej PAN, Ossolineum, Wrocław
9. Holt R (1988) The mills of medieval England. Basil Blackwell, Oxford-New York
10. Główny Urząd Statystyczny (GUS) Central Statistical Office. https://stat.gov.pl/
11. Juniewicz S, Michałowski S (1957) Elektrownie wodne. Państwowe Wydawnictwo Techniczne, Warszawa
12. Hoffmann M (1989) Present state and prospects of development of water-power engineering in Poland with special regard paid to demand for water turbines. Prace Instytutu Maszyn Przepływowych PAN, pp 90–91, Państwowe Wydawnictwo Naukowe, Warszawa-Poznań, pp 51–60
13. Mikulski Z (2004) Rozwój wykorzystania energii wodnej na ziemiach polskich. Gospod Wodna 12:503–509
14. Chudecki M (2017) Małe elektrownie wodne w Gródku i Żurze – cenne elementy krajobrazu kulturowego Wdeckiego Parku Krajobrazowego. Wydawnictwo Unitex, Bydgoszcz. ISBN 978-83-7296-455-7
15. Łaski A (1977) Elektrownie wodne – rozwiązania i dobór parametrów. WNT, Warszawa
16. Kosiński J, Zdulski W (2013) Hydropower potential of the Vistula. Acta Energetica 2(15):38–47. https://doi.org/10.12736/issn.2300-3022.2013203
17. Granatowicz J (2013) The next dam on the Vistula River below Włocławek. Acta Energetica 3(16):99–105. https://doi.org/10.12736/issn.2300-3022.2013307
18. https://pgeeo.pl/Nasze-obiekty/Elektrownie-wodne/Zarnowiec. Accessed 20 Jan 2019
19. https://pgeeo.pl/Nasze-obiekty/Elektrownie-wodne/Zar. Accessed 20 Jan 2019
20. https://pgeeo.pl/Nasze-obiekty/Elektrownie-wodne/Solina. Accessed 15 Jan 2019
21. Decision of the Marshal of the Kuyavian-Pomeranian Voivodeship of February 27, 2012
22. https://www.zzw-niedzica.com.pl/
23. Grządzielski I, Sroka K (2016) Elektrownia wodna Dychów jako źródło rozruchowe do odbudowy zdolności wytwórczych elektrowni cieplnych. Prz Elektrotech 92(10):190–198. https://doi.org/10.15199/48.2016.10.45
24. https://www.tauron-ekoenergia.pl/elektrownie/energia-wodna/ew-krakow/roznow. Accessed 15 Jan 2019
25. https://pgeeo.pl/Nasze-obiekty/Elektrownie-wodne/Debe
26. https://www.tauron-ekoenergia.pl/elektrownie/energia-wodna/ew-jelenia-gora/pilchowice-1#slide2. Accessed 15 Jan 2019
27. https://www.tauron-ekoenergia.pl/elektrownie/energia-wodna/ew-jelenia-gora/pilchowice-1. Accessed 15 Jan 2019
28. European Small Hydropower Association (ESHA). https://www.seehydropower.eu/partners/details.php?id=5
29. Hall DG (2006) Feasibility assessment of the water energy resources of the United States for new low power and small hydro classes of hydroelectric plants. U.S. Department of Energy, Idaho National Laboratory
30. Kosnik L (2010) The potential for small scale hydropower development in the US. Energy Policy 38(10):5512–5519
31. Panić M, Urošev M, Milanović Pešić A, Brankov J, Bjeljac Ž (2013) Small hydropower plants in Serbia: hydropower potential, current state and perspectives. Renew Sustain Energy Rev 23:341–349. https://doi.org/10.1016/j.rser.2013.03.016
32. Paish O (2002) Small hydropower: technology and current. Renew Sustain Energy Rev 6:537–556
33. Huang H, Yan Z (2009) Present situation and future prospect of hydropower in China. Renew Sustain Energy Rev 13:1652–1656. https://doi.org/10.1016/j.rser.2008.08.013

34. Zhou S, Zhang X, Liu J (2009) The trend of small hydropower development in China. Renew Energy 34(4):1078–1083. https://doi.org/10.1016/j.renene.2008.07.003
35. Kong Y, Wang J, Kong Z, Song F, Liu Z, Wei C (2015) Small hydropower in China: the survey and sustainable future. Renew Sustain Energy Rev 48:425–433. https://doi.org/10.1016/j.rser.2015.04.036
36. Yuksek O, Komurcu MI, Yuksel I, Kaygusuz K (2006) The role of hydropower in meeting Turkey's electric energy demand. Energy Policy 34:3093–3103. https://doi.org/10.1016/j.enpol.2005.06.005
37. Kaunda CS, Kimambo CZ, Nielsen TK (2014) A technical discussion on microhydropower technology and its turbines. Renew Sustain Energy Rev 35:445–459. https://doi.org/10.1016/j.rser.2014.04.035
38. Balat H (2007) A renewable perspective for sustainable energy development in Turkey: the case of small hydropower plants. Renew Sustain Energy Rev 11:2152–2165
39. Kucukali S, Baris K (2009) Assessment of small hydropower (SHP) development in Turkey: laws, regulations and EU policy perspective. Energy Policy 37:3872–3879. https://doi.org/10.1016/j.enpol.2009.06.023
40. Malesios C, Arabatzis G (2010) Small hydropower stations in Greece: the local people's attitudes in a mountainous prefecture. Renew Sustain Energy Rev 14:2492–2510. https://doi.org/10.1016/j.rser.2010.07.063
41. Josimovic B, Crncevic T (2012) The development of renewable energy capacities in Serbia: case study of three small hydropower plants in the "Golija" biosphere reserve with special reference to the landscape heritage. Renew Energy 48:537–544
42. Steller J (2013) Hydropower and its development. Acta Energetica 3(16):7–20. https://actaenergetica.org/pl/wp-content/uploads/2014/01/Steller-Hydropower-and-its-development.pdf
43. Pacesila M, Burcea SG, Colesca SE (2016) Analysis of renewable energies in European Union. Renew Sustain Energy Rev 56:156–170. https://doi.org/10.1016/j.rser.2015.10.152
44. Haidar AMA, Senan MFM, Noman A, Radman T (2012) Utilization of pico hydro generation in domestic and commercial loads. Renew Sustain Energy Rev 16(1):518–524
45. Hoffmann M (1992) Małe elektrownie wodne – poradnik. In: Small hydropower – handbook, 2nd ed. Towarzystwo Rozwoju Małych Elektrowni Wodnych, Gdansk
46. Kubiak-Wójcicka K, Kornaś M (2015) Impact of hydrotechnical structures on hydrological regime of the Gwda and Drawa rivers. Quaest Geogr 34(1):99–110
47. Kubiak-Wójcicka K, Muszyńska J (2011) Wykorzystanie odnawialnych źródeł energii do produkcji energii elektrycznej na przykładzie EW Podgaje. Gaz Woda Tech Sanit 10:371–373
48. Kubiak-Wójcicka K, Bąk B (2018) Monitoring of meteorological and hydrological droughts in the Vistula basin (Poland). Environ Monit Assess 190:691. https://doi.org/10.1007/s10661-018-7058-8
49. Hydropower status report (2017) International Hydropower Association. https://www.hydropower.org/sites/default/files/publications-docs/2017%20Hydropower%20Status%20Report.pdf
50. Ankiersztejn I (2013) The Lower Vistula cascade. Acta Energetica 3(16):70–74
51. https://kzgw.gov.pl/index.php/pl/aktualnosci/626-premier-mateusz-morawiecki-w-siarzewie-o-budowie-a-second-stage-to-aqueous-Wisła. Accessed 15 Jan 2019
52. Szydłowski M, Gąsiorowski D, Szymkiewicz R, Zima P, Hakiel J (2015) Hydropower potential in the lower Vistula. Acta Energetica 1(22):18–25. https://doi.org/10.12736/issn.2300-3022.2015102
53. Study of the Vistula River adaptation in the section from the Włocławek to the estuary to the Gulf of Gdańsk to the large and small cascades – modeling (2018) Państwowe Gospodarstwo Wodne Wody Polskie, Warszawa
54. Ustawa z dnia 7 czerwca 2018 roku o zmianie ustawy o odnawialnych źródłach energii oraz niektórych innych ustaw (Dz. U. 2018, poz. 1276)
55. Wiatkowski M, Rosik-Dulewska C (2012) Stan obecny i możliwości rozwoju energetyki wodnej w województwie opolskim. Woda-Środowisko-Obszary Wiejskie 12, 2(38):313–327

56. Zimny J, Michalak P, Bielik S, Szczotka K (2013) Directions in development of hydropower in the world, in Europe and Poland in the period 1995–2011. Renew Sustain Energy Rev 21:117–130
57. Bracken LJ, Bulkeley HA, Maynard CM (2014) Micro-hydropower in the UK: the role of communities in an emerging energy resource. Energy Policy 68(C):92–101. https://doi.org/10.1016/j.enpol.2013.12.046

Chapter 18
Waterways in Poland – The History, Present State and Future

Marcin Gorączko and Katarzyna Kubiak-Wójcicka

Abstract The history of using rivers in Poland for transportation reaches medieval times. Inland navigation utilizing primarily the two main rivers of the country – the Vistula and the Oder – significantly contributed to the expansion and economic growth of cities like Krakow, Warsaw, Wloclawek, Torun, Bydgoszcz, Wroclaw, Szczecin, Elblag, as well as many smaller towns. Polish waterways reached the height of their significance in the period from the 2nd half of the XV century to the 1st half of XVIII century. During that period, Vistula was considered to be, economically, the most important river in the world. The loss of independence due to Partitions of Poland, two World Wars, increased tonnage requirements for transport vessels and the emergence of the railroad as a competitive form of transportation can be seen as the main causes of the stagnation, and then the following recession of the inland water transportation within the country. The gradual process of degradation was not stopped despite the modernization attempts (partial river engineering, construction of new channels). Currently, the overall length of the waterways in Poland is 3655 km, out of which 2417 km are engineered navigable rivers, 644 km channelized river parts, 336 km are channels, and 259 km is on navigable lakes. Around 90% of that length is being utilized by inland navigation. The Polish waterways are elements to three international waterways: E30 – connecting the Baltic Sea with the Danube, E40 – connecting the Baltic Sea with the Black Sea and E70 – connecting the Baltic Sea with the Atlantic Coast. The main factor, however which determines the efficiency of the inland navigation is the compatibility between the ships' sizes and the waterways. Only about 6% of Polish waterways are fit for modern navigation of the international waterway standards (categories IV and V). The remaining 94% (3441 km) of the waterways are of the regional standards (categories I, II and III). The insufficient consideration given to the existing navigable waterways, both in regards to their

M. Gorączko (✉)
Faculty of Civil and Environmental Engineering and Architecture, UTP University of Science and Technology, Kaliskiego 7, 85-796 Bydgoszcz, Poland
e-mail: gorgon@utp.edu.pl

K. Kubiak-Wójcicka
Department of Hydrology and Water Management, Faculty of Earth Sciences and Spatial Management, Nicolaus Copernicus University, Lwowska 1, 87-100 Toruń, Poland
e-mail: kubiak@umk.pl

M. Zeleňáková et al. (eds.), *Management of Water Resources in Poland*, Springer Water,
https://doi.org/10.1007/978-3-030-61965-7_18

nature (channelized rivers, free-flowing rivers, channels) or the navigation parameters (water locks, depth and width of the route, bridge heights) contributes to their low position and marginal role in the current transportation system of Poland (0.4% of the total transported tonnage). Some of their potentials are being utilized by the tourism sector and in the international transport – and it is in these directions that many Polish cities are investing the funds allocated to the development of the inland navigation.

Keywords Inland waterways · Navigation · Transportation system

18.1 Introduction

At the beginning of 2017, Poland has ratified the European Agreement on Main Inland Waterways of International Importance (the AGN Agreement). It is a basic document created by the UN's Economic Commission for Europe which indicates directions for inland waterways development in Europe. Its aim is the introduction of a legal framework that allows for the creation of a coordinated plan for development and construction of inland waterways of international importance based on agreed infrastructural and operational parameters. Accession to the AGN Agreement may allow the inclusion of inland waterways in Poland into the European transportation corridors network TEN-T (Transeuropean Transport Network), a new EU's policy regarding transport infrastructure. It assumes that in the future, in the basis of the current network of roads, railroads, airports and waterways, there will be created a uniform, multimodal system of transeuropean transport network. Ratification of the AGN Agreement is an explicit declaration of adaptation of key sections of the country's waterways to contemporary European standards, which, as it seems, will be a very tough task to complete. Zakłada ona, że w przyszłości na bazie obecnej sieci dróg, linii kolejowych, portów lotniczych i dróg wodnych, zostanie utworzony jednolity, multimodalny system transeuropejskiej sieci transportowej. Undoubtedly, however, it is the problems of inland waterway transport and the concepts of overcoming them in the coming years that will be the main tone in the discussion on strategic directions of water management in Poland from the point of view of the state.

The issue related to inland waterways in Poland was presented in three-time frames: past, present, and future. In the beginning, a short outline of the history of the use of waterways in Poland in historical times and construction of waterway connections based on the existing river network against the background of historical changes of the country were discussed. The second part presents the contemporary state of inland waterway transport and the use of waterways in Poland and its causes. A separate section is the perspective of the use of waterways in Poland, including planned activities in the near future and their possible consequences for the socio-economic development of the country, as well as their significance in the light of the strategies pursued by the European Union. The study uses available literature and materials available on the websites of offices and municipal and government institutions. Based on data collected by the Central Statistical Office [1] information on the degree of use of inland waterway transport in Poland over the last 20 years (1998–2017) has been compiled.

18.2 Mythologized Past – Outline of Waterways Usage History in Poland

The history of using rivers for purposes of transport in the Polish lands dates back to the Middle Ages [2]. Over the past few centuries, the key role in the country played its longest river – Vistula, which, through its network of tributaries, connected main regions, namely: Greater Poland, Kuyavia, Mazovia, Podlachia, Little Poland, Podolia, and Volyn, with the largest Baltic port at that time – Gdańsk. The significance of the Vistula as a transport route for goods has systematically increased from the fourteenth century, from the point of view of the interest of the Polish state for the heyday of the Vistula transport is the period from 1466, when under the Second peace of Toruń Poland gained full control over the Vistula estuary, up to a year 1772, when it was lost as a result of the First Partition. At the turn of the sixteenth and seventeenth centuries, the number of transports recorded on the Vistula in the case of the Rhine was achieved only two centuries later, which clearly shows the importance of the Vistula route in Europe at that time. The Vistula trade was very important for the economy of the former Polish-Lithuanian Commonwealth, contributing to the economic and spatial development of numerous towns along the river, such as Sandomierz, Kazimierz Dolny, Warsaw, Płock, Toruń, Bydgoszcz, and Grudziądz. The goods transported by the river included mainly grain, but also wood, salt, beer, handicrafts, etc. For transport, units with low draft and rowing, sailing or mixed propulsion were used, while wood was floated in the form of rafts. In addition to the Vistula, inland navigation also developed on other rivers, such as Nogat, Pregoła, Dźwina, San, and Oder. The latter river was connected with Havel and Spree already in the seventeenth century. Starting from 1795 (after the Third Partition of Poland), the current main communication artery of the country of Vistula was divided between three partitioners. The upper section of the Vistula was the border of the Austrian partition, the central section ran through the Russian partition, and the lower section was located within Prussia. Vistula was the most important for Prussia. The Bydgoszcz Canal, the missing part of the waterway running through Spree, Oder, Warta, Noteć, Brda, Vistula, Nogat and the Vistula Lagoon, which connected two historical centers of the kingdom, Berlin with Königsberg (now Kaliningrad, Russia) and at the same time two main basins on Polish lands, Oder and Vistula, was built in years 1773–1774. The increase in hard coal mining in the Upper Silesian Basin resulted in the construction of the Kłodnicki Canal in years 1792–1812, which was running from Zabrze (Kłodnica River) to Koźle (Oder), and was replaced with a wider and deeper Gliwice Canal only in years 1933–1939. A complicated system of shipping canals was created in the middle of the nineteenth century in the Iława Lake District. It connected the larger lakes of the region and the cities located on them (Iława, Ostróda) with the seaport in Elbląg. Its main part was the Elbląg Canal, built-in 1844–1860. Starting from the second half of the nineteenth century, the modernization of waterways began. The upper section of the Oder was then canalized between Koźle and Brzeg Dolny (works were completed in 1922). The lower Vistula was regulated within the Prussian partition (from Ciechocinek to Gdańsk) with the estuary (the ditch near Świbno). The Vistula-Oder waterway was thoroughly modernized along with the construction of several new dams on the Noteć River and reconstruction of locks on the Bydgoszcz Canal and construction of the Górnonotecki Canal, which connected the Bydgoszcz Canal with the Gopło Lake, enabling water transport of agricultural and raw materials from the Kuyavia and Greater Poland region. Just before the First World War, a new Oder-Havel Canal was built, the most important connection of Poland's waterways with the West European network. Despite the many hydrotechnical investments carried out, this period was remarked by a decrease in the importance of inland waterway transport in favor of an increasingly fast-developing railway transport. The interwar period turned out to be too short-lived to cause a significant improvement in the condition of obsolete shipping infrastructure in Poland. In addition, the most important waterways of the country were crossed by borders with countries that were politically and economically in conflict (German Reich, Free City of Gdańsk, USSR). As a result, the interwar period brought regression in inland water transport. For example, the Vistula transports in the best years were comparable to those from before the partitions. The remainder of these times is the Gopło-Warta Canal (present Ślesin Canal completed only in 1950) and a number of unrealized canal projects, headed by the Central Canal, which was to connect the Upper Silesian Basin with the Vistula River near Bydgoszcz. This concept soon fell, replaced by the construction of the Herby-Gdynia coal railway in the years 1925–1933. From the point of view of the use of waterways, the establishment of new borders of Poland after the Second World War should be considered very beneficial. Almost all of the Vistula and Oder basins were in its area, and 7 of the 9 largest Polish cities had access to waterways. The most important industrial district of Upper Silesia was connected with the waterways to ports in Gdańsk and in Szczecin. After 1945, several major investments were made that had a direct or indirect impact on improving the conditions for inland waterway transport.

These include:

- the completion of the construction of the Żerań Canal connecting the Vistula River in Warsaw with the Narew (1953),
- the construction of the Łączany Canal (1958) consisting a part of the waterway connecting Oświęcim with Kraków,
- construction of the dam on the Oder River in Brzeg Dolny (1959),
- construction of the dam on the river Narew in Dębe and the creation of the Zegrze Reservoir (1963),
- construction of a dam on the Vistula in Włocławek and creation of the Włocławek Reservoir (1970).

In 2002, the construction of a waterway on the upper Vistula was completed. It lasted from 1949 and was commenced with the construction of four large coal-fired power plants near it (Skawina, Łęg, Nowa Huta, and Połaniec) in order to enable coal to be transported from mines in the vicinity of Oświęcim. However, this goal has not been fully achieved due to the lower than expected parameters of the locks, the narrowing of the river bed left within Krakow, also the construction of only a few of the planned number of dams.

Summing up, it should be noted that the development of inland waterway transport in Polish lands fell at a time when waterways did not require technical intervention in the natural course of rivers. The striving to increase the effectiveness of this means of transport by regulating and channeling rivers and building of large-scale shipping channels in many regions of Europe, in particular in the Rhine and Danube basins, fell on the nineteenth century. At that time, however, Poland was no longer an independent state, sovereign in political and economic terms. From the twentieth century in Europe, the focus has been on the expansion of rail and road infrastructure. Poland was not an exception in this respect compared to other European countries. While it is true that histories have decided to exclude Poland from the pan-European trend of modernization and expansion of waterway infrastructure in the most favorable moment, it would be an abuse to claim that this is the main reason for marginalizing the role of waterways in the contemporary transport system of the country.

18.3 Problematic Present – Current State of Waterways in Poland and Their Usage

The quality of waterways and their spatial arrangement determine the possibilities of development of inland water transport [3]. The total length of waterways in Poland is just over 3600 km, of which the vast majority are freely flowing (about 66%) and canalized rivers (about 18%). Waterways running within lakes constitute about 7% of their total length, while navigation canals – around 9%. Out of the total length of waterways in Poland, just over 90% are exploited. It is worth mentioning here that compared to the 1960s, the length of waterways in Poland decreased by over 20%. In terms of the length of rivers, Poland ranks fifth among European countries, behind Finland, Germany, France, and the Netherlands. Poland is characterized by a fairly significant density of waterways, expressed as an indicator of the length of the network per 1000 km^2 of the area. It is 11.7 km/1000 km^2, with the European average 9.3 km/1000 km^2. Only the Netherlands, Belgium, Finland, Germany, Hungary, and Luxembourg have a higher network density index. For comparison, however, the density of the motorway network (motorways and expressways) in Poland is more than twice, and the railway network twenty times (about 6 km/100 km) larger than waterways. At the same time, the main river network (including channels) of navigational significance consists of:

(a) the Vistula and its basin also Biebrza, Brda, Martwa Wisła, Nogat, Szkarpowa, and Pisa together with the Augustów, Bartnicki, Bydgoski, Elbląski, Jagielloński, Łączański and Żerański canals and lakes such as Drużno, Jeziorak, Szeląg Wielki, Ewingi and the Mazurian Lakes System;
(b) Oder and in its basin Nysa Łużycka, Noteć, Warta, Parnica, Western Oder, Eastern Oder, and Regalica, canals: Gliwicki, Kędzierzyński, Ślesiński and Górnonotecki, ditches Klucz-Ustowo and Parnicki and lakes Dąbie and Gopło.

There are three waterways of international importance across Poland in the sense of the provisions of the AGN agreement of 1996, which define the network of the European waterway system under the TEN-T [4]. It is waterway E30 – connects the Baltic Sea with the Danube in Bratislava, running through the Oder Waterway from Świnoujście through Szczecin and Wrocław to the border with the Czech Republic. Waterway E40 – connects the Baltic Sea with the Black Sea. Waterway leads from Gdansk up the Vistula to Warsaw, and then Narew and Bug to Brest, where it connects with the waterway leading through Polesia to the Dnieper. Waterway E70 – connects the Atlantic coast in Belgium with the Baltic Sea in Lithuania, running in Poland along the Oder, Warta, Noteć, Bydgoszcz Canal, Vistula, Nogat and the Vistula Lagoon to the border of the Kaliningrad Oblast in Russia (Fig. 18.1).

The classification of waterways in Poland (Table 18.1) is almost a direct transposition of the establishments of the European Conference of Ministers of Transport in 2006. Inland waterways are therefore classified according to the allowed size of vessels, taking as the criterion for determining the waterway class are the maximum length, draft and load capacity of a ship and minimum clearance under bridges or other devices crossing the waterway. For waterways or their sections on which there is a variable level of water table, the actual draft value of the ship or pushed set is determined, related to the determined value of the classification parameter, which should not be lower than the draft value occurring for at least 240 days in the average year. The European classification of waterways includes routes of regional importance (classes I–III) and waterways of international importance (classes IV–VII). Class IV parameters, which until recently were considered basic in Europe, are currently considered as minimal for international importance. In many cases, Class IV waterways only play the role of a link between the regional and international networks. Therefore, Class V waterways are accepted as the standard for international transport, which are available for motor barges and pushed sets with tonnage up to 3000 tons. In turn, the separated class VII takes into account the specifics of the lower section of the Danube, where it is possible to operate pushed sets with a capacity from 14,500 to 27,000 tons.

In Poland, requirements for roads of international importance (classes IV and V) are met by only about 6% of the length of waterways (214 km in 2017) (Tables 18.1 and 18.2). The remaining network of waterways consists of regional significance waterways (classes I, II, and III). Waterways with international parameters in Poland include:

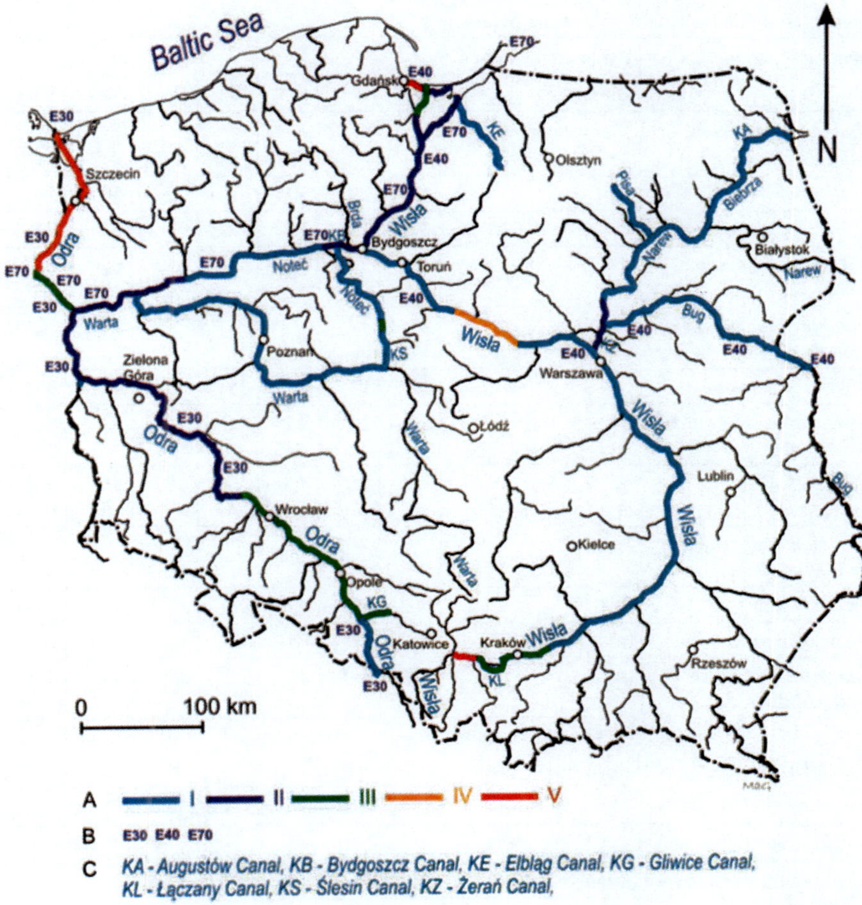

Fig. 18.1 Waterways layout in Poland. A – waterway classes after [5], B – waterways according to AGN, C – main navigable canals

- Vistula from the estuary of Przemsza to the Łączany Canal (class IV on the length of 37.5 km),
- Vistula from Płock to the Włocławek Dam (class Va for 55 km),
- Martwa Wisła (class Vb on the length of 11.5 km),
- Lake Dąbie to the border with internal sea waters (class Vb on the length of 9.5 km),
- Oder from the town of Ognica to Klucz-Ustowo Ditch and further as Regalica to the estuary to Lake Dąbie (class Vb on the length of 44.6 km),
- the Western Oder (class Vb on the length of 36.3 km),
- the river Parnica and Parnicki Ditch from the Western Oder to the border with internal sea waters (class Vb on the length of 6.9 km).

Table 18.1 Inland waterway operating parameters [5]

Operating parameters	Parameters' values							
	Classes	Ia	Ib	II	III	IV	Va	Vb
Minimum dimensions of waterway in a river	Measurement units							
– Waterway width	m	15	20	30	40	40	50	50
– Waterway depth	m	1.2	1.6	1.8	1.8	2.8	2.8	2.8
– Radius of a curve of waterway axis	m	100	200	300	500	650	650	800
Minimum dimensions of a canal								
– Waterway width	m	12	18	25	35	40	45	45
– Minimum water depth in canal	m	1.5	2.0	2.2	2.5	3.5	3.5	3.5
– Radius of a curve of waterway axis	m	150	250	400	600	650	650	800
Minimum dimensions of navigation locks								
– Lock width	m	3.3	5.0	9.6	9.6	12.0	12.0	12.0
– Lock length	m	25	42	65	72	120	120	187
– Depth on the lower step of the lock	m	1.5	2.0	2.2	2.2	3.5	4.0	4.0

Table 18.2 Waterway classification on Vistula and Oder rivers [5]

Waterway stretch	Waterway class	Length (km)	Share (%)
Vistula river		920.8	100
Przemsza river estuary – Łączany Canal	IV[a]	37.5	4
Łączany Canal estuary – Przewóz dam	III	34.3	4
Przewóz dam – Płock	Ib	527.8	57
Płock – Włocławek dam (Włocławek Reservoir)	Va	55.0	6
Włocławek dam – Tążyna river estuary	Ib	43.0	5
Tążyna river estuary – Tczew	II	190.5	21
Tczew – Baltic Sea	III	32.7	4
Oder river		726.8	100
Racibórz – Kędzierzyn-Koźle	Ia	44.4	6
Kędzierzyn-Koźle – Brzeg Dolny	III	187.1	26
Brzeg Dolny – Warta river estuary	II	335	46
Warta river estuary – Szwedt Canal	III	79.4	11
Szwedt Canal – Szczecin Lagoon (Eastern Oder and Western Oder together)	Vb	80.9	11

[a]Waterway class upon modernization completing

Thus the shortcoming of Polish waterways of classes IV and V is not only their total length but a far-reaching fragmentation. Although they meet the conditions imposed on waterways of international importance, their role, due to the lack of connections to the European network, can only be local [6].

18.4 Navigation Traffic on Inland Waterways in Poland

18.4.1 Influence of Water Levels on Navigation Traffic Intensity

Freely flowing rivers, having a predominant share in the total length of waterways in Poland, are subject to seasonal fluctuations in water levels, resulting from their natural hydrological regime. Both periods of too high and too low water levels are a limitation for inland navigation (Figs. 18.2 and 18.3). Floods in the Vistula and Oder basins may have different origins. They are caused by spring and winter thaws (especially in lowland areas), long-lasting and intense rainfall (upland and foothill areas), but also can be caused by sea waters during storms, blocking the free flow of rivers and causing backwater. Too high water levels make it difficult to use waterways, especially in places where they intersect with engineering structures (e.g., bridges, pipelines, power cables, etc.). Elevated water levels, apart from flood hazards, are a significant obstacle to shipping, mainly due to decreased maneuverability of vessels and smaller vertical clearances under bridges [7]. In the case when the clearance between them and the water level at the highest shipping water (WWŻ) exceeds the minimum value for a given waterway class, this causes stopping the navigation. However, relatively high floods are hydrological phenomena occurring relatively

Fig. 18.2 The river bar in the central part of the Vistula river bed in Grudziądz during low water levels in 2015 (K. Kubiak-Wójcicka)

Fig. 18.3 The estuary of the Brda river to the Vistula – a fragment of the international waterway E70 during a low water level in 2015 (M. Gorączko)

rarely and temporarily. Droughts are a much more important problem on Polish waterways.

The analysis of navigation traffic related to water levels on the stretch of waterways E70 and E40 were conducted by several authors [8–12]. Kubiak-Wójcicka and Pokropski [12] analyzed numbers of vessels on lower Vistula lockaged on the Włocławek Dam related to water levels in gauging station Toruń in years 1997–2016. In the case of low water levels and too high water levels, there are restrictions in navigation traffic, which is visible in smaller numbers of lockaged units. Navigation restrictions apply when the water level of Vistula in Toruń is SNW (average low water) – 179 cm [13] or at an emergency level which is close to SWW (average high water) – 650 cm [14]. Based on the Institute of Meteorology and Water Management – National Research Institute data on daily water levels of the Vistula in Toruń, authors [12] estimated the average time of unavailability of the river for navigational purposes, i.e. below the water level of Vistula 179 cm or above 650 cm, below the Włocławek Dam. During the analyzed period 1997–2016, on average, for 10.8% of a year, navigation traffic was impossible due to too low or too high water levels. Low water levels were dominating, and in 2015 they accounted for as much as 41.1% of days during the year, while the highest water levels included 6.3% of days in 2010. Assuming a more rigorous criterion for determining the droughts applied by Gorączko et al. [8], i.e., water level below 240 cm, the time when Vistula cannot be used for the needs of freight shipping may take up to 61% of a year, as it was the case in 2015. In more than 50% of days in navigation season months (from July to October), navigation could not take place due to poor navigational conditions, i.e., water level below 240 cm or above 650 cm. Days with droughts were mainly responsible for that. It should be mentioned that in 2003, 2012, and 2015 in those months, the water levels remained below 240 cm, which meant that the navigation could not take place throughout the whole month [12]. The number of all lockaged units in Włocławek in the analyzed period 1997–2016 varied and fluctuated from 656 vessels in 1997 to only 129 in 2015. It should be noted that the difficulties mainly concerned freight traffic (barges, pushers, and tugs), which require the appropriate depth of the route. These restrictions apply in a lesser extent to tourist traffic, e.g., kayaks, motorboats or other vessels with a small draft.

An ad hoc method of temporarily improving the navigation conditions used on partially canalized rivers in low-water periods is intervention water discharges on dams, calculated for a short-term increase of a waterway's depth. For example, on the lower Vistula river, on the night of 25/26 May 2018, the discharge in Włocławek Dam was increased by 400 m^3 s^{-1}. The wave produced in this way allowed for delivery of a tanker's hull (over 100 m long) from the river shipyard in Płock to Gdańsk. This case, however, aroused great controversy among naturalists, because the discharge resulting in sinking of river islands, was made during the breeding period of the birds, annihilating their large number, including protected species. Habel [15] analyzed the so-called artificial flood waves on the Włocławek Dam in the years 1986–2015, with particular emphasis on the breeding season, from April 1 to September 30. Surfaces of river bars in the highest frontal part of them are raised above the water level when water level on the Fordon gauge equals to 240 cm (water flow 985 m^3 s^{-1}). Artificial floods have a record in the data recorded at water gauges. In the analyzed period, as many as 40 artificial flood waves were recorded in summer months, with intensified events from 2000. It should be noted that in the decade 1986–1995, breeding success was potentially possible almost every year. After 1996, up to now, conditions have deteriorated in this respect, and breeding success was potentially achieved on average once every three years. Interventional discharges have a negative impact on the living conditions and reproduction of water fauna, mainly on the plover birds nesting in the Vistula river bed.

Prevention of such events, but not their complete elimination, is possible through the installation of so-called artificial breeding places in the form of barges or floating piers and islands in river channels (Fig. 18.4). Properly adapted barges as floating "unsinkable" reproductive islands for birds, can be an effective means of reducing the negative impact of such projects. For example, the barge, which is an artificial breeding site since 2000, is located on the lower Vistula near Grudziądz. In recent

Fig. 18.4 Old barge used on the lower Vistula as an artificial breeding island for avifauna (M. Gorączko)

Fig. 18.5 Despite global warming, ice phenomena still constitute an obstacle in the use of waterways in the winter season. Example of the Brda river in Bydgoszcz (M. Gorączko)

years, many artificial islands have been installed, including Brda in Bydgoszcz. Such activities are carried out by landscape parks, ecological associations, and local activists groups.

The consequence of climatic conditions in the country is freezing of rivers and canals in winter. From ancient times, in the history of navigation, periods of ice have been the basis for the separation of individual shipping seasons. The typical time of occurrence of ice phenomena in Poland is usually from the end of November to the middle of March. Although recent research indicates successive shortening of ice-cold periods and a reduction in the occurrence of ice cover in favor of flowing ice, the continuing cyclical occurrence of ice phenomena remains invariably the basic feature of the hydrological regime of Polish rivers [16] (Fig. 18.5).

The importance of inland waterway transport in freight transport in Poland has been marginal for a long time (6.2 million tons of transported loads in 2017, which accounted for 0.2% of the total) and is still decreasing. Of course, such a state is the result of many years of negligence in modernizing infrastructure and unfavorable hydrological conditions, although not only. As it turns out, the very low flexibility of inland waterway transport in the supply chain from the place of shipment to the place of destination is equally important. The solution to this problem is the location of intermodal ports, which combine inland transport with other types of transport, i.e., vehicle and rail along waterways. In case of unfavorable hydrological phenomena (floods, low-waters, ice phenomena) there is a possibility of fast transhipment of goods to other means of transport, as well as their storage, which makes these ports an indispensable link in transport chains [11, 17].

The transport structure is dominated by mass, low processed cargos, such as aggregates (mainly sand and gravel exploited from the bottom of rivers), energy raw materials (hard coal and coke), metal ores, building materials, cement, metal products, fertilizers, agricultural products, waste, etc. (Figs. 18.6 and 18.7). A large group of loads are bulky constructions. The structure of loads in individual regions of the country is different, depending on the dominant profile of production activity.

Fig. 18.6 The structure of freight transport on waterways in Poland is dominated by the transport of aggregates extracted from the river bed (M. Gorączko)

Fig. 18.7 The lock Czersko Polskie located in the estuary section of the Brda river (K. Kubiak-Wójcicka)

In many industrial centers, water transport loses the competition to other modes of transport despite very good communication with the waterway network. An example here can be the Bydgoszcz Water Node, formally one of the most important points on the map of Polish waterways (intersection of E40 and E70). The cargo shipping only takes place on a stretch of a few kilometers – from Vistula where river sand is extracted, upstream of Brda to the port in the eastern part of the city where it is reloaded. The operation of inland waterway transport in this important node depends on the economic situation and the financial condition of the main production plant for which it supplies building material [18, 19].

Unstable hydrological conditions (too low water levels during dry years) combined with changing the economic situation for goods that are mainly transported by waterways, meaning that the volume of cargo transported by inland waterways in Poland is subject to significant fluctuations from year to year. In 2017, 5778 thousand tons of cargo were transported, and 877 million tons of transport work was made, thus noting a decrease by 432 thousand tons in relation to the previous year (about 7%), with a simultaneous increase in transport performance by 45 million t km. In comparison with 2014, this means a decrease in freight transport by 1851 thousand tons (by 24.3%) and an increase in transport performance by 99 million t km (by 12.7%). In the period 2014–2017, the best year was 2015, when the weight of cargo transported by waterways amounted to nearly 12,000 tons [1]. Under these conditions, running a business focused on water transport is characterized by high uncertainty as to its economic viability. No wonder that a significant part of the units belonging to Polish shipowners (fourth fleet in Europe, in terms of the number of barges) performs transport abroad, mainly in Germany. As a result, more than half of the transport of goods carried out by Polish shipowners is foreign transport. The marginal importance of water transport in Poland and lack of prospects for its growth in the near future means that the involvement of financial resources for the maintenance and repair of navigation infrastructure has been much below the needs in this area for many years. This causes a progressive degradation of waterways. According to the report of the Supreme Audit Office [20] of 2014 on the operation of inland navigation in Poland, only the restoration of the national waterways' classification parameters requires expenditures estimated at approximately PLN 14 billion. This is mainly related to the fact that since 1949, the amount of financing for current maintenance of waterways has not been ensuring simple reproduction. It is worth noting here that in the years 2007–2012 the total investment for inland waterways in Poland amounted to only approximately PLN 600 million. The effect is progressive decapitalization of navigation infrastructure (weirs, locks, etc.) and regulation structures.

18.4.2 Passenger Navigation and Touristic Utilization of Waterways

In addition to cargo shipping, passenger transport operates on rivers and canals in Poland. In 2017, the number of transported people amounted to over 1.2 million, showing an upward trend in the last few years. These transports, usually having a tourist and recreational character, are implemented not only on waterways but also on sections of the river network that do not meet the requirements for cargo shipping. It is estimated that the routes for water tourism in Poland amount to approximately 11.6 thousand km, of which 14% are routes of outstanding importance and 58% are very important routes. Nearly 34% of tourist routes are available for all types of boats and ships (sailing yachts, motor yachts, and kayaks), while the remaining part is available only for kayaks. Inland passenger navigation routes in Poland are associated with larger rivers, such as the Vistula, Oder, Warta, Noteć, Brda, Bug, Narew and the canals (Augustów, Elbląg) and the Land of the Great Masurian Lakes. Investments in water tourism, apart from the permanent marking of tourist routes, usually interfere only pointwise in the network of waterways through the construction of marinas and parking spaces, often together with the associated accommodation and catering infrastructure and sanitary and technical facilities. They are most often implemented by local self-governments interested in tourist activation in their regions, with the support of financial resources from EU funds and programs. Among many realizations of this type, one can mention:

- marina and river ports on Brda in Bydgoszcz servicing, among others "Water Tram" (regular passenger vessels within the Bydgoszcz Water Node),
- marina in Nakło nad Notecią,
- yacht harbor on Lake Dąbie in Szczecin,
- river port on the Vistula in Grudziądz,
- yacht marina on the Warta in Poznan,
- passenger ships on the Elbląg Canal.

Of course, this area is also not free from projects that are economically or functionally inaccurate, but the overall direction in the development of infrastructure for users of waterways focused on tourism should be rated positively.

Inland waterways perform a number of non-transport functions that also have significant economic and social value [21]. Over the last 10 years, revitalization of waterfronts has been observed in cities located on larger rivers, e.g., in Warsaw [22], Poznań [23], Bydgoszcz [10, 24], and many others. Comprehensive revitalization of water zones in cities is a must. It changes the image of the city by adding another advantage – a city open to the water, which increases interest in inland passenger navigation or individual development of water sports [19] (Figs. 18.8 and 18.9). An important element of river-bank development is the adaptation of building conditions to minimize the impact on the existing environmental conditions [25].

An additional factor influencing the increased interest in water tourism is the organization of large outdoor events, just to mention, among others, the organization of the Year of Bydgoszcz Canal (2014) or the Year of the Vistula River (2017)

Fig. 18.8 Revitalized river bank of the Brda river in the center of Bydgoszcz (K. Kubiak-Wójcicka)

Fig. 18.9 The river port on the Vistula in Włocławek (K. Kubiak-Wójcicka)

during which numerous canoeing trips, cruises on passenger ships and water trams were organized. The statistical data of the Central Statistical Office confirms that the direction of changes is correct. In 2010, in the entire country, the number of passenger transports was the lowest in the last 10 years, amounting to 879.1 thousand passengers. Since 2011, there has been a steady increase in the number of passengers in inland waterways: in 2014 there were 1038 thousand of them [1], and in 2017, 1262 thousand passengers [26].

18.5 Uncertain Future – Perspectives for Waterways Usage in Poland

Currently, the office of government administration servicing issues related to inland navigation in Poland is the Ministry of Maritime Economy and Inland Navigation. This office was established in 2015, after separating organizational units and employees serving the department of maritime economy from the Ministry of Infrastructure and Development and taking over tasks related to fishing from the Ministry of Agriculture and Rural Development. In the same year, the ministry took over the competence of the Ministry of Infrastructure and Construction in the field of inland waterway transport. At the beginning of 2018, the Ministry of the Marine Environment and Inland Navigation was joined by organizational units and employees of the Ministry of the Environment, so far responsible for water management, thus establishing the Department of Water Management and Inland Navigation. While focusing competences in the management of domestic water resources within one central office seems rational, the subordination of this society's key area to certain group interests may raise concerns about Poland's further respect of the fundamental provisions of the Water Framework Directive.

The fact that this concern is justified may be proved by preparations for the implementation of the Ministry's priority project "Waterway connecting the Vistula Lagoon with the Gdansk Bay." It provides for the construction of a ditch by the Vistula Spit, 1260 m long, 5 m deep, from 25 to 100 m wide, and a waterway connecting the ditch with Elbląg of 22 km length, 5 m depth (with an average lagoon depth of 2.7 m) and width from 60 to 120 m. The value of the investment is estimated at EUR 500 million. The ditch and the waterway will cut Nature 2000 areas: special protection areas of the Vistula Lagoon and Vistula Spit, as well as the area of birds' special protection of the Vistula Lagoon. For this reason, the investment has met with the opposition of many environments. For obvious reasons, numerous non-governmental organizations protest against it, headed by the Polish WWF branch and scientists representing the natural sciences. The purposefulness of the project was also questioned by the Regional Directorate for Environmental Protection in Gdańsk. Only after the director of this institution was forced from the post, a positive environmental decision was obtained for the investor. The project of a waterway through the Vistula Spit and the Vistula Lagoon was also negatively evaluated by the self-government of the Pomeranian Voivodship. The Marshal's Office, apart from objections related to the protection of nature within the scope of the investment's impact, also pointed to the lack of economic justification for its implementation. Finally, the local government authorities of holiday resorts located on the Vistula Spit are concerned with the predicted degradation of local beaches. The matter of the described waterway is also significant because it places Poland in the role of a very uncertain partner in the planned construction of the TEN-T network.

Another controversial project, towards which the Ministry of Maritime Economy and Inland Navigation is inclined, is the creation of a cascade in the Vistula, from Warsaw to Gdańsk. The beginnings of this investment date back to the 1960s. The project involved the construction of dams in Wyszogród, Płock, Włocławek, Ciechocinek, Solec Kujawski, Chełmno, Opalenie and in Tczew together with the accompanying hydropower plants. The growing crisis of the socialist economy in Poland caused that only the dam in Włocławek was finally built. After state system transformation in Poland, the concept of construction of dams on the lower Vistula was repeated several times, but mainly in the context of a dam below the one in Włocławek, as a measure to prevent excessive erosion of the bottom on this section of the river. In recent years, the idea of full cascading of the lower Vistula has been revived, but this time, along with hydro-energetic ones, the benefits that the investment would bring to the development of inland waterway transport are underlined.

Among the in progress investments, there is the construction of a port for a ferry on the Vistula from Czarnowo to Solec Kujawski, which is to be launched in 2019. This is the first step to building a multimodal port in Solec Kujawski, whose task will be to service cargo transported by inland waterway E40 and E70 from the port in Gdansk further into the country. It is assumed that the operation of such a port will contribute to the unloading of road transport and the share of inland waterway transport will increase in the amount of transported cargo. However, an indispensable condition for the launch of a multimodal port is the adaptation of the lower Vistula channel to the requirements of contemporary waterways of international importance, i.e., its channeling.

From the point of view of the planned development of waterways in Poland, the key document is the resolution of the Council of Ministers of 2016 regarding the adoption of "Assumptions for development plans for inland waterways in Poland for 2016–2020 with a prospect until 2030" [27]. It stipulates that the main goal of development of inland waterways important from the transport point of view is their construction or modernization to the parameters of at least navigability Class IV and meeting the requirements of inland waterway transport infrastructure for the TEN-T network. This goal is divided into four priorities covering a total of eleven tasks.

Priority I concerns the Oder Waterway (E-30), assuming that it will achieve an international navigability class and inclusion it in the European waterway network, by eliminating the current bottlenecks, the adaptation of the Oder Waterway to the Class Va parameters, construction of a section of the missing connection between the Danube – Oder – Elbe and the construction of the Silesian Canal (connecting the Vistula and Oder basins in Upper Silesia).

Priority II is connected with the striving for a significant improvement of navigational conditions on the waterway of the Vistula River, through the construction of the Vistula cascade from Warsaw to Gdańsk, modernization of the upper, canalized Vistula and the construction of a dam in Niepołomice.

The third priority area is the connections Oder – Vistula – the Vistula Lagoon and Warsaw – Brest by extension of the E-70 and E-40 waterways. It includes preparation for modernization of the international water route Oder – Vistula – Vistula Lagoon (E-70) and preparation for the construction of the Polish section of the international waterway Vistula – Dnieper (E-40) from Warsaw to Brest.

Priority IV is of organizational nature. It assumes the development of national and international partnership and cooperation for inland waterways. The expected costs of construction and modernization of selected Polish inland waterways in the area of adapting them to the parameters of shipping routes of international importance vary depending on the adopted concept and variants. Initially, it was expected that they would amount to: for the Oder Waterway along with the Gliwice Canal and the Oder-Danube connection – from PLN 16.5 to 22.6 billion, the middle and lower Vistula section from Warsaw to Gdańsk – approximately PLN 31.5 billion, Silesia Canal – approximately PLN 11.0 billion, and the Warsaw-Brest waterway – from 8.1 to PLN 25.5 billion. Currently, the total costs of the above investments are estimated at PLN 60–150 billion.

Shaping the vision of the future state and the importance of waterways in Poland takes place under conditions of acute dispute. It is assumed that it takes place between hydrotechnicians and naturalists, but its nature is more complex. It is worth to notice how divided in this respect is the environment of Polish hydrologists. They fall into both ranks of hot supporters of the reconstruction of waterways as well as their declared opponents. The discussion between the shipping lobby and non-governmental environmental organizations continues.

In the first place, it should be considered whether the current program of development of waterways in Poland takes into account the current state of knowledge about the principles of human management in the natural environment in a responsible and sustainable way. Many hydrotechnical facilities constructed throughout the nineteenth century impress even today with their scale and craftsmanship of technical solutions. This should not, however, obscure the fact that the then human consciousness of the degree of complexity of river ecosystems and synergistic relations between their individual elements was, by the current standards, dim. It is possible to point out, also in Poland, a series of projects that were once undertaken in river valleys, which in the long-term proved to be wrong and even harmful, and whose negative effects we still feel today. For example, one of the arguments for the concept of the Vistula cascade is the possibility of obtaining a water reserve in this way, allowing irrigation of drought-threatened agricultural land in Kuyavia. It is worth noting that the so-called stepping-up of Kuyavia is almost exclusively the result of another great project carried out in an ad hoc manner – the construction of drainage systems in the eighteenth and nineteenth centuries [28]. Previously, it was one of the most waterlogged regions in Poland. Along with the development of civilization, the priorities, goals, and methods for their implementation change, which is always to be remembered. Only a small share of the waterways in Poland meets the requirements for international waterways. It should be obvious that fulfilling the requirements of the AGN Agreement is not so much related to pointwise or section-wise modernization of the existing navigation system (i.e. elimination of so-called bottlenecks) as it is in Western European countries, but to its construction almost from scratch. And in this respect, the project should be considered as contrary to the recommendations of the Water Framework Directive, which recognizes the natural environment as an equal user of water, the same as population, industry, agriculture, inland navigation, etc. According to the WFD, Member States may designate a body of surface water as artificially or heavily changed when changes in hydromorphological characteristics of the body of water that would be necessary to achieve good ecological status could have a significant adverse effect on shipping, including port facilities. There is no doubt that the preservation and even restoration of surface water naturalness remains a priority.

By joining the European Union, Poland pledged to designate Nature 2000 sites on its territory. Their location is closely related to river valleys and wetlands, in most cases. Thus, along the European sections of the E30, E40 and E70 waterways, over 20 areas of special bird protection and over 40 areas of special habitat protection have been designated. Within Poland, these routes, together with the immediate surroundings, are subject to Community protection on almost the entire length. The implementation of the governmental program for the development of waterways in Poland is synonymous with serious interference in Nature 2000 areas and many cases (for example, the planned Vistula channeling) with their physical removal. It is true that both EU and national law assumes in an exceptional case, for the necessary overriding public interest and in the absence of alternative solutions to initiate the procedure of environmental compensation. It is hard to imagine how environmental losses resulting from the location of water reservoirs with a volume of Włocławek Reservoir (permanent reservoir several times wider than the river bed and over 50 km long) could be compensated with a surplus. In general, the implementation of the governmental strategy would undermine the integrity of the entire chain of protected areas in the valleys of the largest rivers in Poland [29] and the permanent destruction of ecological corridors of European importance.

It is anticipated that the cascading of the Vistula, Oder, and other rivers may lead to a permanent transformation of the existing biocenoses, including extermination of migrating fish. Generally, this problem is payed down by the shipping lobby, since currently in the case of surface waters, not only the purity of water is assessed, but also their ecological state, based on phytoplankton, phytobenthos, macrophytes, zoobenthos and ichthyofauna that colonize them. Therefore, it can be anticipated that the implementation of the waterway development program will lead straight to the deterioration of the ecological status of surface waters and economic consequences resulting from failure to meet obligations coming from the WFD.

In light of the above arguments, it can be questioned that inland navigation is the most environmentally friendly mode of transport. One may agree with it only in the case of already existing waterways.

18.6 Summary and Conclusions

One of the basic aims of European transport policy is assuring balanced development in the area of transport. Just talking about the advantages of water transport, its minimal influence on the environment, small costs of transport and small energy consumption, will not contribute to change in inland waterway transport condition. It is a fact that despite the undisputable benefits, inland waterway transport's role in Poland is marginal. Due to the long duration and large costs of investments, in the near future, a significant improvement of navigational conditions on Polish waterways and an increase in the amount of transported cargo should not be expected. This economy sector requires not only significant expenditures from the state budget but also public acceptance of the scope of large-scale activities and their impact on the environment. The lack of a prospective plan for the modernization of waterways and the construction of infrastructure developed in agreement with various social groups only leads to the intensification of existing conflicts.

Possibilities to utilize waterways are sought in their touristic and recreational exploitation. Due to lower transit depths required for touristic and recreational units, there are possibilities for such units traffic, especially in dry years. There is an increasing interest of tourists and residents in the riverside space as a place of rest and recreation.

It seems a top priority matter that on key issues for the country, such as the state of its water resources and their use, the decision-makers would be consensus and cooperation of both opponents and advocates of the inland waterway development program in Poland.

18.7 Recommendations

This chapter draws attention to the multi-faceted problem of development and functioning of inland waterways in Poland. Decision makers at various levels of government and local government should pay attention not only to issues related to the development of inland transport, but also environmental, economic or social issues. The scientific community in Poland should have a significant role to play in the future, especially researchers representing such fields as earth sciences, environmental sciences, socio-economic sciences and technical sciences. It is the results of their research that should form the basis for shaping the transport policy of the country and regions, when drawing up strategic plans and making decisions to implement them.

References

1. Główny Urząd Statystyczny (GUS) Central statistical office. https://stat.gov.pl/
2. Binerowski Z (1982) Transport wiślany w dawnej Rzeczypospolitej. In: Dolina Dolnej Wisły, Wydawnictwo Polskiej Akademii Nauk, Ossolineum, Wrocław-Warszawa-Kraków-Gdańsk-Łódź
3. Wojewódzka-Król K, Rolbiecki R, Rydzkowski W (2007) Transport wodny śródlądowy. Wydawnictwo Uniwersytetu Gdańskiego, Gdańsk
4. Ustawa z dnia 15 grudnia 2016 r. o ratyfikacji Europejskiego Porozumienia w sprawie Głównych Śródlądowych Dróg Wodnych o Znaczeniu Międzynarodowym (AGN), sporządzonego w Genewie dnia 19 stycznia 1996 r. Dz.U. nr 0 poz. 186 z 2017 r
5. Rozporządzenie Rady Ministrów z dnia 7 maja 2002 r. w sprawie klasyfikacji śródlądowych dróg wodnych Dz.U. 2002 nr 77 poz. 695
6. Bolt A, Jerzyło P (2013) Determinants of inland navigation on the Vistula from Warsaw to Gdańsk. Acta Energetica 2(15):56–63
7. Woś K (2011) Analiza nawigacyjna dolnego odcinka Odry objętego obowiązkiem wdrożenia systemu RIS. Logistyka 6:5101–5116
8. Gorączko M, Szyplik J, Pasela R (2013) Wpływ niżówek na warunki funkcjonowania żeglugi w rejonie Bydgoskiego Węzła Wodnego. Geogr Tour 1:69–76
9. Gorączko M (2015) Natężenie ruchu żeglugowego na bydgoskim odcinku drogi wodnej E-70 – stan obecny i perspektywy rozwoju. Geogr Tour 3(1):33–38
10. Kubiak-Wójcicka K (2014) Stopień wykorzystania infrastruktury liniowej w żegludze śródlądowej na przykładzie bydgoskiego odcinka międzynarodowej drogi wodnej E-70. Logistyka 6:12820–12832
11. Kubiak-Wójcicka K (2016) Hydrologiczne uwarunkowania budowy i funkcjonowania intermodalnego portu rzecznego w Solcu Kujawskim. Logistyka 3:129–136
12. Kubiak-Wójcicka K, Pokropski T (2019) Natężenie ruchu żeglugowego na dolnej Wiśle na podstawie analizy śluzowań we Włocławku w latach 1997–2016. Prace Geogr 156:79–99
13. Kubiak-Wójcicka K (2012) Charakterystyka niżówek na Wiśle w Toruniu. In: Marszelewski W (ed) Gospodarowanie wodą warunkach zmieniającego się środowiska. Monografie Komisji Hydrologicznej PTG. Toruń, pp 85–94
14. Kubiak-Wójcicka K (2014) Wezbrania na Wiśle w Toruniu w świetle obserwacji historycznych. In: Ciupa T, Suligowski R (eds) Woda w mieście. Monografie Komisji Hydrologicznej. Kielce, pp 127–134

15. Habel M (2018) Effects of flow regulation and river channelization on sandbar bird nesting availability at the Lower Vistula River. Ecol Quest 29(4):43–53. https://doi.org/10.12775/EQ.2018.028
16. Pawłowski B, Gorączko M, Szczerbińska A (2017) Zjawiska lodowe na rzekach Polski. In: Jokiel P, Marszelewski W, Pociask-Karteczka J (eds) Hydrologia Polski. PWN, Warszawa, pp 195–200
17. Montwiłł A (2013) Węzły transportowe w europejskim systemie transportu i łańcuchach dostaw. Logistyka 6:948–957
18. Taylor Z, Ciechański A (2011) Organizational restructuring and ownership transformation in Poland's inland shipping companies after 1990. Geogr Pol 84(2):77–92
19. Kubiak-Wójcicka K (2016) Uwarunkowania i stan obecny rozwoju żeglugi turystycznej na Wiśle w Toruniu. In: Stankiewicz B, Prochorowicz M (eds) Turystyka wodna jako produkt turystyczny regionu. Wydawnictwo Uczelniane Zachodniopomorskiego Uniwersytetu Technologicznego, Szczecin, pp 67–74. ISBN 97883-7663-229-2
20. NIK (2013) Informacja o wynikach kontroli: Funkcjonowanie żeglugi śródlądowej, KIN-4101-04/2013, nr ewid. 188/2013/P/13/079/KIN
21. Wiegmans B, Konings R (2017) Inland waterway transport. Challenges and prospect. Routledge, New York
22. Zegar T (2015) Rzeka w mieście – perspektywy rozwoju terenów nadwodnych w Warszawie. Prace Stud Geogr 60:139–152. https://wgsr.uw.edu.pl/wgsr/wp-content/uploads/2018/11/Rzeka-w-miescie.pdf
23. Januchta-Szostak A (2011) Front wodny Poznania – Dolina Warty. Rewitalizacja związków z rzeką. Wydawnictwo Politechniki Poznańskiej, Poznań
24. Gorączko M, Pawłowski B, Gorączko A (2014) Problemy transformacji miejskich frontów wodnych na terenach zagrożonych przez powódź. In: Integracja techniki i sztuki w architekturze i urbanistyce. UTP, Bydgoszcz, pp 105–115
25. Kubiak-Wójcicka K, Chodkowska-Miszczuk J, Rogatka K (2017) Integration or disintegration of the ecological and urban functions of the river in the city? A Polish perspective. Transylv Rev Adm Sci 52:59–76. https://doi.org/10.24193/tras.52E.4
26. Transport wodny śródlądowy w Polsce w 2017 roku, Informacje sygnalne (2018) Główny Urząd Statystyczny, Warszawa
27. Strategia rozwoju transportu do 2020 roku (z perspektywą do 2030 roku) (2013) Ministerstwo Transportu, Budownictwa i Gospodarki Morskiej, Warszawa
28. Kaniecki A (2011) Przemiany stosunków wodnych w dolinie górnej Noteci do połowy XIX wieku związane z antropopresją. Bad Fizjogr Pol Zach Ser A Geogr Fiz II(A62):41–58
29. Pawlaczyk P (ed) (2016) Wstępna ocena ryzyka oddziaływania Projektu Banku Światowego P147460 „Ochrona przeciwpowodziowa w dorzeczu Odry i Wisły" na przyrodnicze obszary chronione. Klub Przyrodników, Świebodzin

Chapter 19
The Effects of Plant Irrigation in Poland

Renata Kuśmierek-Tomaszewska and Jacek Żarski

Abstract On the global scale, the main criterion for the advisability of using plant irrigation is the climatic condition – in particular, the amount and distribution of atmospheric precipitation. Therefore, irrigation is mainly applied in regularly or periodically arid climatic zones, because it is in these areas where food production would not be possible without an additional water supply. It is different in Poland, placed in the moderate climatic zone, where irrigation of plants is basically a contingency procedure applicable to supplement the periodic shortages of rainfall in relation to the water requirements of plants. It applies particularly to sandy soils with reduced water retention capacity, situated in the central, lowland part of the country. Droughts in Poland are weather phenomena unfavourable for agricultural production, appearing quite often but irregularly. On the other hand, conditions of fully or excessively moisturized soils during growing seasons occur with a similar frequency to droughts. For this reason, the irrigated area in Poland covers only about 73 thousand hectares. The main indicators of the advisability of applying irrigation and considering its role in domestic agriculture and water management are production effects, i.e., the absolute yield increase obtained under the influence of this treatment in agricultural raw materials production. Irrigation is the most economically effective treatment in horticultural crop production, including orchards, vegetables, and berry plants. In the case of agricultural plants, the indicators of profitability are much less – the only positive financial effect may be obtained in the case of table potato production.

Keywords Water deficit · Supplementary irrigation · Drought · Irrigation efficiency · Poland

R. Kuśmierek-Tomaszewska (✉) · J. Żarski
Department of Agrometeorology, Plant Irrigation, and Horticulture, University of Science and Technology in Bydgoszcz, 6 Bernardyńska Str., 85-029 Bydgoszcz, Poland
e-mail: rkusmier@utp.edu.pl

J. Żarski
e-mail: zarski@utp.edu.pl

M. Zeleňáková et al. (eds.), *Management of Water Resources in Poland*, Springer Water,
https://doi.org/10.1007/978-3-030-61965-7_19

19.1 Introduction

Plant irrigation is a water supply to the soil profile in order to provide or improve the growth, development and yielding conditions of crop plants. Irrigation is considered to be an agrotechnical practice and is one of the very first anthropogenic modifications of the natural environment, originating about 7500 years ago in Mesopotamia.

According to current data published by the International Commission on Irrigation and Drainage [1] and Food and Agriculture Organization of the United Nations [2], around 300 million hectares are irrigated worldwide, which accounts for approximately 20% of the total of agricultural land. It is estimated that about 40% of the global food supply comes from irrigated land. The spatial distribution of the irrigated area, as an essential indicator of water management and food security, is important subject for statistics and scientific research. Recent studies based on vegetation indices used in remote sensing measurements [3] show that the global irrigated area is even higher than that reported in the statistics and possibly amounts to as much as 367 million hectares.

On the global scale, the main criterion for the advisability of using plant irrigation is the climatic condition, in particular, the amount and distribution of atmospheric precipitation. Therefore, irrigation is mainly applied in regularly or periodically arid climatic zones because it is in these areas where food production would not be possible without an additional water supply.

It seems different in Poland (in the moderate zone), where irrigation of plants is basically a contingency procedure, applicable to supplement the periodic shortages of atmospheric precipitation in relation to the water requirements of plants. In terms of the average conditions, these shortages result from too low multiannual average precipitation in relation to the water needs of plants, expressed by mean values of evapotranspiration or optimal precipitation. It particularly applies to sandy soils with reduced water retention capacity, situated in the central lowland part of the country. In selected growing seasons (April–September), or some periods occurring in the middle of them, these shortages result in temporary rainfall scarcity leading to meteorological and agricultural droughts. In the moderate, transitional climate of Poland, these droughts are weather phenomena and are unfavourable for agricultural production, appearing quite often but irregularly. A great example to illustrate the climate risk due to this hazard, in the cultivation of cereal crops and winter rape, (which dominate in the sowings structure), is the occurrence of meteorological drought in the region of Pomorze and Kujawy from May to June between 1981–2010 (Table 19.1). In the analysed multi-annual period in particular cities of the region, meteorological droughts occurred in 7–9 cases, which means a frequency of 23.3–30.0%. This is entirely consistent with the results of research conducted by Łabędzki for central Poland [4].

Conditions of sufficiently or excessively moisturized soils during some periods of plant vegetation, or even throughout the whole growing season, occur in Poland with a similar frequency to droughts. For this reason, among other things, the irrigated area in Poland covers only about 73 thousand hectares according to official data, including

Table 19.1 The occurrence of meteorological droughts in the period of high water needs of cereal crops and rapeseed (May–June) in the years 1981–2010 in the chosen locations of the region of Pomorze and Kujawy (Ed – extremely dry, Vd – very dry, Md – moderate dry, Sd – slightly dry) [5]

Year	Droughts in chosen locations of the region of Pomorze and Kujawy				
	Mochle	Chrząstowo	Toruń	Głodowo	Głębokie
1981	Md	Sd			
1982					
1983	Md	Sd	Sd	Vd	Vd
1984					
1985					
1986	Sd			Sd	
1987					
1988				Sd	
1989	Md	Ed	Ed		Vd
1990	Sd				Sd
1991					
1992	Sd	Md	Sd	Md	
1993					
1994				Md	
1995			Sd		
1996					
1997					
1998					Md
1999					
2000	Vd	Vd	Vd	Ed	Vd
2001					
2002		Sd			
2003	Md	Sd	Vd		
2004					
2005				Sd	
2006		Sd	Sd	Sd	Vd
2007					
2008	Ed	Ed	Md	Vd	Vd
2009					
2010					
Ed	1	2	1	1	0
Vd	1	1	2	2	5
Md	4	1	1	2	1
Sd	3	5	4	4	1

modern pressurized systems (sprinkler irrigation), applied mainly in horticultural crops at only 8.7 thousand hectares [6]. The dimension of these areas places Poland at a distant position from other countries around the world in terms of plant irrigation. According to the ICID record [1], which covers a rank of 46 developed countries with the USA as the leader, Poland maintains the 28th position, while in the classification of all 159 countries, in which China and India are the leaders in terms of irrigation, Poland shifts to the 105th place. A small irrigated area in Poland, covering only 0.5% of agricultural land allows for the conclusion that the importance of irrigation in Polish agriculture is insignificant compared to the area covered by it. This conclusion cannot be changed even by the fact that the actual area of irrigated crops in Poland is much larger than declared by the Statistics Poland [6]. The official data do not take into account systems such as micro-irrigation, especially the drip system, commonly used in horticulture. According to research published by Treder et al. [7], from 80 to 100 thousand hectares of fruit crops in Poland are under irrigation. On the other hand, Lipiński [8] estimates the area of irrigated vegetables in the country at about 45 thousand hectares.

Despite the limitations deriving primarily from economic conditions as well as from the insufficient sources of water and the amount of water available for agriculture, the irrigation of crops in Poland is an important element of water management in the country and the main element of water management in agriculture. This is a promising and perspective solution. The main factors that will accelerate the development of irrigation in the future include: assurance of higher, stable and good quality crop yields, the necessity to increase modernity and make increasingly expanding agricultural farms more competitive, and the expected climate change. In accordance with the theory of global warming, whose main manifestation is an increase in air temperature and evapotranspiration, the frequency of droughts in moderate latitudes is also forecasted to increase [9–11]. Some research results indicate that these changes are already taking place. According to the results of the research presented by Wibig [12], in Poland, one can observe a tendency towards more intense, longer-lasting droughts in the summer season, especially in the areas where water is the main yield-limiting factor.

19.2 Effects of Irrigation on Crop Production

The main indicators for the advisability of applying irrigation and considering its role in domestic agriculture and water management are production effects, i.e., the absolute yield increase obtained under the influence of this treatment in the technology of agricultural raw materials production. The amount and value of production efficiency under irrigation can be discussed in terms of an average approach (multi-annual mean), or in relation to the specific growing season. Positive effects of applying irrigation, such as yield increases, account for the basis of their economic efficiency assessment, which is the main argument for undertaking the investment. Modifications in the quality of yield under the influence of irrigation, which relates to the qualitative features that determine the suitability of the crop for consumption or the processing technology of the raw materials obtained, are also of great importance. An important aspect of irrigation efficiency is also an estimation of the amount of by-product and mass of crop residue, the increase of which improves soil properties.

A direct production effect of plant irrigation, such as an increase in the main yield and changes in its quality, is a consequence of modifications of plant growth and development caused by this treatment. In particular, irrigation has a beneficial effect on the activity of physiological processes, and the morphological and anatomical structure of the plant, and consequently on the canopy structure, and finally on yield components. According to the study by Karczmarczyk [13], irrigation combined with an optimal rate of Nitrogen, Phosphorous, and Potash fertilizers resulted in a significant increase in the intensity of physiological processes occurring in cereals leaves, and thus higher assimilation of carbon dioxide, transpiration and stomatal conductance. The intensification of physiological processes and the prolonged period of the physiological activity of irrigated plants resulted in more robust growth, the formation of taller and thicker stems with longer ears, as well as an increase in the number and weight of grains per ear. Irrigation also contributed to the increase in plant tillering and density of vegetation (the number of blades per unit area). In addition, advantageous changes in yield composition, related to the increased plant density on irrigated plots, higher abundance, and enlarged weight of a single component of the yield, were found in other crops, for instance winter oilseed rape [14], potato [15] and some vegetables [16, 17].

In order to determine the production effects of irrigated crops in Poland, numerous scientific experiments were performed. Then, the results of particular experiments were the subject of numerous syntheses. The most widely known synthesis works are the study by Dzieżyc [18] as well as the multiple-authors monograph edited by Karczmarczyk and Nowak [19]. An analysis of yield modification through the quantitative and qualitative aspects obtained under the influence of irrigation carried out on this basis leads to the conclusion that the yield depends primarily on two factors – rainfall conditions and soil properties.

19.2.1 Atmospheric Precipitation as the Factor Creating Yield

In Poland, in the zone of moderate transitory climate, irrigation treatment supplements the deficiencies of atmospheric precipitation, particularly during the period of high water needs for plants. Therefore, the meteorological factor that has the greatest impact on the production effects of irrigation is rainfall, especially total rainfall and rainfall distribution in time. Since rainfall is characterized by very high temporal variability in particular spans of the growing season, the effects of irrigation might be different in consecutive years. Despite this, the effects significantly correlate with the amount of rainfall during the period of high demand for water of individual species. This relationship, after regression equation transformation, can be generalized according to the formula (1) by Grabarczyk [20]:

$$Q = (P_{OPT} - P_A)k \quad (19.1)$$

where:

Q expected yield increase under the influence of irrigation (kg ha^1),
K yield increase per 1 mm of rainfall deficit supplemented by irrigation (kg ha^1),
P_{OPT} the total of optimal rainfall, i.e., this where no increases in yield due to irrigation are observed (mm),
P_A the total of actual rainfall in the period of the highest need for water of a given species (mm).

Formulas for selected crop species (Table 19.2) were derived on the basis of rigorous field experiments carried out simultaneously in the years 2006–2012 on sandy soil with a compacted subsoil, in the Research Center of the UTP University of Science and Technology located in the village of Mochełek near Bydgoszcz city (Kujawsko-Pomorskie province) [21]. These formulas enable the interpolation of data gained from research from other regions of the country, and for the determination of expected effects of plant irrigation in different zones of rainfall conditions in Poland, assuming congruent soil conditions. In the zone of the lowest rainfall totals during periods of high water needs for plants, which includes central Poland, the expected average yield increases will be the greatest, while in the areas with higher rainfall, in which the totals of atmospheric precipitation in the growing season exceed 400 mm – the increases will be respectively lesser.

The prognostic formulas also provide a determination of the variability of production effects under irrigation in a given area in consecutive vegetation seasons (temporal variation), resulting from uneven rainfall conditions in periods of high water needs of plants in subsequent years. This variability in the example of maize grown for grain in the years 2005–2016 is shown in Fig. 19.1 [22]. For this period, the average production effect of irrigation amounted to 4.63 t ha^{-1}, which constituted an increase in the grain yield of 51%. In particular seasons, the production effects of irrigation depended significantly on the rainfall totals during periods of high water need for maize amounting to 7.28 t ha^{-1} in the dry season, 5.52 t ha^{-1} in periods of

Table 19.2 Regression equations and prognostic formulas of the amount of irrigation effects of selected plant species based on rainfall totals in the period of high water needs [21]

Crop	Type of yield	Period of high water needs	Regression equation	Prognostic formula
Table potato	Tubers	June–July	$Y = -0.15x + 36.1$	$Q = (240 - P_A) \cdot 150$
Maize for grain	Grain	June–July	$Y = -0.04x + 10.2$	$Q = (255 - P_A) \cdot 40$
Spring malting barley	Grain	May–June	$Y = -0.022x + 4.03$	$Q = (180 - P_A) \cdot 22$
Faba bean	Seeds	June–July	$Y = -0.013x + 3.13$	$Q = (240 - P_A) \cdot 13$

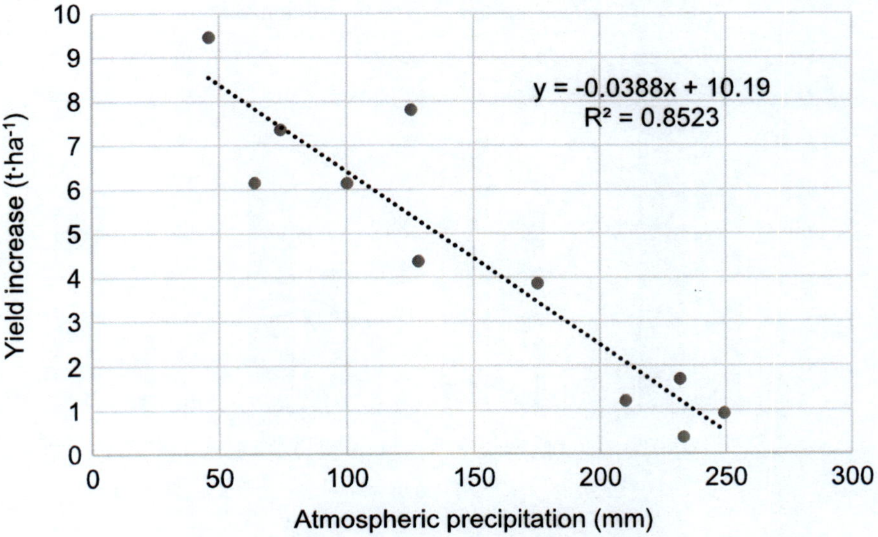

Fig. 19.1 Linear dependence of increase of maize grain yield under irrigation (t ha^{-1}) versus rainfall totals in the period of high water needs of the crop (June–July) [22]

the average rainfall conditions, and 1.05 t ha^{-1} in wet seasons. The regression equation shows that yield increases under irrigation were not significant only when the total of rainfall in the span of June–July exceeded 220–230 mm (Fig. 19.1). Attention should be paid to the high value of the coefficient of determination ($R^2 = 0.85$) describing the significant dependence, even though in the 12 analyzed subsequent growing season distributions of similar precipitation, totals in June–July span were different. Application of additional meteorological parameters and indicators in the formulas (air temperature, evapotranspiration, drought indicators) did not raise the coefficients of correlation and determination against the dependence of irrigation effects on the rainfall totals only.

19.2.2 Soil Conditions Influencing Production Effects

The extent of the production effects of plant irrigation depends significantly on the type of soil, especially its water properties which are usually determined by the degree of compactness. What matters is not only the compactness of topsoil (rich with humus) but also the kind of subsoil (abundant in minerals). According to the plentiful results of field experiments, the absolute and relative yield increases resulting from irrigation as well as from the unitary effects expressed by the yield increase per 1 mm of irrigation water, which are greater for the less compacted the soil is (Fig. 19.2).

The greatest effects of irrigation occur on sandy soils with a deep groundwater table, distinguished by a low water capacity and therefore a low range of ability to

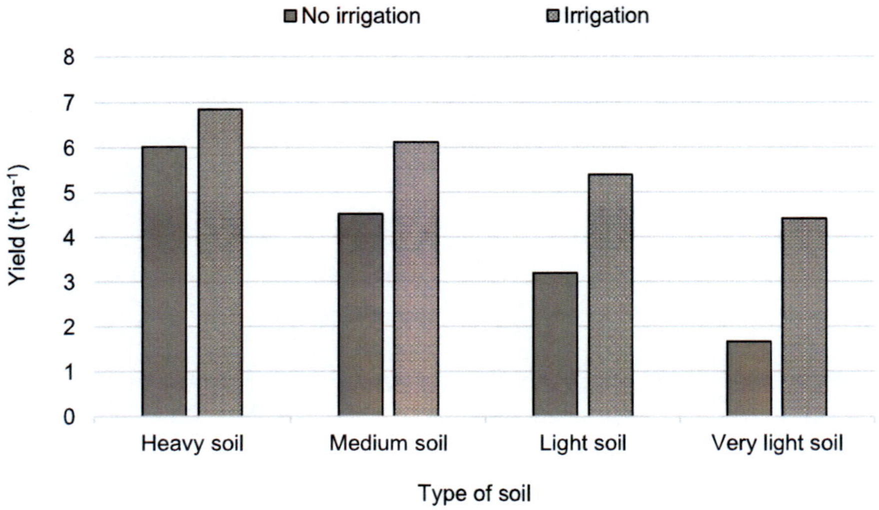

Fig. 19.2 Yield of non-irrigated and irrigated crops depending on the type of soil (based on [23])

perpetually supply plants with water. With the increase of soil compactness and soil water retention, the effects of cereal irrigation quite systematically decrease. In a few experiments carried out on heavy or very heavy soils (fen soil, loess soil), no significant differences were found between the yield of non-irrigated and irrigated crops. What is more, as a result of increased cereals lodging in selected years, the negative effects of irrigation were observed.

When soil factor influence on the crop yield increases is considered, it is extremely important to determine not only their absolute, relative, and unitary values, but also to assess the yield potential of crops grown on different soils under irrigation. It is noteworthy to mention that the yield of crops being irrigated on weak soils is comparable to the yield obtained on better soils without irrigation (Fig. 19.2). Therefore, the use of irrigation on weaker soils (e.g. the good rye, poor rye and very poor rye soil suitability complex related to the classification of complexes of agricultural soil suitability) is a treatment that only compensates production opportunities in relation to better soils (very good rye soil suitability complex and wheat soil suitability complexes).

19.2.3 Irrigation Effectiveness in Crop Groups and Species

Respective groups of crops, species, and cultivars respond to irrigation in different ways. It mainly results from their diverse conditionings of water management, including, e.g. the capability to use water for producing biomass, the properties of the root system and the type of photosynthesis. The different lengths of plant growth stages, as well as the type and the way of yield forming and the yield moisture content are also important.

Comparison of the effectiveness of irrigation in various plants requires the elimination of other impact factors, especially meteorological and soil factors. Experiments should be carried out under the same growing seasons and soil conditions while providing similar agrotechnical systems and ensuring a similar cover of plant water needs by irrigation systems. One of a few experiments that meets the requirements of such comparability of results was carried out in the years 2006–2012 on light soil with compacted subsoil, in the region of Bydgoszcz [21]. The seven-year study of four crop species consisted of growing seasons with different rainfall conditions, which provided considerable data for the average increased yield outputs. The comparison of the results shows that the increased yield outputs were similar in all tested crops and ranged from 37 to 51%, depending on the crop species. The highest relative increase in yield under irrigation was gained in maize for grain cultivation, while the lowest occurred in the raw material production for the brewing industry (malting barley) (Table 19.3). In the case of maize, the result is a consequence of the economic water management by the plant, C4 photosynthesis, and the fact that maize is a monoecious plant, which results in increased water requirements during pollination and fertilization. However, in the case of malting barley, this effect may result from low nitrogen fertilization due to malting production requirements. The absolute average multi-annual production effects of irrigation used in the cultivation of the crop were as follow table potato 12.1 t ha^{-1} of the fresh matter of tubers, maize 3.83 t ha^{-1} of dry matter of the grain, malting barley 1.44 t ha^{-1} of grain and 1.38 t ha^{-1} of faba bean seeds. The effects of unitary production of irrigation expressed in yield increases caused by the use of 1 mm of irrigation water, may be useful in water management optimization in agriculture. The effects amounted to

Table 19.3 Production effects of irrigation of selected agricultural crops on light soil (case of Bydgoszcz region) [21]

Crop	Yield type	Absolute yield increase (t ha^{-1})		Relative yield increase (%)		Unitary yield increase (kg·ha^{-1} mm^{-1})	
		Mean	Range	Mean	Range	Mean	Range
Faba bean	Seeds of 15% moisture	1.38	0.12–3.12	44	2–538	15.0	0.0–34.8
Spring barley for malting	Grain of 15% moisture	1.44	0.0–3.39	37	0–171	18.3	0.0–29.8
Maize	Grain dry matter	3.83	0.32–8.47	51	2–1694	39.7	10.7–70.6
Potato medium-early	Tubers	12.1	0.0–28.7	42	0–204	128	0.0–139

Table 19.4 Average multi-annual production effects of irrigation in selected agricultural and horticultural crops cultivated on very light soil in the village of Kruszyn Krajeński near Bydgoszcz [24]

Agricultural crops	Absolute yield increase t ha^{-1}	Unitary yield increase kg ha^{-1} mm^{-1}	Horticultural crops	Absolute yield increase t ha^{-1}	Unitary yield increase kg ha^{-1} mm^{-1}
Spring wheat	2.27	19.2	Beetroot	17.4	108
Spring barley	2.37	19.5	Carrot	22.6	140
Oat	2.09	16.3	Radish	6.5	100
Sugar beet	19.8	115	Courgette	26.2	104
Early potato	13.1	135	Pumpkin	39.4	323
Late potato	19.8	138	Giant pumpkin	32.9	143
Yellow lupine	1.04	9.5	Black chokeberry	6.2	35
Faba bean	3.07	35.0	Blackberry	8.4	46
Maize (grain)	4.12	28.0	Strawberry	7.2	48

128.1 kg ha^{-1} of potato tubers, 39.7 kg ha^{-1} dry matter of maize grain, 18.3 kg ha^{-1} of barley grain and 15.0 kg ha^{-1} of faba bean seeds.

Production effects of sprinkler and drip irrigation of selected agricultural and horticultural crops, obtained in multi-annual experiments of the Department of Land Reclamation and Agrometeorology of the UTP University of Science and Technology in Bydgoszcz [24] on a very light soil in the location of Kruszyn Krajeński village in the vicinity of Bydgoszcz are presented in Table 19.4 These yield increments are difficult to achieve with limited agrotechnical treatments, such as only drainage or cultivation. The results indicate great prospects for the increase in domestic agriculture productivity, provided that the water factor is optimized. Moreover, they also indicate the potential benefits of sandy soils with low water retention for agricultural use.

19.3 Prospects for Irrigation Development in Poland

Two main factors determine the prospects for the development of plant irrigation in Poland: economic efficiency of this treatment (costs of investment, energy, and water) and the availability of water sources for irrigation purpose without additional investment by farms located near rivers and lakes or having access to easily renewable underground water sources.

Taking this into account, it can be estimated that irrigation development in Poland would primarily concern the large-scale intensive production of vegetables and fruit. In such conditions, it is possible to use micro-irrigation, i.e., low-pressure, precise

Table 19.5 Indicators of economic and financial efficiency of crops production under irrigation [26]

Crop	Profitability indicator
Winter wheat	35–81
Potato	121–217
Sugar beet	153–170
Headed cabbage	241
Onion	364–394
Carrot	419–529
Strawberry	447

irrigation techniques that supply water with drops or streams to the vicinity of plants or directly to the root zone. According to data presented by ICID [1], micro-irrigation is used in 14.4 million hectares throughout the world. This type of system enables fertigation as well as advanced automation based on perpetual monitoring of atmospheric conditions that allows determination of the water needs of the plants [25]. An important advantage of this type of system is that it ensures the economic management of water at the farm scale since there is no need to irrigate the whole field surface, but only precisely chosen parts of the plantation. This argument is particularly important because water resources that can be gained for irrigation purposes should be utilized in Poland in an especially rational manner.

Irrigation of horticultural crops, including orchards, vegetables, and berry plants, is unquestionably an economically effective agrotechnical treatment. This is shown by the profitability indicators for irrigation derived by Jankowiak and Rzekanowski [26] (Table 19.5), as well as the results of the research by Lipiński [8, 27] and Rolbiecki et al. [28]. According to these studies, all indicators of economic and financial efficiency were great; i.e., the capital expenditure for the drip irrigation system of highbush blueberry plantations constituted only 20% of the income increase in the first year of the investment [29].

In the case of irrigation used in agricultural plant production, the indicators of the profitability are much less in comparison to horticultural plants (Table 19.5). In particular, this refers to the most economically important cereal crops, which stand for about 73% of the sown area in Poland, and 61% of the area of irrigated fields worldwide. According to research carried out by Żarski [23], despite the low profitability ratio, cereals should be scheduled on the parts of fields equipped with irrigation systems dedicated primarily to other crop groups. The reasoning that underlies this proceeding is the indispensability of reasonable crop rotation and quality aspects of the yield. For this reason, it is recommended to sprinkle bread-wheat cultivars, but most notably – the malting barley cultivars.

The economic analysis carried out by Kledzik et al. [30], based on the direct surplus growth method, showed that great production effects do not always translate into economic profitability of irrigation. Out of all of the agricultural crops being compared, irrigation caused a positive financial effect only in the case of table potato.

In the production of spring barley and maize for grain, the deployment of this treatment was economically groundless, regardless of the plantation area. The advisability of root crops irrigation, in particular table potato, is proved by the indicators depicted in Table 19.5, or the results of study led by Lipiński [31] as well as the favourable changes in the technological quality and chemical composition of tubers under the influence of this treatment confirmed by Rolbiecki et al. [15] and Wszelaczyńska et al. [32].

Besides the use of irrigation in horticultural and agricultural root crops production, this treatment is also conventional in other areas related to agriculture. Primarly, it is important to mention the forest nurseries and forest planting material productions, irrigation of remediated land – post-mining dumps, dumps and excavations after depletion of surface raw materials, and finally irrigation of backyards, lawns, gardens and recreation areas. Those are relatively recent and perspective areas for fully automated and technically advanced sprinkler systems, often combined with automatic weather stations [24].

19.4 Conclusions

Poland is a country in the EU, which ranks among the last in terms of the availability of water resources per capita (1.6 thousand m^3) with the EU average three times higher. The vast majority of water in Poland comes from atmospheric precipitation (97%) but a large part of it is lost (no storage in drainage systems, reservoirs or rivers) and goes via rivers to the Baltic Sea. This condition, coupled with mild winters, the absence of snow cover in winter, and long periods of high temperatures, which sometimes begin in the early spring and occur in summer, may in some years result in very low water levels in rivers and low level of groundwater table.

On a global scale, agriculture is the largest consumer of drinking water. In Poland, water consumption in plant production is much lower than the average in the world, but as shown by research results and production statistics, plant production is very dependent on water and susceptible to its scarcity and the production effects depend significantly on soil and rainfall conditions over the period of high water needs of crops. Despite this, crops irrigation in Poland has still only a supplementary character.

Poland is an agricultural country with over 14 million hectares of agricultural land, most of which are light soils, with a limited share of organic matter, which consequences into limited water retention capacity. In the last years, the government of Poland has developed several instruments, like water fees, rain tax, as well as a new investment plans in water retention. Water fees are the obligation for the compliance with the EU Water Framework Directive and they also include fees for water taken for irrigation purposes in agricultural production. At present, there have been registered approximately 8% of farms covered by the water fees. Recently, the Polish government has announced the introduction of the Retention Program, which is to be implemented in the years 2021–2027. This program has a budget of EUR 3.11 billion (14 billion PLN), which is to be allocated to the construction of large and small retention reservoirs, especially important for rural areas. Not long ago, the Polish Minister of Agriculture has launched an irrigation program in agriculture. As part of the first call for projects 89 million EUR (400 million PLN) will be allocated to irrigation systems-related investments, and the total amount allocated to the program is 222.15 million EUR (1 billion PLN). The program is aimed primarily at farm owners in the areas most affected by drought in recent years [33]. However, the development of irrigation systems, and thus water consumption in plant production, will be determined by the productive and economic efficiency of agricultural crops.

19.5 Recommendations

The content presented in the chapter is significant for the proper and sustainable management of water resources in agriculture both on a national and regional scale as well as for the design and operation of irrigation systems in family farms or commercial enterprises.

References

1. ICID The International Commission on Irrigation and Drainage (2018) Retrieved from www.icid.org/world-irrigated-area.pdf. Accessed on 21 Nov 2018
2. FAO AQUASTAT Main Database (2018) Food and Agriculture Organization of the United Nations. Retrieved from https://www.fao.org/nr/water/aquastat/data/query/index.html?lang=en. Accessed on 21 Nov 2018
3. Meier J, Zabel F, Mauser W (2018) A global approach to estimate irrigated areas – a comparison between different data and statistics. Hydrol Earth Syst Sci 22:1119–1133
4. Łabędzki L (2007) Estimation of local drought frequency in Central Poland using the standardized precipitation index SPI. Irrig Drainage 56:67–77
5. Kuśmierek-Tomaszewska R, Dudek S, Żarski J, Januszewska-Klapa K (2018) Temporal variability of drought in field crops in the region of kujawsko-pomorskie, Poland. In: The 24th annual international scientific conference "research for rural development 2018", Jelgava, 16–18 May, 2018. Proceedings of the international scientific conference research for rural development 2018, vol 2, pp 62–68
6. Statistics Poland (2017) Ochrona środowiska (Environment). Retrieved from www.stat.gov.pl. Accessed on 19 Nov 2018
7. Treder W, Wójcik K, Tryngiel-Gać A (2011) Development of irrigation of orchard plants reflected by survey investigations. Infrastruct Ecol Rural Area 5:61–69
8. Lipiński J (2012) Efekty produkcyjne i ekonomiczne nawadniania truskawek uprawianych na glebach lekkich. Wiadomości Melioracyjne I Łąkarskie 4:180–183
9. Kundzewicz ZW (2011) Climate changes, their reasons and effects – observations and projections. Land Anal 15:39–49
10. IPCC Intergovernmental Panel on Climate Change (2014) AR5 synthesis report – climate change 2014. Retrieved from https://www.ipcc.ch/report/ar5/syr/. Accessed on 21 Nov 2018
11. Kuchar L, Iwański S, Gąsiorek E, Gąsiorek E (2015) Simulation of hydrothermal conditions for crop production purpose until 2050–2060 and selected climate change scenarios for north central Poland. Infrastruct Ecol Rural Area II(I):319–334

12. Wibig J (2012) Warunki wilgotnościowe w Polsce w świetle wskaźnika standaryzowanego klimatycznego bilansu wodnego. Woda Środowisko Obszary Wiejskie 12(2):329–340
13. Karczmarczyk S (1999) Zbiór prac dotyczących nawadniania. Folia Universitas Agriculturae Stetinensis 193(73):1–220
14. Dudek S, Kuśmierek-Tomaszewska R, Żarski J (2011) The effect of sprinkling irrigation and nitrogen fertilization on the yield of winter rape. Infrastruct Ecol Rural Area 5:193–202
15. Rolbiecki S, Rolbiecki R, Kuśmierek-Tomaszewska R, Dudek S, Żarski J, Cz R (2015) Requirements and effects of drip irrigation of mid-early potato on a very light soil in moderate climate. Fresenius Environ Bull 24(11b):3895–3902
16. Rolbiecki R, Rolbiecki S, Podsiadło C (2011) Porównanie reakcji dyni zwyczajnej odmiany 'Danka' na nawadnianie kroplowe w warunkach opadowo-termicznych Bydgoszczy i Stargardu Szczecińskiego. Infrastruktura I Ekologia Terenów Wiejskich 6:127–135
17. Rolbiecki S, Rolbiecki R, Czop P (2012). Wpływ nawadniania kroplowego i nawożenia azotem na plonowanie cebuli odmiany 'Kutnowska' uprawianej na glebie piaszczystej w rejonie Bydgoszczy. Infrastruktura i Ekologia Terenów Wiejskich 2(I):183–189
18. Dzieżyc J (1988) Rolnictwo w warunkach nawadniania. Powszechne Wydawnictwo Rolnicze i Leśne, Warszawa
19. Karczmarczyk S, Nowak L (2006) Nawadnianie Roślin. Powszechne Wydawnictwo Rolnicze i Leśne, Poznań
20. Grabarczyk S (1987) Efekty, potrzeby i możliwości nawodnień deszczownianych w różnych regionach kraju. Zeszyty Problemowe Postępu Nauk Rolniczych 314:49–64
21. Żarski J, Dudek S, Kuśmierek-Tomaszewska R (2013) Forecasting effects of plants irrigation based on selected meteorological and agricultural drought indices. Annual Environ Prot 15:2185–2203
22. Żarski J, Dudek S, Kuśmierek-Tomaszewska R (2018) Drip irrigation as a factor mitigating drought impact in corn cultivation in central Poland. In: The 8th international scientific conference: rural development. Bioeconomy challenges. Aleksandras Stulginskis University, Kaunas, Nov 23–24, 2017. Proceedings of the international scientific conference rural development. Bioeconomy challenges, pp 182–186
23. Żarski J (2009) The effects of irrigation of cereals in Poland. Infrastruct Ecol Rural Area 3:29–42
24. Rzekanowski C, Żarski J, Rolbiecki S (2011) Potrzeby, efekty i perspektywy nawadniania roślin na obszarach szczególnie deficytowych w wodę. Postępy Nauk Rolniczych 1:51–63
25. Kuśmierek-Tomaszewska R, Żarski J, Dudek S (2012) Meteorological automated weather station data application for plant water requirements estimation. Comput Electron Agric 88:44–51
26. Jankowiak J, Rzekanowski C (2006) Ekonomiczne efekty nawadniania. In: Karczmarczyk S, Nowak L (eds) Nawadnianie Roślin. Poznań: Powszechne Wydawnictwo Rolnicze i Leśne
27. Lipiński J (2015) Efektywność ekonomiczno-finansowa deszczownianego nawadniania ziemniaków jadalnych na glebach lekkich w warunkach produkcyjnego gospodarstwa rolnego. Woda Środowisko Obszary Wiejskie 15(3):61–73
28. Rolbiecki R, Rolbiecka M, Rolbiecki S, Jagosz B, Michalcewicz-Kaniowska M, Zajdel M (2018) Cost-effectiveness of postharvest drip irrigation of chosen North-American cultivars of asparagus grown for green spears in open field technology in the Central Poland. In: International conference "economic science for rural development", Jelgava, LLU ESAF, 9–11 May 2018, Proceedings of the international conference economic science for rural development 2018, vol 49, pp 71–76
29. Kula A, Lipiński J (2017) Produkcyjna i ekonomiczno-finansowa efektywność kroplowego nawadniania borówki wysokiej. Woda Środowisko Obszary Wiejskie, 17,1(57):75–87
30. Kledzik R, Kropkowski M, Rzekanowski C, Żarski J (2015) Economic evaluation of the efficiency of irrigation in selected crops. Infrastruct Ecol Rural Area II:291–303
31. Lipiński J (2016) Efektywność deszczownianego nawadniania ogórków gruntowych w warunkach produkcyjnego gospodarstwa ogrodniczego. Woda Środowisko Obszary Wiejskie 16(3):61–71

32. Wszelaczyńska E, Pobereżny J, Dudek S, Kuśmierek-Tomaszewska R, Żarski J, Pawelzik E (2015) The effects of fertilizers, irrigation and storage on the properties of potato tubers and their constituent starches. Starch/Stärke 67:478–492
33. Website of the Polish Ministry of Agriculture and Rural Development https://www.gov.pl/web/agriculture. Accessed on 14 Aug 2020

Chapter 20
Power Plant Open Cooling System in the Context of the Objectives of the Water Framework Directive

Leszek Pająk, Agnieszka Operacz, and Barbara Tomaszewska

Abstract The maintenance of water using in all Member States of European Union is focused on the rational use and protection of water resources in accordance with the principle of sustainable development. The requirements of water policy are defined in Directive 2000/60/EC of the European Parliament and of the Council of 23 October 2000 establishing a framework for Community action in the field of water policy, also known as the Water Framework Directive. Article 13 of the WFD resulted in the obligation to draw up river basin management plans. As a study case authors selected the conventional power plant with the combined generation of electricity and heat, situated in the Upper Vistula River sub-basin, Poland. On the technological canal behind the conventional power plant, there is a small hydropower plant. The assessment of the general status of a given body of surface water was conducted on the basis of the analysis of the results of the assessment of the ecological status/potential and of the chemical status of the body of surface water in question. The result of the analysis was the inclusion in the Regulation of the Council of Ministers of 18 October 2016 of a derogation resulting from the lack of technical feasibility of achieving the environmental objective.

Keywords Water framework directive · Environmental objectives · Cooling system · Power plant · Poland

L. Pająk · B. Tomaszewska
Department of Fossil Fuels, Faculty of Geology, Geophysics and Environmental Protection, AGH University of Science and Technology, Krakow, Poland
e-mail: pajakl@agh.edu.pl

B. Tomaszewska
e-mail: barbara.tomaszewska@agh.edu.pl

A. Operacz (✉)
Department of Sanitary Engineering and Water Management, Faculty of Environmental Engineering and Land Surveying, University of Agriculture in Krakow, Krakow, Poland
e-mail: a.operacz@urk.edu.pl

M. Zeleňáková et al. (eds.), *Management of Water Resources in Poland*, Springer Water,
https://doi.org/10.1007/978-3-030-61965-7_20

Abbreviations

WFD	Water Framework Directive
EU	European Union
RBMPs	River basin management plans
uRBMPs	Updated river basin management plans
NPG	National power grid
BAT	Best available technics

20.1 Introduction

The European Union's water policy is defined in "Directive 2000/60/EC of the European Parliament and of the Council of 23 October 2000 establishing a framework for Community action in the field of water policy", also known as the Water Framework Directive (WFD) [1]. This document commits all Member States to the rational use and protection of water resources in accordance with the principle of sustainable development (and sets as an overarching objective achieving good status of all waters by 2015. In justified cases, this deadline may be extended until 2021 or 2027 [2, 3]. Some authors suggest that the timescale of the WFD (achieving good ecological status in all surface waters by 2027) is over-ambitious [4].

The overarching strategic objectives of EU water policy are primarily:

- Achieving and maintaining a good status and potential of waters as well as the related ecosystems;
- Ensuring access to water resources in order to meet the needs of the population and of the natural environment as well as socially and economically justified water needs of the economy;
- Mitigating the negative effects of floods and droughts and minimising the risk of emergencies;
- Implementing a system for the integrated management of water resources.

Since the WFD was formally adopted in December 2000, various developments potentially affecting its implementation have occurred [5]:

(a) The importance of specific pressures has been increasingly recognised, including climate change and associated flood and drought risks [6], invasive species [7] and a wide range of emerging pollutants [8];
(b) New perspectives on environmental management have been developed, including ecosystem services, nature-based solutions and adaptive and resilience-based approaches [9–11];
(c) The sustainable development goals have been set, in which many targets are related to water, or affect waters in both positive and negative ways [12].

(d) Other EU policies on biodiversity, renewable energy and floods management have been developed, influencing how we manage aquatic systems.

Among the priority issues in the process of planning investments related to water management is the implementation of WFD assumptions in order to guarantee the protection of natural resources as well as to prevent any deterioration in the status of the natural environment. The above is reflected in ensuring the sustainable development of the Vistula River basin while ensuring that water management needs are met at the same time.

20.2 Legal Framework

20.2.1 River Basin Management Plans

The obligation to draw up river basin management plans (RBMPs) results from Article 13 of the WFD. According to that Article, "*Member States shall ensure that a river basin management plan is produced for each river basin district lying entirely within their territory*" [1].

RBMPs were first developed in Poland in 2009, and they were subsequently approved by the Council of Ministers on 22 February 2011. These plans are updated every six years. Once approved by the Council of Ministers, the updated RBMP (uRBMP), as a strategy paper, is the basis for:

- Making decisions which determine the status of water resources;
- Improving the process for achieving or maintaining the good status of waters and the related ecosystems;
- Determining the need to introduce rational water management principles in the future.

Updated RBMPs include issues related to the achievement of environmental objectives for individual types of surface waters, groundwaters and protected areas. The list of environmental objectives developed during the next planning cycle (2016) included, *inter alia*, more stringent environmental objectives relating to protected areas within which particular bodies of surface water are located. In the first planning cycle, no environmental objectives were indicated, but it was assumed that for natural bodies of surface water those included a good ecological and chemical status, while for heavily modified and artificial bodies of surface water they included a good ecological potential and a good chemical status. For bodies of groundwater, the objective was a good quantitative and chemical status.

Since the publication of the first RBMP, a number of amendments have been introduced into the Polish legislation, which have allowed it to be adapted to water resources management principles which are in line with the requirements of the WFD. The most important of those amendments has been the introduction of a definition of environmental objectives and of provisions which determine the conditions for the application of derogations into the Water Law [13]. The amendment to the Water Law also includes tools which enable the maintenance of water bodies and protected areas in a non-deteriorated state. The introduction of the following provision states: "*the use of waters shall, without limitation, not result in the deterioration of the status of waters and the ecosystems which depend on them, including without limitation the provisions of the RBMP*". This is a very important provision in the context of the possibility of implementing investment projects; among other things. It determines the project owner's ability to obtain a water permit.

20.2.2 Derogations

The obligation to achieve the objectives related to a body of water is not absolute. In special cases, the WFD allows for derogations from these objectives. Such a derogation may only be applied in exceptional circumstances and is subject to the fulfilment of certain strictly defined conditions, which are set forth in Articles 4(4) to 4(9) of the Directive.

The following types of derogations may be applied:

- The extension of the deadline for achieving the objective (derogation conditions are set forth in Article 4(4));
- Establishing a less stringent objective (derogation conditions are set forth in Article 4(5));
- Temporary deterioration in the status as a result of unforeseeable circumstances (derogation conditions are set forth in Article 4(6));
- A failure to achieve the objective as a result of new sustainable human development activities (derogation conditions are set forth in Article 4(7)).

The first two cases relate to a situation where the current state of a body of water is below good, and it is not possible to improve it to a sufficient degree. In such circumstances, the WFD allows for derogations but makes them subject to meeting the conditions set forth in Articles 4(4) and 4(5). Further two cases concern situations where the status deteriorates as a result of unforeseen circumstances which could not have been prevented (e.g. drought) or as a result of new hydromorphological changes introduced by human activity. The conditions for the application of the derogations above are stipulated in Articles 4(6) and (7) of the WFD.

Article 4(4)*WFD: „The deadlines established under paragraph 1 may be extended for phased achievement of the objectives for bodies of water, provided that no further deterioration occurs in the status of the affected body of water when* ***all****of the following conditions are met*:

(a) Member States determine that all necessary improvements in the status of bodies of water cannot reasonably be achieved within the timescales set out in that paragraph for at least one of the following reasons:

(i) The scale of improvements required can only be achieved in phases exceeding the timescale, for reasons of technical feasibility;
(ii) Completing the improvements within the timescale would be disproportionately expensive;
(iii) Natural conditions do not allow timely improvement in the status of the body of water.

(b) Extension of the deadline, and the reasons for it are specifically set out and explained in the river basin management plan required under Article 13.
(c) Extensions shall be limited to a maximum of two further updates of the river basin management plan except in cases where the natural conditions are such that the objectives cannot be achieved within this period.
(d) A summary of the measures required under Article 11 which are envisaged as necessary to bring the bodies of water progressively to the required status by the extended deadline, the reasons for any significant delay in making these measures operational, and the expected timetable for their implementation are set out in the river basin management plan. A review of the implementation of these measures and a summary of any additional measures shall be included in updates of the river basin management plan" [1].

The derogation provided for in **Article 4(7)** is a specific type of derogation, as it does not directly concern bodies of water and the activities which may adversely affect them, either by causing deterioration in their status or by preventing the achievement of environmental objectives. According to Article 4(7):

|*Member States will not be in breach of this Directive when*:

- *Failure to achieve good groundwater status, good ecological status or, where relevant, good ecological potential or*
- *To prevent deterioration in the status of a body of surface water or groundwater is the result of new modifications to the physical characteristics of a surface water body or alterations to the level of bodies of groundwater, or*
- *Failure to prevent deterioration from high status to good status of a body of surface water is the result of new sustainable human development activities and* ***all*** *the following conditions are met:*

(a) *All practicable steps are taken to mitigate the adverse impact on the status of the body of water (Article 4(7)(a) WFD);*
(b) *The reasons for those modifications or alterations are specifically set out and explained in the river basin management plan, and the objectives are reviewed every six years (Article 4(7)(b) WFD);*
(c) The reasons for those modifications or alterations are of overriding public interest and/or the benefits to the environment and to society of achieving environmental objectives are outweighed by the benefits of the new modifications or alterations to human health, to the maintenance of human safety or to sustainable development (Article 4(7)(c) WFD);
(d) *The beneficial objectives served by those modifications or alterations of the water body cannot for reasons of technical feasibility or disproportionate cost*

be achieved by other means, which are a significantly better environmental option" (Article 4(7)(d) WFD) [1].

20.2.3 River Basin Management Plan for the Vistula River Basin District

The RBMP for the Vistula River basin district was approved by the President of the Council of Ministers at the meeting of the Council of Ministers on 22 February 2011 and published in the Monitor Polski Official Journal No. 49 item 549 [14]. The document set forth planning objectives related to water management, which are to be achieved through the implementation of the indicated tasks. It included:

1. A general description of the characteristics of the river basin district in question, including without limitation:
   - The list of bodies of surface water together with their types and the reference conditions established for them;
   - The list of bodies of groundwater;
2. The summary of significant anthropogenic impacts identified and the assessment of their impact on surface water and groundwater status;
3. Lists of protected areas referred to in Article 113(2)(5) together with their graphic representation;
4. The map of the monitoring network together with the presentation of monitoring programmes;
5. The determination of environmental objectives for bodies of water and protected areas;
6. The summary of the results of the economic analysis related to water use;
7. The summary of the measures included in the country's water and environmental programme, taking into account the methods of achieving the environmental objectives set;
8. The list of other detailed programmes and management plans for the river basin district concerning sub-basins, economic sectors, issues or water types, together with a summary of the contents of these programmes and plans;
9. The summary of the public information and consultation measures taken, a description of their results and the changes to the plan made as a consequence;
10. The list of authorities competent for river basin district management.

20.3 Power Plant Open Cooling System

20.3.1 Case Study

The conventional power plant in question, which is situated in the Upper Vistula River sub-basin, is one of the largest in the region. Its operation results in ensuring the stability of power grids, which has a significant impact on the security of the national power grid (NPG) due to that power plant's ability to provide grid recovery services in the event of a blackout. It is one of the utility power plants included in the NPG.Since it features a common steam header, it can produce energy for its own needs as well as for supplying a separate part of the NPG within the framework of so-called "island mode" operation. Owing to its location at the junction of power transmission and power distribution grids, it can play a strategic role within the NPG, in particular in the area covering the Kraków metropolitan area and the city of Skawina. In addition to the combined generation of electricity and heat, this power plant is also a supplier of process steam.

The conventional power plant has an open cooling water circuit, which is fed by a surface intake situated within the power plant canal. Before entering the central pumping station, the water flows through grilles and sieves, and it is subsequently pumped through two collecting pipes from which it is distributed to the condensers of individual turbines and other locations. Spent cooling water from turbine condensers and auxiliary equipment is discharged to a river which is the right-bank tributary of the Vistula River. The hydropower plant converts potential energy of water gained by water in a natural circulation in nature, so this energy should be regarded to be renewable. The doubts about whether energy acquired in a hydropower plant comes from a renewable source should be decided by physics of power processes [15].

According to the RBMP for the Vistula River basin district [14], one of the artificial canals which supplies water from the Vistula River to cool the turbine sets of the conventional power plant is a body of surface water included in the Upper Vistula River region. This canal has the status of an artificial water body. In terms of water quality, its condition is described as bad, which results directly from the quality of water in the Vistula River (which is salinated as a result of drainage from mines). The power plant canal was considered at risk of failing to achieve the environmental objectives set forth in the WFD and was therefore covered by the temporary derogation under Article 4(4)(3) (natural conditions, i.e. conditions which do not allow for improvement in the status of a body of water): "due to the salinity and impact of mine waters".

The body of water receiving spent cooling water from the conventional power plant also has a European body of water code within the water region of the Upper Vistula River assigned. The river, just as the power plant canal, is located within the Eastern Plains Euroregion. In the RBMP, it was defined as a river which is a heavily modified water body with a bad ecological status and was covered by a temporary derogation resulting from Article 4(4)(1): "*in connection with the fact that it directly receives spent CHP cooling waters, the water in the river has an elevated temperature which prevents the achievement of a status which would enable the water to be classified as a useful habitat for cyprinids. The impact of saline waters of the Vistula River*".

20.3.2 Assessment of Water Status and Achievement of Environmental Objectives

As part of the work on the updated RBMP [16], with respect to the body of water receiving spent cooling water from the conventional power plant, the objective was set as follows: achieving good ecological potential and maintaining good chemical status. On the other hand, it was indicated that the water stretch in question was not at risk of failing to achieve environmental objectives. The power plant canal, which supplies water for cooling the turbine sets, was also considered not at risk of failing to achieve environmental objectives.

The criteria for determining the ecological potential and chemical status stipulated in the uRBMP were established in accordance with the requirements in force at the time. They were included in the Regulation of the Minister of the Environment of 9 November 2011 on the method of classification of the status of bodies of surface water and on environmental quality standards with respect to priority substances [17].

The assessment of the general status of a given body of surface water was conducted on the basis of the analysis of the results of the assessment of the ecological status/potential and of the chemical status of the body of surface water in question. The overall status of the body of surface water can only be assessed as good if its chemical status is good and at the same time the ecological status/potential of that body of surface water is at least good.

It should be noted that in the case of bodies of water which are rivers and for which no measurement and control points have been established (as for the power plant canal in question), their status was assessed on the basis of the results of the research carried out within the framework of State Environmental Monitoring. The research should be done accordingly the procedure for mapping the results of water status assessment for unmonitored bodies of surface water on the basis of the results for monitored bodies of surface water using obligatory and optional similarities.

As a consequence, the updated RBMP indicated that the power plant canal in question was not a monitored body of surface water. For the canal, "at least good ecological status/potential" and "good" chemical status were indicated. The current status of the body of surface water was considered "good". It was indicated that the canal had the status of an artificial body of water. With respect to the "assessment of failing to achieve environmental objectives", the body of water was identified as not being at risk. To the canal, derogation (4(4)-3) according to the RBMP was assigned: "*Due to the salinity and impact of mine waters*", in accordance with the provisions of the RBMP and of the national water and environmental programme.

The description in the updated RBMP concerning the power plant canal did not provide for a derogation under Article 4(4) WFD.

On the other hand, it was indicated that the body of water which receives spent cooling water, and which constitutes a separate body of surface water, was subject to monitoring. It was determined that its ecological status/potential was "moderate" and its chemical status was "good", but the current status of the body of surface water was found to be "bad". The river had the status of a heavily modified water body. However, in the "assessment of the risk of failing to achieve environmental objectives" column, it was stated that it was "not at risk". For the river, derogation (4(4)-1) included in the RBMP in force was repeated: "*in connection with the fact that it directly receives spent CHP cooling waters, the water in the river has an elevated temperature which prevents the achievement of a status which would enable the water to be classified as a useful habitat for cyprinids. The impact of saline waters of the Vistula River*". The description in the updated RBMP concerning the river in question did not provide for derogations under Article 4(4) or under Article 4(7) WFD.

20.3.3 Analysis of Possible Derogations

It should be emphasised that in the case of derogations under Article 4(4), the extension of the deadline and indication of the reasons for the derogation should be specifically set out and explained in the RBMP. This means that the updated RBMP should include information on one of the three conditions for the application of the derogation, i.e. whether the derogation was introduced due to:

a. The lack of technical feasibility of achieving good water status promptly;
b. The disproportionate costs of achieving good water status promptly;
c. Natural conditions.

Similarly, the updated RBMP should include a justification for the derogation under Article 4(7) where it is the result of new sustainable human development activities. Such new activities include any investment projects, including a water permit for the use of water or for wastewater disposal. This derogation is acceptable where:

(a) "All practicable steps are taken to mitigate the adverse impact on the status of the body of water;
(b) The reasons for those modifications or alterations are specifically set out and explained in the river basin management plan;
(c) The reasons for those modifications or alterations are of overriding public interest and/or the benefits to the environment and to society of achieving the environmental objectives are outweighed by the benefits of the new modifications or alterations to human health, to the maintenance of human safety or to sustainable development;

(d) The beneficial objectives served by those modifications or alterations of the water body cannot for reasons of technical feasibility or disproportionate cost be achieved by other means, which are a significantly better environmental option" [1].

In the course of the public consultation concerning the updated RBMP, it was argued that the assessment of the fulfilment of conditions under Article 4(7) WFD for the investment projects notified by entrepreneurs would be carried out on the basis of the available data and the documentation provided by those entrepreneurs. This analysis required answers to the following questions:

- Have all possible steps been taken to mitigate the adverse impact on the status of the body of water; if so, what were they?;
- Have the reasons for these alterations or modifications been specifically set out and explained in the river basin management plan?;
- Do the reasons for these alterations or modifications constitute an overriding public interest and/or the environmental and social benefits resulting from achieving water conservation objectives are outweighed by the benefits for human health, maintaining human security or sustainable development resulting from the alterations or modifications; if so, what are they?;
- Can the beneficial objectives served by these alterations or modifications of the body of water not, because of technical feasibility or disproportionate costs, be achieved by means of other measures which would be significantly more beneficial from the point of view of the environment? If so, why?

Fulfilment of the condition contained in the question "*have all possible steps been taken to mitigate the adverse impact on the status of the body of water?*" means that measures have been planned to mitigate the negative impact on the condition of the body of water in question both at the construction and operation stages of the investment project in question. Also an assessment has been conducted whether these measures will ensure its proper protection. These measures may include both design solutions and the manner in which the works are carried out as well as covering the operation stage. These measures should be technically feasible and should not result in any disproportionate costs.

It was important fulfilment of the condition contained in the question "*do the reasons for these alterations or modifications constitute an overriding public interest and/or the environmental and social benefits resulting from achieving water conservation objectives are outweighed by the benefits for human health, maintaining human security or sustainable development resulting from the alterations or modifications?*" This question means that the achievement of this objective is so important that it outweighs the environmental benefits that would result if it were not achieved. This condition was analysed from the point of view of a broadly understood society rather than that of an individual or of a narrow group of entities with specific expectations. According to EC guidelines on exemptions from the achievement of environmental objectives, the concept of overriding public interest covers actions aimed at protecting basic values related to the citizens' lives (health, safety, environment), which are of fundamental importance for the state and society, or actions concerning economic or social activities that fulfil specific public service tasks.

It was also important fulfilment of the condition contained in the statement that the *beneficial objectives served by these alterations or modifications of the body of water cannot, by reason of technical feasibility or disproportionate costs, be achieved by means of other measures which would be significantly more beneficial from the point of view of the environment*. It required a detailed specification of the objective to be achieved and the identification of possible ways of achieving it. Importantly, not only different options for implementing the project planned were considered in this respect, but also entirely different solutions, including non-technical ones.

The final condition for the implementation of an investment project that may have a negative impact on the state of a body of surface water and for the application of the derogation under Article 4(7) WFD is that *the reasons for these alterations or modifications are specifically set out and explained in the river basin management plan*. In the updated RBMP, this condition is considered to be fulfilled for all investment projects that are described in the document.

As indicated above, the waters of the power plant canal are not subject to monitoring. However, this monitoring is carried out on the Vistula River upstream of the location where water is sourced for energy generation purposes. The monitoring demonstrates that the water in the Vistula River in the area in question has both bad status and potential. The main cause of water pollution is the discharge of saline mine waters into rivers within the Upper Silesian Coal Basin. The purpose of draining mines is to ensure that further extraction of coal in active coal mines remains possible. Abandoned mines are drained as well in order to prevent the inflow of water to the neighbouring mines where deposits are still being worked. The national economy is based on energy generation involving the burning of hard coal. More than 50% of national energy needs are covered in this manner. The largest number of mines are situated within the Upper Silesian Coal Basin. In a sense, they have connected vessels due to the hydrogeological and hydrodynamic conditions present there. As coal deposits are extracted, increasingly deep layers are reached, and the mineralisation levels of water flowing into mines increase with depth. The mine waters discharged cause degradation of the environment, which has been reflected in the assessment of the status/potential of the Vistula River bodies of surface water.

A significant number of Vistula bodies of surface water are at risk of failing to achieve good water status/potential, which resulted in their being included in the derogation. With regard to the RBMP (2011), in the Upper Vistula River water region, the temporary derogation covered, *inter alia*, the power plant canal under consideration.

20.3.4 Thermal Conditions of the Water Sourced from the Intake and Discharged

In accordance with the requirements of the water permit, the power plant monitors the quality of water sourced from the intake and also monitors the physicochemical parameters of spent cooling water and wastewater. At the power plant, the temperature of the water sourced from the power plant canal and of that discharged into the river is continuously measured. The lowest temperature difference is observed in winter from December to January. The minimum temperature difference between the water sourced from the intake and discharged during this period is around 1 °C, and the maximum difference is 5.8 °C. A slightly higher difference between the temperature of water sourced from the canal and that of discharged water is recorded in summer when the water is heated by a maximum of 8.6 °C. Taking into account the naturally higher temperature of water in the canal (up to around 24 °C in summer), the temperature of discharged water reaches 28.8 °C. However, it does not exceed the temperature permitted for wastewater discharged into surface bodies of water. On the other hand, the quality of the wastewater discharged into the water body is similar to the quality of the water sourced from the Vistula River via the power plant canal.

However, due to the need to use water from the Vistula River to cool the turbine sets, the water quality of the body of surface water receiving spent cooling water is deteriorated. The monitoring of bodies of surface water carried out by the Provincial Inspectorate for Environmental Protection in Kraków demonstrates that the river exhibits bad water status and potential. The increased electrolytic conductivity of water and chloride content in chlorides in river waters are correlated with the quality of the Vistula River waters used to cool turbine sets. From the analysis of water monitoring results, it appears that the operation of the conventional power plant, which uses an open cooling system based on the Vistula River waters, results in a change in the physicochemical properties of the water body into which it discharges. The changed section of the river is around 3 km long from the outlet through which water is discharged, around the location where it enters the Vistula River. However, the operation of the power plant only increases the temperature of water sourced from the Vistula River and does not change its chemical composition. It should be clearly stressed that the discharge of spent cooling water into the water body does not affect the condition and quality of water in the Vistula River into which the river in question flows. At the same time, the application of the derogation fulfils the condition of Article 4(8) WFD in that it does not permanently exclude or compromise the achievement of objectives in other bodies of water within the same river basin district.

20.3.5 *Assessment of the Economic, Energy and Environmental Aspects of Current and Alternative Solutions*

Taking into account environmental issues and WFD requirements, the feasibility of alternative solutions which could be used in order to reduce the temperature of the spent cooling water discharged has been considered. When considering changes to the cooling system, the impact of these changes on the overall efficiency of the power plant must be taken into account. If possible, efficiency should increase. In the case of temperature-sensitive processes, it must be checked whether the technology used can deliver lower final temperatures while maintaining the same level of safety.

Financial outlays are always among the most important factors influencing the choice of a cooling system, and they can only be estimated for an individual design. The most important types of costs include:

- Investment costs;
- Maintenance costs;
- Operating costs related to the demand for energy (and water);
- Environmental costs such as taxes, fees and waste management costs.

The power plant in question has an open cooling system. In terms of economic, energy and environmental considerations, the existing system has been described as Option 0 (*status quo*). Two alternative options were analysed as well whose implementation would enable: Option 1 – lowering the temperature and discharging spent cooling water back into the power plant canal; Option 2 – no further discharge of water into the river by replacing the open cooling system with a closed system that uses cooling towers. The options listed above meet BAT requirements.

20.3.5.1 Option 0 – Status Quo

Figure 20.1 shows the technological diagram of the power plant circuit including the indication of characteristic points. The specific enthalpy of steam at point 1 is 3470 kJ/kg. The condensation of water vapour behind the turbine takes place (according to data obtained from monitoring the operation of turbine sets) within the pressure range of approximately 2.5 to 4 kPa, which more or less corresponds to vapour condensation temperatures in condensers ranging from 65 to 80 °C.

Figure 20.2 shows points at which, on the basis of actual data, the efficiency of conversion of chemical energy of fuel into electricity was determined for the power plant in question.

Using actual data, on the basis of the graph presented in Fig. 20.2 and with the use of linear interpolation, the equation which makes it possible to determine the efficiency of conversion of chemical energy of fuel into electricity for the power plant as a function of coolant temperature after turbine set condensers (t'') was determined. This relationship is presented below:

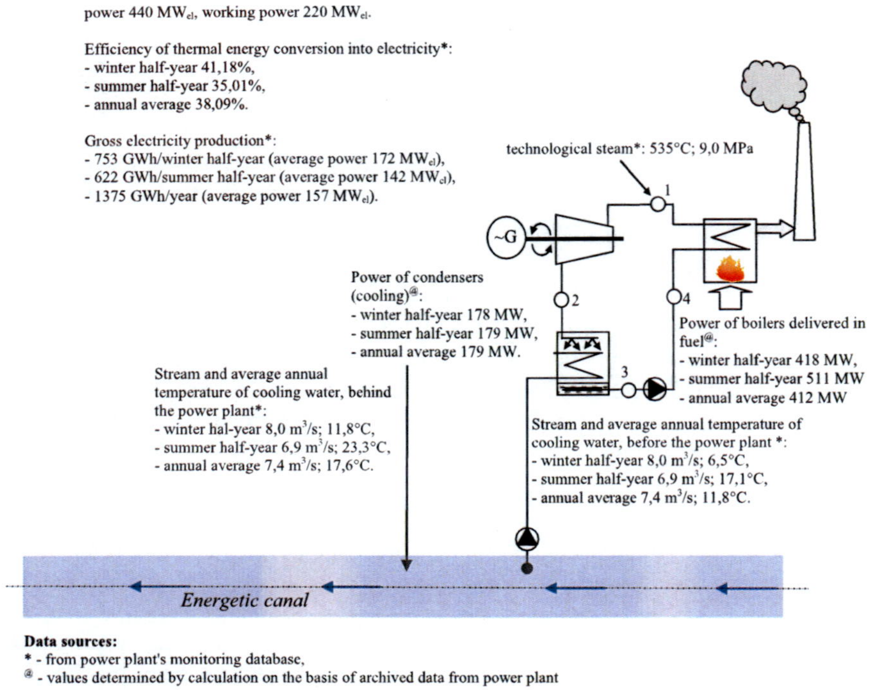

Fig. 20.1 Technological diagram of the power plant circuit including the indication of characteristic points – option 0 – without any changes to power plant operation

$$\eta_{\mathrm{el}}(\mathrm{t}'')[-] = -5.3651 \cdot 10^{-3} \cdot \mathrm{t}''[°\mathrm{C}] + 0.4752 \tag{20.1}$$

where: $\eta_{\mathrm{el}}(\mathrm{t}'')$ – the efficiency of conversion of chemical energy of fuel into electricity as a function of coolant temperature after the turbine set condensers t″ [°C].

This relationship illustrates the dimensionless ratio of electrical energy to the chemical energy of the fuel supplied and is correct for the temperature of the coolant after the turbine set condensers (t'') in °C.

20.3.5.2 Option 1 – Assuming the Additional Cooling of Water After Condensers Before Its Discharge into the Power Plant Canal

The technological diagram of Option 1, which assumes the additional cooling of water after condensers before its discharge into the power plant canal, is presented in Fig. 20.3. This alternative does not change environmental conditions, since the water would be sourced from, and discharged into, the same watercourse. However, this solution would involve additional investment outlays related to the construction of a pipeline which would connect the collecting pipe after power plant condensers with

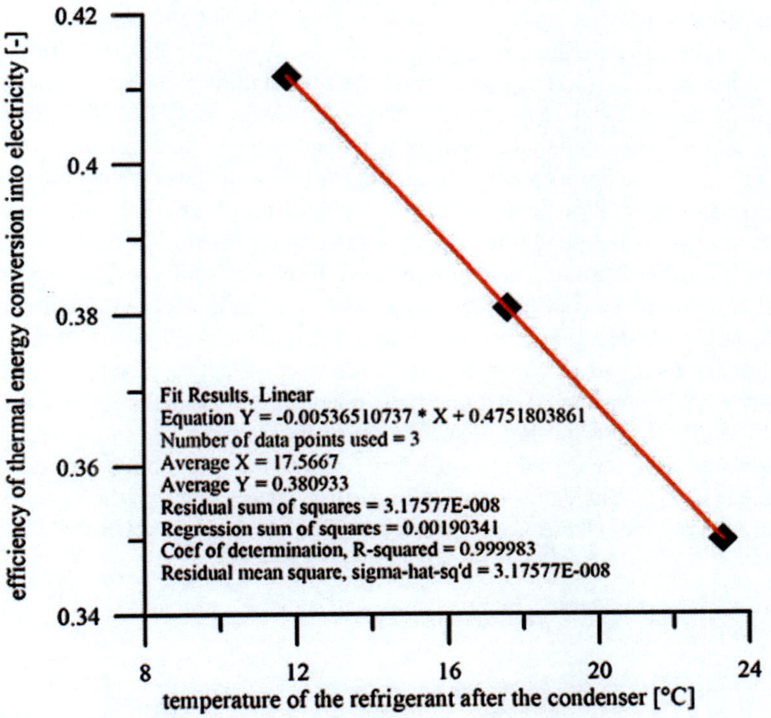

Fig. 20.2 Power plant efficiency as a function of coolant temperature after the condenser (t'')

the power plant canal. The length of the pipeline is estimated at 1,750 m. Assuming an estimated outlay for laying the pipeline at PLN 2000/metre, the total expected capital expenditure (including a reserve for unexpected expenses at the level of 25% of pipeline value) will amount to approximately PLN 4.4 million.

The fan cooler that would be capable of performing the duties set under Option 1 must be an open cooler in which water cooling based on the partial evaporation of the water discharged will be the dominant mode of cooling water after turbine set condensers. The evaporating water flow can be estimated at approximately 46.4 kg/s, which translates to approximately 167 m^3/h. The loss (evaporation) of water discharged into the power plant canal can be replaced by the loss of process water, if this is advisable for some reason. Unfortunately, due to the high cooling capacity required at high ambient temperatures, another method of cooling appears unrealistic.

It has been assumed that the capacity of the cooler would be such so as to ensure that the maximum temperature of the water discharged does not exceed 24 °C. The parameters of the power plant canal were determined on the basis of data from power plant monitoring systems. Extremely unfavourable weather conditions were assumed at which the water temperature after the condensers would be the highest. In this case, the required cooling capacity of the cooler would be 116 MW in order to cool a water flow of 7 m^3/s from 28.6 to 24 °C (see Fig. 20.3). The investment outlay required for a fan cooler is estimated at approx. EUR 57/kW$_c$ = PLN 241/kW$_c$. In the light of the above data, the total capital expenditure for the implementation of the cooling system described above may be estimated at approximately PLN 28 million/EUR 6.4 million. It is assumed that the cooler will operate for 15 years and the costs of its integration with the power plant will amount to 30% of the investment outlay related to the construction of the cooler itself (i.e. PLN 8.4 million/EUR 1.9 million).

The total capital expenditure related to the implementation of the project under Option 1 can be estimated at approximately PLN 40.8 million/EUR 9.3 million (PLN 28 million/EUR 6.4 million for the cooling system + PLN 8.4 million/EUR 1.9 million for integrating the cooler into the system + PLN 4.4 million/EUR 1 million for the pipeline connecting the condenser collecting pipe to the Łączany-Skawina Canal). The annual cost incurred under this option can be estimated at PLN 3.7 million/EUR 0.84 million (depreciation of fixed assets, operation, repairs and maintenance, additional fuel purchase cost). Given the gross amount of electricity produced (1375 GWh/year as stated in Fig. 20.3), the operation of the system would increase the price of electricity generation by PLN 3.93/MWh (EUR 0.89/MWh).

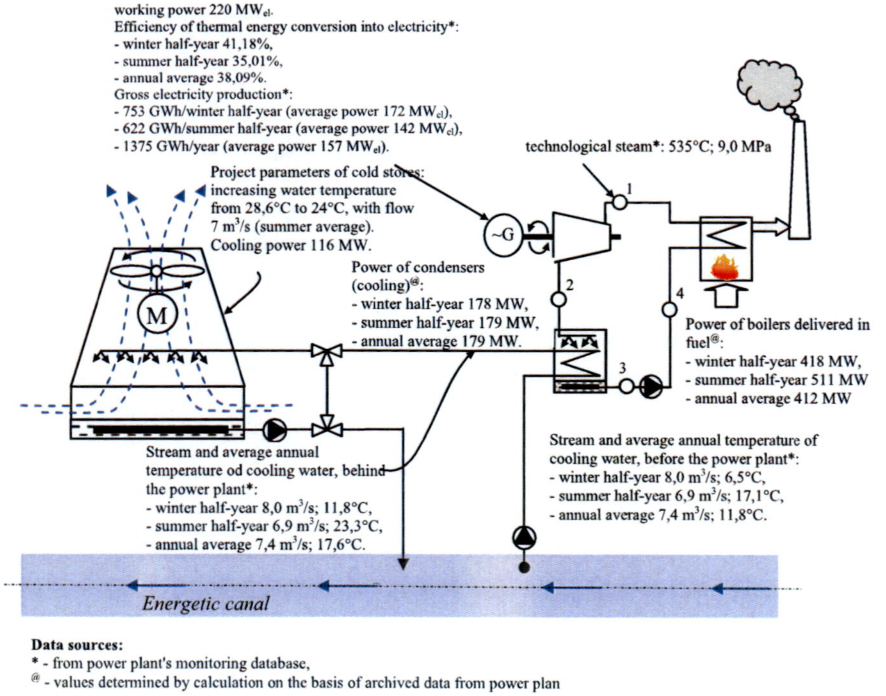

Fig. 20.3 Technological diagram of the power plant after modifications consisting in the installation of an open fan cooler which cools water after the condenser before it is discharged into the power plant canal

Summary for Option 1. The solution analysed appears not to be cost-effective, since for most of the year no problems are present with excessive water temperature after the condensers. Taking into account the results of water monitoring, the number of days per year during which the installation would operate is estimated at approximately 86 (summer months from July to September). The temperature only exceeded 24 °C to an insignificant degree; the maximum temperature after turbine set condensers was 28.6 °C. From the technical point of view, power plant operating conditions should not change, since the cooler will be installed as an additional element before the water is discharged. It will not interfere in any way with the internal turbine-boiler-generator system. The environmental problem potentially associated with the operation of this system would be the use of chemicals: antiscalants used to complex bivalent ions in order to protect the cooler against the formation of secondary deposits. A side effect, however, would be the discharge of water additionally contaminated with chemical compounds into the receiving water body.

20.3.5.3 Option 2 – Assuming the Replacement of the Current Condenser Cooling System with Cooling Towers or Fan Coolers

Figure 20.4 shows a technological diagram, including key operating parameters, for the option under which the current cooling system based on turbine set condensers is abandoned and replaced by evaporative coolers cooled with atmospheric air. It has been assumed that the coolers must enable the generation of electricity even at high (summer) air temperatures reaching 34 °C.

Under these conditions, the efficiency of the cooler decreases significantly and the temperature of the water which cools the condensers increases, which results in a decrease in electricity generation efficiency. If a constant capacity is maintained, this means an increase in the power consumed by burning fuel. The extreme parameters shown in Fig. 20.4 are indicated by the "#" superscript.

The required cooling capacity of a fan cooler or cooling tower should be approximately 390 MW under design conditions (Fig. 20.4). According to the ratio previously adopted, the construction of these facilities should require expenditure of around PLN 94 million. It has been assumed that their integration with the power plant system, including the necessary modifications, would mean capital expenditure at the level of 30% of the capital expenditure on equipment, which results in an additional PLN 28 million. Therefore, the total value of fixed assets required is around PLN 122 million. The annual depreciation of fixed assets would amount to approximately PLN 8.1 million (assuming depreciation of facilities over 15 years). Operating costs and required repairs are estimated at approximately PLN 3.7 million per year (3% of the investment outlay required per year). Therefore, the total operating costs of the new cooling system amount to approximately PLN 11.8 million/year. Given electricity production of 1375 GWh/year, this results in an increase in the unit price of electricity generated by PLN 8.6/MWh.

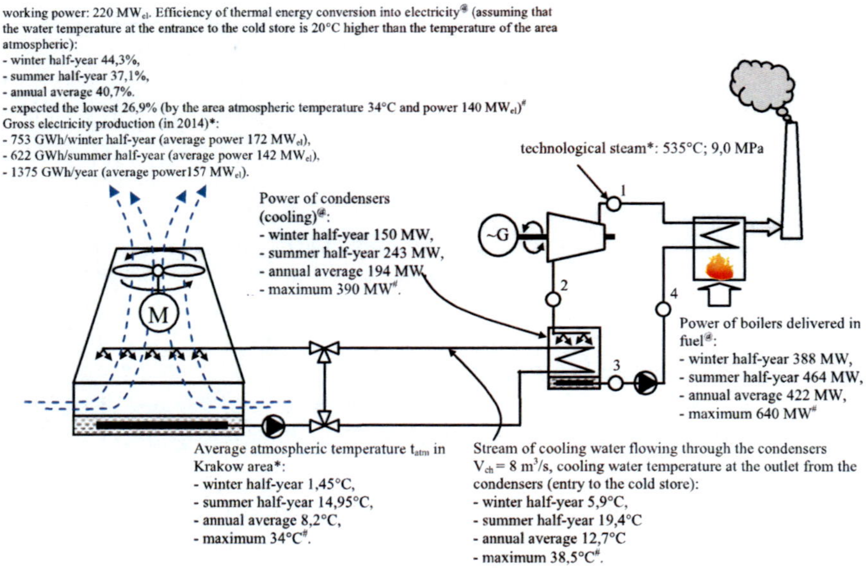

Data soursec:
* - from power plan's monitoring database,
@ - values determined by calculation on the basis of archived data from power plant,
\# - data specified for extreme calculation parameters: temperature of dry thermometer 34 ° C, electric power 140 MW, stream of water in condensers 8 m^3/s.

Fig. 20.4 Option 2 – the current condenser cooling system is abandoned in favour of coolers: process diagram with operating parameters

Summary for Option 2. Achieving such a high cooling capacity given extremely unfavourable weather conditions during the summer forces the use of a cooling system which is based on water evaporation (a system of open cooling towers). Under extreme conditions, the flow of evaporated water, which must be successively replenished with treated water, would amount to 156 kg/s (~562 m^3/h). This high demand for cooling water puts into question the technical feasibility of this option, especially since there is no alternative for cooling turbine set condensers (as water from the power plant canal is no longer used). The expected increase in electricity generation cost is significant as well and will amount to approximately 5%. The actual increase in electricity generation cost will be even higher since the demand for power consumed by burning fuel will increase – a comparison of data in Fig. 20.1 and 4 shows an increase in demand for average annual power consumed by burning fuel from 412 to 422 MW. Given the assumed fuel calorific value at 22 MJ/kg, this translates to an increase in demand for fuel at the level of 14,300 Mg/year. Assuming that the average price of coal dust with this calorific value is PLN 297/Mg, this results in a further increase in electricity generation cost amounting to PLN 4.3 million/year. This results in an increase in the unit price of electricity generated by PLN 3.13/MWh. Therefore, the total increase in electricity generation cost will amount to PLN 11.73/MWh, i.e. approximately 7%. In addition to the increase in electricity generation cost, pollutant emissions will also be higher as more fuel will be burned. For example, the increase in emissions of selected pollutants is estimated as follows: 28,500 Mg/year for CO_2, 110 Mg/year for SO_2, 57 Mg/year for NO_X, and 4 Mg/year for total inhalable dust. The increase in pollutant emissions was determined on the basis of actual data on emissions and fuel consumption at the power plant in question. Therefore, this solution will bring neither economic nor environmental benefits.

20.4 Conclusions

RBMPs were first developed in Poland in 2009 as a result of WFD requirements, and they were approved in 2011. In December 2014, a public consultation was announced on updated RBMPs and the national water and environmental programme, which was concluded by issuing the Regulation of the Council of Ministers of 18 October 2016 [18]. RBMPs are updated every six years. It should be stressed that once they are approved by the Council of Ministers, they constitute the strategy paper which provides the basis for:

- Making decisions which determine the status of water resources;
- Improving the process for achieving or maintaining the good status of waters and the related ecosystems;
- Determining the need to introduce rational water management principles in the future.

The WFD allows for derogations from the achievement of environmental objectives for bodies of surface water, including (1) the extension of the deadline for achieving the objective (Article 4(4) WFD); (2) a failure to achieve the objective as a result of new sustainable human development activities (Article 4(7)). The current updated RBMP analyses and identifies pressures on surface bodies of water and groundwater bodies within the Vistula River basin district, which include, *inter alia*, wastewater discharges into, and sourcing water from, those water bodies.

Under Polish conditions, conventional coal or lignite-fired power plants operate within the NPG and generate electricity continuously. Their electricity generation potential contributes to the stability of local power grid nodes and contributes to maintaining the secure operation of the NPG. In addition to the combined generation of electricity and heat, these power plants are often suppliers of process steam for many enterprises. The cooling solutions used involve both closed and open cooling systems. In the case of the latter, the impact of conventional power plants on surface waters is usually limited to an increase in their temperature by several degrees Celsius. However, each case is different, and multiple aspects related to the water environment must be accounted for. This analysis should be conducted in order to show all the pros and cons of the solutions considered in order to meet the applicable formal and legal requirements.

In order to assess the possibility of avoiding the negative impact of discharges of spent cooling water on the environmental objectives arising from the WFD, it is necessary to evaluate the feasibility of implementing alternative solutions. The analysis presented in this article concerned the operation of a single power plant situated in the Upper Vistula River sub-basin. In terms of economic, energy and environmental considerations, the existing system has been described as Option 0. Two alternative options were analysed as well, and the results of the analysis clearly indicated that with respect to the derogation under **Article 4(7) WFD**, **all** of the following conditions are met [1]:

- Article 4(7)(a): "*all practical steps have been taken to mitigate the adverse impact on the condition of the body of water in terms of the temperature of the spent cooling water discharged, and the non-deteriorated quality of the water sourced from and discharged into, the Vistula River, since the power plant does not change the quality of water sourced from the power plant canal, and the temperature increase in the mixing zone of the receiving body of water does not exceed approximately 6 °C*" [1];
- Article 4(7)(b): plans have been verified, and explanations have been provided for the modification of updated RBMP provisions;
- Article 4(7)(c): it has been demonstrated that the reasons for modifications or alterations are of overriding public interest and the benefits to the environment and to society of achieving environmental objectives are outweighed by the benefits of the new modifications, since the investment project is in line with the constitutional sustainable development principle. The power plant produces both electricity and heat in a combined cycle, which is the most efficient solution. This enables fuel savings of around 25–30% and reduces pollutant emissions;
- Article 4(7)(d): "the beneficial objectives served by those modifications or alterations of the water body cannot for reasons of technical feasibility or disproportionate cost be achieved by other means, which are a significantly better environmental option" [1], which was demonstrated by the assessment of other options related to technical solutions involving the cooling system.

The result of the analysis was the inclusion in the Regulation of the Council of Ministers of 18 October 2016 [16] of a derogation resulting from the lack of technical feasibility of achieving the environmental objective.

Acknowledgements The work presented was financed by the AGH University of Science and Technology, Krakow, Poland, grant 11.11.140.031.

References

1. European Commission (2000) Directive 2000/60/EC of the European parliament and of the council of 23rd October 2000 establishing a framework for community action in the field of water policy. Official journal of the European communities, L327/1. Brussels, European commission
2. Tomaszewska B, Dendys M, Galos K (2017) EIGR – Knowledge base as a tool facilitating the management of groundwater resources in Europe. Min Res Manag 33(4):79–92. https://doi.org/10.1515/gospo-2017-0050
3. Tomaszewska B, Tyszer M, Dendys M (2018) The availability of groundwater information sources in relations to the transposition of the WFD into polish law. Project KINDRA. Therm Sci Eng Process 5:437–443. https://doi.org/10.1016/j.tsep.2018.01.018
4. Hering D, Borja A, Carstensen J, Carvalho L, Elliott M, Feld CK, Heiskanen A-S, Johnson RK, Moe J, Pont D, Lyche SA, van de Bund W (2010) The European water framework directive at the age of 10: a critical review of the achievements with recommendations for the future. Sci Total Environ 408:4007–4019. https://doi.org/10.1016/j.scitotenv.2010.05.031
5. Carvalho L, Mackay EB, Cristina Cardoso A, Baattrup-Pedersen A, Birk S, Blackstock KL, Borics G, Borja A, Feld ChK, Ferreira MT, Globevnik L, Grizzetti B, Hendry S, Hering D, Kelly M, Langaas S, Meissner K, Panagopoulos Y, Penning E, Rouillard J, Sabater S, Schmedtje U, Spears BM, Venohr M, van de Bund W, Lyche SA (2019) Protecting and restoring Europe's waters: an analysis of the future development needs of the water framework directive. Sci Total Environ 658:1228–1238. https://doi.org/10.1016/j.scitotenv.2018.12.255
6. Quevauviller P (2011) Adapting to climate change: reducing water-related risks in Europe – EU policy and research considerations. Environ Sci Pol 14:722–729. https://doi.org/10.1016/j.envsci.2011.02.008
7. Cardoso AC, Free G (2008) Incorporating invasive alien species into ecological assessment in the context of the water framework directive. Aquat Invasions 3(4):361–366. https://doi.org/10.3391/ai.2008.3.4.1
8. von der Ohe PC, Dulio V, Slobodnik J, De Deckere E, Kühne R, Ebert R-U, Ginebreda A, De Cooman W, Schüürmann G, Brack W (2011) A new risk assessment approach for the prioritization of 500 classical and emerging organic microcontaminants as potential river basin specific pollutants under the European water framework directive. Sci Total Environ 409:2064–2077. https://doi.org/10.1016/j.scitotenv.2011.01.054
9. Blackstock KL, Martin-Ortega J, Spray CJ, im Druck. Implementation of the European water framework directive: what does taking an ecosystem services-based approach add? In: Martin-Ortega J, Ferrier RC, Gordon I, Kahn S (Hrsg.) (2015) Water ecosystem services: a global perspective. Cambridge University Press, Cambridge, pp 57–64
10. Grizzetti B, Lanzanova D, Liquete C, Reynaud A, Cardoso AC (2016) Assessing water ecosystem services for water resource management. Environ Sci Pol 61:194–203. https://doi.org/10.1016/j.envsci.2016.04.008
11. Spears BM, Ives SC, Angeler DG, Allen CR, Birk S, Carvalho L, Cavers S, Daunt F, Morton RD, Pocock MJO, Rhodes G, Thackeray SJ, Cadotte M (2015) Thackeray. Effective management of ecological resilience – are we there yet? J Appl Ecol 52:1311–1315. https://doi.org/10.1111/1365-2664.12497
12. Shepherd E, Milner-Gulland EJ, Knight AT, Ling MA, Darrah S, van Soesbergen A, Burgess ND (2016) Burgess. Status and trends in global ecosystem services and natural capital: assessing progress toward Aichi biodiversity target 14. Conserv Lett 9:429–437. https://doi.org/10.1111/conl.12320
13. The Act of 20.07.2017 r. Water law (Dz. U z 2017 r. poz. 1566 – in Polish)
14. Water management plan for the Vistula River Basin. Monitor Polski no 49, position 549, 2011 (in Polish)
15. Operacz A, Tomaszewska B (2016) The review of polish formal and legal aspects related to hydropower plant. Environ Sci Pollut Res 23(19):18953–18959. https://doi.org/10.1007/s11356-016-7466-7

16. Project of uptading the water management plan for the Vistula River Basin, December 2014
17. Regulation of the minister of the environment of 9 November 2011 on the method of classification of the status of bodies of surface water and on environmental quality standards with respect to priority substances (Dz. U. 2011 nr 257 poz. 1545) (not binding now – in Polish)
18. Regulation of the council of ministers of 18 october 2016 (Journal of laws [Dz. U.] of 2016 item 1911

Part V
Conclusions

Chapter 21
Updates, Conclusions, and Recommendations for "Management of Water Resources in Poland"

Katarzyna Kubiak-Wójcicka, Martina Zeleňáková, and Abdelazim M. Negm

Abstract This volume discusses current problems and research topics studied by scientists working on water resources in Poland. Based on many years of research conducted by authors from various research centers from Poland, the main conclusions and recommendations are presented. First of all, surface and groundwater resources as well as institutions responsible for the administration and management of water resources in Poland were discussed. The further part of the book focuses on changes in the hydrological regime of rivers, with particular emphasis on extreme phenomena, i.e. floods and low flows. The remaining part of the study is devoted to the economic importance of water resources for the needs of hydropower, inland waterway, agriculture and industry. This part of the study draws attention to the problems of sustainable use of water resources for the needs of human activities. A set of recommendations regarding future possibilities of using water resources and research was also indicated, which will be in line with strategic topics implemented at the government level, which include projects to counteract the effects of drought and floods.

Keywords Management · Water resources · Floods · Drought · Low flow · Agriculture · Waterway · Hydropower · Sustainable · Poland

K. Kubiak-Wójcicka
Department of Hydrology and Water Management, Faculty of Earth Sciences and Spatial Management, Lwowska 1, 87-100 Toruń, Poland

M. Zeleňáková
Department of Environmental Engineering, Faculty of Civil Engineering, Technical University in Košice, Košice, Slovakia
e-mail: martina.zelenakova@tuke.sk

A. M. Negm (✉)
Water and Water Structures Engineering Department, Faculty of Engineering, Zagazig University, Zagazig 44519, Egypt
e-mail: Amnegm85@yahoo.com; Amnegm@zu.edu.eg

M. Zeleňáková et al. (eds.), *Management of Water Resources in Poland*, Springer Water,
https://doi.org/10.1007/978-3-030-61965-7_21

21.1 Introduction

Water resources are an integral element of the natural environment in which human plays an important role. Modern environmental management, including water, based on the principle of protection through sustainable use favors the preservation of natural resources for future generations. Effective protection of water resources first requires knowing their size and distribution. Having knowledge about the size of water resources and their variability in time, it is possible to properly manage water resources.

Based on many years of research, a summary of the most important findings and conclusions from the research on the size and variability of surface and underground water resources has been presented. Poland belongs to the countries with small water resources. Proper management of water resources indicates the extent of water resources used for economic purposes, which should be in accordance with the principle of sustainable development. A set of recommendations extracted from all articles was also presented to help scientists and decision-makers in the pursuit of sustainable use of water resources in Poland.

The book brings state-of-the-art knowledge that can be effectively used for solving a variety of problems in integrated water resources management as well as the latest developments in the research area. The main goal of the book is put on water resources water management and water structures. The volume presents a summary of the findings, updates, conclusions and recommendations of the studies on water resources management.

21.2 Update

Research carried out by various authors confirms that water resources in Poland are diverse both in time and in space. Compared to other European countries, Poland has relatively small resources of renewable surface waters. In the following the national studies regarding the water resources in Poland concerned with extreme hydrological phenomena – droughts and floods and the use of water resources in the national economy are presented. The brief results of the studies are introduced.

The assessment and management of surface water and groundwater resources have a long tradition in Poland.

Current national surveys of Polish water resources concern the size and variability of water resources and the possibilities of their economic use. According to Kundzewicz [1], the following climate changes, which are observed through warming at various time scales, have important environmental consequences. It is expected that future climate changes will have a significant impact on the hydrological regime of rivers, water resources and their quality in various regions of Poland. In recent years, research into the variability of water resources as a result of global changes and projections of future climate changes has been conducted by various authors [1–5].

Along with climate change, we are increasingly dealing with the occurrence of extreme hydrological phenomena such as floods and droughts [6]. To reduce the risk of flooding, it is necessary to increase retention within the catchment area by building water reservoirs, dry polders, but also by changing land use, which will delay surface runoff [7]. In the coming decades, the risk of drought is expected to increase, which will affect the reduction of flowing water resources and the reduction of groundwater resources. Prolonged drought can lead to water deficits, which are already felt in the area of Kujawy (central Poland) [8]. As a result, they lead to significant losses in agriculture, industry and transport.

Research on Polish water resources has a long tradition. They focus on assessing the size of water resources and their changes under the influence of natural and anthropogenic factors [9]. The research includes the analysis of the impact of selected meteorological elements on changes in river outflow from large river basins, e.g. the Vistula, Oder, Warta, Narew [10–13]. Many researchers in Poland [14–20] dealt with the problems of changes in the hydrological regime of the Polish rivers. From the point of view of the use of water resources in various sectors of the national economy, studies and analyzes regarding maximum and minimum flows for both small and large watercourses in Poland are extremely important [21–23]. An important goal of hydrological research is to understand processes and counteract negative effects. This will allow for the development of watercourse managers' behavior in climate change conditions and the rational use of water resources.

The following are the major updates for water resources management in Poland based on the main book theme.

1. The organizational structure of water management in Poland, was not conducive to the sustainable development of water management. It resulted mainly from two institutions dealing with water management, which functioned in the administrative and sub-basin divisions. The new institution dealing with water management, Państwowe Gospodarstwo Wodne Wody Polskie, has been operating since 2018 and its focus is only on the catchment area.
2. Water resources in Poland (both in quantitative and qualitative aspect) are related to geological structure, surface relief and climatic conditions. The water safety of the country requires the maintenance of the quantity and quality of water resources on a good level and effective protection against the effects of drought and floods.
3. Two usually co-occurring factors are considered to be responsible for the process of the decline of lakes, and therefore also their water resources: (i) water level fluctuations, contributing to changes in the surface area; (ii) aggregation of sediments, can occur through: accumulation of biogenic mass, precipitation of chemical compounds; sedimentation of sediments, effect of aeolian processes.
4. Lakes located in river valleys, created as a result of both natural and anthropogenic processes, are an important and underestimated element of water retention.

5. In Poland, most numerous springs occur in mountainous and upland areas, while the greatest spring discharges are reached in the part of the Tatras built of carbonate rocks. Springs are objects of an exceptional scientific and landscape value.
6. Distribution of available groundwater resources in the country and comparison with the groundwater exploitation is the background of proper, sustainable management of its resources. Different types of mineral and thermal waters are exploited in a number of Spas for therapeutic purposes, and thermal waters are exploited for recreation and heating purposes.
7. Climate changes and the human impact on water relations make the features of the river flow regime change. Rivers with similar changes in flow conditions, except from those with transformation of water relations caused by human activity, also represent similar geographical regions.
8. Two largest Polish rivers follow basically the same course from south to north as they drain the surrounding regions. It is therefore clear that their flow seasonality structure shows spatial differences. However, as the courses of both rivers are similar, greater differences are observed along the rivers than between their basins.
9. The minimum flows are characterized by low seasonality level which generally follows the degree of precipitation continentalism and catchment resources capacity; the exception is the transit rivers Vistula and Oder where seasonality index is almost stable along the whole course.
10. The evaluation of seasonal and multiannual variability, as well as analysis of genetic relationships between selected estimators of hydrological drought, provided new and valuable cognitive insights and made it possible to identify a group of factors the determine drought development.
11. Flood marks have a particularly significant value for the study of historical hydrology, as they allow the recreation of the height and range of floods that occurred prior to the introduction of the systematic measurements of the water levels conducted using special instrumentation.
12. Flood potential index is calculated using long term discharge measurements at hydrological posts located at major rivers of Poland. Highest values of k index represent Oder river mountain tributaries and upper Vistula tributaries. Lowest values of k index are characteristic for the lowland rivers with a high percentage of lakes and bogs in their catchments.
13. Proper development of land exposed to flood, conforming to planning documents, constitutes the imperative goal of flood risk management, and minimisation strategies in Poland. In the European Union countries there are expected the best results in terms of simultaneous achievement of flood risk management objectives and Framework Water Directive environmental goals.
14. From the point of view of flood prevention, buffer and strategic investments are of great importance. Buffer investments mainly relate to the reconstruction of old and construction of new flood embankments, engineering channels and construction of flow regulating devices, strengthening of river banks, as well as the construction of retention reservoirs and development of valley retention.

15. Energy of flowing water is one of the most popular sources of renewable energy all over the World. In Poland there are difficult landscape conditions to build a large hydropower plant so there are only 13 power plants with nominal capacity over 10 MW.
16. Possibilities to utilize waterways are sought in their touristic and recreational exploitation. There is an increasing interest of tourists and residents in the riverside space as a place of rest and recreation in Poland.
17. Plant irrigation in Poland has a supplementary character. The production effects depend significantly on soil and rainfall conditions over the period of high water needs of crops. Production and economic efficiency of agricultural crops condition the development of irrigation systems and hence water consumption whitin the area.
18. The maintenance of water using in all Member States of European Union is focus on the rational use and protection of water resources in accordance with the principle of sustainable development.

21.3 Conclusions

In total, the book "**Management of Water Resources in Poland**" consists of 5 parts. In the first part, there are perface and introduction. The second part discusses water resources in Poland water resources and the third part discusses changes of flows in Polish rivers. The fourth part concerns the economic importance of water resources, while fifth part contains conclusions.

The following findings could be reported based on the chapters of this volume.

1. The lack of integrated management of water resources in Poland resulted from the division of competences between various institutions, which led to conflicts and negligence in water management. It was only the implementation of the new Water Law in 2017 that initiated the reform in water management. As part of this reform, a single institution responsible for water management was established – Państwowe Gospodarstwo Wodne – Wody Polskie (which one may translate as the National Water Holding – Polish Water). This institution started its operations on January 1, 2018. Its division follows the division of the country by the river catchments. Due to the short operation time, the assessment of its activities can be done in a few years. Currently, the most urgent need from the holding is to prepare a generally available hydrological database.
2. Water resources in Poland are extremely low compared to other European countries, which mainly results from climatic and hydrogeological conditions. Water resources are shrinking as a result of water intake for commercial use and for supplying the population for centuries. That is why it is so important to protect the resources and preserve their good quality.

3. In total, there are 7081 lakes in Poland with an area of over 1 ha, and a total capacity of 19.7349 km^3. As much as 99.32% of the lakes' capacity is concentrated in the area covered by the last glaciation (Masurian Lake District, Pomeranian Lake District and Wielkopolska-Kujawy Lake District). The retention of the area north of the boundary line of the maximum range of the last glaciation is 162 mm. This represents only 10% of the average annual rainfall, and the value is about 3.5 times lower than the average runoff from Poland.
4. The lakes located in river valleys are small reservoirs that were created as a result of both natural and anthropogenic processes that took place in the river valleys. These reservoirs are an important and underestimated element of small retention, however hardly explored. Lakes in river valleys are subject to dynamic disappearance processes and require detailed bathymetric, hydrochemical and hydrobiological studies.
5. Special attention should be paid to groundwater outflows based on field observations and review of available literature. The occurrence and efficiency of springs is very important. As well springs' use and needs for their protection in Poland. Particular attention has to be paid to the role of springs in the natural system and culture, as well as to their importance in local water management. Necessary urgent protective and preventive actions were indicated in order to preserve Polish springs as objects of great natural, landscape, economic and cultural significance.
6. Groundwater is the main source of drinking water supply for people in Poland. Over 70% the water demand is covered by that source. The main reason is the abundance of this resource and its good quality. According to the water management policy until 2030, these resources are protected and controlled by a monitoring system. Generally, at a depth of 200–300 m on the Polish Lowlands, there occur saline water of Cl-Na type. These waters are the subject of the growing interest of newly constructed spas and geothermal plants.
7. Five types of hydrological regime are distinguished in Poland. Typological differentiation of the flow regime largely depended on environmental conditions, especially climatic, but also physical-geographical characteristics of river basins. Climate change, along with human impact, will lead to changes in regime features, as evidenced by river responses to changes in the North Atlantic Oscillation.
8. Two largest Polish rivers follow basically the same course from south to north as they drain the surrounding regions. It is therefore clear that their flow seasonality structure shows spatial differences. However, as the courses of both rivers are similar, greater differences are observed along the rivers than between their basins. The river flow shows long-term oscillations due to changes in the water balance components.
9. The spatial, seasonal and long-term variability of low flows in Poland in the years 1951–2015. The minimum flows in Poland in the years 1951–2015 are characterized by low seasonality level which generally follows the degree of precipitation continentalism and catchment resources capacity; the exception is the transit rivers Vistula and Oder where the seasonality index is almost stable

along the whole course. Severe low-flow episodes exceeding maximum credible duration (95 of non-exceedance probability) appeared very often; such image of distribution indicates that hydrogeological limitations, as well as physiographical features of the catchment, are more important than macroscale hydrometeorological conditions.

10. Low flows can be used as indicators of hydrological drought development. In the mountain rivers, low-flow regimes are significantly affected by genetically different summer and winter low flows, while the rest of the country is dominated by summer half-year deficits (July–September). Large transit rivers gradually change the features of their low-flow regime along with the entrance of successive tributaries but retain their regime specificity up to the estuary. The high degree of hydrological drought seasonality occurs when its average seasonal concentration date falls in the summer months. When this date is shifted towards autumn months, the seasonality index decreases as droughts of the winter half-year become more and more important.
11. Flood marks are very numerous in Poland. They are part of the historical heritage and are an important source of hydrological information. The long tradition of their installation should be continued in Poland. Flood markers should play an important role in flood education, especially for people living in areas at risk of flooding. Attention should be paid to the initiative to create a nationwide database available to all Internet users and updated with their participation.
12. The spatial diversity of the flood potential indicator was stated on Polish territory. This indicator is a useful non-dimensional parameter that can be used to compare different river basins. Together with the flood coverage map, it shows flood exposure in various geographical regions of Poland.
13. Proper development of land exposed to flood, conforming to planning documents, constitutes the imperative goal of flood risk management, and minimization strategies in Poland. It was emphasized that full protection against floods is impossible and that their damage and losses can only be reduced to a severe degree and that the risk of flooding should be controlled and managed in advance. In the European Union countries there are expected the best results in terms of simultaneous achievement of flood risk management objectives and Framework Water Directive environmental goals.
14. The high content of substances in the coastal waters of the southern Baltic is mainly due to polluted rivers. The pollution volume depends on the catchment land use and the size and location of the point, linear and area pollution sources. Strategic investments in catchment areas concerned the increase of their retention through afforestation, counteracting water erosion in lowland areas, restoration of riverbed and river banks, increasing retention in urban areas of over 20,000 residents.
15. Increasing the share of renewable energy sources in Poland is an urgent necessity. Hydropower has considerable potential in this area. Construction of hydropower plants is not cheap and fast, but facilities of this type can function for many years without major expenditure. However, it is necessary to draw

the attention of decision-makers to the possibility of using this form of energy production.

16. The inland waterway transport's role in Poland is marginal. The lack of a prospective plan for the modernization of waterways and the construction of infrastructure developed in agreement with various social groups only leads to the intensification of existing conflicts. Possibilities to utilize waterways are sought in their touristic and recreational exploitation. It seems a top priority matter that on key issues for the country, such as the state of its water resources and their use, the decision-makers would be consensus and cooperation of both opponents and advocates of the inland waterway development program in Poland.
17. Plant irrigation in Poland has a supplementary character. The production effects depend significantly on soil and rainfall conditions over the period of high water needs of crops. Production and economic efficiency of agricultural crops condition the development of irrigation systems and hence water consumption within the area.
18. The conventional power plant with a combined generation of electricity and heat, situated in the Upper Vistula River sub-basin, Poland is presented as a case study. The assessment of the general status of a given body of surface water was conducted on the basis of the analysis of the results of the assessment of the ecological status/potential and of the chemical status of the body of surface water in question.

21.4 Recommendations

Research confirms that water resources in Poland are diverse both in time and in space. Compared to other European countries, Poland has relatively small resources of renewable surface waters. In the future, the lack of adequate quantity and quality of water may constitute a barrier to economic development. Therefore, further research is recommended to develop measures coordinating both flood protection and mitigation of effects and drought. Joint activities in the field of water resources protection should be carried out in cooperation with hydrologists, officials, planners and engineers. Practical conclusions, especially regarding regional development barriers, can significantly extend water management aimed at mitigating the effects of drought and floods on a national and regional scale.

This section contain recommendations for water resources management in Poland obtained from the chapters presented in this volume.

1. In the future, the focus should be on research showing current water management in Poland in the opinion of officials, local government officials, stakeholders and society. It will be particularly important to pay attention to the development of effective communication schemes between various institutions. The method of communication plays a key role in making decisions, especially in crisis situations.

2. Measures aimed at water retention in the catchment are recommended. Changes in the conditions of water circulation and outflow should be towards its retention in the soil and bedrock, leading to a decrease in surface runoff and an increase in exploitable resources in catchments.
3. To prevent the decline of lakes is necessary to prevent the intensification of the sedimentation process by preventing a decrease in the water levels and accumulation of sediments.
4. Lakes in river valleys are subject to dynamic disappearance processes and require detailed bathymetric, hydrochemical and hydrobiological studies.
5. There is an urgent need to save and preserve springs and indicate the ways and directions of necessary actions that is interesting for environmental protection services, local authorities and social organizations dealing with the protection of environmental and cultural heritage.
6. The public interest in healing water, thermal water, and highly mineralised water (brines) has been growing considerably for some time. All these factors make scientists and specialists show greater interest in deep water resources, also in the form of a mathematical description of phenomena occurring in the stagnant water zone, which is virtually outside water circulation.
7. The observed transformations of the regime type in the winter North Atlantic Oscillation phases against the background of average conditions are understandable and indicate the possible destabilization of the characteristics of the flow regime of many rivers in Poland in the changing climatic conditions caused by the varying intensity of the macroscale type of circulation.
8. Further research of flow seasonality in two big polish rivers – the Vistula and the Oder should be carried out. It will allow to check whether the introduced changes do not disturb the natural regime of these rivers. These studies will be helpful for decision makers in the Ministry of Maritime Economy and Inland Navigation as well as for ecological organizations monitoring changes in rivers.
9. Good knowledge of low-flows and their determinants may effectively support the tools and strategies of optimal reduction of drought results, its prevention and prediction. Advanced recognition of river low-flow regime is very important for water balance structure and may be crucial for identification of quantity and dynamics of available water resources.
10. Practical conclusions, especially those concerning regional barriers for development and regularities of the duration and range of hydrological drought can significantly expand water management aimed at mitigation of drought effects at the national and regional scale.
11. The "Flood Marks" (www.openhydrology.org/maps/flood_mark) project seems to be a step in the right direction to information about the high water marks during floods. Its strength is the utilization of the popular and still growing internet community of OpenStreetMap users.
12. Flood potential index can be used in regional plans of flood risk management. Flood potential index can be used for comparison of catchment reactions to rain at different regions.

13. As regards effective flood risk management in Poland, it is necessary to work out measures for coordinating both flood protection and flood prevention, and also for mitigating the effects of floods. Further, non-technical methods of protection against floods must be supported with technical methods, for example through legislative and administrative measures.
14. Strategic investments in the catchment areas concerned the increase of their retention through afforestation (comprehensive forest and forestry adaptation projects to climate change), counteracting water erosion in lowland areas, restoration of riverbeds and riverbanks, increasing retention in urban areas of over 20,000 residents.
15. In Poland there is still reserve to build a few large hydropower plants, especially in the lower Vistula and it is possible to build a few hundred of small hydropower plants on smaller watercourses.
16. It seems a top priority matter that on key issues for the country, such as the state of its water resources and their use, the decision makers would be consensus and cooperation of both opponents and advocates of the inland waterway development program in Poland.
17. Significant both on a national and regional scale as well as for the design and operation of irrigation systems in family farms or commercial enterprises is the proper and sustainable management of water resources in agriculture.
18. It should be stressed that the strategy paper approved by the Council of Ministers provides the basis for (i) making decisions which determine the status of water resources; (ii) improving the process for achieving or maintaining the good status of waters and the related ecosystems; (iii) determining the need to introduce rational water management principles in the future.

References

1. Kundzewicz ZW (2011) Nonstationarity in water resources – central European perspective. J Am Water Resources Associat (JAWRA) 47(3):550–562. https://doi.org/10.1111/j.1752-1688.2011.00549.x(ISI)
2. Szwed M, Graczyk D, Pińskwar I, Kundzewicz ZW (2007) Projections of Climate Extremes in Poland. Geographia Polonica 80(2):191–202. https://rcin.org.pl/igipz/Content/107/GP_80_2_cala.pdf#page=185
3. Osuch M, Romanowicz RJ, Lawrence D, Wong WK (2016) Trends in projections of standardized precipitation indices in a future climate in Poland. Hydrol Earth Syst Sci 20:1947–1969. https://doi.org/10.5194/hess-20-1947-2016
4. Szwed M, Karg G, Pińskwar I, Radziejewski M, Graczyk D, Kędziora A, Kundzewicz ZW (2010) Climate change and its effect on agriculture, water resources and human health sectors in Poland. Nat Hazards Earth Syst Sci 10:1725–1737. https://doi.org/10.5194/nhess-10-1725-2010
5. Kundzewicz ZW, Piniewski M, Mezghani A, Okruszko T, Pińskwar I, Kardel I, Hov Ø, Szcześniak M, Szwed M, Benestad RE, Marcinkowski P, Graczyk D, Dobler A, Førland EJ, O'Keefe J, Choryński A, Parding KM, Haugen JE (2018) Assessment of climate change and associated impact on selected sectors in Poland. Acta Geophys 66:1509–1523. https://doi.org/10.1007/s11600-018-0220-4

6. Kubiak-Wójcicka K, Bąk B (2018) Monitoring of meteorological and hydrological droughts in the vistula basin (Poland). Environ Monit Assess 190:691. https://doi.org/10.1007/s10661-018-7058-8
7. Mioduszewski W, Querner EP, Kowalewski Z (2014) The analysis of the impact of small retention on water resources in the catchment. J Water Land Develop 23:41–51. https://doi.org/10.1515/jwld-2014-0028
8. Łabędzki L (2003) Problems of water management in selected catchments in areas of intensive agriculture (in Polish). Woda – Środowisko – Obszary Wiejskie (Water – Environment – Rural Areas) 3:153–172
9. Kowalczak P, Kundzewicz ZW (2011) Water-related conflicts in urban areas in Poland. Hydrol Sci J 56(4):588–596
10. Dubicki A, Malinowska-Małek J, Strońska K (2005) Flood hazards in the upper and middle Oder river basin – a short review over the last century. Limnologica 35(3):123–131. https://doi.org/10.1016/j.limno.2005.05.002
11. Malinowski Ł, Skoczko I (2018) Impact of climate change on hydrological regime and water resources managament of the Narew river in Poland. J Ecol Eng 4:167–175. https://doi.org/10.12911/22998993/91672
12. Kubiak-Wójcicka K (2019) Long-term variability of runoff of Vistula river in 1951–2015. In: Air and water – components of the environment conference proceedings, pp 109–120. Cluj-Napoca, Romania. https://doi.org/10.24193/AWC2019_11
13. Kubiak-Wójcicka K (2020) Variability of air temperature, precipitation and outflows in the Vistula basin (Poland). Resources 9(9):103. https://doi.org/10.3390/resources9090103
14. Dynowska I (1971) The hydrological river regimes (Typy reżimów rzecznych w Polsce in Polish), Zesz. Nauk. UJ, Prace Geogr. p 37, Kraków
15. Fal B (1993) Variability of the outflow from Poland in this century (Zmienność odpływu z obszaru Polski w bieżącym stuleciu in Polish), Wiad. IMGW 16:3–20
16. Fal B (1997) The characteristic flows of the main rivers of Poland in the period 1951–1990 (Przepływy charakterystyczne głównych rzek polskich w latach 1951–1990 in Polish). Bad. IMGW, Seria Hydrol. Ocean, Mat, p 21
17. Wrzesiński D (2016) Use of entropy in the assessment of uncertainty of river runoff regime in Poland. Acta Geophysicavol 64(5):1825–1839. https://doi.org/10.1515/acgeo-2016-0073
18. Wrzesiński D, Sobkowiak L (2018) Detection of changes in flow regime of rivers in Poland. J Hydrol Hydromech 66(1):55–64. https://doi.org/10.1515/johh-2017-0045
19. Pociask-Karteczka J (2011) River runoff response to climate changes in Poland (East-Central Europe). IAHS Publ 344:182–187. https://iahs.info/uploads/dms/16782.33-182-187-344-10-Pociask_JH02_Corrected_jpk.pdf
20. Somorowska U (2016) Changes in drought conditions in Poland over the past 60 years evaluated by the standardized precipitation-evapotranspiration index. Acta Geophys 64(6):2530–2549. https://doi.org/10.1515/acgeo-2016-0110
21. Bartnik A, Jokiel P (2014) Maximum discharges and maximum runoffs in Poland. In: Kobojek E, Marszał T (eds), Natural environment in Poland and its protection in Łódź University Geographical Research, pp 43–73. Łódź University Press, Łódź. https://hydro.geo.uni.lodz.pl/uploads/file/pub_AB/2014_Bartnik_Jokiel_Maximum_discharges_and_runoffs_in_Poland.pdf
22. Nowicka B (2009) Estimation of differentiation extremely high discharges of selected Polish rivers (Ocena zróżnicowania ekstremalnie wysokich przepływów wybranych rzek Polski. Prace i Studia Geograficzne in Polish), vol 43, pp 11–24. https://wgsr.uw.edu.pl/wgsr/wp-content/uploads/2018/11/Nowicka.pdf
23. Kubiak-Wójcicka K, Machula S (2020) Influence of climate changes on the state of water resources in Poland and their usage. Geosciences 10(8):312. https://doi.org/10.3390/geosciences10080312